PATTERNS OF PROBLEM SOLVING

PATTERNS OF PROBLEM SOLVING

Moshe F. Rubinstein

Professor of Engineering and Applied Science
Chairman, Engineering Systems Department
University of California, Los Angeles

PRENTICE-HALL, INC., Englewood Cliffs, New Jersey

Library of Congress Cataloging in Publication Data
Rubinstein, Moshe F., Patterns of problem solving.
Bibliography:
Includes index
1. Problem solving 2. Mathematical models
3. Decision-making 4. Programming (mathematics)
I. Title
QA63.R2 511 74-20721
ISBN 0-13-654251-4

10 9 8 7

Printed in the United States of America

PRENTICE-HALL INTERNATIONAL, INC., *London*
PRENTICE-HALL OF AUSTRALIA PTY. LTD., *Sydney*
PRENTICE-HALL OF CANADA, LTD., *Toronto*
PRENTICE-HALL OF INDIA PRIVATE LIMITED, *New Delhi*
PRENTICE-HALL OF JAPAN, INC., *Tokyo*

To my wife *ZAFRIRA*
and daughters *IRIS and DORIT*

Contents

Chapter 1 PROBLEM SOLVING 1

1-1	Culture, Values, and Problem Solving	1
1-2	Schools of Thought in Problem Solving	2
1-3	Models of Problem-Solving Process	4
1-4	Kinds of Problems	6
1-5	Guides to Problem Solving	7
1-6	Failure to Use Known Information—Difficulty 1	8
1-7	Introduction of Unnecessary Constraints—Difficulty 2	10
1-8	General Precepts as Guides to Problem Solving	14
1-9	Paths to a Solution	19
1-10	Discussing Your Problem	21
1-11	Descartes as Problem Solver	22
1-12	Newton as Problem Solver—Process and Goal	25
1-13	Einstein as Problem Solver—Interaction of Language and Thought	27
1-14	Summary	30
	Exercises	31

Chapter 2 LANGUAGE AND COMMUNICATION 36

2-1	Introduction	36
2-2	The Structure of Language	37
2-3	Communication and Natural Language (Encoding Experience through Language)	40
2-4	Knowledge of the Language and Knowledge of the World	43

2-5 Evolution of Written Language 44
2-6 The Numbers 50
2-7 Gimatria 51
2-8 Language of Statements—Symbolic Logic 53
2-9 Truth Tables 57
2-10 Tautologies and Valid Arguments 61
2-11 Symbols, Operations, and Fundamental Laws of Symbolic Logic 63
2-12 Algebra of Logic and Switching Circuits 64
2-13 Boolean Algebra 66
2-14 Language of Sets—Sets, Subsets, and Operations on Sets 68
2-15 Algebra of Sets 73
2-16 Modern Communication Systems 74
2-17 Use of Redundancy in Communication 75
2-18 Computer Language 81
2-19 Summary 82
Exercises 86

Chapter 3 COMPUTERS: FUNDAMENTAL CONCEPTS 91

3-1 Introduction 91
3-2 A Simple Computer Model 95
3-3 Basic Components of a Digital Computer 99
3-4 A Program to Add *N* Numbers 100
3-5 Flow Chart 104
3-6 Memory (Storage) in the "Real" Computer 105
3-7 Input-Output Media 108
3-8 Computer Language: Language of Zeros and Ones 109
3-9 Conversion of Decimal to Binary Numbers 111
3-10 How a Computer Computes 113
3-11 Programming 120
3-12 Use of Computers 127
3-13 Summary 129
Exercises 131

Chapter 4 PROBABILITY AND THE WILL TO DOUBT 135

4-1 Introduction 135
4-2 Probability and Doubt 136
4-3 Is Mr. X Honest? 138
4-4 Laws of Probability 140

4-5 Bayes' Equation and Relevance of Information 149
4-6 Applications of Bayes' Theorem 151
4-7 More Applications of Bayes' Theorem 156
4-8 Mr. X Revisited 162
4-9 Probability and Credibility 164
4-10 The Concept of Information and Its Measurement 166
4-11 Summary 180
Exercises 183

Chapter 5 MODELS AND MODELING 192

5-1 Introduction 192
5-2 The Purpose of Models 193
5-3 The Nature of Models 196
5-4 Validation of Models 201
5-5 Classification of Models 203
5-6 Models of History 210
5-7 Models of the Universe 212
5-8 Models of the Atom 217
5-9 A Model of the Brain 220
5-10 Models in Engineering 222
5-11 Models in Physical Science and in Human Affairs 227
5-12 More on Mathematical Models 228
5-13 Summary 238
Exercises 241

Chapter 6 PROBABILISTIC MODELS 246

6-1 Probabilistic Models—Preliminary Concepts 246
6-2 Populations and Samples 250
6-3 Probability Distribution Models 253
6-4 Normal Distribution Model 256
6-5 Expected Values of Random Variables and Their Aggregates 264
6-6 Central Limit Theorem and Its Application 269
6-7 Random Walk 273
6-8 From Sample to Population—Estimation of Parameters 277
6-9 Testing Hypotheses—Errors of Omission and Commission 285
6-10 Simulation of Probabilistic Models—Monte Carlo Method 293
6-11 Summary 300
Exercises 304

Chapter 7 DECISION-MAKING MODELS 311

7-1 Introduction 311
7-2 Decision Models 312
7-3 Decision Making Under Certainty 316
7-4 Decision Making Under Risk 317
7-5 Decision Making Under Uncertainty 317
7-6 Utility Theory 322
7-7 Utility Assignments and the Decision Models 335
7-8 Decision Making Under Conflict—Game Theory 337
7-9 Group Decision Making 349
7-10 Summary 350
Exercises 352

Chapter 8 OPTIMIZATION MODELS "Selecting the Best Possible" 356

8-1 Introduction 356
8-2 Linear Functions 357
8-3 Linear Programming—Exposure 361
8-4 Linear Programming—An Application to Dental Practice 366
8-5 Linear Programming—Generalization of Method 371
8-6 Nonlinear Programming 382
8-7 Dynamic Programming 386
8-8 Sequential Decisions with Random Outcomes 391
8-9 Sequential Decisions with Normal Distribution Outcomes 397
8-10 Summary 400
Exercises 403

Chapter 9 DYNAMIC SYSTEMS MODELS 409

9-1 Introduction—An Exposure 409
9-2 Building Blocks in Dynamic System Models 415
9-3 More Recent Models of Dynamic Systems 425
9-4 Homeostasis—Control in Living Organisms 430
9-5 Controllability, and Open vs. Closed Loop Control 434
9-6 Amplitude and Phase in the Response of Dynamic Systems 437
9-7 Characteristics of Feedback Systems 444
9-8 Simulation of Dynamic Systems 455
9-9 A Novel Application in the Making 463
9-10 Summary 466
Exercises 469

Chapter 10 VALUES AND MODELS OF BEHAVIOR 474

10-1	Introduction	474
10-2	Role of Values in Problem Solving	476
10-3	Value Classification	478
10-4	Value Judgment	480
10-5	Knowledge and Values	483
10-6	A Model of Ethical Behavior	489
10-7	Values of the Present	492
10-8	Values of the Future	493
10-9	Dynamic Change in Values and Value Subscription	494
10-10	Cost-Benefit Assessment of Values	496
10-11	Social Preferences—An Axiomatic Approach	497
10-12	Alternate Procedure for Consensus Determination	501
10-13	Metrization of Preferences	503
10-14	The Delphi Method	505
10-15	Use of Delphi Method to Develop an Interdisciplinary Course	508
10-16	Consensus on Values and Consensus on What To Do	513
10-17	Summary	514

TOPICS FOR DISCUSSION WITH FOCUS ON VALUES 516

REFERENCES 522

ANSWERS TO SELECTED EXERCISES 529

INDEX 533

Preface

The material in this book was developed while teaching a campus-wide interdisciplinary course entitled, "Patterns of Problem Solving." The book attempts to provide the reader with tools and concepts which are most productive in problem solving and are least likely to be eroded with the passage of time. Emphasis is placed on developing the proper attitudes for dealing with complexity and uncertainty. A balance is sought between solution techniques and attributes of human problem solvers, so that problem solving is not reduced to a dogmatic and sterile process. To maintain this balance, human values are considered wherever feasible.

Problem solving is presented as a dynamic process, encompassing diverse academic disciplines. Primary objectives of the presentation are:

- To develop a general foundation for problem solving and place in that context tools and concepts of problem solving which the student has acquired.

  To accomplish this, the first five chapters provide a broad framework for consolidating past learning experiences. For example, sets and set theory are discussed in the context of language and as a tool for classification.

- To provide a framework for a better appreciation of the role of tools and concepts which the student may acquire in his continued studies.

  To achieve this, the last five chapters discuss specific classes of problem solving models which have a wide range of applications.

A brief survey of the book's content follows.

Chapter 1 discusses schools of thought and attitudes in problem solving, common difficulties, and general precepts as guides for paths to a solution.

Chapter 2 proceeds from the evolution of written natural language to computer language. Sets, symbolic logic, and modern communications are placed in the context of language as man's most profound tool for learning, thinking, and problem solving.

Chapter 3 introduces the reader to computers, how they are structured, how they work, the concept of programming, areas of computer applications, and the effect of computers on society.

Chapter 4 treats information in terms of its quantitative measurement and the assessment of its relevance and credibility. The basic premise is that a Will to Doubt (an open mind) is most important if information is to be useful.

Chapter 5 is a bridge between the first five chapters which develop a general foundation for problem solving, and the last five which discuss specific classes of problem solving models. Included are models and the modeling process, model classification and evaluation, and limits on what can and cannot be derived from models. Examples are given of models in history, science, and engineering.

Chapter 6 treats probabilistic models, distributions, estimation of parameters, errors of omission and commission, and simulation. The material of this chapter is at an advanced level.

Chapter 7 introduces decision models and the elements of decision theory: decision making under certainty, risk, and uncertainty; attitudes to uncertainty; utility theory and its applications; game theory; and group decisions.

Chapter 8 provides a foundation for optimization using linear, nonlinear, and dynamic programming as a basis for "selecting the best possible."

Chapter 9 is an introduction to cybernetics with emphasis on its ubiquity. Examples explore the history as well as the breadth of applications.

Chapter 10 returns to the problem of problems: understanding values, and forming consensus on values.

To help learn the material, a summary is included at the end of each chapter. The presentation varies in its degree of rigor and sophistication to accommodate students with diverse backgrounds, both at the graduate and undergraduate level. The first part of each chapter is suitable for a general audience of readers. Material that is more suitable for advanced seniors and graduate students is marked by a bold asterisk (**∗**).

The exercises at the end Chapters 1-9 can be classified as:

1) those that emphasize concepts;
2) drill exercises for students whose background is deficient in a particular area.

Answers to selected exercises are given at the end of the book. Notes that precede some exercises contain material that is pertinent to the exercises which follow. See, for example, the note preceding Exercise 4-18 in Chapter 4 or the note preceding Exercise 6-3 in Chapter 6.

Suggested problems within certain sections are meant to serve as possible signals to the reader to pause and reflect. See, for example, the problems at the end of Section 9-1.

Chapters 1 through 5 and selected parts of Chapters 6 through 10 are taught in a campus-wide course entitled, "Patterns of Problem Solving." The course was conceived as a valuable pedagogical and educational experience for all students on campus. This has been realized as evidenced by the following: the students who have taken this course represent more than 30 major fields of study and all levels at the University, from freshmen to graduate students.

Much of the material in Chapters 6 through 10 is taught in a second course entitled, "Applied Patterns of Problem Solving." The two courses are now part of a campus program in Creative Problem Solving.

The experience in teaching the course, "Patterns of Problem Solving," at UCLA led to preparation of a separate *Instructor's Manual*, which includes statements of objectives, suggested allocation of lecture hours to subject matter, and sample quizzes and examinations. A separate *Solution Manual* for all exercises and problems in this book has also been prepared.

In developing the material for these campus-wide interdisciplinary courses, I was motivated greatly by my experiences, for more than ten years, as a lecturer and director of the Modern Engineering Program for Executives. My colleagues, Professors George A. Bekey, Robert S. Elliott, Paul B. Johnson, Walter J. Karplus, Leonard Kleinrock, Bernard Rasof, and C. R. Viswanathan were the principal lecturers in the program. These colleagues, who represent diverse fields of interest, contributed to my learning and understanding of the common patterns of problem solving. I hope to convey sparks of my excitement with the discovery of these patterns to students in all fields.

My colleagues in the Engineering Systems Department at UCLA were helpful in reviewing parts of the text and offering expert advice and constructive suggestions. Professors Bruce L. Miller and Stephen E. Jacobsen reviewed Chapter 8, Dr. Julian Hatcher and Professors Joseph J. DiStefano, Cornelius T. Leondes, and Louis C. Westphal reviewed Chapter 9. I have also benefited from discussions of Chapter 10 with Dr. Norman C. Dalkey and my former student, Gary Gasca.

In preparing the exercises and solutions for the "Patterns of Problem Solving" course, I was assisted by Dr. Julian Hatcher, Dr. Melvin W. Lifson and my Teaching Assistants, Pradeep Batra, Jean Dubinsky, Alex Ratnofsky and Ben Rodilitz. Some exercises were contributed by students in the class as part of their assignment. The students have been an ever-present source of encouragement and, through my interaction with them, the form and content of the manuscript have been shaped.

Professor Harvey S. Perloff, Dean of the School of Architecture and Urban Planning, and Professor Marvin Adelson, Director of the Creative Problem Solving Program, have been continuing sources of encouragement. I have also received strong encouragement to publish this work from more than 100 college teachers from 60 different institutions who participated in my Chautauqua lecture program, "Patterns of Problem Solving," in 1973–74. This program was sponsored by the American Association for the Advancement of Science with support from the National Science Foundation.

The course Patterns of Problem Solving has been approved by the UCLA College of Letters and Science as a body of knowledge which satisfies the Physical Sciences breadth requirement. Professor Russell R. O'Neill, Dean of the School of Engineering and Applied Science, was very helpful in achieving this recognition for the course and has supported it, both in deed and spirit, from its inception.

My wife, Zafrira, did the research for some of the "stories" in the book, and discovered Ahmed of the "Ahmed story" which became the first page of the book. She also sat through a complete offering of the course, when I taught it, and made important suggestions which improved the presentation.

My daughters, Iris and Dorit, acted as a sounding board to many ideas and anecdotes in the book. This they did with sincere interest and remarkable patience.

Edith Corsario and Norma Gear, my Administrative Assistants, displayed an unusual devotion and interest, keeping track of additions and revisions and locating misplaced material. Jean Dubinsky assisted me with a meticulous proof reading of galleys.

Estelle Ratner and Ken Long assisted with suggestions in the production of early versions of the manuscript. Jacki Davis did most of the typing of the manuscript.

I also acknowledge the efforts of Margaret McAbee, production editor, Prentice-Hall.

The idea that led to the development of this material was sparked by Dr. Chauncey Starr, former Dean of the School of Engineering and Applied Science at UCLA.

MOSHE F. RUBINSTEIN

1

PROBLEM SOLVING

1-1 CULTURE, VALUES, AND PROBLEM SOLVING

The Ahmed Episode

One summer my wife and I became acquainted with an educated, well-to-do Arab named Ahmed in the city of Jerusalem. Following a traditional Arabic dinner one evening, Ahmed decided to test my wisdom with his fables. One of them caught me in a rather awkward setting. "Moshe," he said as he put his fable in the form of a question, "imagine that you, your mother, your wife, and your child are in a boat, and it capsizes. You can save yourself and only one of the remaining three. Whom will you save?" For a moment I froze, thoughts raced through my mind. Did he have to ask this of all questions? And in the presence of my wife yet? No matter what I might say, it would not be right from someone's point of view, and if I refused to answer I might be even worse off. I was stuck. So I tried to answer by thinking aloud as I progressed to a conclusion, hoping for salvation before I said what came to my mind as soon as he posed the question, namely, save the child.

Ahmed was very surprised. I flunked the test. As he saw it, there was one correct answer and a corresponding rational argument to support it. "You see," he said, "you can have more than one wife in a lifetime, you can have more than one child, but you have only one mother. You must save your mother!"

Culture, Values, and Problem Solving Differences

Two cultures clashed here, two distinctly different value systems, Ahmed's and ours. The same problem, two different value systems; therefore, two different criteria, different decisions, and different solutions. This is the problem of problems, the subjective element of problem solving and decision making. Man's value system, his priorities, guide his behavior as manifested in problem solving and decision making. Two people, using the same rational tools of problem solving, may arrive at different solutions because they operate from different frames of values and, therefore, their behavior is different. When society faces problems, consensus must be formed to establish a societal value system. The subject of values and value theory is indeed the problem of problems and is treated in the last chapter of this book. See Section 10-2 for a discussion on the role of values in problem solving.

Aftermath of the Ahmed Episode

I told the story to a class of one hundred freshmen and asked for their responses. Sixty would save the child, and forty the wife. When I asked who would save his mother, there was a roar of laughter. No one raised his hand. They thought the question was funny. They were also quite amazed to learn of Ahmed's response.

A group of about forty executives whom I addressed on problem solving and decision making responded as follows: More than half would save the child, less than half would save the wife. One reluctantly raised his hand in response to "Who would save his mother?" (I believe he had an accent. . . .) I promised the group to send the mothers sympathy cards.

The executives were apparently impressed by the story because, at a dinner party that followed the lectures, Ahmed's question was ringing all over the place. Across the table from me sat one of the course instructors and his wife. Both came from Persia and spent the last seven years in the United States. She wanted to know what the conversation about mother, wife, child was all about. Her husband related the story, and she came up with a response immediately: "Of course I would save my mother, you have only one mother." Here her values were a perfect match to Ahmed's culture. But then she turned to her husband and added: "I hope you won't do that." The influence of new values in the USA, or did she mean specifically her mother-in-law. . .?

Most of our friends reacted as if it was natural to save the child. One, an artist, said that she would probably drown before she could ever decide what to do. . . .

1-2 SCHOOLS OF THOUGHT IN PROBLEM SOLVING

There are two approaches to research in human problem solving: the behaviorial and the information processing.

The behaviorists view problem solving as a relationship between a stimulus (input) and a response (output) without speculating about the intervening process. Skinner [1] describes problem solving in terms of behavior as follows: "A hungry man faces a problem if he cannot emit a response previously reinforced with food; to solve it he must change either himself or the situation until a response occurs. The behavior which brings about the change is called problem solving and the response it promotes, a solution.

"A question for which there is at the moment no answer is a problem. It may be solved by calculations, by consulting reference work, or by acting in a way which helps recall a previously learned answer.

"Since there is probably no behavioral process which is not relevant to the solving of some problem, an exhaustive analysis of techniques would coincide with analysis of behavior as a whole."

The *information processing approach* to problem solving is based on the information processing that accompanies the development of computer programs. Here the emphasis is on the process that intervenes between input and output and leads to a desired goal from an initial state. In other words, it seeks to establish the pattern or form of solution as the problem solver is thinking aloud while he proceeds to achieve a desired goal. It attempts to uncover or codify, if possible, the transformation rules that connect input and output, i.e., system identification.

Simon [2], a proponent of the information processing approach, states:

> "The activity called human problem solving is basically a form of means-end analysis that aims at discovering a process description of the path that leads to a desired goal . . . . Given a blueprint find the recipe, given the description of a natural phenomena find the differential equations for processes that will produce the phenomena . . . . Problem solving requires continual translation between the state and process descriptions of the same complex reality . . . ."

As examples of state and process descriptions, Simon cites:

> *State:* "A circle is the locus of all points equidistant from a given point."
>
> *Process:* "To construct a circle, rotate a compass with one arm fixed until the other arm has returned to its starting point."

We *pose* a problem by defining the desired goal in terms of the state description. We *solve* the problem by selecting a process that will produce the desired goal from the initial state. Here the transition from the initial state through the process to the goal will tell us when we have succeeded. The total process, i.e., the solution, may be entirely new although parts of it may not be new. Therefore, we need not resort to Plato's theory* of remembering so we can recognize a solution.

*Plato argues (in the *Meno*) that problem solving is remembering because how else could we recognize the answer to a problem if we did not know it.

1-3 MODELS OF PROBLEM-SOLVING PROCESS

Some psychologists describe the problem-solving process in terms of a simple model that consists of these stages:

Stage 1 *Preparation.* You go over the elements of the problem and study their relationship.

Stage 2 *Incubation.* You "sleep over" the problem. You may be frustrated at this stage because the problem has not been solved yet.

Stage 3 *Inspiration.* You sense a spark of excitement as a solution (or a route to a solution) suddenly appears.

Stage 4 *Verification.* You check the inspired solution against the desired goal.

This model is based largely on the experiences of scientists who have solved difficult problems by inspiration. Such an inspiration is described by Descartes [3] when he discovered the Cartesian coordinates as the solution to the problem of establishing a connection between algebra and geometry.

More sophisticated models of the problem solving process have been proposed. The four stages listed above may, in some situations, take place in parallel rather than in series. The general problem solving theory of Newell, Shaw, and Simon considers the situation in which the problem solver proceeds from the initial state to the desired goal through a set of operations with no distinctly defined intervening stages. Many riddles and the proof of theorems in mathematics are problems of this type. GPS (*G*eneral *P*roblem *S*olver) [4] develops a computer program for the solution of such problems. It uses a means-end analysis, in which a step-by-step procedure examines a description of current status (starting from the initial state) and compares it to the desired end result until a solution is obtained. At each step the degree of misfit between the current state and the desired goal is established. The program gives a number of transformation rules to change a current state to a new one (and also provides a test for the relevance of particular transformations to a given misfit). As an example, the program solves the riddle of the three missionaries and three cannibals who must cross a river by using a rowboat with a capacity of two (Fig. 1-1). Each of the six can row, but the constraint is that under no circumstances should more cannibals than missionaries be present in any location (or else the missionaries will be eaten). The goal is then to have all six cross the river safely, subject to the constraints stated. Try it, and keep track of what you are doing by thinking aloud and recording on paper your step-by-step progress to a solution.

The I-P (*I*nformation *P*rocessing) theory models problem solving in terms of information processing, but not in terms of the more detailed pro-

Figure 1-1 Missionaries and Cannibals

cesses in the nervous system. I-P is a bridge between organized behavior and the function of the nerve system in problem solving. This is analogous to an engineer who works with structural behavior in terms of stress and strain, but without going directly down to the atomic level in his design process.

Artificial Intelligence

We see that problem solving is a subject of research by behavioral and natural scientists. A surge in interest occurred with the advent of the computer. The computer has brought closer diverse disciplines and has caused many interdisciplinary studies to be undertaken. Problem solving is probably the most common ground for all disciplines and is fundamental to all human activities.

Computer scientists are interested in human problem solving because knowledge of the process will enable them to find ways to improve computer programs designed to solve problems. Computer programs have been written to prove mathematical theorems, to solve algebra word problems, to play games such as chess, checkers and tic-tac-toe. Programs have been developed (and are being improved) to translate from one language to another, to compose music, and to generate art forms.

Researchers continue to develop programs which will enable a computer to recognize human handwriting (read) and voice (hear); this is part of an effort in the area of pattern recognition in which rules, guides, and better understanding are sought in the mechanism of pattern identification. The research in the broad area of computer programs for problem solving is known now as the field of *artificial intelligence.*

Research in computer problem solving may provide a better understanding of human problem solving, and vice versa. It is difficult to speculate on the degree of similarity between human and computer problem solving in the future. It is most likely that the two will complement each other, and in the process the computer will become the extension and amplification of man's intelligence as problem solver, just as the industrial revolution became the extension and amplification of man's energy and muscles.

1-4 KINDS OF PROBLEMS

Problem solving can be viewed as a matter of appropriate selection. When we are asked to estimate the number of marbles in a jar, we go through a process of selecting an appropriate number. When asked to name an object, we must select the appropriate word. In performing an arithmetic operation as simple as 8×7, we must select from our store of numbers the appropriate one.

Synthesis

We can distinguish two basic categories of problems. One consists of a statement of an initial state and a desired goal in which the major effort is the selection of a solution process to the desired explicit goal, but for which the process as a whole (i.e., the complete pattern of the solution) is new to us, although the individual steps are not. In such a case, we verify the acceptability of the solution by trying various processes for a solution and eliminating progressively (reducing to zero) the misfit between the desired goal and the results obtained from the trial processes. This kind of problem may be considered as a problem of design or *synthesis* in which a complete solution process is synthesized from smaller steps.

Analysis

The second type of problem focuses more on the application of known transformation processes to achieve a goal. The goal may not be recognized as the correct solution immediately, but can be verified by the process in such a way that no misfit exists between the conditions of the problem (initial state) and the solution. This kind of problem may be considered as a problem of *analysis* in which the solution consists of a transformation or change in representation of given information so as to make transparent the obscure or hidden.

The design of a house is a synthesis problem, while the solution of three simultaneous equations in three unknowns is an analysis problem. Give other examples for each kind of problem.

1-5 GUIDES TO PROBLEM SOLVING

Problem solving is a matter of appropriate selection, selection of ultimate solution, or selection of a process leading to a desired explicitly stated goal. What general rules exist which may guide us to improve our skills as problem solvers? After all, it is not only the work of an executive that consists of continuous chains of problem-solving and decision-making activities. It is a task required of all human beings. Descartes,* Newton, and others devised rules of reasoning in human problem solving. But the topic of problem solving is so broad, so all encompassing, that there is danger in the formalism of rules, at least at this stage of our knowledge. We shall therefore summarize in Secs. 1-6 through 1-10 difficulties and fundamental guides to problem solving that may not be all new; however, as we shall soon indicate, they will prove helpful only if we retrieve them from our storage in the brain when they should be applied. These guides will help develop a general attitude as well as strategies in problem solving that we believe can be taught, or what is more important, can be learned.

Attitude to Problem Solving

It has been said that an education is that which we remember after we have forgotten all we have learned. There is an element of truth in the statement. After we learn something, what we recall with the passage of time is modified in such a way that more and more detail is progressively omitted, but the general idea is reinforced. The reinforcing comes from new experi-

*Sections 1-11 through 1-13 discuss briefly Descartes, Newton, and Einstein as problem solvers. Although these great men dealt with many problems, the concept of *coordinates* is used as a unifying element in discussing all three as problem solvers.

ences accompanied by newly learned material that is compatible with the attitude of the individual to the general idea.

Therefore, the right attitude to problem solving is an asset. Chance, it is said, favors a prepared mind. Being prepared with the right attitude and a frame of useful general ideas will help you recognize new patterns of problem solving at the most unusual times as by serendipity.

1-6 FAILURE TO USE KNOWN INFORMATION—DIFFICULTY 1

The most common difficulty in problem solving is not lack of information, but rather the failure to use information that the problem solver has.

The reason for this common difficulty is explained in terms of the information-processing model of the brain.* The brain, like the computer,† has a storage unit and a processing unit. The storage can hold large amounts of information while the processing unit, which manipulates information, is strictly limited.

The processing unit is a rapid-access storage or short-term memory. Miller [5] states it is limited in capacity to about seven, plus or minus two, unrelated information items which may include such concepts as food, chair, or details such as the digit 3 and the letter B. Another limitation is the time required to transfer an item from the processing unit to the long-term storage, which is about 5 seconds.

For example, look at the following sequence of digits for about 10 seconds:

3 0 7 1 5 9 4 6 2 8 4 5

Now close the book, and write the sequence of digits as you recall them. On the average, people retain and can repeat immediately about seven unrelated digits. Most problems involve many more elements than can be handled by the immediate short-term memory span.

Since problem solving is a process of selection, how do we proceed to select the right elements of information that are relevant to the problem at hand? Considering the vast amount of information in the storage unit of the brain, an exhaustive test of relevance of all information is prohibitive. The processing unit therefore organizes the information in larger blocks or chunks. Information items such as apples, oranges, figs, etc., are chunked as fruit, and at a higher level of chunking, fruits, vegetables, bread, meat, etc., are

*See Sec. 5-9 for a model of the brain.

†Chapter 3 introduces the fundamental concepts and constituent elements of digital computers.

consolidated into food. The concept of food frees the processing unit from keeping all the detail that is in the storage unit under the category of food.

The chunking operation helps retain and manipulate more information in the processing unit. In fact, even the simple example of the sequence of digits will illustrate the point. If you have not done so in the first try, then return to the sequence of digits and try to memorize them in chunks of 3. Take 10 seconds. Then see how much you can recall.

The development of a pattern for seemingly unrelated elements also helps the chunking process.* Diagrams, charts, and equations help to condense information and relieve the processing unit.

To cope with the limitation of the human processing unit, *do not rely on your memory*. Instead:

(a) Write the problem down in its primitive form.
(b) Transform the primitive statement to simpler language. Simpler for you, that is. Package related information in chunks. This does not necessarily mean that you must always shorten the original statement although this may often be the case.
(c) If a verbal problem can be translated to a simpler more abstract mathematical statement, do so.
(d) If the verbal problem can be translated to a simpler more abstract diagram, chart, or graph, do so.

EXAMPLE. Try the above steps on the following problem. The combined ages of a husband and wife are 98. He is twice as old as she was when he was the age she is today. What are the ages of the husband and wife?

Solution Steps:

(a) Original statement.
(b) 1. Husband and wife are together 98 years old *today*.
    2. *Today* the husband is twice the age of the wife T years ago.
    3. *T years ago* he was her age *today*.
(c) Let us use the following symbols to write the verbal statements in the form of algebraic equations.
    $H =$ age of husband today
    $W =$ age of wife today
    $T =$ number of years back when he was her age today
    Using these symbols, the three verbal statements of part (b) take the form of these three equations:

$$H + W = 98$$
$$H = 2(W - T)$$
$$H - T = W$$

*See Sec. 5-9, "A Pragmatic Plan for Remembering."

(d) Past ← T years → Today

husband's age $(H - T)$ husband's age H
wife's age $(W - T)$ wife's age W

relations:

$H + W = 98$, statement 1 in (b)
$H = 2(W - T)$, statement 2 in (b)
$H - T = W$, statement 3 in (b)

1-7 INTRODUCTION OF UNNECESSARY CONSTRAINTS—DIFFICULTY 2

Association Constraints

Consider the following problem. You are given six pencils of equal length and asked to form four identical equilateral triangles with each side the length of a pencil. Try a solution by using six sticks of equal size. Most people find the problem difficult. Some suggest an unacceptable solution of the form

Others try all sorts of configurations on the floor or table. Apparently the word triangle, which is a figure in a plane, constrains the processing unit to test the relevance of solutions in a plane. This is an unstated constraint based on a previously learned association between triangle and plane. Try a spatial solution.

Another example of such a difficulty is observed when people are asked to solve the following problem. Nine dots are arranged in three equally spaced rows and columns as shown. Connect all nine points with four straight lines without lifting the pencil from the paper and without retracing any line. Try the solution before you read ahead.

Most people fail to find the answer because they seem to impose a constraint which leads them to work only within the boundaries of the array, as if the four straight lines must start and terminate only with one of the given nine dots. Perhaps the words *connect*, *dots*, *lines* in the problem statement evoke a previously learned association in which dots are connected by lines with the dots serving as boundaries of the connections.

A third example in which association constraints cause difficulties can be demonstrated in attempts to solve the following problem. You are asked to divide Fig. 1-2 into five equal areas with identical shapes. Try it. If you encounter no difficulty, ask your friends to try it. Can you identify the constraints which caused difficulties in attempting a solution?

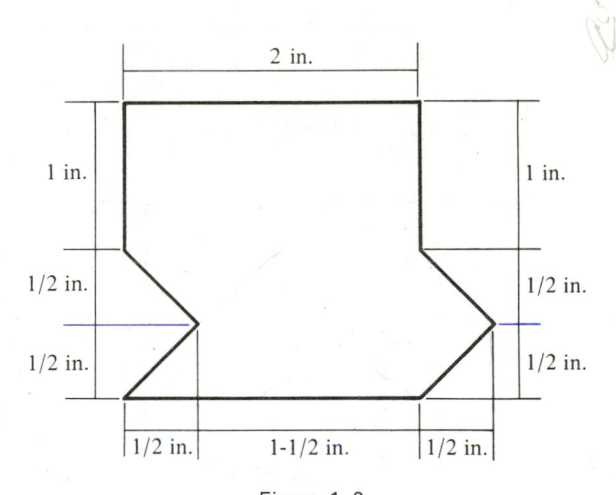

Figure 1-2

Function Constraints

Consider two strings hanging from a ceiling as shown in Fig. 1-3. You are asked to tie the two strings together. However, the strings are too far apart and you cannot hold one string and walk over to grab the other. What would you do? Try a solution before you read ahead.

The difficulty for most people in trying to solve this problem lies in the fact that their previous experiences fixed the function of a monkey wrench (placed in the room) or the function of their shoes, and they fail to see a new use in the form of a weight for a pendulum.

Tie the monkey wrench to one string and swing it so it comes to you while you hold the other string. Presumably, if the monkey wrench were replaced by a lead weight it would expedite the solution process.

Here is another problem employed by Duncker [6] in his experiments:

You are led into a room (Fig. 1-4) which is to be used for visual experiments. You are asked to put three small candles side by side on the door at your eye level so they can be used as a source of light. In the room there is

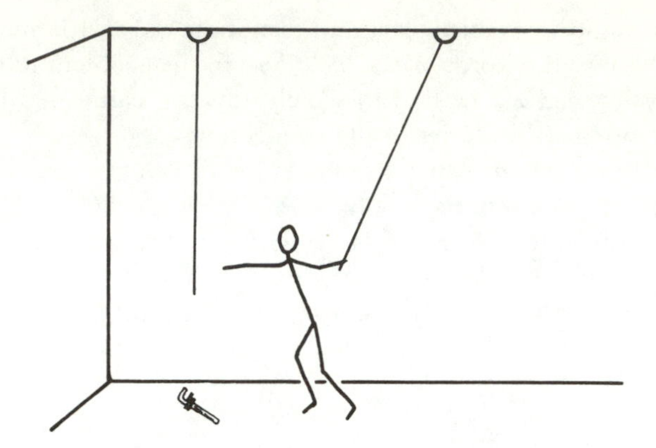

Figure 1-3

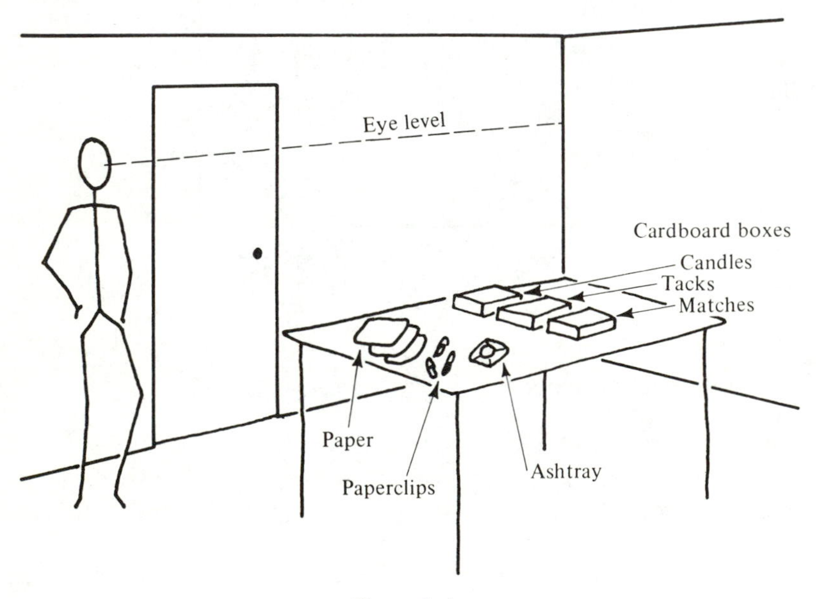

Figure 1-4

a table with various objects which you may use: an ash tray, paper, paper clips, a string, pencils, tinfoil, a small cardboard box with three candles, a small cardboard box with a few dozen tacks, and a third small cardboard box with several boxes of matches. How would you put up the candles? Try it.

When Duncker tried this experiment, 57% failed to solve the problem. In a second experiment he emptied the cardboard boxes, and all subjects solved the problem. In a third experiment when the cardboard boxes were filled with buttons, and the candles, tacks, and matches were left on the table, 87% failed to accomplish the task.

[Note the functional constraint placed by subjects on the cardboard boxes. Think about it and discuss it with a colleague, after you ask him to offer a solution.]

To overcome the introduction of unnecessary constraints, pry loose of previously learned associations and previously experienced functions. As we indicated by examples, the previously learned associations and functions may prevent your processing unit from looking at correct patterns for a solution by blocking them out as part of a chunk of information that is not relevant to the problem at hand. Therefore, look again, do not discard information with firmness, and do it only on a very tentative and temporary basis, in particular in the very early stages of the solution process.

A friend told me how he and his wife were helped when the fan belt of their car ruptured on a desert road, miles from the nearest town. A young man who stopped to help them suggested they use the woman's nylon stocking as a fan belt until they reach the next town. It worked. The young man freed himself from a functional constraint and adapted to the problem at hand. So did Samson when he used the jawbone of an ass to smite the Philistines.

World View Constraint

Churchman [7] relates the story of the mathematics professor who was asked by his students to give the next member in the sequence 32, 38, 44, 48, 56, 60. He was told that the properties of the sequence were well known to him and that the solution is simple. The professor came up with a complicated polynomial after much effort, and gave up when he could not generate a simpler solution. The answer was "Meadowlark," the elevated stop after 60th street in the city subway. The professor rode the subway daily—and got off at Meadowlark.

The professor's world view was mathematics and he related problems to this view. Most professional people have their own world view and attempt to relate problems to this view. It is important to pry loose from the tendency to refer to a fixed world view, and keep a flexibility that will enable the problem solver to explore a number of frames of reference on a tentative basis before becoming deeply committed to one.

Try to overcome the constraints that you attach to functions, those which center on previously learned associations and those which are imposed by your world view. Try also to remove truly imposed constraints on a temporary basis to simplify the problem, see what happens, and then return to the original conditions of the problem. This approach is particularly helpful when the simplification reduces the problem to one you know how to solve already. For example, the fourth order equation $x^4 - 18x^2 + 81 = 0$ can be reduced to the second order equation $y^2 - 18y + 81 = 0$ in which $y = x^2$. The solution is $y = 9$ and $x = \pm 3$. Can you think of other examples?

1-8 GENERAL PRECEPTS AS GUIDES TO PROBLEM SOLVING

The following precepts constitute a brief summary of some general guidelines to problem solving. They are generalizations of results from laboratory experiments by experimental psychologists and the studies of other researchers in the field of problem solving. As is always the case in forming conclusions on the basis of limited information, these precepts are plausible guides. We therefore maintain that they are probably true, but should be used with an open mind.

Total Picture

Before you attempt a solution to a problem, avoid getting lost in detail. Go over the elements of the problem rapidly several times until a pattern or a total picture emerges. Try to get the picture of the forest before you get lost in the trees.

I normally start the reading of technical material by going over everything with the exception of the actual text (the preface, table of contents, titles, subtitles, etc.) so as to form a total picture first. The details fall in place more easily after the total picture has been formed.

When I first started my university studies in the U.S.A. my English was only fair. I recall the frustration of reading text material with the aid of a dictionary, trying to understand each word as I proceeded. I only got more deeply lost in detail because each word was explained in terms of new words. This sent me searching for the meaning of the meaning, and from there to the meaning of the meaning's meaning. . . . I soon learned that I must try to move ahead fast several times, gleaning whatever meaning I could. After several passes, a total picture of the chapter emerged. Then the meaning of paragraphs and even single words started falling into place.

Withhold Your Judgment

Do not commit yourself too early to a course of action. You may find it hard to break away from the path, find it may be the wrong one. Search for a number of paths simultaneously and use signs of progress to guide you to the path that appears most plausible.

Models

Verbalize, use language to simplify the statement of the problem, write it down. Use mathematical or graphical pictorial models. Use abstract models such as symbols and equations, or use concrete models in the form of objects. For example, you will find it easier to work the cannibal-missionary problem by using 3 objects of one kind (pencils) and 3 of another (say, paper clips) to represent the travelers as they are moving between the two ends of your desk. You will also find it easier to use 6 sticks of equal length to construct 4 equilateral triangles with each side the length of a stick, rather than trying to sketch the figures on paper.

A model is a simpler representation of the real world problem; it is supposed to help you.

Change in Representation

Problem solving can also be viewed as a change in representation. The solutions of many problems in algebra and mathematics in general consist of transformations of the given information so as to make the solution, which is obscure, become transparent in a new form of representation. Most mathematical derivations follow this route.

Changes in representation by a transformation may form an auxiliary step that simplifies the solution process. The use of logarithms reduces multiplication and division to the simpler operations of addition and subtraction, respectively. In a similar manner, the Laplace transformations reduce the solution of differential equations to the solution of algebraic equations. Representation of a spatial or time function in terms of frequency components using a Fourier transformation is a similar change in representation employed by engineers in many fields to make the obscure transparent.

Consider the game, number scrabble (2), in which two players draw from nine cards numbered from 1 to 9. The cards are placed in a row face up. The players draw alternately one at a time, selecting any card remaining in the row. The goal is to be first to select 3 cards with a total value of exactly 15. If all 9 cards are drawn with neither player securing 15 in a set of 3 cards, it is a draw. How can this game be represented to help the player form a strategy? Try it.

Another problem that can be solved by an appropriate change in representation comes from chess. Two black and two white knights are positioned as shown in Fig. 1-5. Only the nine squares shown can be used in moving the knights.

Devise a plan to move the knights one at a time until the black and white knights trade positions. A knight can move three squares in the sequence (1,2) or (2,1), in which the first number indicates the number of squares in one direction, and the second is the number of squares in a perpendicular direction. A square can be occupied by one knight only. Knights can skip over each other in making the moves. For example, the knight in the upper left corner can move (2,1) and land in the square in row two column three, or in row three column two. The same results can be obtained by the move (1,2). The objective is to achieve a trade in positions of the knights, using the minimum number of moves.

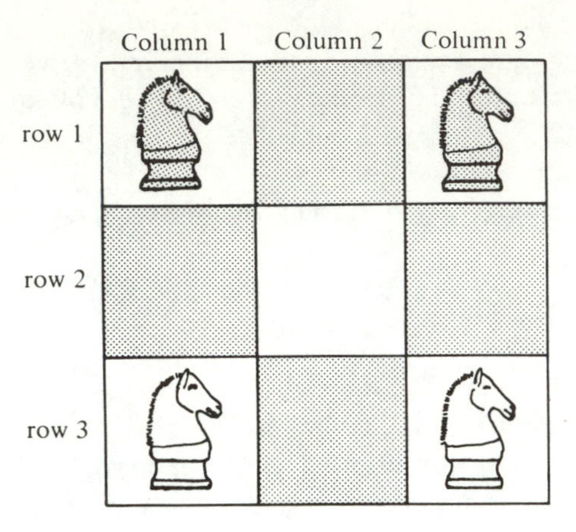

Figure 1-5

4	9	2
3	5	7
8	1	6

Figure 1-6

In the preceding problem of number scrabble, does the array of Fig. 1-6 help you construct a strategy? This array is also called a magic square in which the numbers in each row, column, and main diagonal add up to 15. Will the transformation to a game of tic-tac-toe be helpful?

Asking the Right Questions

Language in all its forms is a most powerful tool in problem solving. Asking the right question, uttering the correct word, or hearing it, may direct your processing unit to the appropriate region in your long-term storage to retrieve complete blocks of information that will guide you to a successful solution. The right question and the appropriate word may cause you to change the frame of reference as you are searching your brain for relevant information and patterns for a solution.

In the case of the two strings (Fig. 1-3), if instead of asking how you can get to the free string while you are hanging onto the other, you ask how you can get the string to come to you, you effectively solve the problem.

Hammurabi in Babylon changed the course of history by asking the right question and changing the frame of reference when dealing with the problem of water. Instead of asking how to get the people to the water, he asked how to get the water to the people. This led to canals. The rest is history. A similar situation arose when, in recent history, it was necessary to design hangars so that airplanes could be removed quickly in an emergency. It was difficult initially to achieve a successful solution with conventional designs. An outstanding creative solution was finally obtained by asking how the hangars could be rapidly removed from the airplanes so they could take off. A folding-type mobile configuration for the hangars offered the design solution.

Asking the right question may even change a negative to a positive response, as was the case with a young clergyman who was a heavy smoker. When he asked his superior whether he could smoke while praying, the response was a strong no. A wise friend suggested that next time he ask whether he could pray while smoking. . . .

Will to Doubt

The president of a large research organization related to me the following interesting development. A division of the company was awarded a government contract to study the reasons why elementary school children in a Mexican-American community could not read English. After preliminary investigation it became apparent that the basic premise was wrong. The children could read rather well; the difficulty was traced to the testing procedure. The tests used instructions such as "*underline* the word that *corresponds* to the picture." While the children could read the given list of five words and knew that the word cat described the picture, they had never encountered the words *underline* or *corresponds* before.

Have a will to doubt. Accept premises as tentative to varying degrees, but be flexible and ready to question their credibility, and, if necessary, pry yourself loose of fixed convictions and reject them. Rejection may take the form of innovation, because to innovate is, psychologically, at least, to overcome or discard the old if not always to reject it outright. Einstein overcame the 200-year reign of Newtonian mechanics, and Freud overcame the theories that dwelt on neurology and psychology to innovate in their respective fields of physics and psychology. Edison solved the problem of electric light by rejecting the direction pursued by other experimenters in his field. While they tried to get light out of current by reducing the resistance of the conductors, Edison tried it with increased resistance. While others studied the effect of a current through a filament in air, Edison tried it in a vacuum.

Keeping the premises tentative is deeply rooted in the scientific method. We do not prove hypotheses to be true in science, we simply continue to subscribe to them on a tentative basis until proven wrong. To retain a flexibility that permits us to discard an old theory in the face of new evidence, we must have the will to doubt. In a later chapter we shall use probability theory to show that when a man has no will to doubt, then it is a waste of time to present him with evidence that should reduce the credibility of a premise to which he subscribes. He will not change his mind.

The will to doubt is an important attitude in problem solving. It provides the flexibility for imagination and innovation. Hanging onto false premises and assumptions in the face of evidence of error is one of the greatest obstacles to problem solving.

We must caution, however, that the rejection of a premise must also be on a tentative basis so you can return to it when new evidence justifies it. It was probably this attitude that inspired the psalmist in the book of Psalms to exclaim: "The stone which the builders rejected became the corner stone."

An interesting experiment illustrates the differences between the flexible and rigid problem solver. A picture of a cat is shown on a screen. Subjects are asked to identify the picture. Then the figure on the screen is changed progressively in stages until the cat is transformed to a dog. The subjects are asked to identify the figures at each stage in the transformation. Those who have a will to doubt are usually quick to reject their original identification as soon as a reasonable amount of evidence is shown. The rigid subjects with no will to doubt continue to insist that they see a cat on the screen long after no dog would bark at it. . . .

When you are given the following logical argument:

Premise 1: All poisonous things are bitter
Premise 2: Arsenic is not bitter
Conclusion: Therefore, arsenic is not poison

Do not worry only about the logic. Question the credibility of the premises.

In a famous anecdote, a rabbi tells a young student about two men who, while working on a roof, fall down the chimney. One man's face is clean, the other's dirty. The rabbi asks, "Who will wash his face?" The student first answers that the one with the dirty face will. The rabbi disputes this by arguing that the reverse is true, because the clean one sees the dirty one and thus concludes he is also dirty, so he washes his face.

The rabbi repeats the question to the student a few days later. This time the student answers that the clean one will wash his face. The rabbi disagrees again on the basis that the clean man tells the other that his face is dirty, so he washes it.

At this point the student gives up, and the rabbi says: "Two men fall down through a chimney. Is it possible that one man's face is clean and the other's dirty?"

Question the premises!

1-9 PATHS TO A SOLUTION

The general guides of this chapter are not to be followed in any particular sequence and become a prescribed route to problem solving. In fact, the successful problem solvers avoid getting stuck in any prescribed direction. They explore many routes, maintain an "open mind," and a flexibility to abandon and return to various routes. There are a number of guides that are helpful in the search for a solution, once the total picture has been formed and no solution is in sight.

Working Backwards

Do not start at the beginning and follow systematically step by step to the end goal. For example, when I ask my students how many tennis matches of single eliminations must be played by 1025 players before a winner is declared, most of them start from the beginning by considering 512 games with one waiting out and so on. It is much more direct and general to start from the end and view one winner and 1024 losers and recall that each loser played one losing game in a single elimination match. This way, not only is the answer obtained, but you derive a new path to a solution. The solution path is as important as the answer and, in problems where the goal is specified, the path is the solution. You now have the solution for any number of players N.

Try to work backwards to find a solution path to the following problem. How can you get exactly 6 quarts of water from a river by using two containers with capacities of 9 and 4 quarts?

Generalize or Specialize

In the problem of the tennis matches, the number of players was not relevant, and in the solution when working from the end the problem was effectively generalized to N players. Insight is often gained by the reverse process of specializing, namely, instead of 1025 players, try 3, then 4. Both approaches are valid and helpful in the path to a solution.

Explore Directions That Appear Plausible

Do not hesitate to embark on a new direction on the basis of partial evidence. If a route appears plausible, pursue it. It is a mistake to disregard a conclusion because it is only probable and not certain. This is the basis of heuristic reasoning. Neglecting heuristic reasoning will block your progress to a solution. Of course, heuristic conclusions may lead to disappointing results, but then they are only probable and not certain to begin with. So follow routes based on heuristic reasoning, but keep your will to doubt. Heuristic reasoning is also known as plausible or inductive reasoning. It helps us go from limited information to a generalization in the form of a classification as we do in statistics where a sample is used to identify an entire population. It is also helpful in selecting satisfactory alternatives for problems in which the alternatives are not specified from the outset.

Stable Substructures

In complex problems it helps to proceed in a way that permits you to return to your partial solution after interruptions. Stable substructures that do not collapse or disappear when you do not tend to them will serve this purpose.

Simon [2] relates the story of two watchmakers who made the same watch, consisting of 1000 parts. One of them used a pattern which required that he assemble all 1000 parts without interruption because intermediate constructions were unstable and collapsed when he was disturbed by telephone calls. The second watchmaker used a construction pattern in which 10 parts were assembled as a stable substructure that could be set aside. Ten such substructures were assembled into a stable substructure of 100 parts, and then ten of these resulted in a watch. You can speculate whose business was a success.

Simon [2] also relates the fact that Lawrence of Arabia employed the same philosophy to organize a successful Arabian revolt against the Turks. He used the existing socially stable building blocks of the tribes to form his army.

Analogies and Metaphors

Use an analogy whenever you can think of one. An analogy provides a model which serves as a guide to identify the elements of a problem as parts of a more complete structure. It also helps recognize phases as elements of a complete process. For example, if your problem is a need to convey understanding of the structure and function of a digital computer to an audience which has had no exposure to it, you may devise a simple model as the one described in Sec. 3-2. The analog describes the structure and function in terms familiar to the reader.

In many cases metaphors help convey understanding. For example, a farmer who has never seen a ship in the ocean will be helped if told that the ship plows the water as it sails through it. Similarly, on a more sophisticated level, the metaphors of particle and wave in modern physics are of great help in understanding the concepts of theoretical physics.

Emotional Signs of the Right Path

When we work intently on a problem we are frustrated by slow progress and are elated by a rapid pace to the goal. Follow your emotions. If you feel elation in a certain thought, pursue it. Follow your hunch when a route "looks good to you." Here heuristics come into play. Of course, the signs of euphoria may be deceptive, but in general it is wise to follow them.

1-10 DISCUSSING YOUR PROBLEM

Talk

When you are stuck after an intensive effort to solve a problem, it is wise to take a break and do something else. It is also helpful to talk about your problem at various stages in your search for a solution. Talking to someone may help you pry loose of the constraints we mentioned, because your colleague may have a different world view and he may direct you to new avenues of search when he utters a word or asks a question. In fact, you may find an exciting inspiration while you are talking, before your listener has a chance to say a word.

Talking to someone about your problem helps at all stages: When you are very vague about the problem and find it difficult to verbalize clearly; when you have a good grasp of the problem, but consider it difficult to explain to others; or when you have a clear picture of the problem and can communicate it clearly to your listener.

The three stages above remind me of the story that is related in *The Philosophy of Niels Bohr* by Aage Peterson. In the story a young man was sent from a small isolated village to the big city to hear a famous rabbi speak. When he returned, the villagers gathered in the synagogue, anxious to hear the young man's report. "The first talk was brilliant; clear and simple. I understood every word. The second was even better; deep and subtle. I did not understand very much, but the rabbi understood all of it. The third was the finest, a great and unforgettable experience. I understood nothing and the rabbi himself didn't understand much either."

Talk about your problems at all three levels. The rabbi did too. . . .

Listen

Listen constructively to the ideas of others. It may help you get new ideas of your own and enhance your innovation and creativity in problem solving. Hyman [8] tested this precept on a group of engineers. He asked them individually to solve an automatic warehousing problem. First, the group was given solutions that were proposed earlier by others. Half the group was asked to list reasons why the proposed solutions could not work, and the other half was asked to list the strong points. Then each was asked to generate his own solution. When the results were rated for creativity by a committee of experts, those who were asked to evaluate the ideas of others constructively (strong points) had the better, more creative solutions.

While a positive attitude to the ideas of others is generally beneficial, be critical of your own ideas, tear them apart, look for all the weak points. Do not stop after you have produced one solution. Try another, and compare them critically. Push yourself to the limit. Hyman [8] asked a group of people to offer a solution for maintaining the quality of higher education in the face of an increase in student-faculty ratio. After some time interval the same people were asked to reconsider the problem and offer a second solution. Relevant statistical data was supplied in both attempts. After another time interval Hyman *pushed the subjects to the limit.* He gave them a copy of their solutions and a list of the most commonly offered solutions and asked that they offer a solution which did not make use of any of these ideas. About 25% of the subjects failed to produce a new solution. All of the others came up with acceptable new ideas; a third of these were outstanding creative solutions.

1-11 DESCARTES AS PROBLEM SOLVER

Sections 1-11 through 1-13 briefly discuss Descartes, Newton, and Einstein as problem solvers. Although these giants dealt with many problems, the concept of *coordinates* is used as a unifying element in discussing all three as problem solvers.

René Descartes (1596–1650), great philosopher and mathematician, was very much interested in the process of problem solving. In his *Rules for the Direction of the Mind* he started a plan to derive a universal method for problem solving. He never completed this effort. Descartes tells that, as a young man, he used to go through the exercise of reinventing in his mind ingenious inventions that he had heard about. This gave him the opportunity to perceive that in the process of so doing he was making use of certain rules. Descartes tried to formalize these rules in the process of problem solving. Here is what he had to say about the mental operations that will relieve our short-term memory in the process of problem solving and help us see the forest (i.e., total picture) when we deal with complex problems:

> "If I have first found out by separate mental operations what the relation is between magnitudes A and B, then that between B and C, between C and D, and finally between D and E, that does not entail my seeing what the relation is between A and E, nor can the truths previously

learned give a precise knowledge of it unless I recall them all. To remedy this, I would run them over from time to time, keeping the imagination moving continuously in such a way that while it is intuitively perceiving each fact it simultaneously passes on to the next; and this I would do until I had learned to pass from the first to the last so quickly that no stage in the process was left to the care of memory, but I seemed to have the whole intuition before me at the same time. This method will relieve the memory, diminish the sluggishness of our thinking and definitely enlarge our mental capacity."

Descartes and Cartesian Coordinates—an Inspiration?

Descartes introduced what is known as the Cartesian coordinate system. As Descartes relates, he had a remarkable vision in which the Angel of Truth appeared to him one night in 1619, while he was serving in the army as a young student from Paris, and blessed him with a supernatural capacity in science. Inspired by this vision and following a short period of study and verification, he introduced the Cartesian coordinates as the leading step in the process of fusing geometry and algebra into one, i.e., analytic geometry. Until that time the two disciplines were two separate sciences. But from then on, algebraic equations could be visualized in terms of geometric constructions and vice versa.

Cartesian coordinates consist of three mutually perpendicular rectilinear axes that emanate from a common point of origin 0 as shown in Fig. 1-7. The axes are labeled x, y, z. The position of any point P in space can be specified by giving the three coordinates of the point that are the three projected distances x, y, z from the origin to the point. For example, if we designate the coordinate of point P_1 as (0,3,2), in which the numbers correspond to x, y, z, respectively, then we locate point P_1 by walking 3 units along y, then 2 units parallel to z as shown in Fig. 1-7.

The equation of a line in the yz plane intercepting the y axis a distance c/a from the origin and intercepting the z axis a distance c/b from the origin is given by

$$ay + bz = c$$

The equation of a circle of radius r in the plane containing axes y and z and with the center at the origin is given by

$$y^2 + z^2 = r^2$$

The equation of a sphere of radius r with the center at the origin is

$$x^2 + y^2 + z^2 = r^2$$

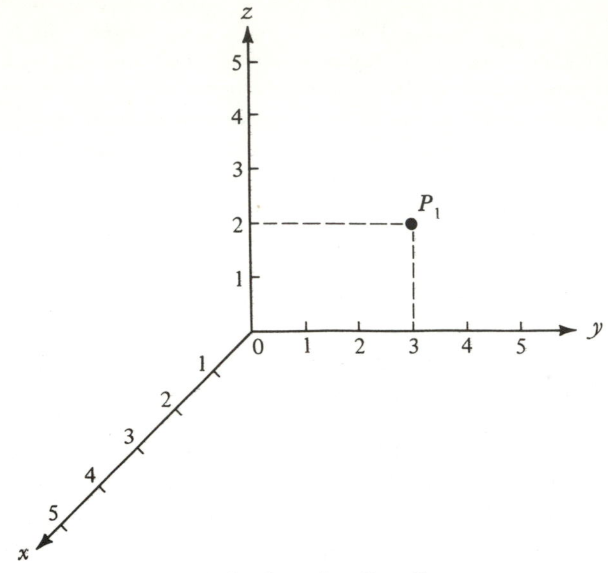

Figure 1-7 Cartesian Coordinates

Equations can be written for a parabola, hyperbola, ellipse, or a configuration such as a hyperbolic paraboloid, which is a sophisticated architectural form used in various parts of the world. The equation for the hyperbolic paraboloid shown in Fig. 1-8 is

$$z = kxy$$

in which k is a constant.

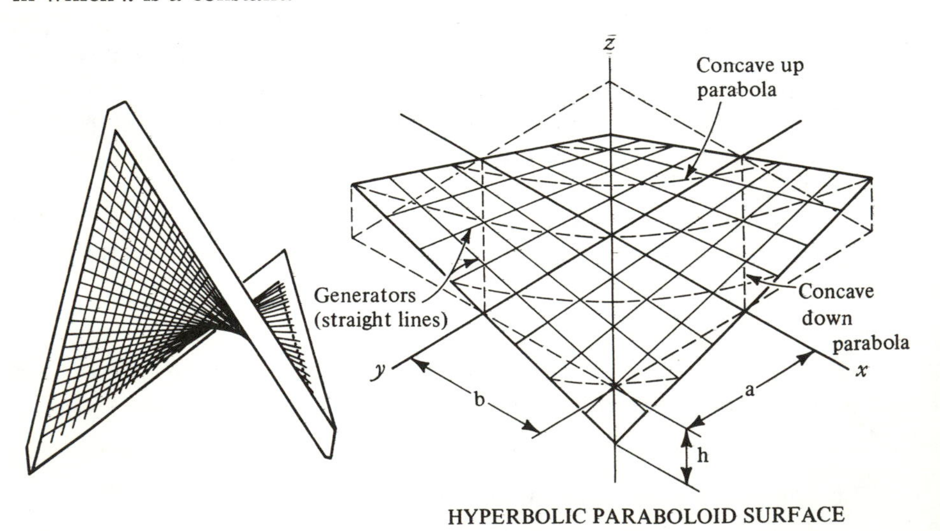

Figure 1-8 Hyperbolic Paraboloid

The Cartesian coordinates and analytical geometry provided the springboard for Newton that led to his discovery of the Law of Gravitation.

* **Problem.** Show that the hyperbolic paraboloid can be generated by straight lines, i.e., can be built from wires connected to a frame of four straight beams as shown in Fig. 1-8. Also show that when this form is intersected by a plane perpendicular to the xy plane and at a 45° angle to the x and y axes, the intersection has the shape of a parabola. In contrast, when the form is intersected by a plane parallel to the xy plane, the intersections yield hyperbolas. Can you now speculate on the origin of the name hyperbolic paraboloid?

1-12 NEWTON AS PROBLEM SOLVER—PROCESS AND GOAL

The work of Isaac Newton (1642–1727) in mechanics, his famous three laws of motion,* and his discovery of the Law of Gravitation,† guided scientific search for natural knowledge for more than 200 years. Newton's writings were first published in 1687 in the work that was entitled *Philosophiae Naturalis Principia Mathematica* (Mathematical Principles of Natural Philosophy) and commonly referred to as the *Principia* in short. *Principia* explained for the first time how a single mathematical law accounts for the motion of objects on earth (fall of an apple), the movement of heavenly bodies (motion of the planets around the sun, orbit of moon, etc.), and the phenomenon of tides.

Newton was also profoundly interested in religion, but he kept his convictions and thoughts on this matter secret, just as he did with his experiments in alchemy (transformation of basic metals to gold). In fact, Newton published very little and had to be persuaded by his colleagues to complete *Principia.*

Why Mechanics?

Why did Newton, a man with great desire to explain the world, become obsessed with the study of mechanics, the motion of objects? True, it led to a unified exploration that encompassed the free fall of objects, the oscillation of a pendulum, the motion of the planets, and it set the stage for modern

---

*Newton's three Laws of Motion:

1. A particle not acted on by an outside force moves in a straight line with constant velocity.
2. When a force acts on a particle, the particle moves in the direction of the force, and the force equals the time rate of change of momentum of the particle.
3. For every action (force) there is a precisely equal opposite reaction.

†Newton's Law of Gravitation: Every particle attracts every other particle with a force that is proportional to the product of their masses and the inverse square of their separation. (See Eq. 5-1).

astronomy. But there were other problems in natural philosophy that Newton could have studied. It was common for the early scientists-philosophers to pursue a host of problems in their quest for knowledge and desire to understand and explain natural phenomena, as Leonardo da Vinci (1452–1519) did. Why then Newton's almost total devotion to mechanics?

Newton had a goal and he considered the study of mechanics as the most appropriate process to arrive at his desired goal. Newton believed that through mechanics he could find certainty about the existence of *absolute space.*

Why Absolute Space?

Consider an object in motion. To specify its position, direction of travel, and velocity we need a frame of reference. We can use a Cartesian coordinate system with the origin placed at a certain location on earth, and the x axis pointing to a given direction, say, north. We could select the origin somewhere in New York City, in Los Angeles, or as fixed in the center of a train that is traveling at a *constant velocity* on a straight line rail connecting L.A. and N.Y.C. The choice of location for the origin of the coordinates and their orientation, i.e., direction of x axis, is arbitrary. The position of an object will change, depending on which coordinate system is used, but position is a relative concept. The velocity of the object will also be different when a moving frame of reference is employed. But then velocity, too, is relative.

Consider all conceivable coordinate systems that move with constant velocity relative to each other, *but do not accelerate or rotate.* These are known as Newton's *inertial* or *natural coordinate systems.* They are more convenient for describing mechanical phenomena than other coordinate systems. What is remarkable about the inertial frames is that the laws of mechanics appear the same in all of them. Namely, *the form* of the physical laws is independent of the coordinate system.

But can we find a point at which to fix a natural coordinate system? After all, the earth rotates in space, and the sun, one of millions of stars in the periphery of the Milky Way galaxy, also rotates, and so does the Milky Way, which is a nebula among a multitude of nebulae. All of these heavenly bodies are rotating and accelerating. With all these masses accelerating and swirling in space, is there anywhere an absolutely stationary point, a point at which the net resultant of the gravitational forces exerted by all masses in the universe is zero? Only one point—the center of gravity of the entire universe—is such a point in view of the gravitation hypothesis. Newton's contention was that the center of gravity of the universe must exist, and in it we can fix the origin of the one and only most royal member of the natural coordinate systems, the *absolute space.*

Let us compare the characteristics of absolute space, according to Newton, with the attributes ascribed to the concept of God in the Middle

Ages [3]:

Absolute Space:	*God:*
Eternal, infinitely large, cannot be created or destroyed, present everywhere, superior to all its content because it can exist without matter, but matter cannot exist without space; it is beyond the comprehension and imagination of human beings, all emanates from it, all effects are measured with respect to it.	Eternal, one, simple, immobile, complete, independent, existing through Himself, untransformable, necessary, uncreated, unlimited, omnipresent, immaterial, all penetrating, all embracing, all is emanating from Him. He is the cause of all causes.

These attributes are remarkably similar. Newton searched for absolute space. Was this his desired goal, or was it God he was searching for? The process he employed was the study of mechanics. Is it reasonable to match the supreme goal with the process? Newton's own words give a clue:

> "The business of natural philosophy is to argue from phenomena and to deduce causes from effects, until we come to the very first Cause which is certainly not mechanical."

1-13 EINSTEIN AS PROBLEM SOLVER—INTERACTION OF LANGUAGE AND THOUGHT

Existence of Absolute Space and Mechanics

Newton never succeeded in proving or gaining certainty about the existence of absolute space. He did not reach his desired goal, but his process of solution was new and rich in discoveries to a degree that changed and continues to change man's life on earth.

Albert Einstein (1879–1955) caused the search for absolute space to be discontinued when he showed that the existence of absolute space cannot be verified through experiments in mechanics.

Consider observers A and B in Fig. 1-9 who are moving with respect to each other parallel to the length of their vehicles. A may claim that he is moving and B is stationary; B may claim the reverse. For all we know, either claim may be true, or both may be moving, and yet in all cases the same relative velocity will be observed by A and B. The Laws of Nature will be the same for A and B if each will fix a frame of reference in his vehicle. Hence, A could be situated at the center of gravity of the universe, but never know it; neither would B be able to verify this.

Einstein introduced the four-dimensional space-time (x, y, z, t) which is neither eternal nor untransformable and does not have the attributes of God. Thus Einstein drove absolute space, which cannot be observed or its

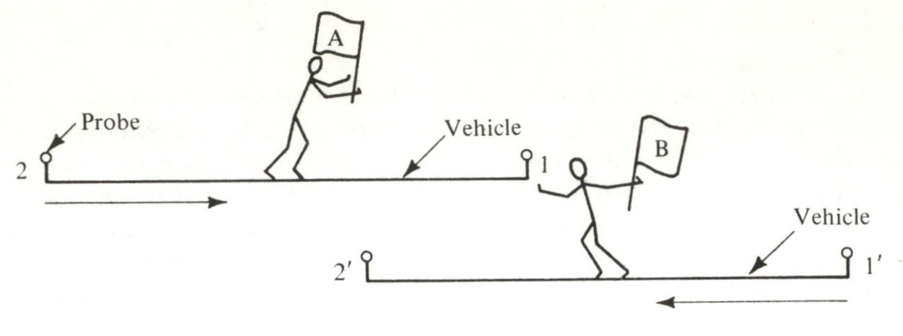

Figure 1-9

existence verified, from physics back into the domain of the supernatural or metaphysics (i.e., the domain in which empirical verification is not possible).

The next chapter is devoted to Language and Communication. It is therefore appropriate to discuss at the end of this chapter the interaction of language and thought which led Einstein to the discovery of relativity.

Interaction of Language and Thought

Einstein discovered relativity in the process of trying to understand the meaning of the abstract concept of *simultaneity* in the frame of space and time. He reduced his problem to a metaphor involving a train and the observation of bolts of lightning striking at two different places. The lightning is detected by two observers, one on the moving train and one alongside the track. Mirrors helped the observers to see both points. Einstein described it as follows in a 1916 paper [9, p. 164]:

> "If we say that the bolts of lightning are simultaneous with regard to the tracks, this now means: the rays of light coming from two equidistant points meet simultaneously at the mirrors of the man on the track. But if the place of my moving mirrors coincides with his mirrors at the moment the lightning strikes, the rays will not meet exactly simultaneously in my mirrors because of my movement." (Einstein is on a moving train.)
>
> "Events which are simultaneous in relation to the track are not simultaneous in relation to the train, and vice versa. Each frame of reference, each system of coordinates therefore has its special time; a statement about a time has real meaning only when the frame of reference is stated, to which the assertion of time refers.
>
> "Similarly with the concept of simultaneity. The concept really exists for the physicist only when in a concrete case there is some possibility of deciding whether the concept is or is not applicable. Such a definition of simul-

taneity is required, therefore, as would provide a method of deciding. As long as this requirement is not fulfilled, I am deluding myself as a physicist (to be sure, as a non-physicist, too!) if I believe that the assertion of simultaneity has a real meaning. (Until you have truly agreed to this, dear reader, do not read any further.)

"After some deliberation you may make the following proposal to prove whether the two shafts of lightning struck simultaneously. Put a set of two mirrors, at an angle of 90° to each other, at the exact halfway mark between the two light effects, station yourself in front of them, and observe whether or not the light effects strike the mirrors simultaneously."

Thus Einstein arrived at a relation between space and time. Each coordinate system has its special time, and events are described in terms of four quantities x, y, z, and t.

* A simple extension of the train story can be employed to show how objects contract in travel [10]. Consider the travelers A and B, each in the center of his vehicle moving alongside each other (Fig. 1-10). The ends of the vehicles are equipped with probes that emit a spark, like the bolt of lightning in the train story, when a probe of A just passes one of B's. Suppose the vehicles of A and B were of equal length when they took off for a space journey which finds them now passing each other in parallel. Now, as probes

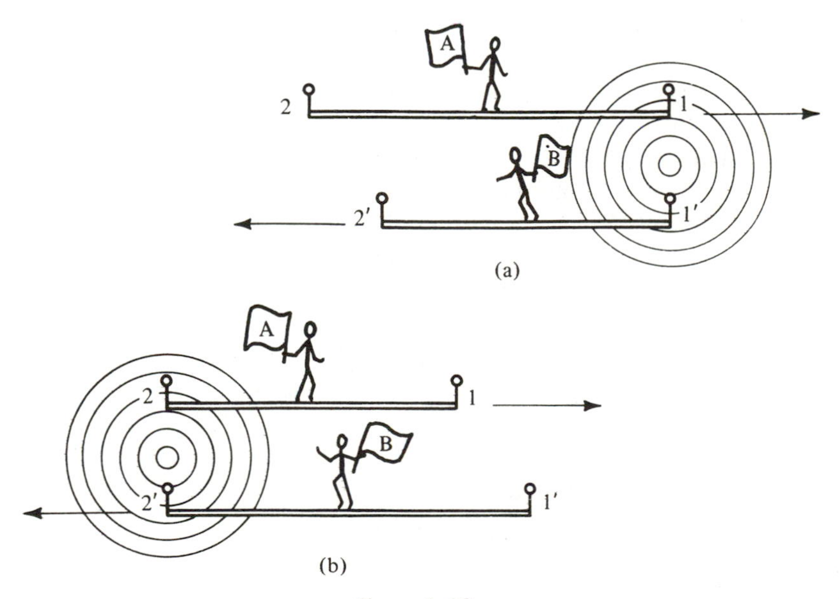

Figure 1-10

1 and 2 of A coincide with probes 1′ and 2′ of B, respectively, spherical waves of light are emitted from each end. But A is traveling away from the origin of the spark at end 2 and toward the origin of the spark at 1; hence, he will receive the light from 1 first. On this basis he will conclude that B's vehicle is shorter, as shown in Fig. 1-10(a), because that is how he will account for the later arrival of the light from the spark at probe 2.

B's reasoning will lead him to conclude that A's vehicle is shorter on the basis of the same observations, as shown in Fig. 1-10(b).

1-14 SUMMARY

There are two basic schools of thought in problem solving: *behaviorist* and *information processing*. The behaviorist approach deals with the stimulus (input) response (output) aspects of problem solving. The information-processing approach deals with the process that intervenes between input and output and leads to a desired goal from an initial state. While the models of problem solving by the two schools of thought are therefore different, they may complement each other in the quest for better understanding of human problem solving, and pave the way for new accomplishments in artificial intelligence, i.e., computer problem solving, so as to amplify human intelligence.

We distinguished *two basic kinds of problems*. The *first* consisted of a statement of an initial state and a desired goal and requires a process to lead to the goal from the initial state. The *second* consisted of an initial state that requires the application of a known transformation to achieve a goal. The first may be viewed as a *synthesis problem*, the second as an *analysis problem*.

Guides were offered to problem solving. These included a discussion of common difficulties such as

Difficulty 1: Failure to use known information, Miller's seven plus or minus two

Difficulty 2: The introduction of unnecessary constraints: association, function, and world view

General precepts were listed as guides:

1. Get the total picture
2. Withhold your judgment
3. Use models
4. Change representation
5. Ask right questions
6. Have a will to doubt

A number of concepts were introduced as milestones and paths to generating a solution:

1. Work backwards
2. Generalize or specialize

3. Explore directions when they appear plausible
4. Use stable substructures in the solution process
5. Use analogies and metaphors
6. Be guided by emotional signs of success

A problem can be discussed effectively if you take a break when you are stuck and

1. Talk about it,

provided you also learn to

2. Listen constructively to the ideas of others.

In the last three sections of the chapter we discussed Descartes, Newton, and Einstein as problem solvers. The concept of coordinates was used as a unifying element which was treated by each of these three giants of history in their creative human problem-solving activities.

The introduction to the chapter discussed the importance of culture and values in problem solving. Culture, values, and consensus regarding values are at the very heart of human problem solving and form the frame of reference for human behavior.

EXERCISES

1-1 Select a topic from this list or create one of your own.

1. Distribution of wealth
2. Enrollment procedures at a university
3. Population control
4. Transportation
5. Communication
6. Privacy and community
7. Technology and society

Now identify and state a specific problem associated with the topic of your choice. Exercises 1-2 to 1-11 inclusive refer to this problem.

1-2 Describe the present state, and the future state if no action is taken.

1-3 Describe the desired state.

1-4 Discuss values of society which could affect your approach to a solution.

1-5 What difficulties may arise due to conflict between the values of the individual and of society?

1-6 Select two value systems and describe how they might lead to different solutions.

1-7 Describe a process for achieving the desired state.

1-8 Discuss "synthesis" and "analysis" in the context of your problem.

1-9 Identify constraints on problem states, solution process, and on your activities in solving your problem.

1-10 What models, in the form of diagrams, charts, equations, or others, can you use to help you work on your problem?

1-11 Discuss how you might use the following guides:

(a) Total picture
(b) Withhold judgment
(c) Change representation
(d) Work backwards
(e) Generalize and specialize
(f) Explore alternatives
(g) Use stable substructures (modules)
(h) Analogies
(i) Chunking (aggregation)
(j) Remove unnecessary constraints
(k) Talk about the problem

1-12 From your own experience, briefly describe:

(a) a problem constrained by association constraints.
(b) a problem constrained by function constraints.
(c) a problem constrained by world-view constraints.

How do these constraints affect the attainment of the desired goal state?

1-13 Why is artificial intelligence an important field of research?

1-14 What are some of the implications of an age of "intelligent" machines?

1-15 *The 15-Tile Puzzle.* This game consists of a 4×4 grid on which there are tiles (numbered with integers from 1 to 15) and one empty space (see Fig. 1-11). In the initial state of the game, the tiles are scrambled, with the space in an arbitrary location on the grid. In the desired goal state, the tiles appear in numerical order across the rows of the grid, as shown below.

The grid is constructed such that the only way to change the order of the tiles is to slide (NOT lift) a tile into the empty space: this constitutes a "move." The transition from the initial state to the desired goal state is accomplished by a series of such moves. The objective is to achieve the goal state using the minimum number of moves.

10	14	2	9
1		12	7
15	8	3	5
6	11	4	13

(a)

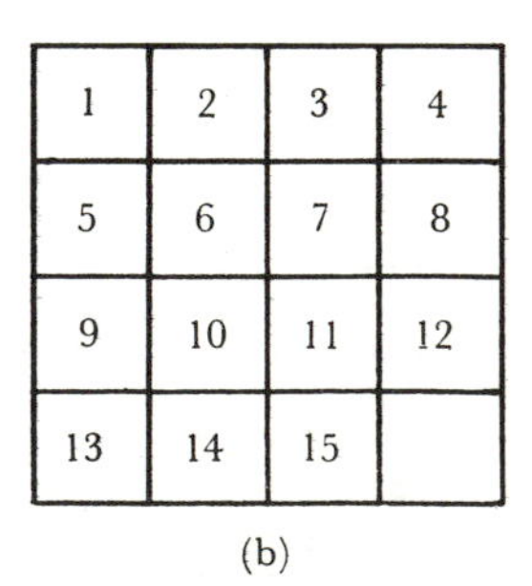

(b)

Figure 1-11 Fifteen-Tile Puzzle: (a) Initial State, (b) Desired Goal State

(a) Considering the initial state given in the figure, which constraints govern the first move?
(b) Based on the moves available, what criteria might you use to choose the "best" move? Is this the ONLY "best" move? Explain.
(c) Why is it important to conduct an ordered search for "good" moves?
(d) What problems could be encountered in such an ordered search?

NOTE: *Exercises 1-16 through 1-22* are fun problems for your enjoyment. In attempting to find solutions, try to discover your own problem-solving style. Use the guides discussed in Chapter 1; for example, work backwards, identify and remove unnecessary constraints, etc.

1-16 Two locomotives, each with 3 cars, are facing each other on a curved track, as shown in Fig. 1-12. They must pass each other and maintain their cars in the initial order. The curved main track can hold as many cars as we want, but the straight siding can hold only 3 cars, or 2 cars and 1 locomotive at one time. The locomotives A_0 and B_0 can move forward or backward, and the cars can be coupled or uncoupled at will.

Show how the goal can be accomplished, using an abstract representation of the process.

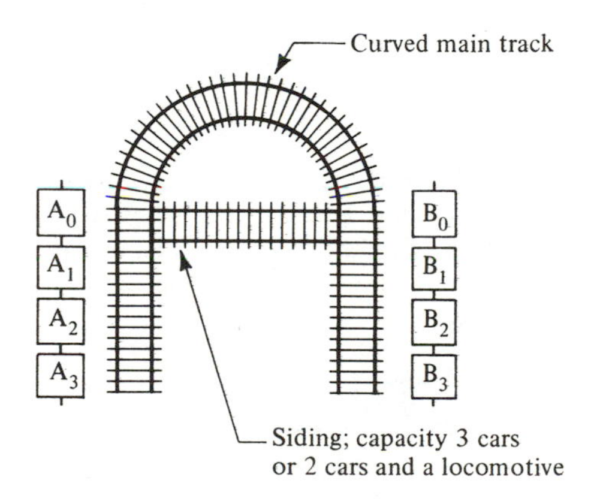

Figure 1-12

1-17 How can you derive each of the numbers 1, 2, 3, 4, 5, and 6, using the digit *4* exactly *four* times in sums, powers, etc.?

1-18 How can you divide an area bounded by a circle into ten parts with only three lines?

1-19 A rocketship is to go from Earth to Planet X. It is a five-day trip, but the rocket holds only enough fuel for 3 days' flight. Fuel stations 1, 2, and 3 are 1, 2, and 3 days' flight from Earth, respectively. The fuel stations are empty, but Earth has an unlimited supply of fuel. The rocketship can leave fuel at stations 1, 2, and 3. What is the minimum number of days required to reach Planet X from Earth?

1-20 A farmer had a square property with 24 trees, as shown in Fig. 1-13. In his will he stated that each of his 8 sons should receive the same amount of land and the same number of trees. How would you divide the land?

Figure 1-13

1-21 *Tower of Hanoi Problem*

Eight discs rest on pin 1. What is the minimum number of steps in which you can move the eight discs to either pin 2 or pin 3? A step is defined to be the movement of one disc from one pin to another. You can remove only one disc at a time, and a larger disc can never rest on top of a smaller disc, as shown in Fig. 1-14.

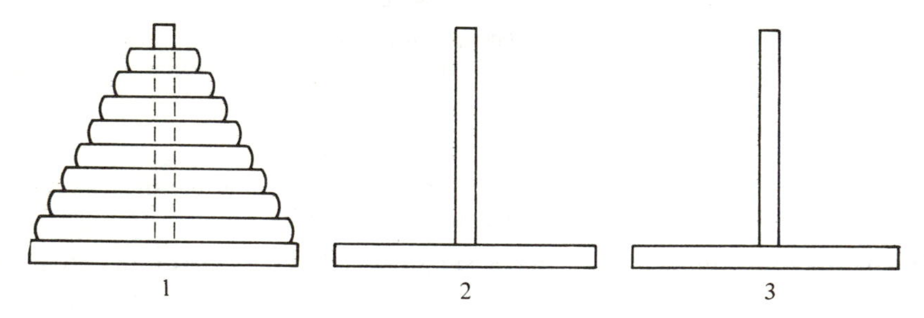

Figure 1-14 Tower of Hanoi Problem

1-22 Many years from now, two readers of this book meet on the street. The following is part of their discussion:

Man 1: Yes, I'm married and have three fine sons.
Man 2: That's wonderful! How old are they?

Man 1: Well, the product of their ages is equal to 36.

Man 2: Hmm. That doesn't tell me enough. Give me another clue.

Man 1: O.K. The sum of their ages is the number on that building across the street.

Man 2 (after a few minutes of thinking with the aid of pencil and paper): Ah ha! I've almost got the answer, but I still need another clue.

Man 1: Very well. The oldest one has red hair.

Man 2: I've got it!

What were the ages of the three sons of Man 1? (Hint: All ages are integers.)

2

LANGUAGE AND COMMUNICATION

2-1 INTRODUCTION

Man's tools have had a marked influence on the artificial world which he has created, irrespective of how primitive or sophisticated the tools were: stone, wood, metal, mechanical tools, or electronic tools. But of all tools, language is the most important and marvelous tool man has devised.

Language has had the most profound effect on man as a self-thinking social being. Language is man's tool for thinking and the basis for his communication. The definition of a community is intimately connected with the concept of communication. It has been said that a community is a geographic region in which news travels from one end to the other in one day. In the past, communities were small on the basis of this definition. Today the entire world is one community. In fact, it is quite common to hear and read the expression *world community* these days.

Language is the tool we use to form concepts, generate ideas, manipulate information, and solve problems. We form models of the physical world, as well as models of abstract unreal worlds, by using language. Through language we can communicate to others thoughts, feelings, and perceptions that they have not experienced.

The authors of the Bible tell us that the first tool used by God was the spoken word: "And God said: 'Let there be light.' And there was light." [Genesis 1:13]. In the same manner, according to the Bible, man's first act was to devise names for all the creatures around him as they were created: "And the man gave names to all cattle, and to the fowl of the air, and to every beast of the field." [Genesis 2:20].

The power of language as a tool in creating a man-made world was understood by the authors of the Bible, as can be inferred from the story of the tower of Babel:

> "And the whole earth was one language . . . And they said: 'Come let us build us a city, and a tower, with its top in heaven. . . . .' And the Lord said: 'Behold they are one people, and they all have one language; and now nothing will be withholden from them, which they purpose to do. Come let us confound their language, that they may not understand one another's speech.' So the Lord scattered them, and they left off to build the city. Therefore was the name of it called Babel;* because the Lord did there confound the language of all the earth."

This is one of the early stories of recorded history in which a communication breakdown resulted in the collapse of a mighty creation of man: the tower of Babel. The critical factor was not a physical tool, a material or human resource, but rather the most marvelous of all the gifts of man—his language.

Language enters all phases of problem solving and decision making, and it is our most profound tool for learning and thinking. It is not surprising, therefore, to learn that the many aspects of language have been studied in various fields. The humanist is concerned with the aesthetic aspects of language. The psychologist is interested in the mechanism that makes language use possible, the role of language in behavior, and the use of language to represent experience in symbolic form. The linguist treats language as a complex structure based on rules of logic and organization which permit the communication of meaning. The computer scientist studies natural language with a view of getting a better understanding of language in general so that more powerful and efficient artificial languages can be created to program computers.

In the following sections we discuss the structure of language, the origin of written language, language as a symbolic process in communication, the interaction of language and thought, and, finally, symbolic logic and the language of computers.

2-2 THE STRUCTURE OF LANGUAGE

A language may be considered as a collection of rules that specify how a particular set of symbols may be combined to form a statement which conveys meaning. A language has a vocabulary of words that are coded by means of an alphabet or a finite set of characters. The words of the vocabulary can be used to form sequences of words which are known as *strings*. Considering a vocabulary V and the set of all strings N which can be formed from

*The word *Babel* is derived from the Hebrew word *Balbel* which means "confuse."

V, then a language is a subset of N. The strings which belong to this subset are the sentences of the language.

Semiotics

The American philosopher Charles Morris [11,12] has developed a science of signs which he calls *semiotics*. Semiotics permits a classification of three major categories of rules in the structure of language.

1. Syntactics: A set of rules to determine whether a string (a sequence of words) belongs to the language.
2. Semantics: A set of rules to establish the meaning of a sentence.
3. Pragmatics: A set of rules to discern the relationship between the language and its user.

For example, the sentence *The book is good* is a syntactic assertion because the sentence belongs to the English language in terms of its structure; hence, it is syntactically correct. The sentence '*The book is good*' *is true* is a semantic assertion; and the sentence *I think the book is good* is a pragmatic assertion.

If we now change the word *good* to *bad* in the above three sentences, then the syntactic assertion remains true, the semantic assertion becomes false (negation of original statement), and the pragmatic assertion cannot be decided because it depends on what the user thinks.

Grammar and Semantic Theory

Grammar provides the rules for generating the sentences of a language. These rules generate infinitely many strings, some of which, although being syntactically correct, have never been uttered by a speaker. Grammar generates also a structural description with each sentence. Such structural description specifies the following: the constituent elements of the sentence, the grammatical relations between these elements and between the higher substructures of the sentence, the relation between the sentence and other sentences of the language, and the ways the sentence is syntactically ambiguous together with an explanation of why it is ambiguous in these ways.

Semantic theory of language takes over the explanation of the speaker's ability to produce and understand new sentences at the point where grammar leaves off.

To appreciate how much of understanding sentences is left unexplained by grammar, we compare the grammatical and semantic characterization of sentences. When we do so, we note that in the sentences *The cat eats the meat* and *The man eats the bread*, grammar provides an identical structural description of the form shown in Fig. 2-1 for two sentences with different meanings.

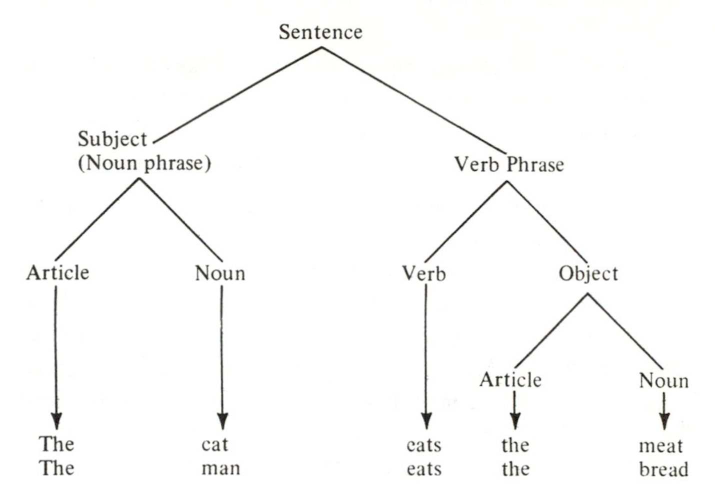

Figure 2-1 Grammatical Structure of Sentence

The structural description is essentially given by

(noun phrase, verb phrase)

Or in more detail

(article, noun, verb, article, noun)

On the other hand, sentences which have identical meaning may have different grammatical structures. For example, *The dog bit the boy* and *The boy was bitten by the dog.*

Language and Metalanguage

There are levels of semantic meaning as conveyed by corresponding levels of a language. Object language talks about objects. For example, *The book is big* is object language. We can substitute an actual book for the word *book* in the sentence and retain the meaning conveyed by the original proposition.* But we could not substitute an actual book in the sentence *'Book' has four letters,* because here the name of the object is considered rather than the object itself. Such a sentence is *metalanguage.* In a similar manner, the sentence *'Book has four letters' has four words* is a *meta-metalanguage* because here metalanguage is the object of conversation.

*It is interesting that in the Hebrew language the word *davar* means both word and thing, as if to remind us that the fundamental level of language is object language in which a thing can replace a word, and vice versa, without loss of meaning in the process of communication. In fact, the Hebrew words that deal with talking are derived from the root of the word for thing, i.e., *davar.*

In summary, the following semantic levels are associated with language:

semantic level zero: object language; talks about objects
semantic level one: metalanguage; talks about names of objects
semantic level two: meta-metalanguage; talks about metalanguage or the names of names

Higher semantic levels are possible by further extension of the remoteness from the object language.

In object language, belief and disbelief are expressed by the absence or presence of the word *not*. For instance, *The book is good*, and *The book is not good*. In metalanguage, belief and disbelief are expressed by the words *true* and *false*: *It is true that the book is good* or *It is false that the book is good*. Hence, *true* and *false* are metalanguage because they refer to statements about things. They have no meaning when referred directly to objects. Sentences such as *Chalk is false* and *House is true* have no meaning.

2-3 COMMUNICATION AND NATURAL LANGUAGE (ENCODING EXPERIENCE THROUGH LANGUAGE)

Language gives us the capacity to represent experience in a symbolic form, namely, we can encode our experience through the use of language. Language serves an important role in the operations of our mental processes known as cognition, and it is our medium for communication. Communication requires a receiver and a transmitter; we serve as both alternately, as we communicate with others. But communication also requires that encoding of experience via language have a meaning socially shared by both receivers and transmitters. Effective communication requires that we encode our experience in symbols that are compatible with the receiver's encoding rules.* This is not always simple, because the symbolic encoding of experience differs with the age level, and depends on the total being and experience of man. This includes genetic, environmental, and cultural factors. We illustrate these differences in the following discussion. First, the influence of the limited experience of the young.

Private Encoding versus General Encoding

Robert M. Krauss [13] relates an interesting anecdote from his research. In a study of linguistic communications of four-year-old children he formed teams of two, a speaker and a listener seated at two ends of a table

*This is why metaphors are so helpful. Thus, for example, when you are trying to communicate to a farmer your experience of observing the motion of a ship, you may say that the ship plowed the water.

with an opaque screen separating them. Each child was given the same set of blocks. The speaker was asked to describe to the listener the design which he saw on each block as he was stacking them. The listener was asked to identify the blocks as they were being described and stack his blocks the same way as the speaker. One block had the numeral 3 on it. The speaker kept referring to the figure as *sheet*. When asked why he called it *sheet*, he responded: "Have you ever noticed when you get up in the morning the sheet is all wrinkled? Well, sometimes it looks like this." It is true that sometimes it does, but this encoding is not shared by many and as such it is a poor encoding when communication is the goal. As a consequence, the listener found it impossible to decode the message *sheet*, and the two were not communicating. We may view the encoding of *sheet* for the experience of seeing the numeral 3 as a private encoding of the transmitter. Being private, no communication takes place because it is incompatible with the receiver's encoding of such an experience and, therefore, he cannot decode it, i.e., translate the code to the experience of the transmitter.

Words and Concepts

The concepts associated with words such as chair, table, and pen contain the most general characteristics, and as we utter these words the receiver decodes them to images stored in his mind. These produce the general configuration void of detail such as shape (round, square), dimensions (short, long, high, low), material, color, etc. Hence, the words are an abstraction of the objects achieved through a generalization that conveys only the basic features.

The concept of a number is more complex, and the concept of zero requires further sophistication. Zero was first mentioned in the works of a Hindu mathematician, Aryabhatta (c. 500 A.D.). It came only after the idea of showing values by position evolved, and gave us our positional number system.

Definitions

The concepts of force, gravity, and mass are even more difficult to form. To learn such concepts requires that we first learn other underlying concepts. When concepts are defined in terms of other concepts we call such definitions *circular definitions*. Consider, for example, the following definitions in geometry:

Point: the intersection of two *lines*
Line: the trace of a moving *point*

It is something like a cat chasing its own tail (Fig. 2-2).

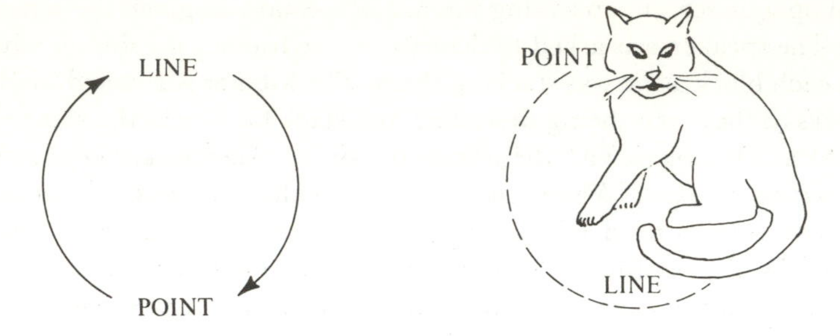

Figure 2-2 Circular Definitions—Point, Line

In a similar manner the equation which is derived from Newton's Second Law,

$$\text{Force} = (\text{Mass})(\text{Acceleration})$$

can be viewed as a circular relationship between the concepts mass, force, and acceleration,* in which each concept can be defined in terms of the two remaining concepts in the equation. Thus, for example, mass is a property of an object that reveals itself through the relation between the force applied to the object and the resulting acceleration.

Definitions of words in natural language are circular because any effort to define all words in a language must be in terms of the words in the language. We therefore conclude that a dictionary cannot define a single word because it attempts to define all words; it must go in circles.

Definitions, therefore, are possible only in terms of concepts already introduced. In constructing a hierarchy of concepts in the field of mathematics, the first concept or concepts in the hierarchy cannot possibly be defined. Such basic concepts that are undefined are referred to as "primitive elements" in mathematics. We may therefore consider "line" and "point" as possible undefined terms in geometry.

Culture and Language†

Culture has a marked influence on the way we perceive the world. It affects the way we represent experience, store it in our memory, and encode it through language in the process of communication. Walker [14] gives an example of the language spoken by the Wintu Indians of California. In Wintu you describe Fig. 2-3 as "The table bumps." In English you describe the same figure as "The book is on the table." The Wintu sentence is strictly topological. It shows that the Wintu-speaking Indians have a strong orientation to

*Acceleration can also be defined in terms of the concepts in the statement: acceleration is the rate of change of velocity.

†There are about 3000 languages in the world. Each of only 100 of these have more than a million speakers. Each of 13 have more than 50 million speakers.

Figure 2-3 The Book is on the Table (English Language). The Table Bumps (Wintu Language).

thought process in terms of the topology of form. This overshadows the relationship between the position of the book with respect to the table, and the presence of two objects as reflected in the English encoding of the same experience.

Studies [15] have indicated that different cultures have different divisions of the spectrum of color from purple to red. In Bassa, a language spoken in Liberia, the spectrum is divided into two: (I) purple, blue, and green are all represented by one word, and (II) yellow, orange, and red by another, as shown in Fig. 2-4.

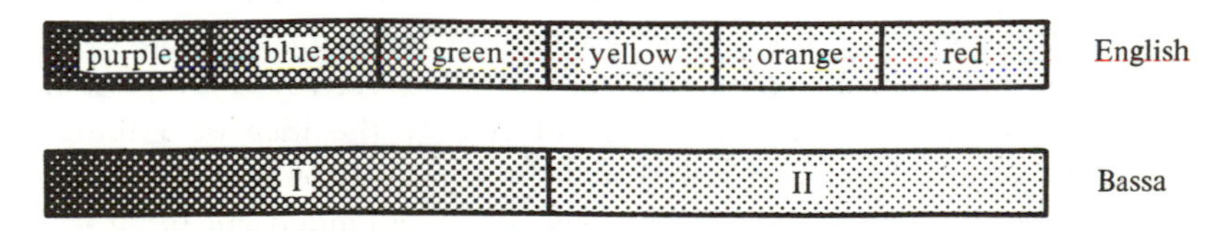

Figure 2-4 Partition of Color Spectrum in English and Bassa [15]

Here we see that two languages use different ways of mapping names given to colors, and vice versa. Red, yellow, blue, etc. are concepts, just as chair and table are. When we say that we remember seeing a red carpet, we have retained *red* in our memory. But it may be difficult for us to identify accurately the exact shade of red in the spectrum because the word *red* encodes a broad portion of the spectrum in much the same way as *chair* encodes the idea of chair, hence the entire family of chairs in an abstract way.

2-4 KNOWLEDGE OF THE LANGUAGE AND KNOWLEDGE OF THE WORLD

The ability to communicate in a language requires the knowledge of the encoding-decoding process of the language. An additional feature that helps convey meaning is our knowledge of the world, in particular when ambiguities are encountered in natural language. The following example illustrates this point. The store which carries the sign: *We sell alligator shoes* may convey two different meanings. However, here our knowledge of the world, in addition to our knowledge of the language, helps resolve the ambiguity. We know that alligators do not wear shoes, but shoes are made from alligator skin. Hence, the sentence cannot mean shoes for alligators, but rather shoes from alligator skin. But now take the sentence: *We sell horseshoes.* Could it mean that the shoes are made from the hide of a horse? Or does the knowledge of the world tell us something else? Besides, what kind of a store carries the sign?

In some cases knowledge of the world will not help. For example, in the sentence *The bill is large*, *bill* may refer to a debt or to the beak of a bird. Here the ambiguous meaning of the sentence may be resolved only in context, from its relation to another sentence. For example, *the bill is large, but it must not be paid.* Here are some more examples of ambiguous meanings: *Time flies* (noun, verb; or verb, noun), *date rocks*, *spring in the air*, *I hit the man with the stick*, *A man eating lobster*. You may think of many more.

Semantic Overtones

Our knowledge other than that of the language as such causes us to associate with some words more meaning than the words should convey. For example, when we see the words *white house* most of us in the U.S.A. see something by far more specific than the general ideas conveyed by *house* and *white*. Similarly, *fourscore* does not register merely the idea of a number. Instead, Gettysburg, Lincoln, Civil War are retrieved from our brain's long-term storage unit. Such added, and not necessarily intended, meaning which we attach to words is known as *semantic overtones*. Semantic overtones are usually connected with a repeated use of a word in a particular context. The communication media tend from time to time to propagate the repeated use of certain words as connected with certain events. For example, during the Johnson administration, whenever I heard the word *escalation*, the added words *war* and *Vietnam* came to my mind. It is interesting that such words enter even the foreign press and carry the same overtones outside the U.S.A. In a similar manner, such words as *drugs* and *space* carry semantic overtones.

2-5 EVOLUTION OF WRITTEN LANGUAGE

We are so familiar with alphabetical symbols as used to remind us of sound and to form words that we seldom stop to ask how the letters got their shapes. Instead, we attempt to develop better techniques for speed reading so that we become less conscious of words, and much less so of letters, as we extract meaning from written language.

Culturally, written language is late, but historically it is early because recorded history could not exist without the written language to record it. Primitive early cultures had no use for an alphabet and, therefore, early languages were oral languages. The written language evolved in five basic stages that resulted in an alphabet as we know it. These stages will be discussed in this section. The history of the alphabet goes back to 1300 B.C., but its origin is obscure. However, some strong evidence has been preserved in language itself which permits a plausible hypothesis on the origin of our alphabet.

The Hebrew language is the only language in which the names of all the letters in the alphabet have meaning. These meanings provide the clues to the origin of the symbols. Many dictionaries tell us that the word alphabet comes from the Greek alpha and beta. These are the names of the first two letters in the Greek alphabet; the symbol for alpha is α and for beta β. But the words alpha and beta have no meaning in Greek. On the other hand the words *Aleph* and *Bet*, which are the names of the first two letters in the Hebrew alphabet, have definite meanings that go back to Biblical times and are equally relevant in modern Hebrew. The same holds true for the names of other letters in the Hebrew alphabet. It is this meaning of the name of each letter in Hebrew which we shall use to show how the symbols of the alphabet evolved.

The Stages in the Evolution of Writing

The following five stages were involved in the development of written language:

Stage 1	Thing picture
Stage 2	Idea picture
Stage 3	Word-sound picture
Stage 4	Syllable-sound picture
Stage 5	Letter-sound picture

Examples of stage 1 are shown in Fig. 2-5, in which the picture of a warrior and an old man were each symbolic written representations of the corresponding object, i.e., the thing itself.

Stage 2 represented a higher level of abstraction in which the picture of the thing was augmented by the idea that a symbol such as a line might indicate one more of the same kind of thing. In Fig. 2-6, the six vertical lines indicated six warriors. This was the early form of the idea of a number, because the same six lines could have been shown with a picture of an ox, a child, or any other object.

Stages 3, 4, and 5 represented a revolutionary departure from the philosophy which underlaid the written language of stages 1 and 2. While stages 1 and 2 attempted to communicate directly the visual experience of things,

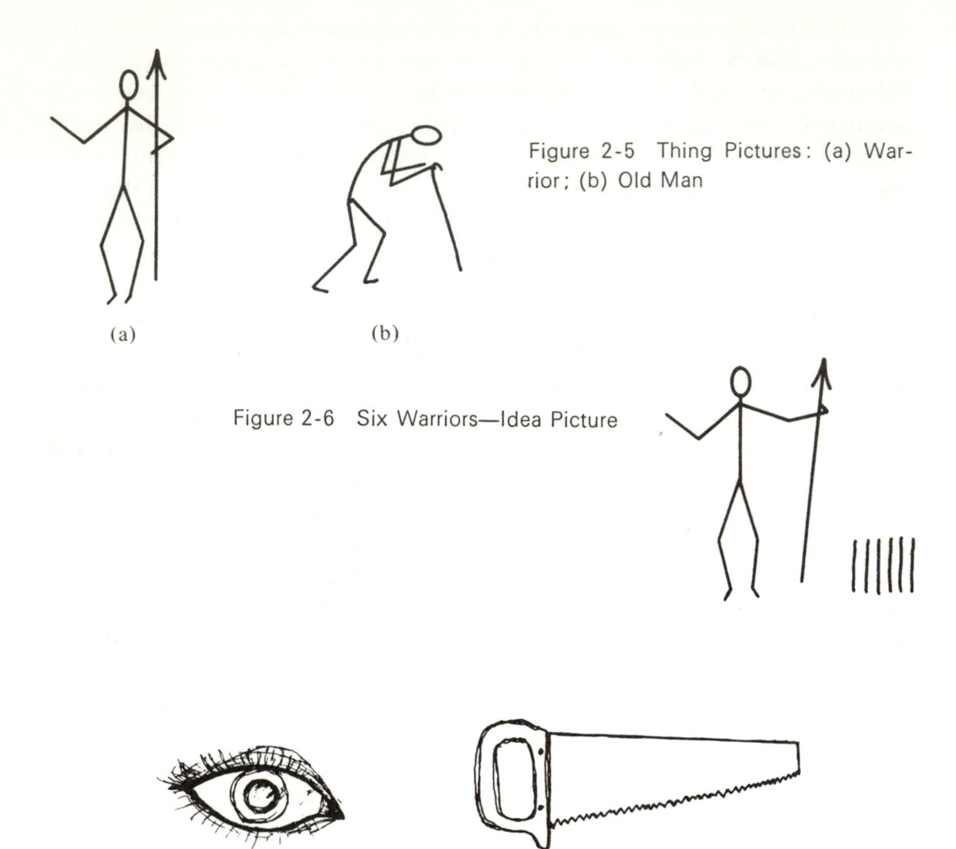

Figure 2-5 Thing Pictures: (a) Warrior; (b) Old Man

Figure 2-6 Six Warriors—Idea Picture

Figure 2-7 Word-Sound Pictures—I Saw.

stages 3, 4, and 5 attempted to communicate the same experiences through the sounds (names) that the oral language symbolically attached to things. Thus, in stage 3 a picture was created to convey the sound of a word. For example, the pictures in Fig. 2-7 could be the first two words in the sentence *I saw a man*. Here the picture of an eye was to remind you of its name, and the sound of the name conveyed the meaning *I* as the context might reveal. Of course, there was room for ambiguity, but remember this is a primitive written language. Besides, doesn't modern natural language leave room for ambiguity? (Have you forgotten: *Time flies*, or *I saw a man eating shrimp*?)

Stage 4 was an attempt to economize the number of different pictures when it was recognized that words were synthesized from syllables and that the same syllables appeared in more than one word. Stage 4 was marked by syllable-sound pictures, as shown in Fig. 2-8.

In stage 5 it was realized, apparently, that the sounds of a language could be synthesized from a small number of building blocks—the letters. However, the basic philosophy of stages 3 and 4 was retained. A picture of a known object represented the sound of the word that was the name of the object. However, now the picture was supposed to convey a single fundamental sound; therefore, only the first sound of the word was conveyed. The complete word became the name of the letter and the first sound of the word became the sound of the letter. Let us clarify this by tracing the origin of some letters.

The Origin of the Letter A

The Hebrew word for bull, ox, or champion in the Bible, as well as in modern Hebrew, is Aluf. The picture of an aluf (ox) could be recognized from an abstract figure in which the basic features of a head with horns were retained, as shown here:

The name of the symbol was changed slightly to *Aleph* and its sound was the sound of A, the first sound in its name. This form was used by the Phoenicians and the Hebrews in ancient times. The Greeks took the letter, tilted it to the right  (α) and named it *Alpha.* The Romans took the Greek symbol, tilted it another 90° clockwise,  , and that is how A was born.

The Hebrew *Aleph* remains faithful to its origin with its shape still showing the head and the horns:

The Origin of Other Letters

Two more detailed examples will suffice to demonstrate the origin of the symbols and sounds of letters. The Hebrew letter *Zayin* comes from the word Zayin which means weapon. The letter has the shape of a sword  in printed form, and a dagger in written form . It is rather easy to speculate how the Greeks got their *Zeta*  both in name and shape.

The Hebrew letter *Sheen* comes from the word Shen which means tooth. Note the teeth in the shape of the letter  . The corresponding Greek letter is *Sigma* Σ (again the Hebrew letter is turned to the right).

Table 2-1 indicates the meaning associated with the other letters of the alphabet in the Hebrew language [16]. *Beth* in Hebrew comes from the word Ba'yit which means house. The shape reflects it. *Gimel* comes from Gamal which means camel. *Daleth* from Delet which means door, and so on. In most

Figure 2-8 Syllable-Sound Pictures: (a) Chairmanship; (b) Penmanship; (c) Carpentry; (d) Carpetbag

Table 2-1 Origin of Hebrew Alphabet

Letter Name (Hebrew)	*Meaning of Name* (Hebrew)	*North Semitic* 1300–1000 B.C.	*Early Phoenician* 1300–1000 B.C.	*Modern Hebrew* Cursive	Print
Aleph	Ox-Head	𐤀	𐤀	א א	א
Beth	House	𐤁	𐤁	ב	ב
Gimel	Camel	𐤂	𐤂	ג	ג
Daleth	Door	𐤃	𐤃	ד	ד
Hey	Window (?)	𐤄	𐤄	ה	ה
Vav	Hook	𐤅	𐤅	ו	ו
Zayin	Weapon	𐤆	𐤆	ז	ז
Heth	Fence (Barred window)	𐤇	𐤇	ח	ח
Teth	Cross in circle	𐤈	𐤈	ט	ט
Yod	Hand	𐤉	𐤉	י	י
Kaf	Palm of hand	𐤊	𐤊	כ	כ
Lamed	Goad	𐤋	𐤋	ל	ל
Mem	Water	𐤌	𐤌	מ	מ
Nun	Fish	𐤍	𐤍	נ	נ
Samech	Support (post)	𐤎	𐤎	ס	ס
Ayin	Eye	𐤏	𐤏	ע	ע
Pe	Mouth	𐤐	𐤐	פ	פ
Tzade	Fisher (?)	𐤑	𐤑	צ	צ
Kof	Monkey [Back of head (?)]	𐤒	𐤒	ק	ק
Resh	Head	𐤓	𐤓	ר	ר
Shin-Sin	Teeth	𐤔	𐤔	ש	ש
Tav	Mark	𐤕	𐤕	ת	ת

cases the shapes of the letters can be traced to their origin as reflected by the name.

2-6 THE NUMBERS

At this point you may be curious to find out how the symbols for numbers evolved. You may be surprised to discover that the Arabic numerals we use are not of the same shape as those used in the Arab world. The Arabic symbols corresponding to

0 1 2 3 4 5 6 7 8 9

are ٠ ١ ٢ ٣ ٤ ٥ ٦ ٧ ٨ ٩

Table 2-2 Origin of Numbers

(Hindu 300 B.C.)

(Hindu 876 A.D.)

(Hindu 11th century)

(West Arabic 11th century)

(East Arabic 1575)

(European 15th century)

(European 16th century)

The number system is the same, i.e., a positional number system (see Chapter 3). There are various speculations as to how the shapes of our number symbols evolved from the Hindu originators through the Arabic culture of the middle ages and then 16th century Europe. See Table 2-2. I shall offer only one explanation which may easily be interpreted as the product of someone's rich imagination after the fact and no more. According to this explanation, the shape of each number contains a corresponding number of angles:

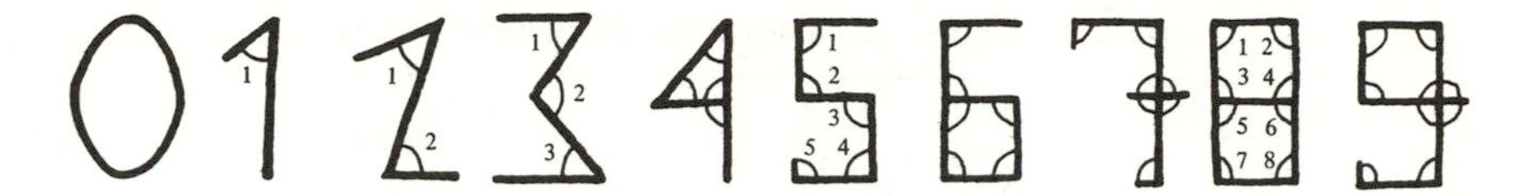

Note that 7 and 9 are stretching the theory to the limit!

Ancient symbols for numbers around 3400 B.C. had the simple form which reflected the use of reed and clay tools of those days, as shown in Table 2-3.

Table 2-3 Number Symbols c. 3400 B.C.

	1	2	3	4	5	9	10
Sumerian	V	V V	VVV	∨∨∨ ∨	∨∨∨ ∨∨	∨∨∨∨∨ ∨∨∨∨	<
Hieroglyphics	Λ	Λ Λ	ΛΛΛ	ΛΛΛΛ	ΛΛΛΛΛ	ΛΛΛΛΛΛΛΛΛ	Λ

2-7 GIMATRIA

The Hebrews, the Greeks, and the Romans used the letters of their alphabets to represent numbers. Aleph was one, Beth two, and so on. To the present day, Hebrew uses letters to designate dates, thus (Hebrew is written from right to left):

יום ב י׳א אדר ה״תשלא

5731 ADAR 11 2 DAY

There is no number in the above statement which stands for: *Monday (day 2) 11th day in the month of Adar year* 5731* (the year is from creation, according to Hebrew tradition). The numerical equivalents of the 22 Hebrew letters are shown in Table 2-4.

The use of the alphabet for numbers is not convenient in performing arithmetic calculation. It is therefore speculated that this may be one of the reasons why neither the Greeks nor the Hebrews contributed much to number theory in the field of mathematics. Instead, the use of letters for numbers, coupled with the fact that the Hebrew language is fundamentally a consonant language,† led to the development of *Gimatria.*

Gimatria is a form of numerology in which the sum of the numbers represented by the letters of a word is the *number of the word*, and two words

*The symbol " above the ה in ה״תשלא signifies that ה, which ordinarily stands for 5, designates 5000 here.

†The vowels were invented in the 7th century and appear in the form of lines and dots below the letters or above them, so that the written structure of the Bible could be preserved. It is alleged that Michelangelo placed horns on the forehead of Moses in his famous statue because he misinterpreted the Hebrew word KRN which has different meanings, depending on vowels. Keren means *horn.* Karan means *was shining.*

Table 2-4 Numerical Values of the 22 Hebrew Letters

30	ל	1	א
40	מ	2	ב
50	נ	3	ג
60	ס	4	ד
70	ע	5	ה
80	פ	6	ו
90	צ	7	ז
100	ק	8	ח
200	ר	9	ט
300	ש	10	י
400	ת	20	כ

are considered equivalent if they add up to the same number. Gimatria is still part of the curriculum for Hebrew scholars of today. Here are a number of examples of Gimatria.

The name of God, Jehovah, or YHVH as it appears in Hebrew scripture, has the following form using Hebrew letters (remember Hebrew is written from right to left)

H V H Y

יהוה

Here is how Gimatria explains the origin of this name. Consider the Hebrew sentence

היה הוה ויהיה

and shall be is was

which means: Was, is and shall be. Certainly a most concise and appropriate description of the attributes of God. Using Table 2-4, you will find that the numerical value of the sentence is 72. Now generate the name of God by using a triangle that begins with the first letter and adds letters progressively until the complete name is formed as shown:

י

יה

יהו

יהוה

The numerical value of the triangular structure is. . . (you guessed it).

Do not argue, we do not propose that this is how the name of God, YHVH, was generated, but this is how Gimatria explains it. Here is a more complex scheme. The Hebrew word Makom (means place) is another name for God. In Hebrew the word is

מקום

(ם is the same letter as מ; some letters have a different shape at the end of a word). The numerical value of the word is 186. This is equivalent to the sum of the squares of the numerical values of each of the letters in YHVH, namely,

מקום $= (\text{ה})^2 + (\text{ו})^2 + (\text{ה})^2 + (\text{י})^2$

$$40 + 6 + 100 + 40 = (5)^2 + (6)^2 + (5)^2 + (10)^2$$

The difference in numerical value between the Hebrew phrase for *this world* and *the world to come* equals nine. This is one-half the numerical value for the word *life*. Draw your own conclusions. The late Samuel Joseph Agnon, Nobel prize winner in literature, pointed out in one of his writings that *male and female* and the word *heaven* have the same numerical value. Did this inspire the phrase *matches are made in heaven*? Your guess is as good as mine. Scholem [17] discusses *serpent* and *Messiah* which have the same value.

Christian theology also made use of Gimatria to interpret the past and forecast the future on the basis of the Hebrew scripture. Peter Bungus, a theologian who lived in the days of Luther, wrote a book on numerology along the lines of Gimatria. In it he found Luther's name equivalent in numerical value to the number for Beast of Revelation; this, he claimed, proved Luther to be an Antichrist. Luther interpreted the same number as equivalent to the duration of the Papal regime. Many examples of Gimatria are also found in Greek mythology. The superiority of Achilles was attributed to the large numerical value of his name, 1276; while Patroclus and Hector were 87 and 1225, respectively [18].

2-8 LANGUAGE OF STATEMENTS—SYMBOLIC LOGIC

Symbolic logic is a language that attempts to overcome the difficulties of ambiguity, semantic overtones, and emotional response of user which are encountered in natural language. It is suitable for abstract reasoning and has many useful applications such as checking validity of arguments, simplifying circuit design, and in the logic design of computers.

The fundamental primitive element or building block in symbolic logic is a proposition. A proposition is a statement in which something is affirmed or denied of a subject in the form of a verbal or written assertion such as "The book is good," "Snow is white," or "I am taking the course Logic 141." A single proposition is a pragmatic assertion and as such we cannot prove or disprove it. We shall therefore not be able to say anything about the truth or falseness of a single proposition; however, the logical operations in symbolic logic will enable us to determine the truth or falseness of combinations of propositions.* Such combinations are referred to as *compound statements.* A single proposition is referred to as a *simple statement* or just a *statement.*

The basic axioms about propositions are:

1. A proposition is either true or false.
2. A proposition cannot be simultaneously true and false.

These axioms are the basis of a two-valued logic because a statement can be true, T, or false, F. This is distinct from a three-valued logic in which a statement can also be undecidable.

Basic Logical Operation

Let the letter p stand for the statement "I am taking Logic 141," and let q stand for: "I am taking Economics 101." The symbols T and F will be assigned to a true and false statement, respectively. The basic logical operations in compound statements to determine their truth value, i.e., T or F, are:

AND, OR, NOT

The names of these operations and the symbols that designate them in symbolic logic are:

Operation	*Name*	*Symbol*
AND	Conjunction	$\wedge$
OR (inclusive) at least one	Disjunction	$\vee$
OR (exclusive) only one, not both	Disjunction	$\underline{\vee}$
NOT	Negation	$\sim$

OR, Inclusive, Exclusive. The symbol $\vee$ comes from the Latin conjunction *Vel* which means "one or the other or both" and is the *inclusive OR* operation.

*This is similar to the definition of words in a language, or terms in mathematics and geometry (recall line and point); primitive elements are undefined. A single proposition, therefore, should be viewed as a primitive concept.

Another Latin conjunction *Aut* means "one or the other but not both." This is the operation *exclusive OR* and has the symbol $\underline{\vee}$.

Using the statements p and q above, then

$p \wedge q$	stands for:	I am taking both Logic 141 and Economics 101
$p \vee q$	stands for:	I am taking at least one of the two courses (i.e., Logic or Economics, or both)
$p \underline{\vee} q$	stands for:	I am taking one course only, Logic or Economics
$\sim p$	stands for:	I am *not* taking Logic 141
$\sim q$	stands for:	I am *not* taking Economics 101
$\sim(p \wedge q)$	stands for:	Negation of the compound statement $p \wedge q$, namely, I am *not* taking both Logic and Economics.

The last statement, $\sim(p \wedge q)$, implies three possibilities: I am taking Logic and not Economics; I am not taking Logic and am taking Economics; I am not taking either course. Using symbolic logic operations, these three compound statements can be written as $p \wedge \sim q$, $\sim p \wedge q$, $\sim p \wedge \sim q$, respectively. These last three could be summarized as not p OR not q, i.e., $\sim p \vee \sim q$. Hence, $\sim(p \wedge q)$ is equivalent to $\sim p \vee \sim q$.

Tree Diagrams

Considering two statements p and q, each with two possible values of T or F, then we can have a total of four combinations for the values of compound statements for p and q. This can be demonstrated by a tree diagram which is a very useful tool for listing all logical possibilities in symbolic logic. Figure 2-9 shows in row 1 the two possibilities, T and F, for p. Each of these can be combined with two possibilities for q, as shown in row 2. At the end of each branch of the inverted tree (the root is on top, and the end branches on the bottom) we record the sequence of T and F symbols as we read them from

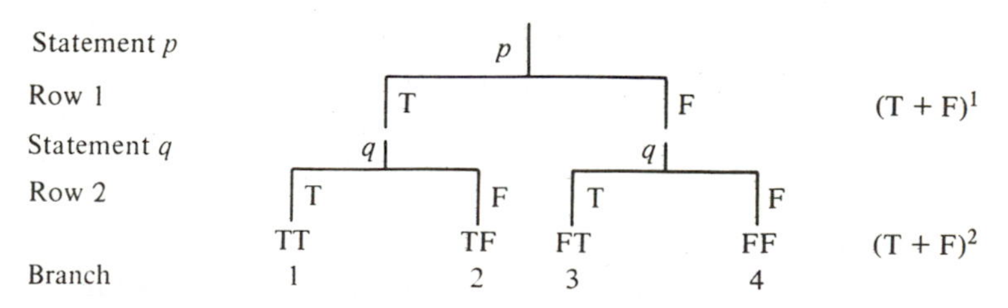

Figure 2-9 Tree to List Logical Possibilities for Two Statements

the top down along the branch. The tree may be constructed by starting with either p or q.

Figure 2-10 shows a tree diagram for three statements p, q, r. There are eight logical sequences of T, F values as listed at the ends of the branches. Again, the order of p, q, r in the rows is immaterial. We see then that the addition of a fourth statement, say, s, will appear in the form of two branches T, F at the end of each of the 8 sequences in Fig. 2-10, and thus double the

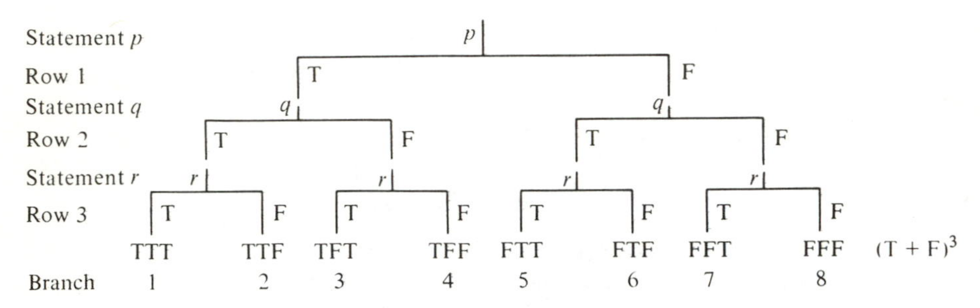

Figure 2-10 Tree Diagram for Three Statements

number of logical possibilities. The same will hold true in going from n to $n + 1$ statements. The reader can verify that there are 2^n logical possibilities for n statements. If we consider each row in the tree to represent the outcome of tossing a coin, then the tree of n rows will represent the possible outcomes of n tosses, with each outcome consisting of a string of n symbols of T's and F's with F standing for *head* and T for *tail.*

In Fig. 2-10 it can be seen that the first statement, p, has a value of T for the first four branches, and F for the remaining four. The second statement, q, has a value of T for the first two branches, F for the next two, T for the following two, and F for the last two. The last statement, r, has the highest frequency of changes from T to F. In general, for n statements, the first statement will have a value T for the first half of the 2^n possibilities, and F for the last half. The last statement will alternate between T and F in succession. The intermediate statements in the tree will alternate with increasing frequency by a factor of two from the first with two (2^1) "waves" of T's and F's (a "wave" stands here for a string of uninterrupted T's or F's); the second, four waves (2^2); the third, eight waves (2^3); the fourth, sixteen waves (2^4); and finally 2^n waves for the last. Note, however, that each statement contributes $(1/2)(2^n)$ T's and $(1/2)(2^n)$F's to the total number of logical possibilities, irrespective of how many waves it used to achieve this. It, therefore, is immaterial how we order the statements. What we are discussing here has to do with how to generate systematically all logical possibilities, once we have selected an arbitrary sequence for the given simple statements.

2-9 TRUTH TABLES

The description of the tree structure in terms of waves of uninterrupted symbols of T's and F's with an increasing frequency of twofold provides a convenient scheme for recording all logical possibilities in a table called a *truth table.* The truth table enables us conveniently to determine the truth value of compound statements on the basis of the truth values of the simple statements or building blocks. Consider Table 2-5 for two statements, p and q. Under the first statement, p (on the left), we list in a column 2^1 waves of T's and F's, and under statement q, we list 2^2 waves. In Table 2-6 we list 2^1 waves for p, 2^2 for q, and 2^3 for r. The shortest wave consists of a single

Table 2-5

Row No.	Statements		logically true	$p \vee q$	$q \rightarrow p$	p	$p \rightarrow q$	q	$p \leftrightarrow q$	$p \wedge q$	$\sim(p \wedge q)$	$\sim(p \leftrightarrow q)$	$\sim q$	$\sim(p \rightarrow q)$	$\sim p$	$\sim(q \rightarrow p)$	$\sim(p \vee q)$	logically false
Column No.	p	q	1	2	3	4	5	6	7	8	9	10	11	12	13	14	15	16
1	T	T	T	T	T	T	T	T	T	T	F	F	F	F	F	F	F	F
2	T	F	T	T	T	T	F	F	F	F	T	T	T	T	F	F	F	F
3	F	T	T	T	F	F	T	T	F	F	T	T	F	F	T	T	F	F
4	F	F	T	F	T	F	T	F	T	F	T	F	T	F	T	F	T	F

Table 2-6

Row No.	p	q	r	1	2	3	4	5	6	7	8	251	252	253	254	255	256
1	T	T	T	T	T	T	T	T	T	T	T	F	F	F	F	F	F
2	T	T	F														
3	T	F	T														
4	T	F	F														
5	F	T	T	T	T	T	T	T	T	T	T	F	F	F	F	F	F
6	F	T	F	T	T	T	T	F	F	F	F	T	T	F	F	F	F
7	F	F	T	T	T	F	F	T	T	F		F	F	T	T	F	F
8	F	F	F	T	F	T	F	T	F	T		T	F	T	F	T	F

symbol T or F, the longest wave has an uninterrupted string of $(1/2)(2^n)$ T's or F's; n is the number of simple statements.

The sequences

TT
TF
FT
FF

beneath p and q in Table 2-5 and the sequences

TTT
TTF
TFT
TFF
FTT
FTF
FFT
FFF

beneath p, q, and r in Table 2-6 correspond to the sequences generated in the trees of Figs. 2-9 and 2-10 respectively.

Interpretation of Columns 1–16 in Table 2-5

In Table 2-5 there are four entries (four rows) for each column because we have four logical possibilities for two statements. Each symbol in the four positions of columns 1 through 16 represents a truth value (F or T) for one of the four compound statements TT, TF, FT, FF for p and q in the four rows beneath p and q in the table. There are $2^4 = 16$ different arrangements (permutations) of symbols T and F in a column of four rows, and this is why we have sixteen columns. These permutations are generated systematically by considering a tree with four rows. The waves of T's and F's are recorded in the four rows because the possibilities are listed in columns. (This is the reverse from what we did for p and q.) The first row for the sixteen columns in the table has two waves, the second has four waves, the third eight, and the fourth sixteen.

Now let us interpret the significance of each column:

Column 1: Logically true regardless of the truth values of p and q. For example, if "I am taking Logic 141" (p), and "I am taking Economics 101" (q), then "December is the last month of the year" (1). The statement (1) is true regardless of the truth values of p and q.

Column 2: $p \vee q$. This compound statement is true when at least one of p or q is true, and is false otherwise (i.e., when both are false).

Column 3: $q \longrightarrow p$. This signifies that q implies p. Consider Fig. 2-11. Whenever "switch Q is closed" (we designate this as statement q), "light P goes on" (p). Whenever "switch R is closed" (r), "light P goes on" (p). Light P goes on when Q or R or both are closed; namely,

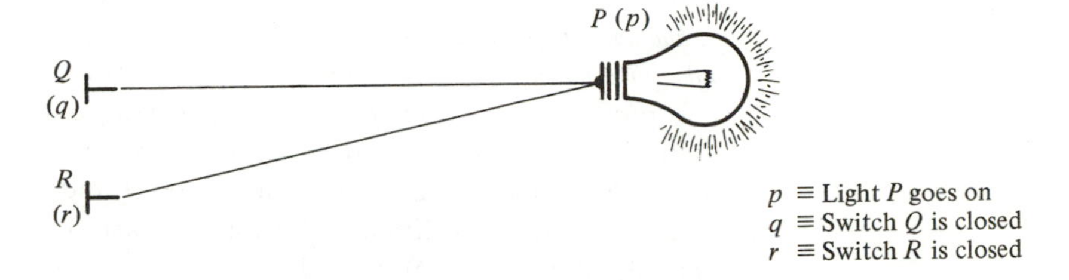

Figure 2-11 $q \longrightarrow p$ (q implies p). Light P is on when Switch Q is Closed.

p is true when $q \vee r$ is true. However, focusing our attention on statements p and q, we can say that in this arrangement p is true whenever q is true; namely, *the truth of q implies the truth of p*. We denote the last statement symbolically in the form $q \longrightarrow p$. Note that all we ascribe to the symbols $q \longrightarrow p$ is: If q is true, then p is true. This requires more discussion which follows.

Let us inspect, one at a time, the entries in column 3. In the first row we have values TT for pq; $q \longrightarrow p$ has the value T because light P is on and switch Q is closed. In the second row we have TF; light P is on and switch Q is open. This is possible because switch R could cause light P to be on. Hence, $q \longrightarrow p$ is still true because it does not violate the fact that if q is true, then p is true. In row three, FT indicates that light P is not on and switch Q is closed. This makes $q \longrightarrow p$ false because its truth implies that P is on whenever Q is closed. Hence, row three represents a condition for which the statement $q \longrightarrow p$ is false. For the fourth row, FF, light P is not on and switch Q is open. This does not violate the truth of $q \longrightarrow p$ because it is possible for both switches Q and R to be open, in which case light P is not on.

$q \longrightarrow p$ can be stated in words as "q implies p" or "if q, then p" or "a *sufficient condition* for p to be true is for q to be true." Note that the condition is *not necessary* because p can be true when q is false, i.e., when r is true. The operation $\longrightarrow$ is named the *conditional*.

Column 4: Merely a restatement of column p.

Column 5: $p \longrightarrow q$. Verify this column of entries T and F. To help you visualize the significance of the entries, interchange the symbols P and Q in Fig. 2-11, i.e., let q be "light Q is on" and p "switch P is closed," and test all four possibilities.

Column 6: Merely a restatement of column q.

Column 7: $p \leftrightarrow q$. The operation symbolized by the two-headed arrow is called the *biconditional*. It can be stated as: *if, and only if, p is true, then q is true*; or *a necessary and sufficient condition for q is p*; or *p and q are equivalent*. The positions of p and q can be interchanged in all these statements. As an example, consider the arrangement of Fig. 2-11 in which only Q and P are included and R is eliminated. In such an arrangement, light P is on (p) if, and only if, Q is closed (q), and vice versa. It can also be said that a *necessary and sufficient* condition for p is q, and vice versa. $p \leftrightarrow q$ is true only when both p and q are true or both are false. This is equivalent to $\sim(p \underline{\vee} q)$. Verify.

Column 8: $p \wedge q$; the statement is true only when both p and q are true.

Column 9: $\sim(p \wedge q)$; negation of column 8.

Column 10: $\sim(p \leftrightarrow q)$; negation of column 7. Can also be written symbolically as $p \underline{\vee} q$. Verify.

Column 11: $\sim q$; negation of column 6.

Column 12: $\sim(p \rightarrow q)$; negation of column 5.

Column 13: $\sim p$; negation of column 4.

Column 14: $\sim(q \rightarrow p)$; negation of column 3.

Column 15: $\sim(p \vee q)$; negation of column 2.

Column 16: Negation of column 1. Logically false, regardless of truth values for p and q. For example, if "I am taking Logic 141" (p), and "I am taking Economics 101" (q), then "December is the first month of the year" (16).

Note that the last eight columns are the negations of the first eight. Thus, column 16 negates column 1, column 15 negates column 2, etc. Any column number j is negated by column number $(16 - j + 1)$.

Table 2-6 considers three statements p, q, and r and all $2^8 = 256$ possible columns of T and F values in strings of 8. You may wish to consider the first 3 columns and the last 3 columns, discuss their interpretation, and convince yourself that column number j is the negation of column number $(256 - j + 1)$. Can you see the analogy to Table 2-5? Also, can you see how a similar table could be generated for p, q, r, s, t, u? How many columns would be necessary?

2-10 TAUTOLOGIES AND VALID ARGUMENTS

Tautologies

A tautology is a compound statement that is true regardless of the truth values of the constituent simple statements. For example, the compound statement

$$(p \longleftrightarrow q) \longleftrightarrow [(p \longrightarrow q) \land (q \longrightarrow p)]$$

is a tautology. To confirm this, we construct a truth table and determine the truth values of the compound statements in the parentheses in succession, then for the square braces, and finally for the entire statement as shown in Table 2-7.

Table 2-7 Verification of the Tautology $(p \longleftrightarrow q) \longleftrightarrow [(p \longrightarrow q) \land (q \longrightarrow p)]$

Column $\longrightarrow$		1	2	3	4	5
p	q	$p \longleftrightarrow q$	$p \longrightarrow q$	$q \longrightarrow p$	$2 \land 3$	$1 \longleftrightarrow 4$
T	T	T	T	T	T	T
T	F	F	F	T	F	T
F	T	F	T	F	F	T
F	F	T	T	T	T	T

In column 4: $2 \land 3$ is $(p \longrightarrow q) \land (q \longrightarrow p)$
In column 5: $1 \longleftrightarrow 4$ is $(p \longleftrightarrow q) \longleftrightarrow [(p \longrightarrow q) \land (q \longrightarrow p)]$

Syllogisms and Valid Arguments

A syllogism is a valid argument with two premises (propositions) and a conclusion. For example,

Major premise:	All men are mortal.
Minor premise:	Socrates is a man.
Conclusion:	Therefore, Socrates is mortal.

The function of logic is to check the validity of an argument, namely, the assertion that a *conclusion* follows from the premises (two or more). *An argument, therefore, will be valid if, and only if, the conjunction* ($\land$) *of the premises implies the conclusion.* Stated another way, if the premises are true, then the conclusion must be true, or

$$(\text{Premise 1 true}) \land (\text{Premise 2 true}) \land \ldots \land (\text{Premise } n \text{ true}) \longrightarrow \text{Conclusion true}$$

Note that the procedure does not propose to question the credibility of the premises. This is all right for exercises in logic, but, as pointed out in Chapter 1, *have a will to doubt and question the premises in problem solving.*

EXAMPLE 1. The validity of arguments can be checked by using truth tables. For example, consider the following arguments by referring to Fig. 2-11.

premise 1: $q \rightarrow p$
premise 2: q
conclusion: p

This argument is valid because the light P is on, i.e., p is true when both $q \rightarrow p$ and q are true. Namely, $(q \rightarrow p) \wedge q$ implies p; i.e.,

$$[(q \rightarrow p) \wedge q] \rightarrow p$$

Using a truth table, the validity of the argument is checked as shown in Table 2-8. The appearance of all T's in the last column, which represents the argument, confirms its validity.

Table 2-8 Truth Table for Checking Validity of Argument

p	q	$q \rightarrow p$	$(q \rightarrow p) \wedge q$	$[(q \rightarrow p) \wedge q] \rightarrow p$
T	T	T	T	T
T	F	T	F	T
F	T	F	F	T
F	F	T	F	T

EXAMPLE 2. Using a truth table (Table 2-9), we can also verify that the following argument is not valid (again refer to Fig. 2-11).

premise 1: $q \rightarrow p$
premise 2: p
conclusion: q

Namely,

$$[(q \rightarrow p) \wedge p] \rightarrow q$$

This is false because in Fig. 2-11 we cannot argue that "switch Q is closed" (q) on the ground of the two premises: "q implies p," ($q \rightarrow p$) and "p is true," (p). The two premises could be true with switch R closed and Q open.

Table 2-9

p	q	$q \rightarrow p$	$(q \rightarrow p) \wedge p$	$[(q \rightarrow p) \wedge p] \rightarrow q$
T	T	T	T	T
T	F	T	T	F
F	T	F	F	T
F	F	T	F	T

The last column second row of Table 2-9 confirms this reasoning and shows that the conclusion (q) can be false when the premises are true.

EXAMPLE 3. Consider a more involved argument. Let the symbols g, w, e stand for the following premises:

g = Logic 141 is a *good* course.
e = The course is *easy*.
w = The course is *worth* taking.

Here is the argument (try to translate it into English):

premise 1: $g \rightarrow w$
premise 2: $e \vee \sim w$
premise 3: $\sim e$
conclusion: $\sim g$

Table 2-10 shows the check for the validity of the argument:

$$[(g \rightarrow w) \wedge (e \vee \sim w) \wedge \sim e] \rightarrow \sim g$$

or

premise 1 $\wedge$ premise 2 $\wedge$ premise 3 $\rightarrow$ conclusion

Table 2-10

				1	2	3	← Premise Number		
e	g	w	$\sim w$	$g \rightarrow w$	$e \vee \sim w$	$\sim e$	$1 \wedge 2 \wedge 3$	$\sim g$	$(1 \wedge 2 \wedge 3) \rightarrow \sim g$
T	T	T	F	T	T	F	F	F	T
T	T	F	T	F	T	F	F	F	T
T	F	T	F	T	T	F	F	T	T
T	F	F	T	T	T	F	F	T	T
F	T	T	F	T	F	T	F	F	T
F	T	F	T	F	T	T	F	F	T
F	F	T	F	T	F	T	F	T	T
F	F	F	T	T	T	T	T	T	T

2-11 SYMBOLS, OPERATIONS, AND FUNDAMENTAL LAWS OF SYMBOLIC LOGIC

Summary of Symbols and Operation in Symbolic Logic

Operation	*Name*	*Symbols*
AND	Conjunction	$p \wedge q$
OR	Disjunction	$p \vee q$
Exclusive OR		$p \underline{\vee} q$
NOT	Negation	$\sim p$
IMPLY (If p, then q)	Conditional	$p \rightarrow q$
EQUIVALENCE (If, and only if, p then q)	Biconditional	$p \leftrightarrow q$

Fundamental Laws or Equivalent Forms in Symbolic Logic

Using truth tables, we can show that the equivalence statements in the following laws hold true. The reader should verify this.

1. Commutative Laws:

$$(p \wedge q) \leftrightarrow (q \wedge p)$$
$$(p \vee q) \leftrightarrow (q \vee p)$$

2. Associative Laws:

$$(p \wedge q) \wedge r \leftrightarrow p \wedge (q \wedge r)$$
$$(p \vee q) \vee r \leftrightarrow p \vee (q \vee r)$$

3. Distributive Laws:

$$p \wedge (q \vee r) \leftrightarrow (p \wedge q) \vee (p \wedge r)$$
$$p \vee (q \wedge r) \leftrightarrow (p \vee q) \wedge (p \vee r)$$

4. De Morgan's Laws:

$$\sim(p \wedge q) \leftrightarrow (\sim p \vee \sim q)$$
$$\sim(p \vee q) \leftrightarrow (\sim p \wedge \sim q)$$

5. Idempotent Laws:

$$(p \wedge p) \leftrightarrow p$$
$$(p \vee p) \leftrightarrow p$$

In the following, F and T stand for *false proposition* and *true proposition*, respectively.

6. Complement Laws:

$$(p \wedge \sim p) \leftrightarrow \mathrm{F}$$
$$(p \vee \sim p) \leftrightarrow \mathrm{T}$$
$$\sim\sim p \leftrightarrow p$$

7. Identity Laws:

$$(p \wedge \mathrm{T}) \leftrightarrow p$$
$$(p \wedge \mathrm{F}) \leftrightarrow \mathrm{F}$$
$$(p \vee \mathrm{T}) \leftrightarrow \mathrm{T}$$
$$(p \vee \mathrm{F}) \leftrightarrow p$$

2-12 ALGEBRA OF LOGIC AND SWITCHING CIRCUITS

The laws of Sec. 2-11 can be used in an algebra of symbols x, y, z, etc., in which each symbol can assume a truth value of T or F. Thus x could represent a switch in an electric circuit; T and F could stand for *switch is closed* and *switch is open*, respectively. The algebra of logic could then be employed to simplify the structure of complex circuits.

Consider, for example, the circuit of Fig. 2-12. A signal will flow from 1 to 3 if the following is true:

$$(x \vee y) \wedge (x \vee z)$$

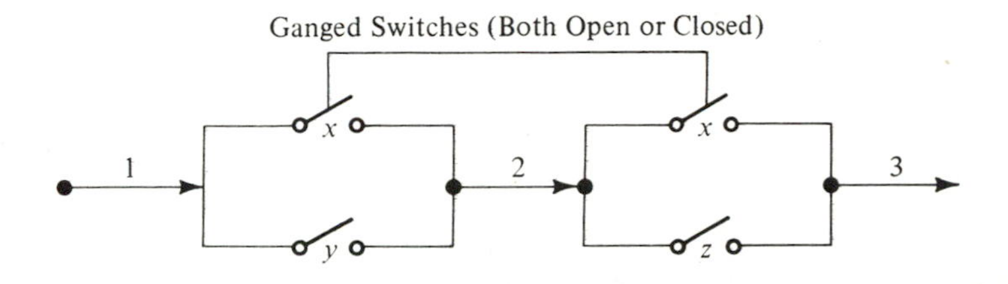

Figure 2-12 $(x \vee y) \wedge (x \vee z)$

$(x \vee y)$ guarantees that there will be a signal from 1 to 2. $(x \vee z)$ guarantees that a signal will pass from 2 to 3. To pass a signal from 1 to 3, both must be guaranteed. From the distributive laws, the above can be written as $x \vee (y \wedge z)$. This expression can be translated to a simpler circuit if we recognize that $\vee$ indicates a parallel connection of the elements (switches) on each side of the disjunction, and $\wedge$ indicates a connection in series for the elements on each side of the conjunction. The simplified circuit is shown in Fig. 2-13.

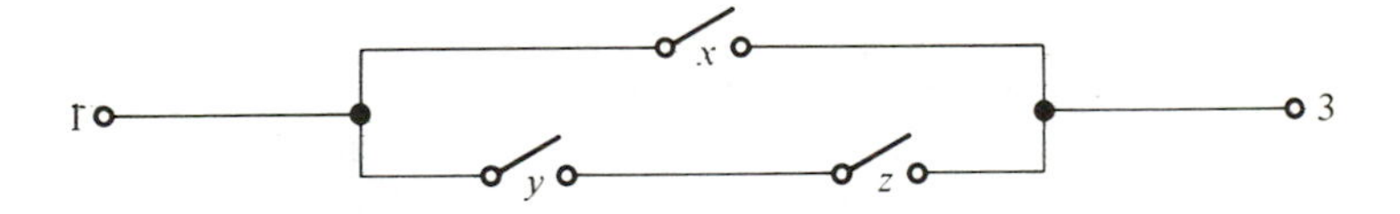

Figure 2-13 Simplified Structure for the Circuit shown in Fig. 2-12

AND, OR, and NOT Circuits

We can construct circuits that will function like truth tables for logical operations. Three such circuits, referred to as "boxes," are shown in Fig. 2-14.

In Fig. 2-14 (a), the light is on only when $p \wedge q$ is true; in (b) for $p \vee q$; and in (c) for $\sim p$. In both (a) and (b), p and q stand for closed switches. In (c), switch p is an assembly of rigidly connected segments that can rotate on a pivot. In the figure, switch p is shown in its open position, i.e., $\sim p$. When the switch is closed (p) by pushing it to the right [the same way as for the switches in (a) and (b)], then the segment below the pivot swings to the left

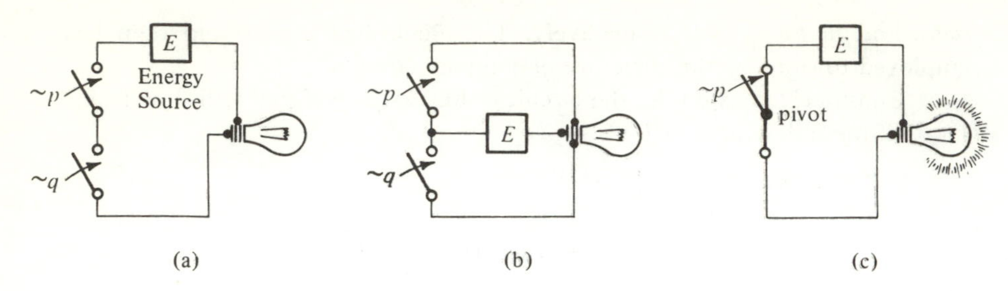

Figure 2-14 (a) AND Box—Light on when $p \wedge q$ True; (b) OR Box—Light on when $p \vee q$ True; (c) INVERTER Box (NOT Box)—Light on for $\sim p$, Light off when p True

and opens the circuit, turning off the light.* Hence, in (c) the light is on for $\sim p$ and off for p. This arrangement explains the origin of the name *NOT box* or *INVERTER box* for this circuit.

We shall make use of the AND, OR, and NOT (or INVERTER) boxes of Fig. 2-14 in Chapter 3 when we discuss the logic design of the computer arithmetic unit.

2-13 BOOLEAN ALGEBRA

If the symbols T and F are replaced by 1 and 0, respectively, so that in the algebra of logic each variable x, y, z, etc., can assume *only one* of these values, and if, in addition, the symbols $\wedge$, $\vee$, and $\leftrightarrow$ are replaced by a dot $(\cdot)$, a plus sign $(+)$, and an equal sign $(=)$, respectively, then the algebra of logic takes the form of Boolean algebra.† We must remember that the plus sign $(+)$ and multiplication sign $(\cdot)$ have the interpretation of disjunction and conjunction of logic, respectively. Thus, the seven fundamental laws of symbolic logic (Sec. 2-11) take on the following form in Boolean algebra:

1. Commutative Laws:

$$x \cdot y = y \cdot x$$
$$x + y = y + x$$

2. Associative Laws:

$$(x \cdot y) \cdot z = x \cdot (y \cdot z)$$
$$(x + y) + z = x + (y + z)$$

---

*This leads to the following configuration for the switch:

†George Boole, an English mathematician, published in 1854 his classical book, *An Investigation of the Laws of Thought on Which are Founded the Mathematical Theories of Logic and Probabilities*. This was the foundation of Boolean algebra.

3. Distributive Laws:

$$x \cdot (y + z) = x \cdot y + x \cdot z$$
$$x + (y \cdot z) = (x + y) \cdot (x + z)$$

4. De Morgan's Laws:

$$\sim(x \cdot y) = (\sim x + \sim y)$$
$$\sim(x + y) = (\sim x \cdot \sim y)$$

5. Idempotent Laws:

$$x \cdot x = x$$
$$x + x = x$$

6. Complement Laws:

$$x \cdot \sim x = 0$$
$$x + \sim x = 1$$
$$\sim\sim x = x$$

7. Identity Laws:

$$x \cdot 1 = x$$
$$x \cdot 0 = 0$$
$$x + 1 = 1$$
$$x + 0 = x$$

Using Boolean algebra in truth tables for the AND, OR, and NOT boxes of Fig. 2-14, we can construct Table 2-11.

Table 2-11 Truth Table for AND, OR, NOT Circuits of Fig. 2-14

		AND box	OR box	NOT box
p	q	$p \cdot q$	$p + q$	$\sim p$
1	1	1	1	0
1	0	0	1	0
0	1	0	1	1
0	0	0	0	1

Table of Boolean Algebra Functions Analogous to Table 2-5

Table 2-11 is a portion of Table 2-12, which is analogous to Table 2-5 and includes all 16 possible functions of two variables x and y in Boolean algebra. The functions represented by the first eight columns are:

Column 1: 1
Column 2: $x + y$
Column 3: $y \rightarrow x$ or $x \geq y$

Column 4: x
Column 5: $x \rightarrow y$ or $y \geq x$
Column 6: y
Column 7: $x \leftrightarrow y$ or $x = y$
Column 8: $x \cdot y$

Columns 9 to 16 are the negations or complements of columns 1 to 8 in such a way that column j is the complement of column $(16 - j + 1)$.

Table 2-12 Table of Boolean Algebra Functions Analogous to Table 2-5

		1	$x + y$	$y \rightarrow x;\ x \geq y$	x	$x \rightarrow y;\ y \geq x$	y	$x \leftrightarrow y;\ (x = y)$	$x \cdot y$	$\sim(x \cdot y)$	$\sim(x = y)$	$\sim y$	$\sim(x \rightarrow y);\ \sim(y \geq x)$	$\sim x$	$\sim(y \rightarrow x);\ \sim(x \geq y)$	$\sim(x + y)$	0
x	y	1	2	3	4	5	6	7	8	9	10	11	12	13	14	15	16
1	1	1	1	1	1	1	1	1	1	0	0	0	0	0	0	0	0
1	0	1	1	1	1	0	0	0	0	1	1	1	1	0	0	0	0
0	1	1	1	0	0	1	1	0	0	1	1	0	0	1	1	0	0
0	0	1	0	1	0	1	0	1	0	1	0	1	0	1	0	1	0

2-14 LANGUAGE OF SETS—SETS, SUBSETS, AND OPERATIONS ON SETS

One of the most basic and important activities in problem solving is the classification and subsequent aggregation or chunking of elements of a problem. This was discussed in Chapter 1, Sec. 1-6, and is also treated in Chapter 5, on Models and Modeling, Sec. 5-3. Elements are aggregated or chunked together by virtue of their common attributes, the strong connections of function, form, or interactive connections. First, the relevant elements are separated from those that are not relevant to the solution of the problem at hand. Next, the relevant elements are classified and aggregated in chunks, or sets, to reduce the amount of detail and transform what may be initially a complex and unmanageable problem, laden with detail, to a manageable representation of the problem.

In this section we discuss sets, which represent aggregates of objects or elements, and consider the language of sets. In the context of problem solving with the need for classification and chunking, sets and the language of sets take on a meaning and relevance which are often missing in exposures to this subject.

A set is a well-defined collection of objects. All the students at UCLA are the *elements* of a set, say, set A. A set that consists of some elements of another set is called a *subset*. Thus, if set B is the set of all students at UCLA taking Economics 101, then B is a subset of A. The original complete set A is called the *universal set* $\mathscr{U}$, and the set that contains no elements is called the *null set* or *empty set* $\varnothing$.

In the operations of symbolic logic we generated new compound statements from given statements. In an analogous manner we shall generate new sets from given sets. There is a close connection between sets and statements, and between the operations on sets and those on statements. To illustrate the analogy between sets and statements, we introduce the symbols and operations for sets, and compare them with their counterparts in the preceding sections on statements.

Symbols and Operations

A, B, C	Symbols for sets A, B, C
a, b, c	Elements of sets
$a \in A$	a is an element of set A
$x \notin A$	x *is not* an element of set A
$\mathscr{U}$	Universal set. $\mathscr{U}$ is arbitrary and depends on the "universe" of conversation or discourse.
$\varnothing$	Null set, or empty set, contains no elements. Serves a function similar to zero in the theory of numbers.
$B \subset A$	B is contained in A, or B is a subset of A.
$X \not\subset A$	X *is not* a subset of A
$E \cap L$	The *intersection* of E and L; namely, the subset containing all elements which are both in E and L. For example, let E be the set of all UCLA students taking Economics 101, and let L be the set of all UCLA students taking Logic 141. Then $E \cap L$ is the subset of the students who are taking both courses. When two sets are disjoint, i.e., they have no elements in common, then their intersection is the null set $\varnothing$.
$E \cup L$	The *union* of sets E and L. The set of elements which belong to E or L or both.
$\tilde{B}$	The complement of a set B. For example, if our universal set $\mathscr{U}$ is the set of all students at UCLA and B is the set of those who are taking Economics 101, then $\tilde{B}$ is the set of students who are not taking Economics 101. The union of a set and its complement is the universal set $\mathscr{U}$. The intersection of a set and its complement is the null set.
$\{\ \}$	The elements of a set are recorded between brackets. For example, the set of boxes in Fig. 2-14 could be designated as $\{a, b, c\}$.

Analogy of Operations on Sets and Statements

Language of Statements	*Language of Sets*
p	P
q	Q
$\sim p$	$\tilde{P}$
$p \wedge q$	$P \cap Q$
$p \vee q$	$P \cup Q$
$p \rightarrow q$	$P \subset Q$
$p \leftrightarrow q$	$P = Q$

To illustrate the analogy by an example, consider the symbols p, q, P, and Q to represent the following:

p: I am taking Logic 141
q: I am taking Economics 101
P: Set of all students at UCLA taking Logic 141
Q: Set of all students at UCLA taking Economics 101

Venn Diagrams

The same four logical possibilities which apply for statements p and q can be constructed for sets P and Q. In sets we are concerned with the two possibilities of an element belonging or not belonging to a given set P. This is analogous to the two truth values of a statement p. The four possibilities for P and Q can be identified from a *Venn diagram*, as shown in Fig. 2-15.

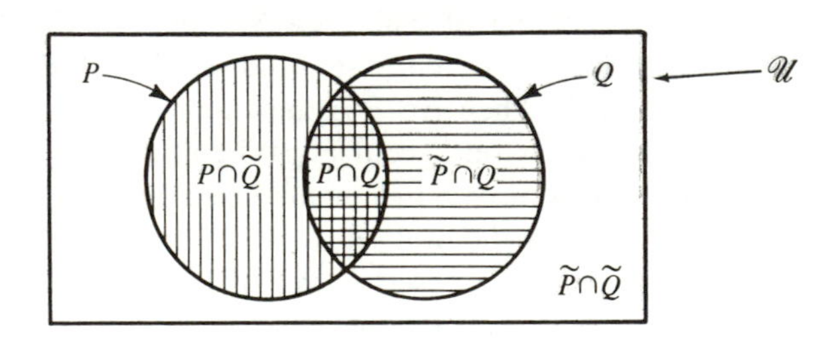

Figure 2-15 Venn Diagram for Two Sets: *P, Q*

The universal set $\mathscr{U}$ in the diagram consists of all elements in the rectangle, say, in our case, all students at UCLA. The circles P and Q identify the sets P and Q. If we now ask all the students at UCLA to stand in the appropriate region of the figure, which is the subset to which they belong, then they would position themselves as follows:

region $P \cap Q$: all students taking both Logic 141 and Economics 101 region $P \cap \tilde{Q}$: all students taking Logic 141 and not taking Economics 101

region $\tilde{P} \cap Q$: all students not taking Logic 141 and taking Economics 101

region $\tilde{P} \cap \tilde{Q}$: all students not taking Logic 141 and not taking Economics 101

Note the correspondence between the regions in Fig. 2-15 and the logical possibilities in the tree of Fig. 2-9 at the branch ends, or Table 2-5:

$P \cap Q$ TT
$P \cap \tilde{Q}$ TF
$\tilde{P} \cap Q$ FT
$\tilde{P} \cap \tilde{Q}$ FF

Similarly, for three sets P, Q, R (say, R is the set of students taking Engineering 11), the Venn diagram is shown in Fig. 2-16. (The symbol $\cap$ for the intersection has been dropped and will be considered to exist between all symbols.)

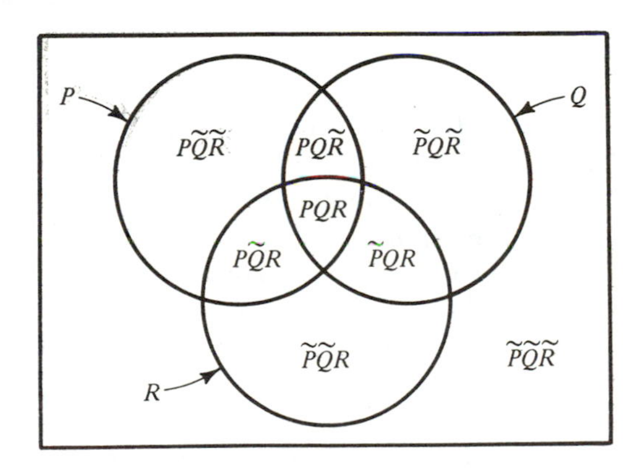

Figure 2-16 Venn Diagram for Three Sets: *P, Q, R*

The correspondence between the regions of Fig. 2-16 and the logical possibilities in the tree of Fig. 2-10 or Table 2-6 is:

PQR TTT
$PQ\tilde{R}$ TTF
$P\tilde{Q}R$ TFT
$P\tilde{Q}\tilde{R}$ TFF
$\tilde{P}QR$ FTT
$\tilde{P}Q\tilde{R}$ FTF
$\tilde{P}\tilde{Q}R$ FFT
$\tilde{P}\tilde{Q}\tilde{R}$ FFF

The frequency of waves of T's and F's in the language of statements corresponds to the frequency of absence and presence of the symbol ~ over the letters in sets.

For *P*, *Q*, *R*, *S*, sixteen regions can be identified in Fig. 2-17. Elements that belong to *P* are contained in the region above the diagonal line from the

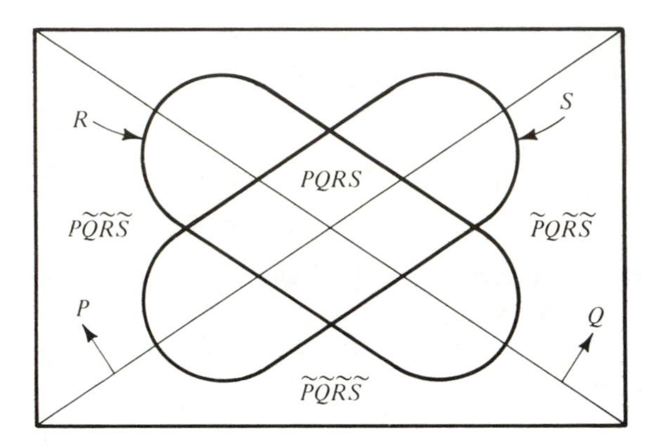

Figure 2-17 Venn Diagram for Four Sets: *P*, *Q*, *R*, *S*

upper right to the lower left corner. For *Q*, it is the region above the second diagonal line. *R* and *S* are confined to ellipses. Four of the sixteen regions are identified. Try to identify the remaining twelve in terms of a sequence *PQRS* with the symbol ~ over the appropriate letters. Figure 2-17 is due to the late Warren S. McCullough who described this diagram in a modern engineering seminar at UCLA in 1963.

Venn diagrams are not useful beyond four sets, but the regions can be generated by using trees or truth tables the same way as we generated the logical possibilities with statements.

Venn diagrams can be used for enumerating the logical possibilities of statements and, hence, also for determining the validity of arguments. For

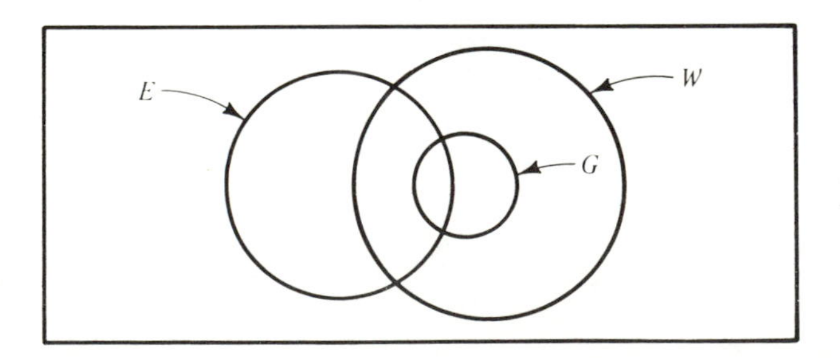

Figure 2-18

example, the first premise $g \longrightarrow w$ or $(G \subset W)$ in the argument, whose validity is confirmed in Table 2-10, takes the form of Fig. 2-18 when Venn diagrams are used. Since the second premise is $E \cup \tilde{W}$ and the third $\tilde{E}$, it is apparent that the conclusion $\tilde{G}$ is valid because $(E \cup \tilde{W}) \cap \tilde{E}$ implies $\tilde{W} \cap \tilde{E}$ and, therefore, $\tilde{G}$. See laws 3 and 7 in the algebra of sets that follows. Again, Venn diagrams are not useful beyond four premises.

2-15 ALGEBRA OF SETS

The seven sets of equivalent forms or laws in the operations on statements can be translated to the following corresponding laws for the algebra of sets (confirm their validity by using Venn diagrams):

1. Commutative Laws:

$$P \cap Q = Q \cap P$$
$$P \cup Q = Q \cup P$$

2. Associative Laws:

$$(P \cap Q) \cap R = P \cap (Q \cap R)$$
$$(P \cup Q) \cup R = P \cup (Q \cup R)$$

3. Distributive Laws:

$$P \cap (Q \cup R) = (P \cap Q) \cup (P \cap R)$$
$$P \cup (Q \cap R) = (P \cup Q) \cap (P \cup R)$$

4. De Morgan's Laws:

$$\widetilde{(P \cap Q)} = \tilde{P} \cup \tilde{Q}$$
$$\widetilde{(P \cup Q)} = \tilde{P} \cap \tilde{Q}$$

5. Idempotent Laws:

$$P \cap P = P$$
$$P \cup P = P$$

6. Complement Laws (Here $\varnothing$ and $\mathcal{U}$ correspond to F and T, respectively, in the language of statements.):

$$P \cap \tilde{P} = \varnothing$$
$$P \cup \tilde{P} = \mathcal{U}$$
$$\widetilde{(\tilde{P})} = P$$

7. Identity Laws:

$$P \cap \mathcal{U} = P$$
$$P \cap \varnothing = \varnothing$$
$$P \cup \mathcal{U} = \mathcal{U}$$
$$P \cup \varnothing = P$$

2-16 MODERN COMMUNICATION SYSTEMS

Communication underwent a most revolutionary change in recent years in terms of speed and distance of communication. As stated earlier, the world can be viewed as a single community in terms of our ability to transmit and receive information. Early in our civilization man communicated over distance by shouting. He could communicate at the rate of about one word per second over a distance of less than one mile. Then came coded smoke and fire signals, later, flag signals from ships (first used in the 17th century) and, next, mechanical apparatus consisting of arms on towers for signaling (semaphores) about the 18th century. Tzar Nicholas I in Russia (19th century) had a network of semaphores linking Moscow, Warsaw, and St. Petersburg. Telegraph came in the middle of the 19th century, then radio, TV, microwave towers, satellites, and lasers. Now we can communicate information at the rate of 10,000 words per second over distances which link us with outer space. Thus we can send or receive a signal from the moon in about one second.

The fundamental elements of a communication system are shown in Fig. 2-19. In oral communication between a speaker and a listener, the

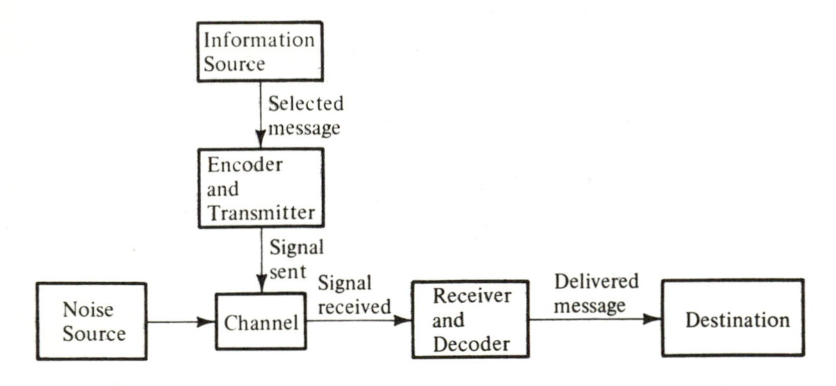

Figure 2-19 Schematic Diagram of a Communication System

information source is the brain, the *encoder* is the speaker who encodes his experience into language, and the voice mechanism is used as a *transmitter*. The varying sound pressure produced by the speaker as he utters the words is the *signal sent*. The signal is transmitted through air which acts as the *communication channel*. In the process of transmission, distortions of sound take place, which were not intended by the information source (the speaker and his brain). These unwanted additions to the signals sent are called *noise*. The source of noise is in the environment of the communication system. The *signal received* by the listener is a combination of the signal sent and the unwanted noise. The listener, who acts as *receiver*, *decodes* the received signal by translating it to an experience on the basis of his knowledge of the language and the world. This experience becomes the delivered *message* which registers in the *destination*, i.e., the brain of the receiver.

For telephone communication, the selected message is changed by the telephone *transmitter* from sound pressure of voice to varying electrical current. The *channel* is a wire or microwave, and the signal is the varying electrical current. The decoding is done at the other end.

Transmitted messages can be coded in various ways, provided receiver and transmitter have agreed on the encoding rules. Thus we can use a binary code to transmit decimal numbers or English words if we wish. The amount of information* we transmit depends on the amount of distortion, i.e., noise that corrupts the intended message as it makes its way from transmitter to receiver. Fortunately, redundancy exists in natural language. Yes, you read it correctly: "redundancy exists in natural language." Our written language is based on complex figures that were abstracted to the form of letters of the alphabet, as discussed earlier. A letter can be recognized even when it is only partially seen because the essential figure which it intended to convey was not abstracted to the point where no redundancy is left.

There is also redundancy in written words: Mst ppl cn rd ths sttmnt wth lttl dffclty. Now you can see how Hebrew managed with no vowels until the 7th century, and still does. Redundancy also appears in sentences; words can be omitted without loss of meaning. The same holds for paragraphs and so on.

2-17 USE OF REDUNDANCY IN COMMUNICATION

When we transmit a message by a binary code, there is also a possibility of corruption because of noise. For example, suppose we wish to transmit information in the form of message units that consist of four binary digits (bits). Consider a pulse train as shown in Fig. 2-20 as the signal sent (representing 1011). A clock controls the fixed time spacing between bits. During the pulse interval, a pulse is generated when a bit 1 is sent, and no pulse is generated when a bit 0 is sent. Noise in the transmission process may cause the sent signal of Fig. 2-20 to be transformed into a received signal of the form shown in Fig. 2-21. The second bit in the received signal is about midway between 0 and 1. Which is it? There are clever ways to code the information by using redundancy so that errors due to noise can be detected and corrected.

---

*Measurement of information is discussed in Chapter 4.

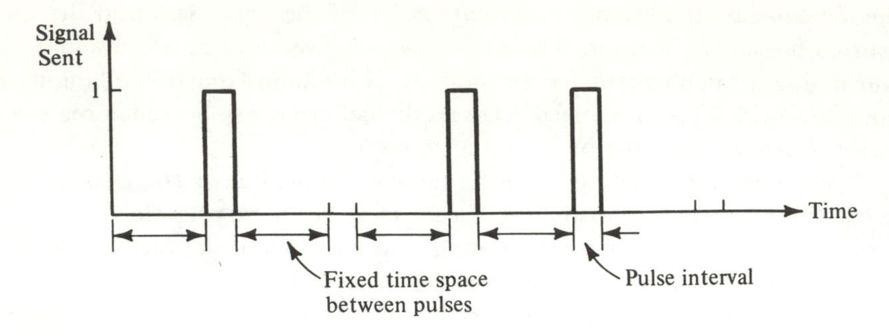

Figure 2-20 Signal Sent: 1011

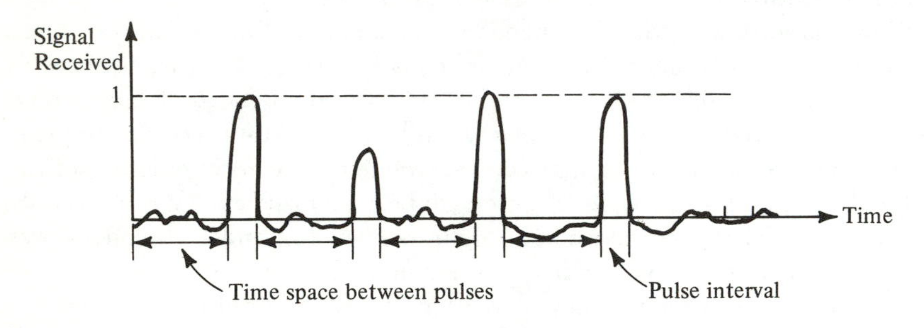

Figure 2-21 Signal Received: 1011 or 1111 ?

In the following discussion we develop the Hamming* code [132] which is a communication code that will detect and correct a *single* error in one bit out of a pulse train of seven bits in which four constitute the desired message and three are extra digits for redundancy. The development will demonstrate a novel application of Venn diagrams.

Hamming Code

Consider a string of seven bits designated by letters a_i in which the subscript indicates the position of the bit:

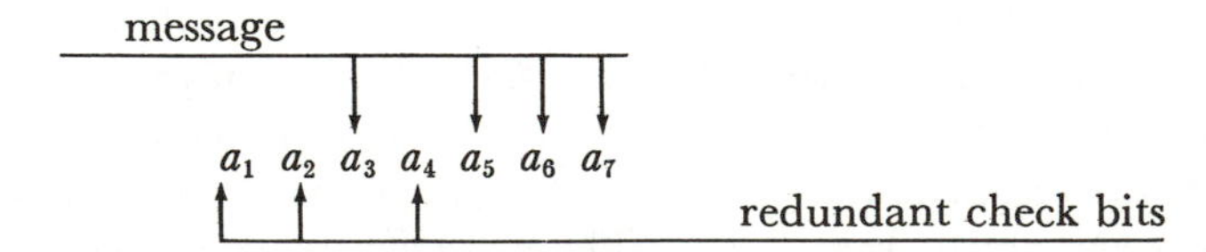

Each letter a_i can assume the values of 0 or 1. The string $a_3a_5a_6a_7$ constitutes the message, and the bits $a_1a_2a_4$ are the redundant bits added to

*Originated by R. W. Hamming [132].

the message as check bits in order to detect and correct a single error in the transmitted string of seven bits. The three check bits are a function of the four bits in the message, as will be shown. Let us observe the transmitter and the receiver at work while communicating with this code.

Transmitter. Suppose the transmitter wishes to send the message 1011; i.e., $a_3a_5a_6a_7 = 1011$. Before sending the message, he extends it by adding check bits $a_1a_2a_4$ whose values are selected using the Venn diagram of Fig. 2-22. Each of the seven regions in the area bounded by the circles I, II, and

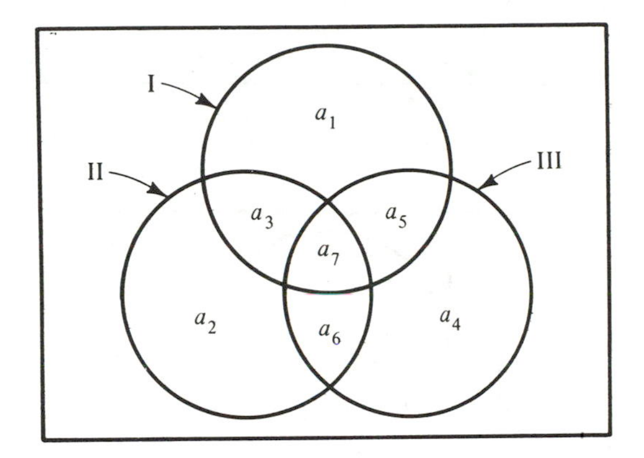

Figure 2-22 Venn Diagram For Hamming Code

III is identified by a letter a_i. The transmitter inserts the values for $a_3a_5a_6a_7$, i.e., 1, 0, 1, 1 in our case, and then he selects 0 or 1 for each check bit $a_1a_2a_4$ so that the sum of the bits in the four regions of each circle adds up to an even number. This is called a parity check. In the present case, $a_1a_2a_4$ are, respectively, 0, 1, 0, as shown in Fig. 2-23. The transmitter sends the signal of seven bits 0110011.

Receiver. Let us suppose that noise causes the transmitted bits 0110011 to be received as 0100011. We now follow the receiver as he decodes the received signal. Each bit is entered in the appropriate region in the Venn diagram of Fig. 2-22, as shown in Fig. 2-24. He now adds the bits in each circle and records 0 for a sum which is an even number, and 1 for an odd number. The resulting three bits are listed in the order III, II, I, corresponding to the results obtained with each circle. In the present case, the receiver obtains 011. This is the binary equivalent* of the decimal number 3. Hence, there is an error in position three; namely, $a_3 = 0$ in the received signal should be changed to 1. Now, after the correction, the parity check will yield 000 which signifies that there is no error.

*See Chapter 3 for the binary number system.

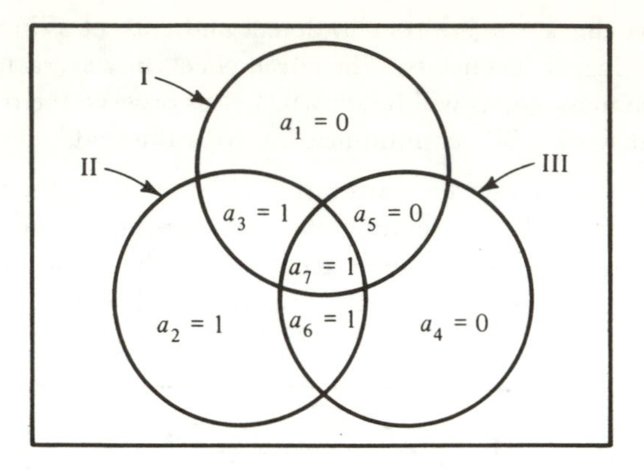

Figure 2-23 Hamming Code For Message 1011

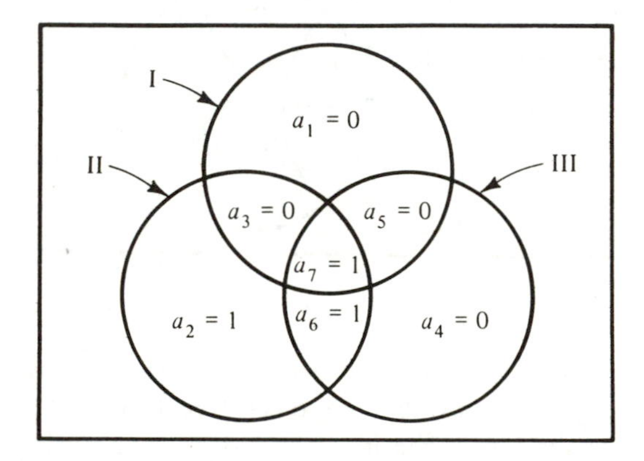

Figure 2-24

The Hamming code not only detects the error, but tells you where it occurred. This is accomplished by the special arrangement of the message and check bits, and the parity check as performed by using Venn diagrams with circles and regions designated, as shown in Fig. 2-22. The parity check yields three binary digits which can stand for the numbers 0 to 7 and represents no errors (000) or the position where a single error occurred by identifying a_1 (001) to a_7 (111).

Why Does the Hamming Code Work?

Let us consider messages of four bits with the stipulation that an error can occur in only one bit. To identify which of the four bits in the message is wrong, we require a *check* code of three bits for the numbers zero to four.*

*The number zero (000) in the check code signifies "no error".

But now an error may occur not only in the message, but also in the check bits. Restricting ourselves to only one error, we can still use the three check bits to identify seven positions, (001, 010, 011, . . . , 111), four for the message and three for the check bits, as well as the number zero (000) to signify "no error." To make the Hamming code work as described above, we arrange the string of seven bits a_1 to a_7, as shown in the Venn diagram of Fig. 2-22, with the message identified by $a_3a_5a_6a_7$ and the check bits $a_1a_2a_4$. Each check bit is placed so that it is contained in one circle only. This way, each check bit can be entered independently for any selected message because, first, the message bits are recorded in the appropriate regions, and then the check bits are added to yield an even sum of bits in each circle separately. To check for a single error in the transmission, we add the bits in the regions of each circle. Suppose an odd number of bits is obtained in each circle. We then write III II I = 111, which is the number seven. Hence, a_7 is wrong. Note that a_7 is at the intersection of all three circles and, therefore, the error affects all of them. That is why a_7 was recorded there in the first place. Similarly, an error in position one (a_1) is identified as III II I = 001 and, therefore, only circle I is affected; that is why a_1 is in the region which is contained in circle I only. An error in a_5 is identified by III II I = 101; therefore, a_5 is at the intersection of circles III and I outside of II. You can now justify on the same basis the assignments of all a_i in Fig. 2-22.

Since this is not simple and may require rereading and independent reflection, let us repeat the above by starting with the check bits. If we have three check bits, we can identify eight distinct numbers or regions: 000 will signify no error; 001, 010, 100 will signify the three possible errors in the check bits a_1, a_2 and a_4, respectively. We have four numbers left out of the total of eight and, therefore, the message can consist of no more than four bits if we insist on identifying an error in the message. These four identification numbers are:

$$011, 101, 110, 111$$

These are, respectively, the decimal numbers 3, 5, 6, 7. Note also that the sequences of the three bits III II I in the parity check identify the eight regions of Fig. 2-22; a one indicates that the region is in the corresponding circle, a zero indicates the region is outside the corresponding circle. For example, 000 identifies the region outside all circles, and 111 identifies the intersection of the three circles.

You may now wish to satisfy yourself by reasoning why four check bits can be used in a Hamming code for a message of eleven bits, 5 check bits for a message of 26 bits, 6 for 57, and in general n check bits for a message of $[2^n - (n + 1)]$ bits.

The positions of the check bits and message bits can be identified from the following table in place of a Venn diagram:

CIRCLE OR ROW			COLUMN NUMBER 1	2	3	4	5	6	7
I	:	0	1	0	1	0	1	0	1
II	:	0	0	1	1	0	0	1	1
III	:	0	0	0	0	1	1	1	1
			a_1	a_2	a_3	a_4	a_5	a_6	a_7

check bits (a_1, a_2, a_4)

To check for an error in a received message, we proceed as follows: For any $a_i = 1$, enter its identification from the above table in column i; for example, for $a_7 = 1$, place three ones in column seven. If $a_i = 0$, place three zeros in column i; for example, for $a_7 = 0$, place three zeros in column 7. Then add the bits in each row. Record 1 for an odd sum, and a zero for an even sum. The sequence of bits III, II, I in the three rows identifies where a single error occurred. When transmitting a message, we record the a_i of the message and assign check bit a_1 a value of 0 or 1 as required to make the sum of bits in row I even. Similarly, we assign check bits a_2 and a_4 to make the sums in rows II and III even.

For example, suppose we receive the sequence 0100011. Then we construct this table:

I:	0	0	0	0	0	0	0	1
II:	0	0	1	0	0	0	1	1
III:	0	0	0	0	0	0	1	1
		a_1	a_2	a_3	a_4	a_5	a_6	a_7

Row I: odd sum, I = 1; Row II: odd sum, II = 1; Row III: even sum, III = 0. Thus, III II I = 011 and there is an error in a_3; namely, it should be 1 instead of 0 and the correct sequence is 0110011.

To transmit the message 1001, for example, we construct this table ($a_1 = 0$, $a_2 = 0$, $a_4 = 1$ for this message):

I:	0	0	0	1	0	0	0	1
II:	0	0	0	1	0	0	0	1
III:	0	0	0	0	1	0	0	1
		a_1	a_2	a_3	a_4	a_5	a_6	a_7

For a message of 11 bits with 4 check bits, the table takes the following form with the location of error identified by the four-bit sequence IV, III, II, I:

ROW															
I:	0	1	0	1	0	1	0	1	0	1	0	1	0	1	0	1
II:	0	0	1	1	0	0	1	1	0	0	1	1	0	0	1	1
III:	0	0	0	0	1	1	1	1	0	0	0	0	1	1	1	1
IV:	0	0	0	0	0	0	0	0	1	1	1	1	1	1	1	1
		a_1	a_2	a_3	a_4	a_5	a_6	a_7	a_8	a_9	a_{10}	a_{11}	a_{12}	a_{13}	a_{14}	a_{15}

check bits (a_1, a_2, a_4, a_8)

The Venn diagram that corresponds to this table is shown in Fig. 2-25. As an exercise, identify the location of all a_i in the 15-bit sequence. (Some are identified in the figure.) Also, generate the table for a Hamming code for

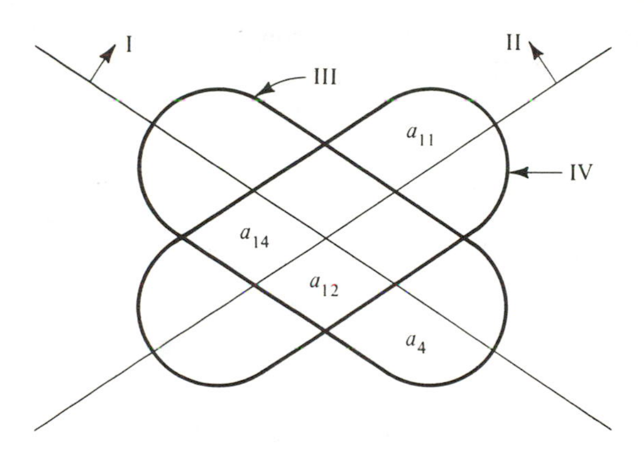

Figure 2-25 Venn Diagram for the Hamming Code of Eleven-Bit Message and Four Check Bits

a message of 26 bits, and explain how it can be used by first coding a message of your choice (adding the appropriate check bits), then introducing an arbitrary single error, reconstructing the table as a receiver, and detecting the error.

2-18 COMPUTER LANGUAGE

The language of the computer is basically a binary language which employs two fundamental symbols 0, 1. In this form the language is referred to as *machine language*. To simplify communication with the computer (programming), higher levels of artificial languages have been developed. It is simplest to illustrate the level of sophistication of these higher-level languages by examples. Consider the statement:

> "Clear the calculator and add to it the content of computer memory register number 64."

This statement appears as follows in the computer languages listed below:

Machine Language (30-bit word):
011011 000000 000000 000001 000000
Assembly Language:
CLA 00064
Assembly Language, Symbolic:
CLA SUM
Here CLA stands for *Cl*ear and *A*dd, and SUM is the address of the register, or the name of the number in the register.
Compiler Language:
PRINT SUM (Print the content of register SUM)
SUM = STF + COF + FEM (add contents of registers STF, COF, and FEM, and place the result in register SUM)

Examples of a program in machine language and the procedure involved in using a compiler language are discussed in more detail in Chapter 3. Therefore, consider this section as a brief exposure to the subject of computer languages.

2-19 SUMMARY

Language is man's tool for thinking and the basis for his communication. Through language we can communicate to others thoughts, feelings, and perceptions which they have not experienced.

Language may be considered as a collection of rules that specify how a particular set of symbols may be combined to form a statement which conveys meaning.

Semiotics, the science of signs, has three major classifications for rules in the structure of language: *syntactics*, *semantics*, and *pragmatics*.

Grammar provides the rules for generating the sentences of a language. *Semantic theory* of language takes over the explanation of a speaker's ability to produce and understand sentences at the point where grammar leaves off.

The following semantic levels are associated with language:

object language:	concerned with objects
metalanguage:	concerned with names of objects
meta-metalanguage:	concerned with names of names

The higher the meta level, the more remote we are from object language.

Effective communication requires that the *transmitter* and *receiver* encode their experiences in symbols in a compatible form. At times, when a *private encoding* differs from the *general encoding*, communication fails to take place. (Remember *sheet* and the number three?)

Words are abstract models or representations of objects, concepts, and experiences. Culture has a marked effect on the way experience is represented and, therefore, influences language.

Knowledge of the world helps to extract meaning from statements that may otherwise be ambiguous. Natural language is attended by *semantic overtones* which may lead to ambiguities and misunderstanding.

Culturally, written language is late, but historically it is early because recorded history depended on language to record it. Written language evolved in five stages:

Stage 1	Thing picture
Stage 2	Idea picture
Stage 3	Word-sound picture
Stage 4	Syllable-sound picture
Stage 5	Letter-sound picture

The origin of our alphabet can be traced by these stages.

The Hebrews, Greeks, and Romans used letters to represent numbers. This form of representation was not productive for progress in the theory of numbers, but it led to a form of numerology called *Gimatria* in the Hebrew language.

Symbolic logic is a *deductive* language which is void of ambiguity and semantic overtones, and is not user dependent. The basic axioms about propositions in symbolic logic are:

1. A proposition is either true or false (two-valued logic)
2. A proposition cannot be simultaneously true and false

The basic operations of symbolic logic are: AND, OR, NOT, IMPLY, EQUIVALENCE. Conclusion regarding the truth or falsehood of compound statements can be derived from a *truth table* by exploring all the possible ways in which the compound statement can occur. A *tree diagram* may be used to generate the set of possible compound statements.

A *tautology* is a compound statement which is true regardless of the truth values of the constituent simple statements.

A *syllogism* is a valid argument with two premises and a conclusion.

An *argument is valid* if, and only if, the conjunction (intersection $\wedge$) of the premises implies the conclusion:

$$(\text{Premise 1 true}) \wedge (\text{Premise 2 true}) \wedge \ldots \wedge (\text{Premise } n \text{ true}) \rightarrow \text{Conclusion true}$$

Boolean algebra is an algebra of symbols (say, x, y, z, etc.) in which each symbol can assume a value of 0 or 1, analogous to F or T in symbolic logic. The basic operations of Boolean algebra are analogous to the operations of symbolic logic. The algebra of logic can be used to simplify switching circuits, to generate ways of monitoring various activities, and in general to deduce logical possibilities of compound statements and reduce them to the simplest form.

A *set* is a well-defined collection of objects. The objects may be characterized in terms of any attribute. The objects may be physical or abstract. A set which contains no elements is called a *null set* or *empty set* $\varnothing$. The complete set of all elements of interest is the *universal set* $\mathcal{U}$. When all the elements of a set B are contained in set A, then B is a *subset* of A.

The operations on sets are analogous to the operations on statements in symbolic logic. Tables 2-13 and 2-14 show the analogy of symbols and laws for operations in symbolic logic, Boolean algebra, and sets.

Table 2-13 Analogy of Symbols for Operations in Symbolic Logic, Boolean Algebra, and Sets

	Symbolic Logic proposition p proposition q	*Boolean Algebra* variable x variable y	*Sets* set P set Q
NOT	$\sim p$	$\sim x$	$\tilde{P}$
AND	$p \wedge q$	$x \cdot y$	$P \cap Q$
OR	$p \vee q$	$x + y$	$P \cup Q$
IMPLY	$p \longrightarrow q$	$x \leq y$	$P \subset Q$
EQUIVALENCE	$p \longleftrightarrow q$	$x = y$	$P = Q$

Venn diagrams can be used to identify logical combinations of attributes from two sets (Fig. 2-15), three sets (Fig. 2-16), and four sets (Fig. 2-17). Beyond four sets, tree diagrams or tables or algebraic statements must be employed.

Early in civilization man communicated at a rate of one word per second over a distance of less than a mile. At present, using modern communication systems, the rate is 10,000 words per second over distances which link Earth with outer space.

A *modern communication* system consists of these elements: *Information source, encoder and transmitter, communication channel, receiver and decoder, destination* for *message* from information source. These elements can be identified in oral communications. The speaker acts as *information source* (his brain), as *encoder* of experience into language, and as *transmitter*. The air is the *communications channel*. The listener is the *receiver* and *decoder* and is also the *destination* of the *message*.

Noise is a form of unwanted distortions that enter the message as it makes its way from source to destination. Redundancy can help overcome

Table 2-14 Analogy of Fundamental Laws or Equivalent Forms in Symbolic Logic, Boolean Algebra, and Sets

Laws	Symbolic Logic	Boolean Algebra	Sets
1. Commutative	$(p \wedge q) \longleftrightarrow (q \wedge p)$	$x \cdot y = y \cdot x$	$P \cap Q = Q \cap P$
	$(p \vee q) \longleftrightarrow (q \vee p)$	$x + y = y + x$	$P \cup Q = Q \cup P$
2. Associative	$(p \wedge q) \wedge r \longleftrightarrow p \wedge (q \wedge r)$	$(x \cdot y) \cdot z = x \cdot (y \cdot z)$	$(P \cap Q) \cap R = P \cap (Q \cap R)$
	$(p \vee q) \vee r \longleftrightarrow p \vee (q \vee r)$	$(x + y) + z = x + (y + z)$	$(P \cup Q) \cup R = P \cup (Q \cup R)$
3. Distributive	$p \wedge (q \vee r) \longleftrightarrow (p \wedge q) \vee (p \wedge r)$	$x \cdot (y + z) = x \cdot y + x \cdot z$	$P \cap (Q \cup R) = (P \cap Q) \cup (P \cap R)$
	$p \vee (q \wedge r) \longleftrightarrow (p \vee q) \wedge (p \vee r)$	$x + (y \cdot z) = (x + y) \cdot (x + z)$	$P \cup (Q \cap R) = (P \cup Q) \cap (P \cup R)$
4. De Morgan's	$\sim(p \wedge q) \longleftrightarrow (\sim p \vee \sim q)$	$\sim(x \cdot y) = (\sim x + \sim y)$	$\widetilde{(P \cap Q)} = \tilde{P} \cup \tilde{Q}$
	$\sim(p \vee q) \longleftrightarrow (\sim p \wedge \sim q)$	$\sim(x + y) = (\sim x \cdot \sim y)$	$\widetilde{(P \cup Q)} = \tilde{P} \cap \tilde{Q}$
5. Idempotent	$(p \wedge p) \longleftrightarrow p$	$x \cdot x = x$	$P \cap P = P$
	$(p \vee p) \longleftrightarrow p$	$x + x = x$	$P \cup P = P$
6. Complement	$(p \wedge \sim p) \longleftrightarrow F$	$x \cdot \sim x = 0$	$P \cap \tilde{P} = \varnothing$
	$(p \vee \sim p) \longleftrightarrow T$	$x + \sim x = 1$	$P \cup \tilde{P} = \mathscr{U}$
	$\sim(\sim p) \longleftrightarrow p$	$\sim \sim x = x$	$\widetilde{(\tilde{P})} = P$
7. Identity	$(p \wedge T) \longleftrightarrow p$	$x \cdot 1 = x$	$P \cap \mathscr{U} = P$
	$(p \wedge F) \longleftrightarrow F$	$x \cdot 0 = 0$	$P \cap \varnothing = \varnothing$
	$(p \vee T) \longleftrightarrow T$	$x + 1 = 1$	$P \cup \mathscr{U} = \mathscr{U}$
	$(p \vee F) \longleftrightarrow p$	$x + 0 = x$	$P \cup \varnothing = P$

distortion introduced by noise. Redundancy exists in written natural language and can also be introduced in binary codes of modern communication alphabets. The *Hamming code* is an example of redundancy at work to detect and correct a single error in a message. Redundancy reduces the rate of message transmission.

The language of the digital computer is basically a binary language which employs two fundamental symbols, 0 and 1. Higher levels of artificial languages have been developed to simplify communication between the user of natural language and the computer and its basic language of binary symbols.

EXERCISES

NOTE. Exercises 2-1 through 2-4 refer to the problem you chose in Exercise 1-1.

2-1 Refer to Fig. 2-19, the schematic diagram of a communication system. Identify elements of your problem which correspond to the elements of the diagram: information source, encoder, transmitter, etc.

2-2 (a) How can you test for and identify noise in the channel you used to acquire information in the solution of the problem?
(b) How might you introduce redundancy in the channel? What effect would this have?

2-3 What can be done to ensure that encoding is not private?

2-4 Identify areas where you anticipate you might be able to use tree diagrams, truth tables, and Venn diagrams in the solution. Did you anticipate the use of any of these in your answer to Exercise 1.10?

2-5 Suppose that the first three questions on a Gallup poll are:

(a) Are you from the North, South, East, or West?
(b) Are you Democrat, Republican, or other?
(c) Are you younger than 21, or either 21 or older?

Draw a tree diagram to enumerate the possible replies to the three questions.

2-6 Two-valued logic, in which logical variables have one of two values (e.g., 0 or 1, T or F), was described in Sec. 2-8. Suppose we want to use a three-valued logic system in which a statement can be true, false, or undecided. Draw a tree to list the logical possibilities for 2 statements.

2-7 There are six children at a party. They belong to two families of three children each. Each child has either brown or blue eyes, and brown or blond hair. Children in one family may differ by at *most* one characteristic.

Tom—blue eyes, brown hair; John—blue eyes, blond hair
Jim—brown eyes, blond hair; Mary—blue eyes, brown hair
Jane—brown eyes, brown hair; Sue—blue eyes, blond hair

Which children are brothers and sisters?

2-8 A man is put in a room with 2 doors. If he walks through one, he will be executed. If he goes through the other, he will be set free. There is a guard in front of either door. One always lies, and the other always tells the truth. The man does not know which guard is which. He is permitted to put *one* yes-or-no question to *one* guard only. What should he ask and what will the answer mean? (*Hint:* Use a truth table.)

2-9 Three blindfolded men each choose a hat from a barrel containing three black and two white hats. They stand in a line, with C at the head of the line, B behind C, and A behind B. The men remove their blindfolds and remain facing forward. A can see the colors of B's and C's hats, B can see only C's hat, and C cannot see any hats.

> A says, "I don't know what color my hat is."
> B says, "I don't know what color my hat is."

What color is C's hat? Show this by using a tree diagram.

2-10

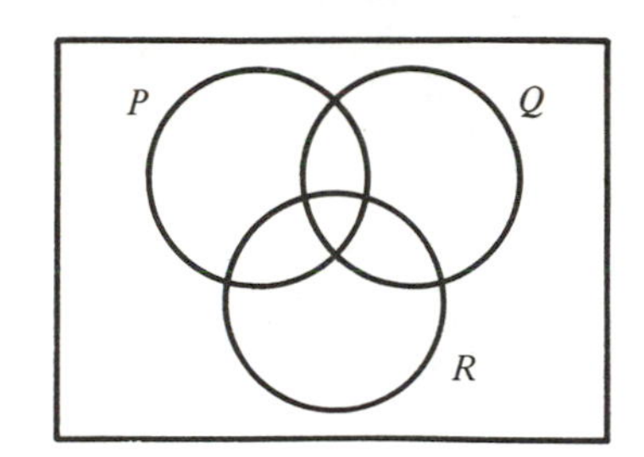

Given: $P \cap Q \cap R = 3$
$Q \cap R = 7$
$P \cap R = 9$
$(P \cup R) \cap Q = 9$
$Q = 13$
$P \cup Q = 24$
Compute $P \cap \tilde{Q} \cap \tilde{R}$

2-11 Determine the validity of the following:

> Premise 1: If the sun is shining, it is not raining.
> Premise 2: The sun is not shining.
> Conclusion: It is raining.

2-12 You receive the following binary message: 0 0 1 1 1 0 1. The message includes three check bits.

(a) Which bit was erroneously received?
(b) What was the correct number (in base 10) sent?

2-13 A message of 11 bits will require four redundant check bits to detect a single error; you can write 16 numbers, zero to decimal 15, using 4 bits. Try to use the diagram of Fig. 2-17 to devise a scheme for such a Hamming code. Label the regions and test your scheme.

2-14 A message of 26 bits will require five check bits; you can write 32 numbers, zero to decimal 31, using 5 bits. Can you devise a Hamming code to detect a single error for such a signal without resorting to a Venn diagram? First, generalize the code for the message of 4 digits discussed in the chapter, by resorting to a table instead of a Venn diagram.

2-15 Three men, Mr. A, Mr. B, and Mr. C, p. ıy the following game. Each holds a hoop and two balls. Each player, in turn, may throw his ball anywhere he wishes. If a ball is thrown through a player's hoop, that player is out of the game. The last one left in the game wins.
C never fails to throw the ball through the hoop at which he is aiming, B sometimes misses, and A is even less accurate than B.
The game starts with Mr. A throwing his first ball, then Mr. B, and then Mr. C (if they are still in the game). Then the cycle repeats with the second ball of each player still in the game.
Draw a tree diagram to show possible outcomes if, on his first throw, Mr. A:

(a) Throws his ball at Mr. B's hoop
(b) Throws his ball at Mr. C's hoop
(c) Purposely misses both opponents' hoops.

(*Hint:* To reduce the size of the trees, assume that since C never misses, the other players will always try to get him out first [except for A's first ball in part (a) above]; also, since B is more accurate than A, if C is given the choice between aiming at A or B, he should aim at B.)

2-16 Explain, in your own words, how and why the Hamming code works.

2-17 Using a truth table, check the validity of the following argument:

> If this is a good book, then it is worth purchasing. Either the book has a well-known author, or the book is not worth buying. But the book is not written by a well-known author. Therefore, it is not worth purchasing.

2-18 Show that $a \wedge (b \vee c) \longleftrightarrow (a \wedge b) \vee (a \wedge c)$.

(a) Using a truth table
(b) Using a Venn diagram

2-19 People are classified according to their blood type by testing a blood sample for the presence of three proteins (antigens) called A, B, and Rh. The presence or absence of antigens A and B determines blood type, while the presence or absence of antigen Rh determines the "Rh factor" as follows:

Antigen(s) in blood	Blood type
A and B	AB
A only	A
B only	B
neither A nor B	O

Antigen in blood	Rh factor
Rh	Rh positive (+)
no Rh antigen	Rh negative (−)

If we let A, B, and Rh denote the sets of people whose blood contains the A, B, and Rh antigens, respectively, then all people are classified into one of the 8 possible blood type-Rh factor categories.
Suppose a laboratory technician reports the following statistics after testing blood samples of 100 people:

Number of samples	Antigens in blood
50	A
52	B
40	Rh
20	A and B
13	A and Rh
15	B and Rh
5	A, B, and Rh

According to the above statistics, how many people of the 100 tested were of type A and Rh negative?

2-20 For the Venn diagram (Fig. 2-27), identify each of the regions.

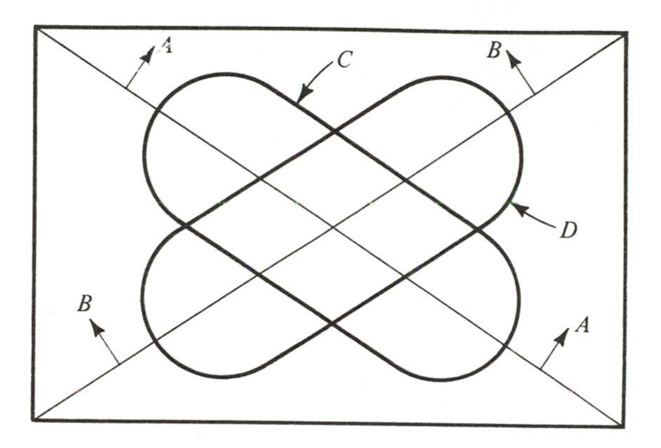

Figure 2-27

2-21 Earthquakes occur at a particular location at a rate of one a month. The earthquakes can be classified into two categories: those of magnitude M or greater, and those of magnitude less than M. Use a tree diagram to enumerate all possible sequences of earthquake magnitudes in a four-month period.

2-22 A person wants to travel from A to B (Fig. 2-28). Assuming that he can travel only North or East, find the total number of distinct paths that he may take. Note: A path is distinct if there exists no other path that contains exactly the same arcs (that is, line segments connecting two points or nodes). (*Hint:* Use a tree diagram.)

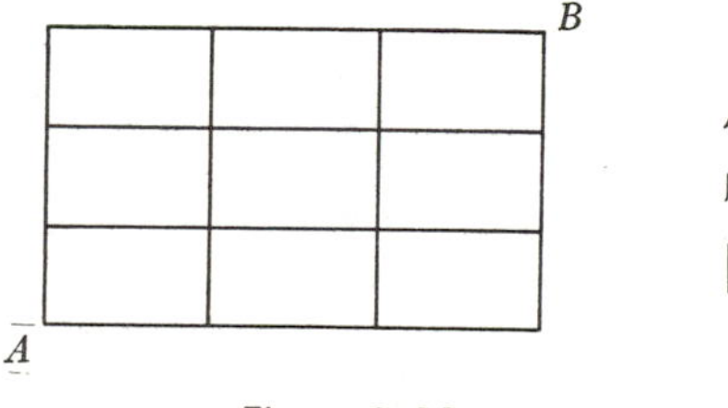

Figure 2-28

2-23 Two persons play the following game. They take turns flipping a fair coin, and the first person to get a head wins. Draw a tree diagram to represent the possible outcomes of the game.

2-24 Verify symbolic logic laws 4 and 7 by using truth tables.

2-25 Verify Boolean algebra laws 1 and 6 by using truth tables.

2-26 In the following list, identify words that are related, and order them according to levels of aggregation (or abstraction).

society	cell
swordfish	group
organism	automobile
animal	transportation
energy	education
force	school
atom	book
temperature	teacher

3

COMPUTERS: FUNDAMENTAL CONCEPTS

3-1 INTRODUCTION

The computer has entered into almost every phase of our lives. It is used in banking, finance, industry, government, education, and research, to name some areas; and applications continue to expand. The computer is destined to produce a profound effect on the life of man because it will serve to amplify man's intelligence. It will most likely surpass the effect produced by the invention of the steam engine which brought about an amplification of man's energy. The engine opened the way to the industrial revolution; the computer opened the way to a new information revolution with an impact that is so difficult to foretell that we are still groping for an appropriate choice of words to describe it. The pace of new developments in computer technology is amazing when we consider that the first electronic computer was commercially available only in 1950.

In the early stages of computer technology the emphasis was on *hardware*—the physical makeup of the computer; next came the surge in *software* studies—programming languages for effective computer use by man. While developments in both hardware and software are continuing, emphasis is now shifting to the user—to you, me, and all of us. We shall ultimately be direct users of computers. As writing was once the magic tool of priests and is now almost universally shared by man, so will computer programming cease to be the exclusive knowledge of a new cult of "priests," the computer programmers. We shall learn a simple programming language some day that will enable us to use a computer from our homes the way we use a telephone. We shall be able to interrogate library material for information, communicate with our bank account, obtain government statistics on health, education, welfare, and politics, by using our programming language on the keyboard of a home-installed console.

We shall also be able to file away information in the computer storage, manipulate it, and retrieve it as we please. A television-like screen, or possibly a three-dimensional cube, will provide us with visual displays of information in two and three dimensions (holography).

The computer provides the most revolutionary aid to problem solving and decision making. In Chapter 1 we discussed the great interest in artificial intelligence, the development of heuristic programs for problem solving and decision making; this interest is spreading to many disciplines and is bringing about unprecedented cooperative interdisciplinary efforts in the use of computers for problem solving.

We shall need all the help we can get from the computer in problem solving in order to cope with the problems which the computer is creating for society by virtue of its existence. The emphasis is now on society, the user, as we said earlier. How can we protect privacy? Can everyone interrogate our bank accounts? Are our health records open to all? How is the computer, coupled with the new communication technology, going to affect public decision making? What about education and learning? Voting, elections, work, leisure, business, world trade? The list is long; in fact, it embraces all of society and the artificial world man has created. Perhaps new consensus of values and moralities will emerge to help society cope with a new world. Values and moralities are showing strong evidence of change, and the pace of change is increasing. No longer do father and son share the same values, and the future will increase the gulf between two successive generations unless man adapts to change in values as he adapts to changes in his artificial man-made world.

Brief History

The history of the computer is indeed brief. Ideas and devices that started paving the way to the modern digital computer appeared in the 17th century. In 1617 John Napier, who invented logarithms, developed a device for multiplication. Then came the slide rule which is an analog computing device in which multiplication is analogous to the addition of distances on the sticks of the slide rule. Blaise Pascal built an adding machine in 1642, using numbered wheels geared together in a way similar to a mileage indicator in an automobile.

A breakthrough in concepts came from a different field altogether. In the 18th century, a Frenchman, Joseph Marie Jacquard, developed a machine that represented the breakthrough in the philosophy of computer concepts. He built a loom that could automatically weave a pattern in cloth. The instructions to the loom to weave a specific pattern were provided by a code of holes on punched cards that were fed in chain form to the loom. In 1820, an English mathematics professor, Charles Babbage, fell upon the idea that a pattern of holes on punched cards such as used by Jacquard could also be used to specify and control a prescribed sequence of arithmetic operations on an arithmetic calculator. Babbage built a small model of a "difference engine" to evaluate polynomials by a method that uses the difference between numbers.* About 1833 he proposed to build an "analytical engine" which had the basic features of the modern computer. The plan called for a *storage unit* in which numbers could be stored, an *arithmetic unit* for calculations, and a *control unit* to decode instructions from the punched cards and execute them. The mechanisms required for the analytical engine were too sophisticated for the technology of that time, and Babbage died frustrated with his dream unfulfilled.

The punched card appeared again in 1890 in the U.S.A. when Herman Hollerith used it in a tabulating machine to register census data. Modern card-reading machines came next.

Developments in electromechanical devices led to desk calculators of various kinds. Then in 1938 came the MS thesis of Claude E. Shannon at M.I.T., in which the connection was established between switching circuits and Boolean algebra. This opened the door for using electronic devices with an *on-off* two-valued action in place of mechanical or electromechanical devices.

In 1945, Professor John Mauchly and his graduate student, J. Presper Eckert, Jr., built the first computer based on electronic technology. Their computer, called the ENIAC (Electronic Numerical Integrator and Computer), had 18,000 electron tubes and generated so much heat that it was

*For example, the following tables of differences can be used to generate the squares and cubes of numbers ($\nabla^1, \nabla^2, \nabla^3$ designate first, second, and third differences in the numbers of the preceding column). To generate 6^2 in the first table, add $25 + 9 + 2 = 36$. For 7^2, record 2 beneath ∇^2, add to 9, and record 11 beneath ∇^1, and $11 + 25$ under 25. Now, add $36 + 11 + 2 = 49$. Similarly, to obtain 7^3 in the second table, add $216 + 91 + 30 + 6 = 343$. For 8^3, start with 6 under column ∇^3, add to 30 and record 36 under ∇^2, 127 under ∇^1, and $127 + 216 = 343$ under n^3. $343 + 127 + 36 + 6 = 8^3$.

n	n^2	∇^1	∇^2	∇^3
1	1			
		3		
2	4		2	
		5		0
3	9		2	
		7		0
4	16		2	
		9		
5	25			
6				
7				
8				

n	n^3	∇^1	∇^2	∇^3	∇^4
1	1				
		7			
2	8		12		
		19		6	
3	27		18		0
		37		6	
4	64		24		0
		61		6	
5	125		30		
		91			
6	216				
7					
8					

likened to a furnace. One of the main limitations of the ENIAC was the need to specify the instructions to the computer by means of a plug board. The board had to be rewired for each problem, and this was a tedious job.

In 1946 came another breakthrough in concept. John von Neumann and his colleagues proposed the idea of the *stored program.* The idea called for the program, which instructs the computer to perform the specific operations in the solution to a problem, to be stored in the computer memory just as the data are. The program could then be changed internally, if desired, the same way as data are manipulated and changed internally. In fact, the program could cause parts of itself to change on the basis of results which are obtained as the calculations proceed. The first stored-program computer was built in 1949, and the first such commercial electronic computer was the UNIVAC, installed at the Bureau of the Census in Washington in 1950.

The invention in 1947 of the transistor which can switch in a few billionths of a second, the development of integrated circuits (IC) in which hundreds of electronic components can be placed on a "chip" of silicon measuring a fraction of an inch, the further development of large-scale integration (LSI), a further miniaturization, made it possible to build complex computers with enormous logical capabilities, increased speed, reliability, and drastic decrease in cost. The computers of the 1970's make the UNIVAC appear like an abacus by comparison; they can perform a million operations in a second and it costs cents to perform a million calculations. Costs are still coming down while capabilities are improving. It is now feasible to prepare a concordance (index of words in a text and the sections in which they are mentioned) of the Bible, to interrogate the computer on who was the authentic author of disputed literary work, to schedule commercial flights, to reserve passenger space, to control manufacturing processes, to perform medical diagnosis, and to monitor the vital functions in intensive care units. The list can be continued.

The computer has produced a marked effect on the economy. The value of installed computers in the United States is expected to surpass 50 billion dollars by 1975, and the computer industry is expected to surpass the oil and automobile industry in size in the next decade. But its most profound effect will be on our life style and our values. It is, therefore, important that every member of our society learn the fundamental concepts of computers just as he learns the alphabet of language.

This chapter discusses the digital computer, in particular the fundamental concepts incorporated in elementary building blocks to synthesize a complex and sophisticated tool that can perform the staggering multitude of functions described here. Starting with a symbolic hypothetical computer model, we develop an appreciation for the basic components and the method of operation of the modern digital computer as well as aspects of automatic programming.

Analog and Digital Computers

Before proceeding with our discussion of digital computers, let us point out the difference between the two basic types of computers: analog and digital.

In the analog computer, data are represented by a continuous variable, such as voltage (or current), which is read from a dial, or by the length of a line that is read from a graduated scale. The accuracy in the analog computer is limited by the accuracy of the measuring instruments employed to assess the value of the continuous variable. A slide rule is a simple example of an analog computer in which numbers are represented by their corresponding logarithms on a continuous scale, and the accuracy depends on the ability to measure the length of a line.

In the digital computer, on the other hand, all data are represented in a discrete form by a sequence of digits. For instance, the number 1/9 will be represented in the digital computer by a discrete finite number of digits 11111 . . . in which the decimal point is known to precede the first digit. The accuracy depends on the number of discrete digits employed to represent a number. Theoretically, an infinite number of digits are required to represent the value of 1/9.

Although an analog computer is very useful in the solution of many specific engineering problems involving differential equations [19, 20], the digital computer has the advantage of much greater flexibility for multipurpose use and is more accurate [21, 22].

There are also *hybrid computer* systems which contain analog as well as digital hardware. Our discussion here is limited to digital computers.

3-2 A SIMPLE COMPUTER MODEL

We begin with a simple computer model that has conceptually the basic features of a modern computer. The heroes in our story for the simple model are a boss and his secretary. The boss runs a computer service business with the assistance of the secretary. The computer is housed in a single room and consists of the following components (see Fig. 3-1):

1. A wall with wooden compartments
2. A wall on which is inscribed a code of operations and their meaning
3. A typewriter
4. A calculator
5. A dial for setting integer numbers from 0000 to 9999

Each compartment (*register*) in Fig. 3-1 is connected to the wall and has a fixed *address* or *location* in the form of four digits above it. For instance,

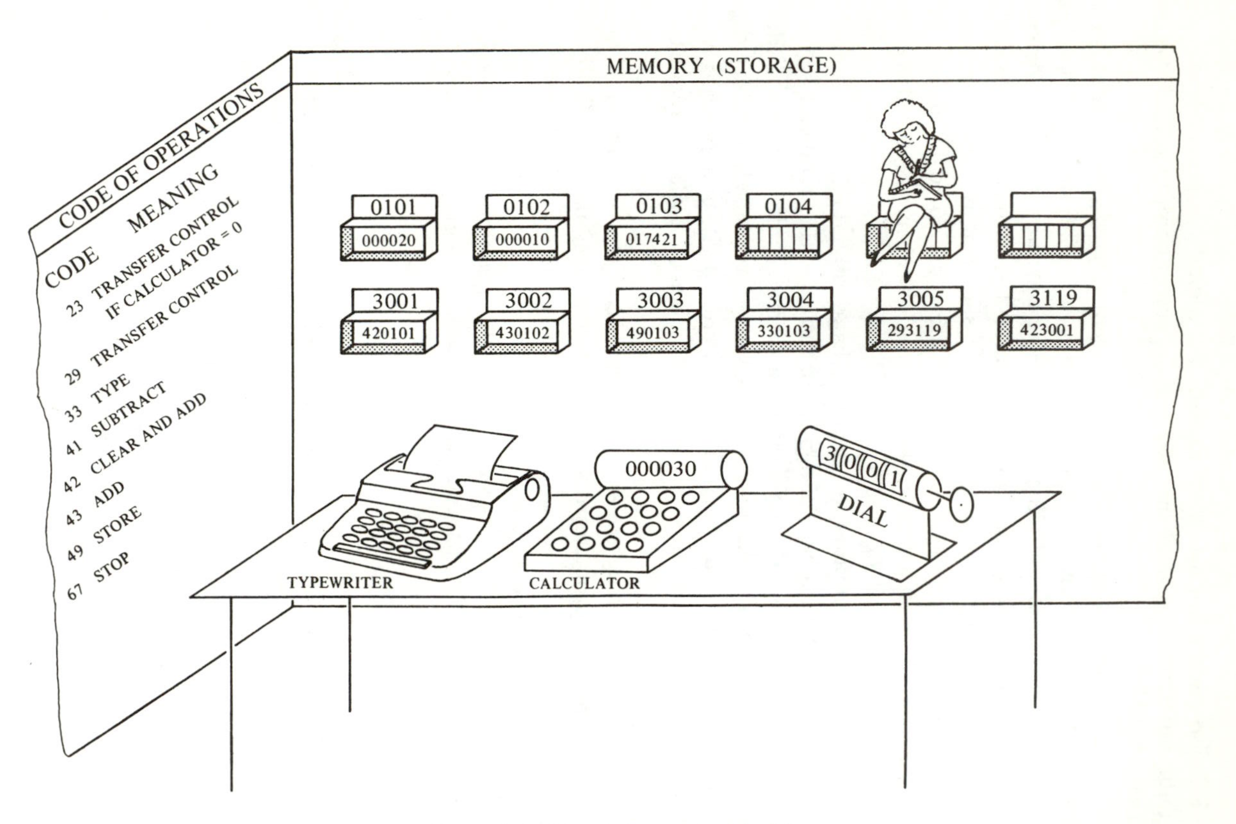

Figure 3-1 Simple Computer Model

the address 3119 locates the compartment at the lower right corner of the wall very much as a house is located by its address. The inside back wall of each compartment is a blackboard surface divided by vertical lines into six equal spaces. One digit can be written in each space with a total of six digits per compartment.

The secretary is equipped with a secretarial pad, a pencil, and a piece of chalk. Before assuming her responsibility in the computer room, the secretary was trained for one full day. The boss found her acceptable because she had the following desirable qualifications: Good eyesight, not inquisitive, and willing to follow instructions with perfect accuracy like a robot without ever questioning their wisdom.

Let us follow the secretary as she enters the computer room in Fig. 3-1 at 8:00 A.M. As soon as she enters she walks over to the desk, picks up her pad, pencil, and chalk, and looks at the dial where she sees the digits 3001. The dial always tells the secretary where to go (to which address) for her next instruction. Consequently, she goes to the compartment with the address 3001 (the compartment at the lower left of the wall), looks inside it and records in her pad the digits 420101 which she sees. She is now ready to execute the first computer instruction which has the form: 420101.

First Computer Instruction 420101. In interpreting an instruction she has been trained to identify the first two of the six digits as an operation code and the remaining four as an address of a compartment. From the wall with the code of operations she finds the digits 42 to mean *clear and add.* With this information available, she looks again at the digits 420101 and proceeds to execute the following operation: She *clears* the calculator to zero *and adds* to it the number which she finds in the compartment with the address 0101 that is specified by the last four digits of the instruction. This number is 000020. Consequently, the calculator will read 000020. The first instruction has been completed. She walks over to the dial and advances it by one number to read 3002. She always advances the dial by one number after the completion of an instruction unless she is specifically instructed to do otherwise.

Second Computer Instruction 430102. As indicated earlier, the dial tells her where to go for her next instruction. Hence, she goes to address 3002 and reads the six digits 430102 which she finds at this address. Again the first two digits are an operation code and the last four represent an address of a compartment that contains the number she will operate on. Following this interpretation, she walks over to the calculator and adds (43 is the code for add) the number that she finds in compartment 0102. This number is 000010 and, therefore, after the execution of this instruction, the calculator reads 000030. She now walks back to the dial and advances it by one number to read 3003.

Third Computer Instruction 490103. Next, she goes to address 3003 and reads the digits 490103. This she interprets as: store (49 ≡ store) the number that you have in the calculator in the address specified by the last four digits of this instruction, that is, 0103. She promptly copies the number 000030 from the calculator and records it with chalk in compartment 0103 after erasing the digits 017421 which are currently shown there. Then she sets the dial to 3004. Note that this is the first time she has erased the inside of a compartment. This is always implied in the instruction *store* which instructs to replace the existing six digits with six new ones.

Fourth Computer Instruction 330103. With the dial set at 3004 she proceeds to this address and reads the instruction 330103 which she executes as follows: She types (33 ≡ type) the number which she reads in address 0103. Since in executing the third instruction she stored there the number 000030, this is the number she reads there now and subsequently types. The dial is set to 3005.

Fifth Computer Instruction 293119. The next instruction at location 3005 reads 293119. This is executed as follows: She goes to the dial, which controls the sequence of operations by specifying the address of the next instruction, and sets it to 3119. Hence, the instruction 293119 is interpreted as follows: Transfer the control of operations (29 ≡ transfer control) to the address specified by the last four digits 3119. This is the one case, then, when she is specifically instructed to set the dial to a particular number instead of advancing it by one as she usually does. With the dial set at 3119 she proceeds to location 3119 to read her next instruction.

Sixth Computer Instruction 423001. In compartment 3119 she reads 423001. She accurately interprets this instruction as: clear the calculator to zero and add to it (42 ≡ clear and add) the number which is presently stored in compartment 3001. Following the execution of this instruction, the calculator reads 420101. The dial is advanced to 3120 and the secretary goes to this address to read her next instruction.

In executing the sixth instruction, 423001, the digits stored in compartment 3001 were treated as a number (four hundred and twenty thousand one hundred and one), whereas the first thing this morning after reading the number 3001 on the dial, the secretary walked over to this very same compartment and interpreted the very same six digits 420101 as an instruction (see First Computer Instruction). Is anything wrong with this? Absolutely not! The information within a compartment is always in the form of six digits, referred to as a *machine word.* Whether these digits are to be interpreted as a number (data) or a meaningful instruction depends on how they are being used by the boss (the *programmer*) who used a piece of chalk (*input medium*) to write a sequence of coded instructions (*program*) and data on the back walls of the compartments (*storage or memory*). As will be shown later, there may be a good reason for using a machine word both as a meaningful instruction and a number.

This, then, is how the *stored program* idea works. The program is indistinguishable from the data in the computer, and it can cause parts of itself to change as the calculations proceed. Before proceeding to analyze what has been accomplished by our computer so far, let us relate what the boss had in mind when he prepared and stored two programs in our computer. In the first program, he wanted to add the numbers 10 and 20, store the result in location 0103, and type it on the typewriter. To accomplish this he wrote a *program* that consists of a sequence of coded instructions and stored it in locations 3001 to 3004, inclusive. The data he stored in locations 0101 and 0102. He then set the dial to 3001 and proceeded to write a second program. The details of this program are not given, but the boss decided to store its first instruction in location 3119. To be sure that the secretary will proceed to 3119 after she completes the execution of the first program, he placed in location 3005 the instruction to transfer control to 3119. Now we compare the plans of our boss with what has been accomplished by the computer and find complete agreement. The number 30, which is the sum of 10 and 20, appears in location 0103 and on the typewriter; the computer has just finished executing the first instruction of the second program and is proceeding to the location 3120 for its next instruction.

3-3 BASIC COMPONENTS OF A DIGITAL COMPUTER

Now let us relate some terms used in connection with a modern electronic digital computer to the components of our model.

The wall with the wooden compartments is the *computer memory* with its registers where information (instructions and data) is retained as long as it is not erased. The calculator used by the secretary is the *arithmetic unit* of the computer. The chalk used by our boss to enter the information into the computer is the *input medium*. The typewriter used by the secretary to type out the information which is stored in the computer memory is the *output medium*. The secretary, assisted by the dial and the operations code inscribed on the wall, acts as the computer *control unit*. The control unit reads and stores information, interprets instructions, and causes all operations to be performed in the proper sequence. The dial takes the form of an electronic *command counter* (instruction counter) and the pad with the operation code becomes a *command register* where the present instruction is interpreted in terms of operation and associated address, and subsequently is executed.

Figure 3-2 shows the analogy between the components of our simple computer model and the terms describing the basic components of a digital computer. The arrows in the diagram show the direction in which informa-

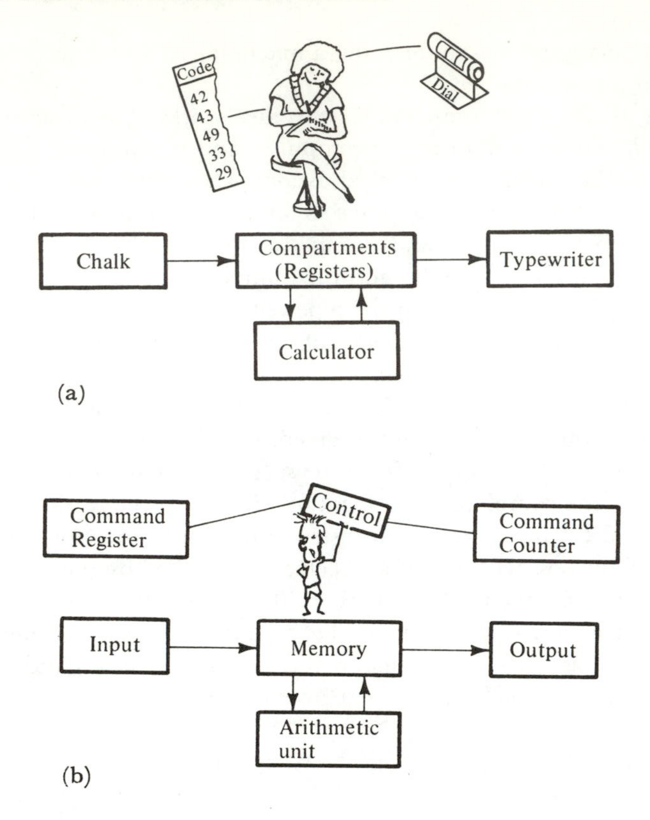

Figure 3-2 (a) Components of Our Simple Computer Model; (b) Components of a Digital Computer

tion (numbers, instructions) flows. We shall discuss the components of the "real" digital computer in later sections, but before we do so, let us get a better understanding of the concepts introduced in Sec. 3-2 and develop some new ones by following the computer of Fig. 3-1 as it executes a program designed to add N numbers.

3-4 A PROGRAM TO ADD N NUMBERS

A program to add N numbers is shown in Table 3-1. The program is assigned to registers (locations) 0200 to 0214, inclusive, as shown in the left-hand column of the table. The program instructions on the right-hand column of the table are headed by OPADDR to remind us that the first two digits of an instructions signify an OPeration code and the last four represent an ADDRess.

To be more explicit, let us use the computer of Fig. 3-1 and the program of Table 3-1 to add 700 numbers ($N = 700$). The data (the 700 numbers) will be stored in (locations) 1001 to 1700, inclusive, as shown in Table 3-2.

Table 3-1 Program to Add N Numbers

Address	OPADDR
0200	424000
0201	495000
0202	425000
0203	431001
0204	495000
0205	426000
0206	414001
0207	230213
0208	496000
0209	420203
0210	434001
0211	490203
0212	290202
0213	335000
0214	670000

← Counter (0205–0208)

← Address Modification (0209–0212)

Table 3-2 Data for Program of Table 3-1

Address	Data
1001	000075
1002	000010
1003	000024
.	.
.	.
.	.
1700	002050

The programmer selected the address 5000 as the register where the answer (sum of the 700 numbers) will be stored. He also selected register 6000 to store in it the number 000700 so the computer can be told how many numbers it is expected to add. He also found it necessary to store 000000 and 000001 in the computer. The reason for this will become apparent later. He selected registers 4000 and 4001 for 000000 and 000001, respectively. The last four pieces of information are designated, for ease of reference, as auxiliary information and are summarized in Table 3-3 for future reference.

We now input the program of Table 3-1 into the computer registers 0200 to 0214, inclusive, then input the data of Tables 3-2 and 3-3. With the

Table 3-3 Auxiliary Information

Address	Information
4000	000000
4001	000001
5000	Answer
6000	000700

program and data in the computer, we now set the dial to 0200 (location of first instruction), and start executing the program (*computing*). The operations are interpreted from the code of operations in Fig. 3-1 and after each instruction the dial is advanced by one unless instructed otherwise. The program is executed as follows.

Clearing the Calculator and Register 5000

Instruction in 0200: *424000.* The calculator is cleared and the number 000000, which is in register 4000, is added to it.
Instruction in 0201: *495000.* The number 000000, currently in the calculator, is stored in 5000, replacing any six digits that are currently there.

Adding the First Number

Instruction in 0202: *425000.* The calculator is cleared and the number 000000 in register 5000 is added to it.
Instruction in 0203: *431001.* The number 000075 in 1001 is added to the calculator.
Instruction in 0204: *495000.* The number 000075 currently in the calculator is stored in 5000.

At this point, location 5000 has the number 000075, which is the first number to be added.

The next four instructions constitute a counter which is designed by the programmer to tell the computer how many more numbers must be added, and when to stop.

The Counter

Instruction in 0205: *426000.* The calculator is cleared and the number 000700 in 6000 is added to it.
Instruction in 0206: *414001.* The number 000001 in 4001 is subtracted from the calculator (the calculator now reads 000699).
Instruction in 0207: *230213.* This instruction is a conditional transfer of control and permits the computer to make a decision based on the results of the preceding computations. The interpretation is as follows: If the calculator reads 000000 (zero), transfer control to 0213 (the address specified by the last four digits of the instruction) and proceed to execute the instruction in 0213. If the calculator does not read zero, advance the dial by one and proceed to the next instruction (in 0208). Since the calculator reads

currently 000699 (see last instruction), the computer proceeds to 0208.

We note here, however, that the calculator now reads 000699, because only one number has been added, but will read 000000 after all 700 numbers are added. At that time the computer will proceed to 0213 and type the answer which will be at 5000, then go to 0214 and stop.

Instruction in 0208: 496000. The number 000699 currently in the calculator is stored in 6000, replacing the number 000700 currently there.

So far only the number 000075 in 1001 has been added; therefore, we must go back and add the next number which is in 1002, then the next one in 1003, and so on, until we add the last one which is stored in 1700. To avoid writing the instruction of addition 700 times, each time specifying a different address, we use the next four instructions to modify the address portion (last four digits) of the instruction in 0203 and cause the control to transfer back to 0202, so that, when the instruction in 0203 is executed, the number added to the sum will be read from location 1002. This will be repeated until all 700 numbers are added. Here is how this is accomplished:

Address Modification

Instruction in 0209: 420203. The calculator is cleared to 000000 and the number 431001 in 0203 (see Table 3-1) is added. Note that earlier 431001 was treated as a meaningful instruction; now it is used as a number.

Instruction in 0210: 434001. The number 000001 in 4001 (see Table 3-3) is added to the calculator. The calculator now reads 431002.

Instruction in 0211: 490203. The number 431002 currently in the calculator is stored in 0203, replacing the digits 431001 which are currently there (see Table 3-1).

Instruction in 0212: 290202. Control is transferred to 0202; that is, the dial is set to 0202 so that the computer goes there for its next instruction. From here the program is executed again as described earlier.

The instructions in registers 0202 to 0212, inclusive, are called a *loop* because they are performed repeatedly many times, proceeding in a loop from 0202 to 0212 and back to 0202. So far, we have completed one cycle of the loop. Before proceeding with a second cycle from 0202 to 0212, inclusive, we note that at this stage the content of registers 0203, 5000, and 6000 appears as in Fig. 3-3.

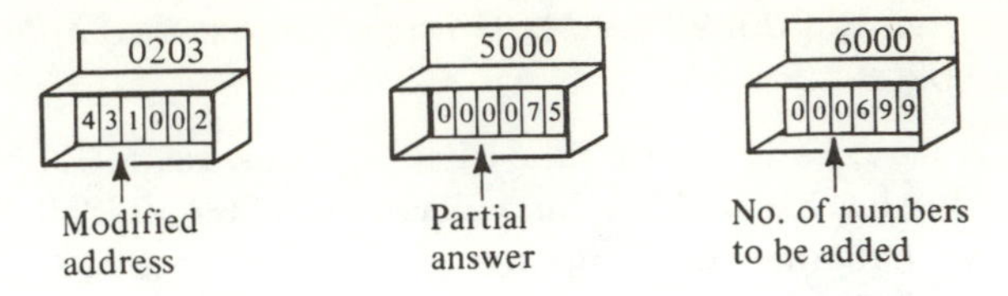

Figure 3-3 Contents of Registers 0203, 5000, and 6000 after Executing One Cycle of the Loop 0202 to 0212, Inclusive.

We suggest that the reader follow the computer as the second and third cycles are executed. After each of these cycles, the contents of registers 0203, 5000, and 6000 appear as shown in Table 3-4.

Table 3-4

Register	After Second Cycle	After Third Cycle
0203	431003	431004
5000	000085	000109
6000	000698	000697

3-5 FLOW CHART

When a program involves a large number of operations, it is convenient first to draw a diagram describing the flow of the various blocks of operations in the proper sequence. Such a diagram is called a *flow chart.* The flow chart for the program of Table 3-1 is shown in Fig. 3-4. Each block of operations,

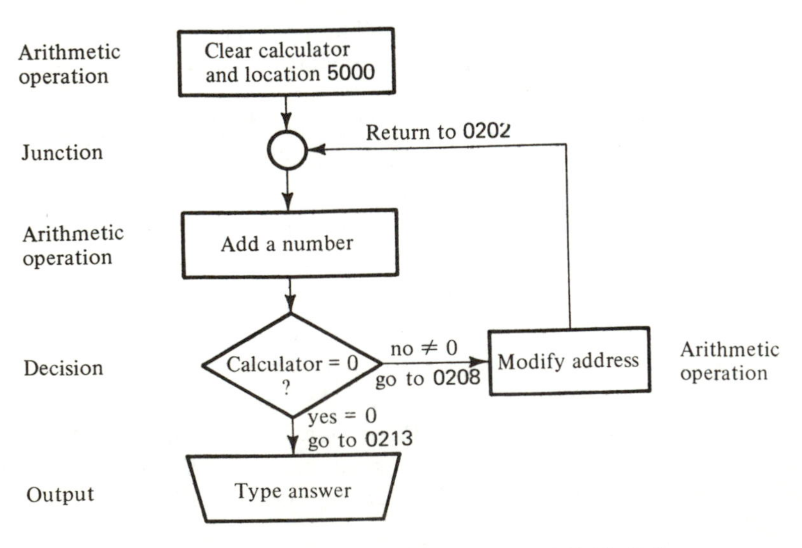

Figure 3-4 Flow Chart for the Program of Table 3-1

such as address modification, is shown in a rectangular box. The diamond-shaped box represents a decision. In our program this represents the instruction in register 0207, but it also contains adjustment of the counter which is stored in 6000, because the decision in 0207 depends on the value in 6000.

Each block of operations in the flow chart can, in turn, be broken into other flow charts describing the operations in more detail. For ease of reference, we identified in the flow chart of Fig. 3-4 some of the locations in the program of Table 3-1. In general, the flow chart can be initially more abstract; for example, instead of the arrow identified by "return to 0202" we could initially state "continue calculations," and only later identify specific location in the expanded version of the flow chart.

3-6 MEMORY (STORAGE) IN THE "REAL" COMPUTER

Actually, we have been discussing a real computer all along. Ours, however, was a simple model. Now we describe some basic components of the more sophisticated modern digital computer, beginning with the computer memory in this section.

Serial Access Memory

The function of a computer memory is to store data and instructions, referred to as *machine words*, in properly addressed positions, so that they can be retrieved when necessary. In our simple computer model, a machine word consisted of a sequence of six decimal digits, and it was stored in a properly addressed compartment. Let us now see how a machine word and an address appear in a modern digital computer. Figure 3-5 shows two types of computer memories. The drum is about 6 in. in diameter and about 8 in. long;

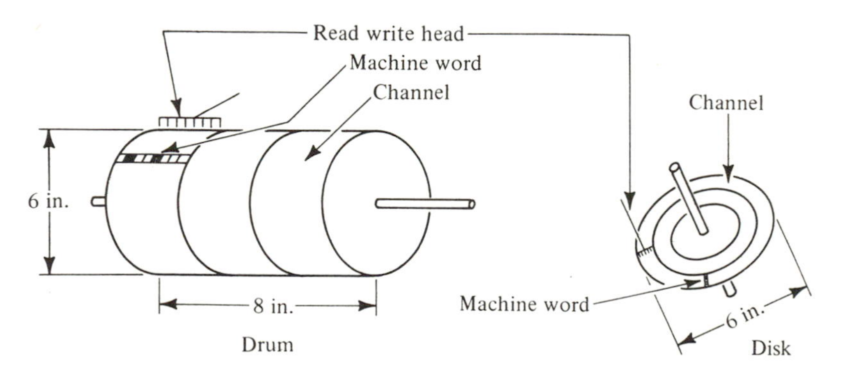

Figure 3-5

the disk is about 6 in. in diameter.* Both are mounted on shafts that rotate when the computer is in operation (approximate speed, 2000 rpm). The disk and drum are coated with a magnetic surface which is divided into channels. Each channel consists of a serial sequence of positions that are identified by *addresses*. Each such position corresponds to one compartment in our simple computer model and consists of a sequence of minute magnetic surfaces occupying the width of a channel. Each of these minute surfaces can be polarized (magnetized) or depolarized independently so that a machine word can appear as shown in Fig. 3-6(a) in which the black square indicates

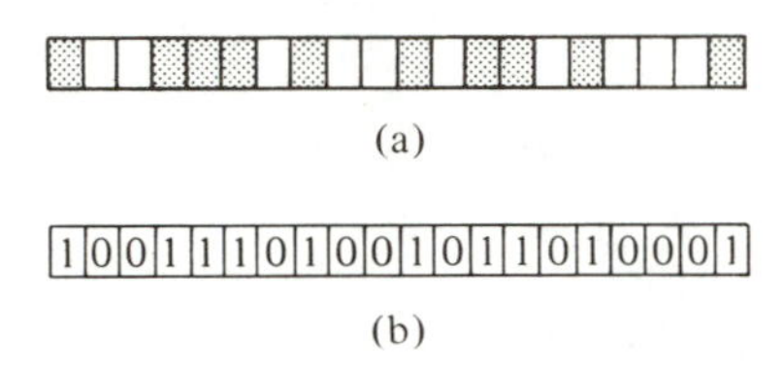

Figure 3-6 (a) Machine Word in the Computer; (b) Symbolic Representation of a Machine Word

polarization. Since each square can assume one of two states, we can choose the digits 1 and 0 to represent polarization and no polarization of a square, respectively.† The machine word of Fig. 3-6(a) appears, then, symbolically as shown in Fig. 3-6(b). Any machine word in the computer appears in the form of a sequence of the digits 0, 1. Each digit is called a *bit* (contraction of binary digit) and represents the most fundamental unit of information, 0 = No and 1 = Yes, in responding to a question. The number of bits in a common-length machine word is 32. A machine word is stored (written) or retrieved (read) from the drum or disk through the use of a *read-write head* (also referred to as read-in and read-out head) shown in Fig. 3-5. One such head is positioned above each channel and can polarize or depolarize (write) each bit in any machine word in the channel as this machine word passes underneath it while the disk or drum is rotating. It can also detect (read) the state of polarization of the bits in any machine word. The writing and reading by the read-write head is aided by the input-output media which we shall discuss later.

We observe that the information is read or written on the drum or disk-type memories in a sequential or serial fashion. The read-write head must wait for the drum or disk to rotate until the particular location is

*These dimensions vary for different computers.

†Normally there are two polarized states, depending on the direction of the magnetic field. One of these represents 1 and the other 0. The introduction of the two states as polarized and depolarized was done symbolically for simplicity.

directly beneath it so its content can be processed. Because of this, the disk and drum are referred to as *serial access memories.*

Random Access Memory

A different type of memory, called *magnetic core memory*, is shown in Fig. 3-7. This memory consists of a matrix of doughnut-shaped core elements of ferromagnetic material. Each core has two energizing coils, x and y. The

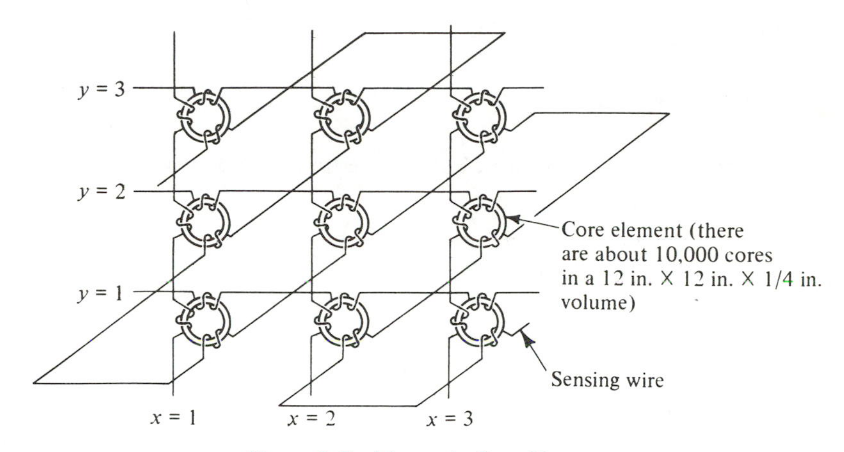

Figure 3-7 Magnetic Core Memory

signals are such that one coil x or y is not sufficient to energize a core element. Only when a signal enters simultaneously through x and y will the corresponding core be energized. For instance, the upper left core element in Fig. 3-7 will be energized by a signal from $x = 1$ and $y = 3$ simultaneously. The coordinates x and y establish the address of each bit. Information is read from the core elements through a sensing wire.* The core memory has the advantage of being free of any rotating element and being capable of storing and reading information at any location in memory with equal speed for all practical purposes.† It is therefore called a *random access memory*. A machine word in this type of memory also consists of a sequence of bits.

As in our simple model, a machine word can be interpreted as an instruction or a number. The first portion of the instruction (for instance, the first 8 bits) serves as an operation code and the remainder specifies an ad-

*To read the content of position (x,y) the corresponding coordinates (x,y) are energized. If a 1 is stored in (x,y), this will lead to no change, and the sensing wire will not respond. But if a 0 is stored in (x,y), the reading will change it to a 1, and the sensing wire will respond. Then the 0 must be restored after the reading.

†The speed depends on the speed of signal travel and the location of the machine word being processed.

dress. It is easy to see how an operation code can be specified by a predetermined sequence of bits. If only two-bit positions are available for operation codes, then the following four (2^2) distinguishable codes are possible: 00, 10, 01, 11. With three-bit positions, $2^3 = 8$ distinguishable codes are possible and, using eight bits, 2^8 distinctly distinguishable codes are possible. But how about the address portion of an instruction, or how about a machine word representing a number (data)? What does it signify when in the form of a sequence of zeros and ones, and how does the computer compute with numbers in this form? We shall discuss these matters later, but, first, let us describe briefly input-output media.

3-7 INPUT-OUTPUT MEDIA

The principal types of input-output media in a modern digital computer are (see Fig. 3-8):

1. Typewriter (or printer)
2. Punched cards
3. Punched tape
4. Magnetic tape, magnetic disk
5. Console, keyboard, and light pen (photoelectric pencil)

It is not difficult to visualize how depressing a key on a typewriter can cause a circuit to be activated and, consequently, cause a "one" bit to be

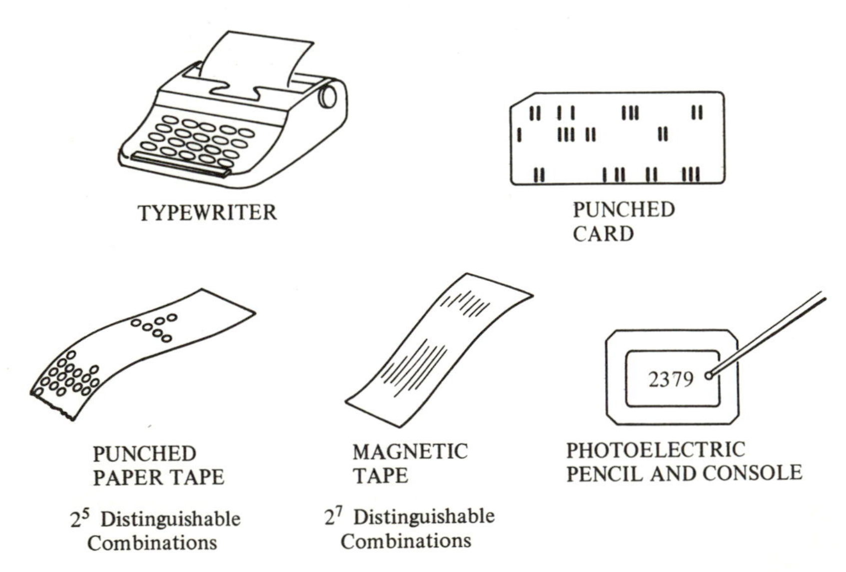

Figure 3-8 Input-Output Media

stored in the computer by the read-write head or by the *xy* coils in the core-type memory.

Similarly, mechanical fingers scanning a punched card will cause "one" bits to be stored when holes are detected. No holes will store "zero" bits.

Punched tapes may also be scanned mechanically or may be used with a photoelectric device. The tape moves next to a light source and, when a hole is present in the tape, the light goes through and activates a circuit that causes a "one" bit to be stored. No hole will store a "zero" bit.

Magnetic tapes and magnetic disks are the most compact forms of storage and as input-output media can be read faster than the other three storage types mentioned earlier. Detection of one state of polarization causes a "one" bit to be stored, and the second state causes a "zero" bit to be stored.

The reasoning used to describe the foregoing devices as input media can also be used to describe them as output media; that is, detection of one polarized state (a "one" bit) in memory will activate a circuit that causes a typewriter key to be depressed, or a hole to be punched in a card or tape, or an equivalent polarization on magnetic tape or disk.

A light pen and console are used for man-computer communication, following basic concepts similar to those discussed above.

3-8 COMPUTER LANGUAGE: LANGUAGE OF ZEROS AND ONES

In any written language there are about 60 to 100 different characters. In the modern digital computer, the fundamental characters are the bits 0 and 1. These are combined to form distinguishable codes signifying various operations and control commands that are part of the *computer language*. They may also be used to represent the digits 0, 1, . . . , 9, or any number, for that matter. To see how this is done, let us briefly discuss number systems.

The most familiar of all number systems is the decimal system. Here is how we use this system (or convention) to interpret the number 252.2. The value represented by each digit is interpreted according to its position relative to the decimal point, so that

$$252.2 = 2 \times 10^2 + 5 \times 10^1 + 2 \times 10^0 + 2 \times 10^{-1}$$

in which 10 is called the *base* or *radix* of the system. For convenience, the foregoing interpretation can be obtained from the following table:

. . .	10^4	10^3	10^2	10^1	10^0	10^{-1}	10^{-2}	10^{-3}	. . .
			2	5	2	2			

Each decimal digit of the second row in the table is multiplied by the value above it. Any other number can be evaluated by placing it in the second row of the table with the decimal point positioned as indicated. In the decimal system, the base is 10 and there are ten digits 9, 8, 7, . . . , 1, 0. The highest digit 9 is the first integer number smaller than the base 10.

We can devise an *octal* system with base 8 and eight digits 7, 6, . . . , 1, 0 (again the highest digit is the first integer number smaller than the base). By analogy to the decimal system we can assess the value of octal numbers from the following table:

. . .	8^4	8^3	8^2	8^1	8^0 •	8^{-1}	8^{-2}	. . .
					•			

For instance, the number $252.2_{\boxed{8}}$ (the number in the box designates the number system) in the octal system has the value

$$252.2_{\boxed{8}} = 2 \times 8^2 + 5 \times 8^1 + 2 \times 8^0 + 2 \times 8^{-1} = 170.25_{\boxed{10}}$$

When only the two digits 0 and 1 are used in a number system, by analogy the base must be 2 and the number system is referred to as the *binary system*. Binary numbers are evaluated from the following table:

. . .	2^3	2^2	2^1	2^0 •	2^{-1}	2^{-2}	. . .
				•			

In general, for a number system with base B the integers used are $(B - 1), (B - 2), \ldots, 1, 0$ and the numbers are evaluated from the following table:

. . .	B^3	B^2	B^1	B^0 •	B^{-1}	B^{-2}	. . .
				•			

Any value B^n in the top row of the table is the value of the digit 1 associated with the corresponding position relative to the decimal in the second row. Table 3-5 summarizes the information relevant to number systems.

Table 3-5 Number Systems

System	Base	Digits	Values of the Digit 1 Associated with Position							
Decimal	10	9, 8, 7, . . . , 1, 0	. . .	10^3	10^2	10^1	10^0 •	10^{-1}	10^{-2}	. . .
Octal	8	7, . . . , 1, 0	. . .	8^3	8^2	8^1	8^0 •	8^{-1}	8^{-2}	. . .
Binary	2	1, 0		2^3	2^2	2^1	2^0 •	2^{-1}	2^{-2}	. . .
B	B	$(B - 1), \ldots, 1, 0$	. . .	B^3	B^2	B^1	B^0 •	B^{-1}	B^{-2}	. . .

3-9 CONVERSION OF DECIMAL TO BINARY NUMBERS

We shall now show how to convert decimal numbers to the computer language of binary numbers. The reverse (binary to decimal) can be accomplished from Table 3-5.

Let us start with the decimal number 25. To convert it to its binary equivalent we divide it by the base 2 repeatedly as follows: The number 25 is divided by 2; the result 12 and a remainder of 1 are recorded beneath it:

$$\begin{array}{r|ll} 2 & 25 & \text{remainder} \\ & 12 & 1 \end{array}$$

Next, the result 12 is divided by 2 and the result 6 and remainder 0 are recorded beneath it:

$$\begin{array}{r|lc} 2 & 25 & \text{remainder} \\ 2 & 12 & 1 \\ & 6 & 0 \end{array}$$

This process is continued until finally 1 is divided by 2 to yield 0 and a remainder 1, which are recorded in the last row.

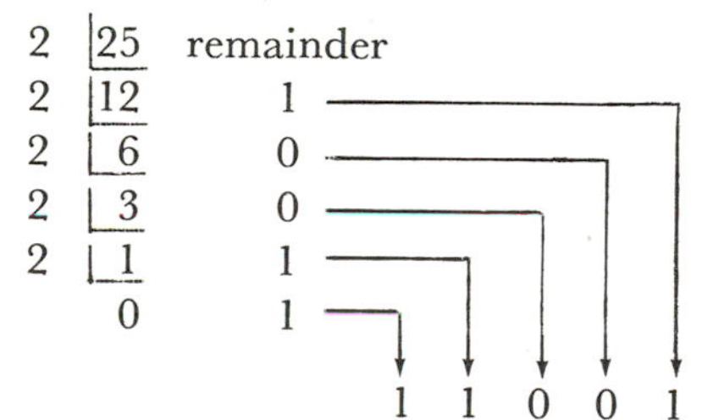

The binary equivalent of 25 is the column of remainder digits read from the bottom up or rotated to the right as shown. Hence,

$$25_{\boxed{10}} = 11001_{\boxed{2}}$$

A similar procedure is followed in converting from decimal to other bases. For instance, the octal equivalent of the decimal number 25 is:

$$\begin{array}{r|lc} 8 & 25 & \text{remainder} \\ 8 & 3 & 1 \\ & 0 & 3 \end{array}$$

$$3\ 1_{\boxed{8}} = 25_{\boxed{10}}$$

Conversion of Noninteger Decimal Numbers to Binary Form

To convert a noninteger decimal number to its binary equivalent we multiply it by the base 2 repeatedly as follows.

Consider the decimal number $0.625_{\boxed{10}}$. To convert it to its binary equivalent we multiply it by 2 and record the result 1.25 in two parts: the leading digit 1 to the left of the decimal and the remaining portion .25 to the right of the decimal as shown.

```
Leading digit        0.625
     ↓                 ×2
                     -----
     1                 .25
```

Next, the portion to the right of the decimal, .25, is multiplied by 2 and the result is again recorded in two separate parts of a leading digit 0 and the remaining portion .50 to the right of the decimal

```
Leading digit        0.625
     ↓                 ×2
                     -----
     1                 .25
                       ×2
                     -----
     0                 .50
```

This process is continued until finally the part to the right of the decimal is zero. In the present example this is accomplished when .50 is multiplied by 2:

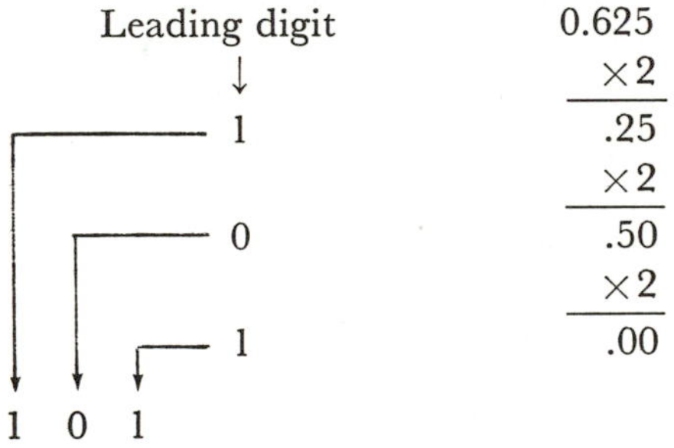

```
        Leading digit        0.625
              ↓                ×2
                             -----
  ┌────────── 1                .25
  │                            ×2
  │                          -----
  │  ┌─────── 0                .50
  │  │                         ×2
  │  │                       -----
  │  │  ┌──── 1                .00
  ↓  ↓  ↓
  1  0  1
```

The binary equivalent of 0.625_{10} is the column of leading binary digits read from the top down preceded by a decimal point:

$$0.625_{10} = 0.101_{2}$$

Binary Coded Decimal Numbers

An alternate way to represent decimal numbers by binary digits is to convert each decimal digit independently into its binary equivalent by using four bits, as shown in Table 3-6.

A decimal number that is translated (or coded) by this procedure is called a *Binary Coded Decimal* number (BCD number). For instance, the decimal numbers 25 and 11 are coded as

$$25_{10} = 0010 \quad 0101$$
$$11_{10} = 0001 \quad 0001$$

Note that this conversion is entirely different from that discussed earlier; it is simpler in the sense that everything required for conversion is given by Table 3-6.

Table 3-6 Binary Coded Decimal Digits

Decimal Digit	Binary Equivalent
0	0000
1	0001
2	0010
3	0011
4	0100
5	0101
6	0110
7	0111
8	1000
9	1001

Now that we know how operations and numbers can be translated into the computer language of zeros and ones, we can conceive of devices that can effect such translation. What remains to be shown is how the computer computes in its own language of zeros and ones.

3-10 HOW A COMPUTER COMPUTES

Since numbers in the computer are in binary form, we expect the computations to be performed in the binary system. The fundamental rules for adding binary digits are shown in Table 3-7. For convenience, we identify the results of the four operations by a sum digit S and a carry digit C as indicated in the table. These rules can be applied to add any two numbers, A and B, as follows:

Table 3-7 Rules for Adding Binary Digits

	Case I	Case II	Case III	Case IV
$a =$	0	0	1	1
+	+	+	+	+
$b =$	0	1	0	1
C: Carry digit, S: Sum digit → $\boxed{C}\,\boxed{S}$	00	01	01	10

$$
\begin{array}{rl}
\text{Carry digits} & C \longrightarrow 1\ \ 1\ \ 1\ \ \\
 & A \longrightarrow \ \ 1\ \ 0\ \ 1 \\
 & + \\
 & B \longrightarrow \ \ 1\ \ 1\ \ 1 \\
\text{Result} & \longrightarrow 1\ \ 1\ \ 0\ \ 0
\end{array}
$$

We begin at the right-hand column of digits by adding 1 1 and recording a sum digit of 0 in the last row and a carry digit of 1 at the top of the second column from the right. Now we add the digits of the second column, $1 + 0 + 1$, and record a sum digit 0 and a carry digit of 1 above the next column. Next, we add $1 + 1 + 1$ and record the sum digit 1 and carry digit 1 which is also recorded in the result.

In general, then, for any two binary numbers $A = a_1\, a_2\, a_3\, a_4$ and $B = b_1\, b_2\, b_3\, b_4$, we have

		[0]	[1]	[2]	[3]	[4]	← Column numbers
Carry digits	$C \longrightarrow$	C_1	C_2	C_3	C_4		
	$A \longrightarrow$		a_1	a_2	a_3	a_4	
		+					
	$B \longrightarrow$		b_1	b_2	b_3	b_4	
Result	$\longrightarrow$	C_1	S_1	S_2	S_3	S_4	

C_j indicates the carry digit resulting from the addition operation in column j. For example, the following results are obtained by using the above general form for adding binary numbers. The reader should verify these results.

Carry digits	$C \longrightarrow$ 1111100	1111	10011111
	$A \longrightarrow$ 1011101	1111	11010111
	+	+	+
	$B \longrightarrow$ 110110	111	10011111
Result	$\longrightarrow$ 10010011	10110	101110110

From the discussion to this point it becomes apparent that if we can add three binary digits which consist of a carry bit C_j from a preceding addition in column j, and two bits a_{j-1} and b_{j-1} from the numbers A and B in column $j - 1$, then we can add any two numbers. Of course, once we can add any two numbers, we can add n numbers by adding the sum of the first two to the third, the new sum to the fourth, and so on. We must, therefore, show how the computer can add three bits $a_{j-1} + b_{j-1} + C_j$, record the sum digit, shift the carry digit to the next column, and repeat the operation. We show this in two stages. First, we discuss a *half-adder* which can perform the operations of Table 3-7. Next, we combine two half-adders into a *full adder* which can add any two binary numbers.

It can also be shown that the operations of multiplication, division, extracting roots, etc., can be reduced to the most fundamental operations of addition. If this is so, then if the computer can perform addition, it can also perform multiplication, division, extraction of roots, etc.

Half-Adder

Figure 3-9 shows diagrammatically a half-adder. The digits ⓐ and ⓑ to be added (see Table 3-7) are entered in the form of signals from the left,

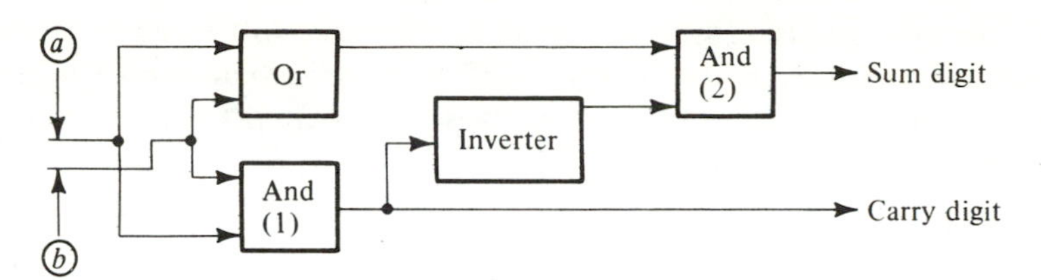

Figure 3-9 Half-Adder

and the answer is recorded in the form of signals on the right. The directi of signal flow is indicated by the arrows. A signal will come through ⓐ only when a polarized spot is detected where ⓐ is stored, signifying the digit 1 for ⓐ; similarly for ⓑ. The locations of the sum digit and carry digit will be polarized, signifying the storage of the digit 1, only when a signal gets there; otherwise, a zero is stored. The components that will effect this information transfer consist of two *AND boxes*, an *OR box*, and an *INVERTER box*. (These were discussed briefly in Chapter 2, but are repeated here for completeness.)

AND Box. An AND box is a device that has two inputs and one output. It will output a signal if, and only if, signals enter through the two inputs simultaneously. A simple example of an AND box is shown in Fig. 3-10(a).

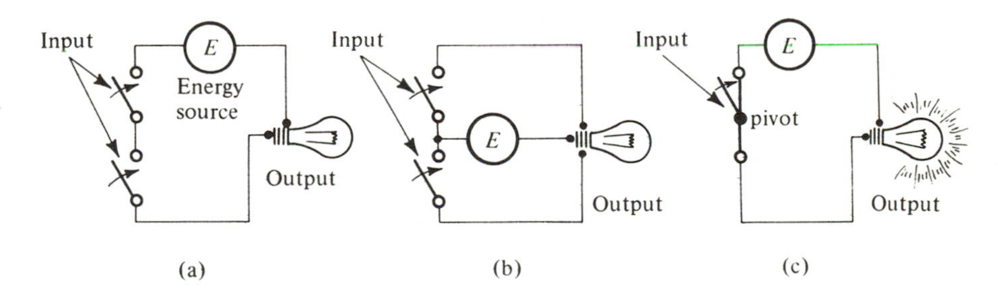

Figure 3-10 Circuits for AND, OR, and INVERTER Boxes: (a) AND Box (Input Signal Closes Switch); (b) OR Box (Input Signal Closes Switch); (c) INVERTER Box (Input Signal Opens Switch)

In the circuit of this figure a switch closes when an input signal excites it. Consequently, two input signals entering simultaneously will close the circuit and cause an output signal represented by the light on.

OR Box. An OR box is a device that has two inputs and one output. It will transmit a signal if at least one signal enters the box. [See Fig. 3-10(b).]

INVERTER Box. An INVERTER box has one input and one output. It will transmit *no signal* when a signal enters, and will transmit a signal when no signal enters the box. An INVERTER box circuit is shown in Fig. 3-10(c) in which the entry of a signal causes the switch to open. When no input signal enters, the switch is closed.

In early computer models AND, OR, and INVERTER boxes were constructed with electromagnetic relays. These were replaced later by vacuum tubes and then by transistors.

Let us follow the half-adder diagram of Fig. 3-9 as it executes case (IV) of Table 3-7. Since, in this case, $a = 1$ and $b = 1$, two signals enter AND box (1) and the OR box; consequently, each of these boxes outputs a signal. From AND box (1) the signal enters the INVERTER box and, at the same time, enters the carry digit position to store 1. AND box (2) receives a signal from the OR box, but no signal from the INVERTER box; consequently, it transmits no signal so that a 0 is stored in the sum digit position. This agrees with case (IV) of Table 3-7. (Verify the other cases.)

The operations of the half-adder of Fig. 3-9 performing the additions of Table 3-7 can be summarized in a truth table, shown in Table 3-8.

Table 3-8 Truth Table for Half-Adder of Fig. 3-9

Input		Output	
a	b	C Carry Digit	S Sum Digit
1	1	1	0
1	0	0	1
0	1	0	1
0	0	0	0

Using the notation of Boolean algebra, we can write the following equations for the carry digit C and the sum digit S as functions of a and b:

$$C = a \cdot b$$
$$S = a \cdot b' + a' \cdot b$$

The prime is another way of designating *not* or *complement;* namely, it is the same as the symbol $\sim$. An overbar, as in $\bar{a}$, is also used to designate the same thing. Note that the function for S is equivalent to an *exclusive OR* statement, $a \underline{\vee} b$; namely, S is 1 when a is 1 or b is 1, but not both.

* Full Adder

Figure 3-11 represents a *full adder*, which consists of two half-adders and an OR box connected as shown. The input-output for each of these three elements is as follows:

Element	*Input*	*Output*
Half-adder (1)	Bits a_j and b_j of numbers A, B of column j	Sum digit $S(1)$ Carry digit $C(1)$
Half-adder (2)	Carry digit C_I resulting from addition in the preceding column of digits (i.e., column $j + 1$).	Sum digit $S(2)$ recorded as S_O (result of addition in column j)
	Output $S(1)$ from half-adder (1)	Carry digit $C(2)$
OR box	$C(1)$ and $C(2)$	Carry digit C_O resulting from addition in column j

C_O, i.e., the output carry digit from the OR box in the addition performed in column j, is not recorded but delayed one unit of time by a clock that controls the operations, and is entered as an input digit C_I with the

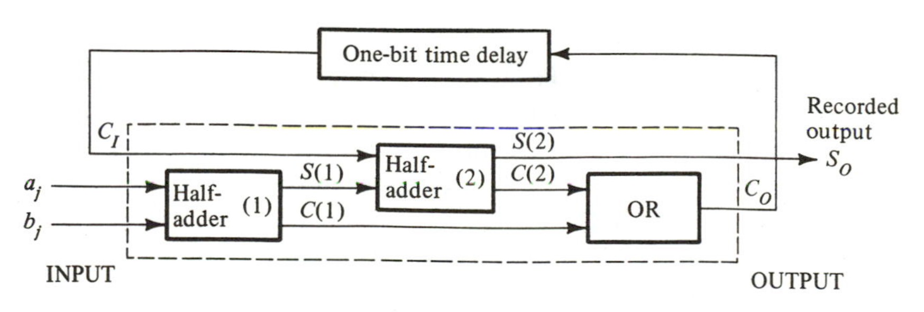

Figure 3-11 Full Adder

Table 3-9 Truth Table for Full Adder of Fig. 3-11

Row Number ↓	Input			Output	
	a	b	C_I	C_O	S_O
1	1	1	1	1	1
2	1	1	0	1	0
3	1	0	1	1	0
4	1	0	0	0	1
5	0	1	1	1	0
6	0	1	0	0	1
7	0	0	1	0	1
8	0	0	0	0	0

digits a_{j-1} and b_{j-1} of column $j - 1$. This is designated by the 1-*bit time-delay box* in Fig. 3-11.

The operation of the full adder can be summarized in a truth table, Table 3-9.

Using the notation of Boolean algebra, we can write the following equations for C_O and S_O in terms of a, b, and C_I:

$$C_O = a \cdot c + a \cdot b + b \cdot c$$
$$S_O = a' \cdot b' \cdot c + a' \cdot b \cdot c' + a \cdot b' \cdot c' + a \cdot b \cdot c$$

It is constructive to verify the outputs for C_O and S_O in Table 3-9 by following the operations in Fig. 3-11. We suggest you try it.

EXAMPLE. Let us follow the full adder of Fig. 3-11 as it performs the following addition:

	[0]	[1]	[2]	[3] ← Column numbers
C_O, C_I →	1	1	1	
$a_1a_2a_3$, A →		1	0	1
+				
$b_1b_2b_3$, B →		1	1	1
$C_1S_1S_2S_3$, Result →	1	1	0	0

Column 3: $a_3 = 1$, $b_3 = 1$, and $C_I = 0$ are entered at the input to the full adder. From the second row of Table 3-9, we have: $S_3 = S_O = 0$, $C_O = 1$

Column 2: $a_2 = 0$, $b_2 = 1$, and $C_I = 1$. From row 5 of Table 3-9, we have: $S_2 = S_O = 0$, $C_O = 1$

Column 1: $a_1 = 1$, $b_1 = 1$, and $C_I = 1$. From row 1 we have: $S_1 = S_O = 1$, $C_O = 1$

Column 0: $a_0 = 0$, $b_0 = 0$, and $C_I = 1$ From row 7: $S_0 = S_O = 1$, $C_O = 0$

The result of the addition is given by the values of S_O, $S_0S_1S_2S_3$ which is also the sequence $C_1S_1S_2S_3$ because $S_0 = C_1$

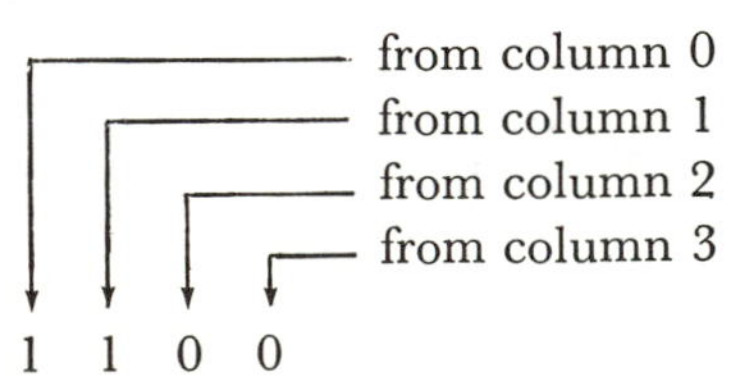

The train of pulses entering the full adder (INPUT) in the above addition and the train of output pulses have the form of Fig. 3-12. The signals

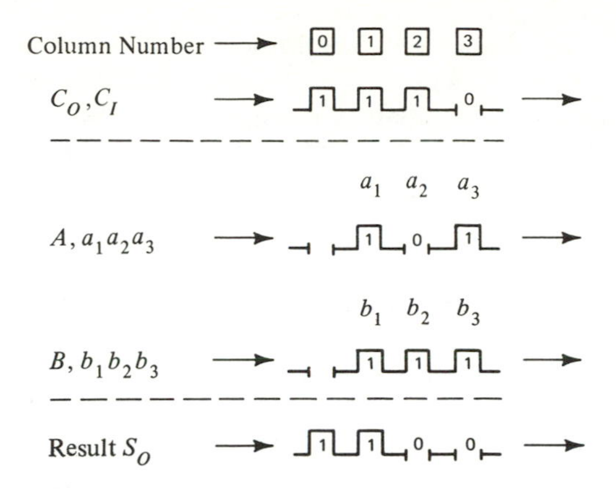

Figure 3-12 Train of Pulses at Input and Output of Full Adder in the Addition of A = 101 and B = 111

for C_O, C_I, and A, B enter in the direction of the arrows (to the right), namely, starting with the least significant digit on the right; and the output signal S_O exits the same way with the least significant digit first.

* Parallel Adder

The adder of Fig. 3-11 is part of the computer arithmetic unit (calculator) and it can add two binary numbers in serial fashion, proceeding in sequence from the right-most column to the left-most. This, of course, requires time delays to shift the carry bit from each column j to column $j-1$ on its left. It is possible to speed up the operation of addition by employing a number of full adders in parallel, as shown in Fig. 3-13. The adder of

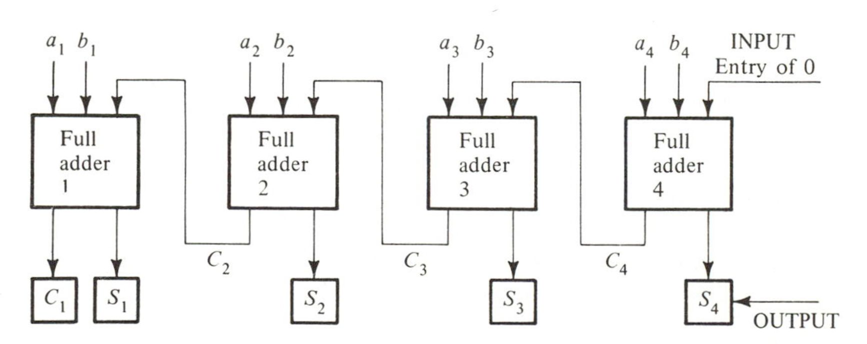

Figure 3-13 Parallel Adder (for 4-Bit Adder; n-Bit Adder has n Full Adders)

Fig. 3-13 performs simultaneously the addition of two numbers, each consisting of four binary digits. The result is a five-digit binary number, $C_1S_1S_2S_3S_4$, in which the most significant digit is the carry digit, from the addition in column 1, which is designated by C_1. Note that the output carry digit C_4 of full adder 4 becomes input to full adder 3, and so on until we get to C_1. The carry digits are generated almost instantaneously so that the final sum appears after a very short delay. To add two binary numbers of n digits will require n full adders linked in parallel as shown in Fig. 3-13.

Verify the addition of 1101 and 1110 by following Fig. 3-13. Use Table 3-9 to determine the output from the full adders.

Computer Control

The control of a computer has the functions of causing information to be stored in memory or retrieved from it, and causing a program to be interpreted and executed in a proper sequence.

In our simple model of Sec. 3-2 the secretary, aided by a wall chart and a dial, constituted a control unit. In the modern digital computer the control contains a *command register* where operations are interpreted (analogous to the wall chart in Fig. 3-1) and a *command counter* which controls the sequence of operations (corresponding to the dial in Fig. 3-1). To channel information in the computer, and into and out of the computer, the control makes use of AND, OR, and INVERTER boxes described earlier.

Types of Digital Computers and Their Key Features

Some digital computers are special-purpose computers designed to solve specific problems. Others are general-purpose computers capable of solving a great variety of problems. The general-purpose computers are not all the same; they may vary in the number of fundamental operations they perform, their speed of operations,* storage capacity, input-output media, and, of course, cost. In the light of our discussion in this chapter so far, we can state that no matter how many complex functions a digital computer can perform, these functions are combinations of the fundamental operations of detecting, transferring, and manipulating information in the form of bits.

3-11 PROGRAMMING

Programming a computer in machine language (its language of zeros and ones) consists of writing a sequential set of instructions, specifying addresses where machine words (data, instructions, and answers) will be stored.

*Some computers can add more than 10 million numbers in 1 second at a cost of less than 1 cent.

All these are in binary forms in the computer. The task of programming in machine language can become very involved although it is very efficient in exploiting the capabilities of the computer [23, 24, 25, 26]. In recent years much has been done to make the task of programming simpler so that communication with the computer will be more direct insofar as the user is concerned. This brings us to *automatic programming* which provides an automatic translation from "almost" the English language and the language of mathematics to computer language.

Automatic Programming

The program of Table 3-1 is an example of a program written in a machine language (the language of our simple computer model). In writing this program we had to specify location registers for the program, the data, and the answer, and to consider each step of the operations. All this is simplified when automatic programming is used. Automatic programming involves the use of a *compiler* program (which is a computer program) that translates a simple language of programming to machine language and automatically assigns registers to the program, the data, and the answers. The use of a compiler causes a less efficient utilization of the computer, but this sacrifice is outweighed, most of the time, by the benefits of the simple programming language. An example of an automatic programming language is the IBM FORTRAN (short for FORmula TRANslator) [27, 28]. This language is remote from the computer language and is very close to the English language and the language of mathematics. For instance, the words

```
DO    READ    PUNCH    PRINT
```

instruct the computer to do, read, punch, and print; and the equation

```
X = A + B - C * D/E
```

instructs the computer to execute the operations* on the right of the equal sign and assign the result to the variable X on the left. To get a better appreciation of this language, let us apply it to the program of Sec. 3-4 and follow all the steps in its development.

FORTRAN Program to Add *N* Numbers

The program of Table 3-1 is written in FORTRAN in Table 3-10. Each line in the table is called a FORTRAN statement. Using punched cards as input media, one statement is punched per card. Cards are punched by depressing keys on the keyboard of the punching equipment, with the keys correspond-

*In FORTRAN the asterisk(*) and slash (/) designate multiplication and division, respectively.

ing to the symbols (letters, numbers, etc.) that appear in the FORTRAN statements. The statement number preceding each statement is optional. A statement number is required only for those statements that are referred to by other statements of the program, as we shall see later. After the program is written, the programmer checks it for errors in the programming language, such as spelling, omission of commas, parentheses, equal signs, and the like. Then the deck of FORTRAN cards is checked automatically by the computer (*precompiled*) to detect any further errors insofar as the FORTRAN language is concerned. When errors are detected, special typed signals indicate the nature of the errors so that they can be corrected.

Table 3-10 FORTRAN Program to Add N Numbers*

Statement Number	FORTRAN Statement
1	SUM = 0.0
2	READ N
3	DO 4 I = 1,N
4	READ A(I)
5	DO 6 J = 1,N
6	SUM = SUM + A(J)
7	PRINT SUM

*The program should also specify the form of the data and answer (how many significant figures, floating point or fixed point) and also the number of A(I) terms. These details are eliminated for the sake of clarity.

After the FORTRAN program, also referred to as *source program*, has been precompiled and no errors detected, it is placed in the computer for translation into the computer language (*compilation*). The compilation produces a new deck of punched cards that represent the source program in machine language. The program in machine language is referred to as the *object program*.

At this point, all data are punched on cards and arranged in the order called for by the program. (In the program of Table 3-10, N appears first and is followed by the 700 numbers to be added.) The data cards are placed behind the object program cards, and the entire deck of cards is placed in the computer. When the start button is depressed, the computer proceeds as follows: First the object program is stored in the computer memory, then it begins to execute the program. Note that the data have not yet been stored in the computer memory and will be stored only when the program is being executed and the word READ is encountered. Let us follow the computer as it executes the program with the value of $N = 700$ and the data as given by Table 3-2 (ignore the addresses specified in this table), using the concept of the compartments of Sec. 3-2.

Program Execution

Statement 1. The computer assigns the value zero to the inside of a compartment which it has named (identified by an address) SUM.

Statement 2. The computer reads the first number (700) in the data cards and assigns it to a compartment with address N. The data card with the number 700 has now been processed through the input media, and the card with the next piece of data (75) is ready to be read.

Statement 3. DO 4 I = 1,N The computer assigns to I the value of 1 and executes all the following statements down to and including Statement 4. Then I is assigned the value 2 and the process is repeated, until finally Statement 4 has been executed the number of times specified by the number in address N (which is 700 after Statement 2 has been executed). At this point the computer proceeds to Statement 5. Statements 3 and 4 are called a *DO loop*. This is indicated for our own reference by the line to the left of these statement numbers. As a result of executing this DO loop, the computer stores in the compartments with addresses A(1), A(2), . . . , A(700) the data from the data cards, (see Fig. 3-14), so that the number 75 is in A(1), 10 is in A(2), 24 is in A(3), and finally 2050 is in A(700). (See Table 3-2 for data.)

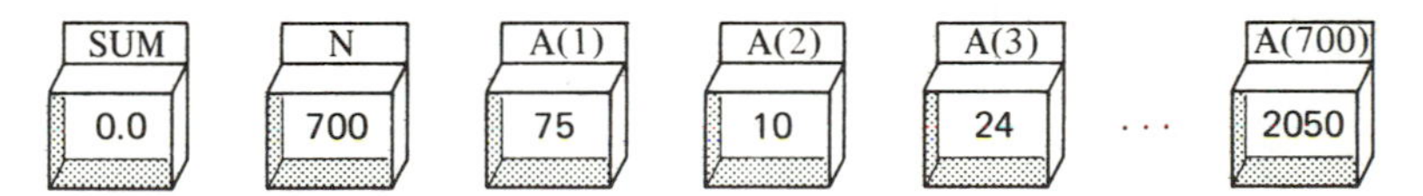

Figure 3-14 Content of Memory Compartments After the DO Loop of Statements 3 and 4 is Completed

Statements 5 *and* 6 form another DO loop. First, the computer sets J = 1, then goes to Statement 6, which will now be SUM = SUM + A(1), and proceeds to execute the instruction on the right of the equal sign as follows: It goes to compartment SUM and reads the number inside it; next, it goes to compartment A(1) and reads its content. Then it adds these two numbers. (The present content of these compartments is shown in Fig. 3-14.) Next, it stores the result in the compartment specified on the left of the equal sign;* that is, it stores 75 in SUM, replacing the 0.0.

The DO loop has now been satisfied for J = 1. Now the computer sets J = 2 and returns to Statement 6 which now reads

SUM = SUM + A(2)

Proceeding as before, the computer will add 75 and 10, and store the result of 85 in SUM, replacing its former content (75). Then J is set equal to 3,

*Note that the equal sign (=) in FORTRAN does not have the conventional meaning.

and so on until, finally, J = 700 (the value of N) at which point SUM contains the sum of the first 699 data numbers and Statement 6 appears as

SUM = SUM + A(700)

After this is executed, SUM contains the desired sum of all 700 numbers. The computer proceeds to Statement 7.

Statement 7. The computer prints the content of SUM which, of course, is the desired answer.

The flow chart for the program of Table 3-10 is shown in Fig. 3-15. Of course, in general, the flow chart is prepared before the computer program is written and serves as a guide for generating the program. You can

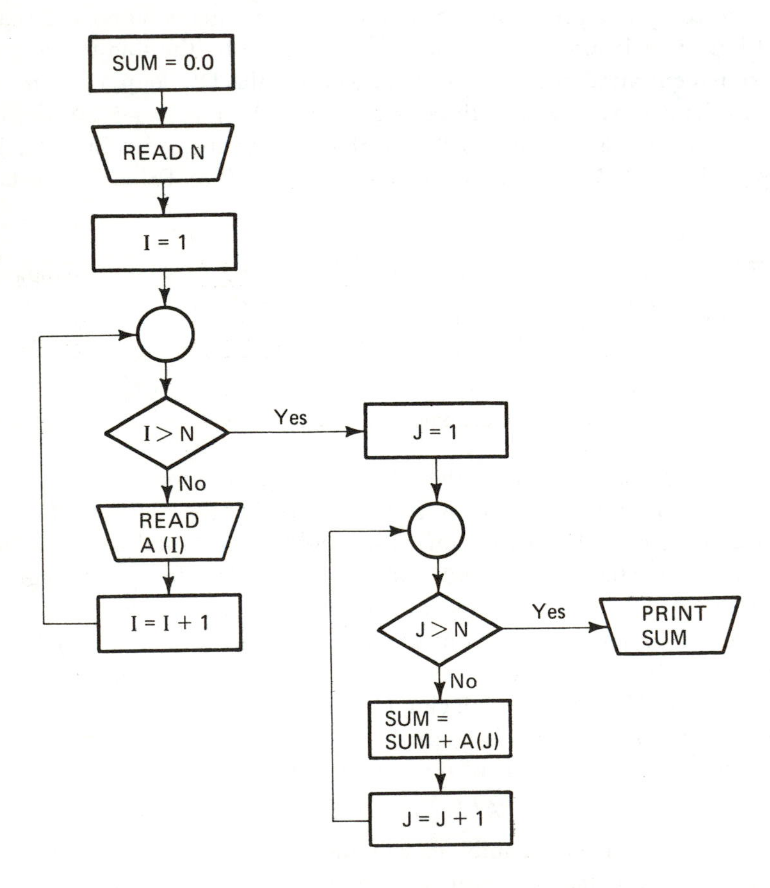

Figure 3-15 Flow Chart for FORTRAN Program in Table 3-10

identify the two DO loops in the flow chart by noting the flow of arrows. The basic building blocks commonly used for preparing a flow chart are as follows (these symbols were also used in the flow chart of Fig. 3-4):

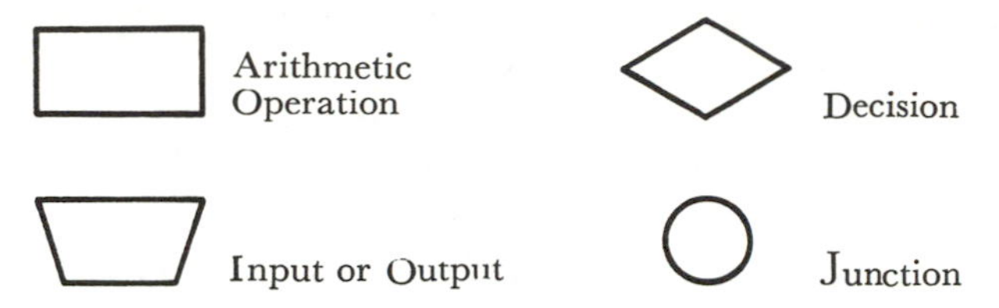

A More Efficient FORTRAN Program to Add *N* Numbers

The program of Table 3-10 requires 700 data storage compartments, one for each piece of data. We observe, however, that after a number has been added to the sum, it is not needed any more. Also, we are adding one number at a time. This suggests the more efficient program of Table 3-11, which will save a lot of computer memory space. In this program there is a single compartment with address A for all data. There is also a single DO loop (Statements 3 to 5 inclusive). When the DO loop is executed for I = 1, the computer first reads the data card with the number 75, stores it in address A, then executes Statement 5. Now I is set equal to 2, and when the computer encounters Statement 4 it reads the next data card (which has the number 10) and stores it in A, replacing the 75. Then Statement 5 is executed and causes the value of 85 to be stored in SUM. This is repeated until the DO loop (3 to 5) is executed 700 times (the value of N). We suggest that the reader follow the execution of this program by picking five numbers, for instance (storing 5 in N), and drawing compartments for SUM, N and A.

Table 3-11 More Efficient Program to Add *N* Numbers

Statement Number	FORTRAN Statement
1	SUM = 0.0
2	READ N
┌ 3	DO 5 I = 1,N
│ 4	READ A
└→ 5	SUM = SUM + A
6	PRINT SUM

Subroutines

It is possible to economize in program writing by using symbolic names to refer to subprograms that were previously developed. For example, consider 600 numbers arranged in an array of 20 columns and 30 rows. Such an array is called a *matrix*. The matrix can be denoted by a single letter A with each element (number) in row I, column J, designated as A(I, J). Table 3-12 gives a program that will add each number A(I, J) from matrix A which has M rows and N columns to a corresponding number B(I, J) from a matrix B which also has M rows and N columns, and store the results in a matrix C. This program can become a part or a *subprogram* of a larger program called the *main program*. The subprogram is also called a *subroutine* and it is identified by a name. Here MATADD(A,B,C) is the name for the subroutine that specifies that matrix A is to be added to matrix B and the result stored in matrix C.

Table 3-12 Program to Add Two Matrices

Subroutine MATADD(A,B,C)	
1	DO 3 I = 1,M
2	DO 3 J = 1,N
3	C(I,J) = A(I,J) + B(I,J)

Suppose the main program requires the addition of two matrices RX and RY with the result stored in matrix RZ. Then all we need write is the following one line in the main program:

```
CALL MATADD (RX,RY,RZ)
```

Similarly if matrix RX must be multiplied* by matrix RY and the result stored in matrix RZ, the main program will require a statement of the form:

```
CALL MATMPY (RX,RY,RZ)
```

This statement will refer to a subroutine of the form given by Table 3-13 which is stored in the computer and executes the required operation. The statement RETURN (statement 6 in Table 3-13) causes computer control to return to the main program after the subroutine has been executed.

Table 3-13 Subroutine MATMPY(A,B,C)

1	DO 5 I = 1,M
2	DO 5 J = 1,N
3	C(I,J) = 0.0
4	DO 5 K = 1,L
5	C(I,J) = C(I,J) + A(I,K)*B(K,J)
6	RETURN
7	END

*Matrix multiplication is a procedure in which the numbers in each row of the first matrix are multiplied by corresponding numbers in the columns of the second matrix.

3-12 USE OF COMPUTERS

The purpose of this chapter was to introduce fundamental concepts and key features of digital computers. In so doing, we did not consider the great detail and complexity required to design these sophisticated machines. Familiarity with fundamental concepts, however, should enable the reader to embark on a program of study of an automatic programming language [27, 28] so that he will be able to exploit the vast capabilities of the digital computer. We hope that an appreciation of the fundamental concepts, which are the basis of the structure and operation of the computer, will dispel the anxiety that results from the mystery that surrounds computers. The need to dispel the anxiety becomes evident when we consider the ever-increasing use of computers in our society, as mentioned briefly in the introduction to this chapter and expanded in the following paragraphs.

Use of Computers [29]

The use of computers has expanded enormously since their commercial introduction in the early 1950s, and new applications continue to emerge. Early applications concentrated on file maintenance, data sorting, payroll accounting, inventory status information, sales records, accounts receivable, invoicing, and other information and data summaries in administrative and management applications. With time, applications penetrated other areas of private business, government, and the military. A partial list includes the following applications:

Files and Information Retrieval. Master files in banks, industry, business, and government can be kept current with changes, deletions, and additions made by the computer. For example, many businesses use the services of a credit check which is a computerized check of a file. When a customer wishes to pay by using his credit card, the sales clerk can telephone a request for a credit check; the response is instantaneous. Library reference files can be searched by the computer and information printed for the user. Here, the computer is used to store information, keep it current, and retrieve it on demand.

Data Manipulation. Data can be sorted, classified, and printed out in a variety of desired forms: tables, graphs, matrix arrays, and the like. For example, income tax returns can be printed out in the same format as that required by the state or federal government.

Calculations. The computer can perform calculations from the most elementary to the most sophisticated evaluations of mathematical functions and complex algorithms. It can perform the symbolic logic operations of Chapter 2, the decision theory computations of Chapter 7, the optimization calculations of Chapter 8, etc. The field of numerical analysis deals with models for computer calculations.

Simulation. The computer can be used to generate models of experience by simulating the behavior of physical systems such as transportation, pollution, flood control, bridge response, airplane, automobile, and the like. Simulation can also be applied to human behavior such as acceptance of a new product, a new law, a candidate for office. Both deterministic and probabilistic models can be considered. In all simulation studies we ask questions of the sort "What if . . ." and get quick responses, say, in minutes, regarding phenomena which involve years in real life. (See Chapter 9 for simulation models of a city.)

Control. In Chapter 9 we discuss dynamic systems in which control to achieve a desired level of performance is a key feature. For example, in a chemical plant, where a particular process is in operation, the output is sensed and a computer can conduct the appropriate computations as well as cause a signal to correct rapidly for any deviation from a desired level of performance. This function of the computer in control is more evident and explicit in guiding the flight of a space vehicle. The direction of travel is monitored, and any deviation from the desired course triggers rapid computer calculations to effect a correction in direction. Here the speed of calculation is extremely important because elapsed time, or delay, between observation and correction may make the correction ineffective or even detrimental.* (See Chapter 9 for a discussion of feedback control.)

Pattern Recognition. The computer can be used to recognize letters, numbers, and words, for sorting mail or in keeping checking account records current in banks. Computer pattern recognition can be used in classification of patterns in scientific work such as biology, zoology, anthropology, and in engineering models of complex systems. Here the computer is given limited data and the program generalizes by synthesizing or identifying a pattern. This is a use of computers in an inductive process.

Heuristic Problem Solving and Artificial Intelligence. The computer can be used to discover proofs in mathematics, to generate new combinations of moves in playing chess, and it can employ means-ends analysis procedures in general problem solving (GPS), as discussed in Chapter 1. This area of computer use is among the most novel and exciting. It opens the door to effective programs for such applications as computerized weather prediction and to adaptive learning to deal with complex problems of a social, socio-economic, and socio-technical character. The potential for extending adaptive computer learning to model the human learning process may prove the most far-reaching advance in the history of man. (See Chapter 9 for a discussion of Adaptive Control.)

---

*True, computer calculations can account for some time delay; however, effective control requires that the delay be small.

These categories are, of course, not exhaustive. You may wish to expand the list of applications to specific areas, such as education and medicine, and consider how the computer can be used (or is being used) effectively in these areas. This partial list of applications reminds us that the computer is widely used and that a fundamental knowledge of its characteristics and potential uses must be part of modern-day education.

3-13 SUMMARY

A modern electronic digital computer consists of these components: *input media, memory, arithmetic unit, control unit, and output media. The input and output media* are, respectively, the elements for entering and retrieving information in communicating with the computer. The *memory* is the computer storehouse of information in the form of instructions or data. The *arithmetic unit* is the computation center. The *control unit* reads and stores information, interprets instructions, and causes all operations to be performed in the proper sequence. A *command counter* in the control unit keeps track of the instructions and acts as a clock for timing and sequencing operations. The interpretation of instructions takes place in the *command register* of the control unit.

A *flow chart* is a schematic diagram that describes the flow of various blocks of operations in the plan to solve a particular problem. The flow chart can be constructed to various levels of detail in the descriptions of the operations. A flow chart is helpful as a first step prior to developing the details of a computer program to solve a problem.

A *computer program* consists of a sequence of coded instructions for the solution of a problem.

Computer memories are of the *serial access* or *random access* type. The *serial access* memories consist of rotating disks or drums with a magnetic surface (or tapes), and information can be stored or retrieved by using a write-read head which is positioned above the proper channel. The access time depends on the position of the desired location in memory; the closer it is to the write-read head, the shorter the access time. The *random access* memory consists of a matrix of core elements of ferromagnetic material. It is free of any rotating elements and each location in it has the same access time for all practical purposes.

Input-output media can have the form of a printer, typewriter, punched cards, punched tape, magnetic disk, magnetic tape, or console with keyboard and light pen.

The computer language employs binary numbers. Numbers can be converted from one base to another. Binary numbers can be used to code the decimal digits in the decimal number system by coding each of the digits 0 to 9 separately and independently, thus preserving the positional value of the decimal system. Decimal numbers coded this way are called *Binary Coded Decimal Numbers.*

The computations in the digital computer are performed in the binary number system, using AND, OR, and NOT (INVERTER) circuits or boxes based on symbolic logic or Boolean algebra. A *half-adder* is a device for adding two binary digits. The half-adder consists of two AND boxes, one OR box, and one NOT (INVERTER) box. The operation of the half-adder can be described by a truth table or the following equations of Boolean algebra in which a and b are the two binary digits to be added:

Carry digit, $c = a \cdot b$
Sum digit, $s = a \cdot b' + a' \cdot b$

A *full adder* can add any two binary numbers. It consists of two half-adders, one OR box, and a ONE-BIT DELAY box. It proceeds by adding a digit a_i from binary number A to corresponding digit b_i from binary number B, plus the carry digit c from the preceding stage which involved a_{i+1} and b_{i+1}. (Numbers are represented in the form $a_1a_2 \ldots a_n$.) The operation of the full adder can be described by a truth table or the following equations of Boolean algebra; c designates the carry digit from the previous stage:

$$\text{Carry digit, output } C_O = a \cdot c + a \cdot b + b \cdot c$$
$$\text{Sum digit, output } S_O = a' \cdot b' \cdot c + a' \cdot b \cdot c' + a \cdot b' \cdot c + a \cdot b \cdot c$$

A *parallel adder* consists of an array of full adders to permit parallel operations. There are n full adders for the parallel addition of binary numbers with as many as n bits per number.

Automatic programming provides an automatic translation from "almost" natural language (say, English) and the language of mathematics to computer language. Automatic programming involves the use of a *compiler* which is a computer program designed to translate a simple programming language into the computer language (called *machine language*) and assign registers to the instructions and data. IBM FORTRAN is an example of an automatic programming language. This language is remote from computer language and close to the English language and the language of mathematics. For example, the words DO, READ, PUNCH, PRINT, and the equation

```
X = A + B - C * D/E
```

are legitimate statements in the language of FORTRAN.

The steps in developing and using a FORTRAN program are: First, the desired operations of a flow chart are translated to the FORTRAN language. The punched cards representing the FORTRAN program are checked automatically by the computer to detect errors in the FORTRAN language. This phase is called *precompilation.* After the FORTRAN program, which is called the *source program*, has been precompiled successfully, i.e., no errors detected, it is placed in the computer for *compilation*, i.e., translation into computer language. Compilation produces a new set of cards called the *object program*. The *object program* represents the original source program in machine language.

Subroutines are subprograms which can be referred to and used by a *main program*. Subroutines are identified by a name and can be retrieved from memory by reference to that name.

The use of computers has expanded enormously since their introduction in the early 1950s. Computers are used in banking, finance, industry, government, education, and research in science and the humanities, to name some areas. Specific task applications include files and information retrieval, data manipulation, calculations, simulation of technical and social systems, process control, pattern recognition, heuristic problem solving, and artificial intelligence.

EXERCISES

3-1 Write a flow chart for the FORTRAN program in Table 3-11. How does this compare with Fig. 3-15 (the flow chart for the less efficient program)?

3-2 Using the computer model discussed in this chapter, draw a flow chart for a program which generates and prints out the Fibonacci sequence, 1, 1, 2, 3, 5, 8, 13, . . . , where each element is the sum of the two preceding elements. Have the program stop after the 100th element of the sequence.

NOTE. Exercises 3-3 through 3-6 refer to the following grade assignment problem. At the end of the term, a professor assigns the students grades based on the number of points they scored on the tests. First, he must select cutoff totals T_A, T_B, T_C, and T_D which are the lowest point totals for which he will give the grades A, B, C, and D, respectively.

3-3 One possible scheme is to make T_A equal to 85% of the total possible points, $T_B = 65\%$ of the total, $T_C = 35\%$ of the total, and $T_D = 15\%$ of the total. Draw a flow chart for a program which would assign T_A, T_B, T_C, and T_D. Assume the total possible $= T_{\text{MAX}}$.

3-4 Another plan would be to use a rule similar to that described in Exercise 3-3, except to take the percentages from the highest point total achieved by a student (T_{HIGH}) rather than from the maximum total possible points. Draw a flow chart for a program that would assign T_A, T_B, T_C, and T_D by using this new plan. (*Hint:* You must first find out what T_{HIGH} is. Assume the number of students in the class is 200.)

3-5 Assuming that T_A, T_B, T_C, and T_D have been assigned, draw a flow chart for the program which would count the number of A's, B's, C's, D's, and F's that were given out, and would compute the average point total of the scores of 200 students.

3-6 What are the advantages to the professor in using the computer to assign grades? What are the disadvantages?

NOTE. Exercises 3-7 through 3-9 refer to the following character identification problem: When Samuel F.B. Morse designed his code, he tried to assign shorter codes for letters most often used and longer codes for letters and symbols least often used. (This is why "•" stands for "e" and "•••-" stands for "v".)

3-7 If Morse were attempting the task now, he might find the use of a computer quite helpful. Draw a flow chart for a computer program that would read the first 50,000 characters of this book ("read" by means of an input device, Fig. 3-8) and count how many times the letter "t" is used. (Characters include letters, punctuation, numbers, and spaces.)

3-8 Draw a flow chart for a computer program that would read the same characters and count how often the letter combination "th" is used.

3-9 Suggest some other application of the computer's ability to perform this kind of character identification.

3-10 Suppose it is required to add only the first three numbers in Table 3-2, type the result, and stop the computer. What change(s) must be introduced in Tables 3-1 and 3-3, if any? Follow the execution of the program until the computer is stopped by the program.

3-11 You are a spy with an agent in Switzerland who sends you top secret information. The information is preceded by a three-digit binary number. From time to time your agent will send you false information to keep enemy spies off balance. An electronic circuit is devised which uses AND, OR, and NOT elements (as in a half- or full adder); this circuit takes the binary number and tells you if the information is true. If the binary number is (ABC), then the document is real if the following Boolean statement is true:

$$(A \wedge \bar{B}) \vee (\bar{A} \wedge B \wedge C) \vee (A \wedge \bar{C})$$

(a) Diagram such a circuit.
(b) Assume you no longer have any OR elements. Now construct the circuit.

3-12 Concerning the problem you chose in Exercise 1-1, discuss how the use of a computer could aid you in some aspect of the solution (for example, in data collection, sorting, arithmetic calculations, etc.).

NOTE. Exercises 3-13 through 3-23 are drill exercises intended to give you a better understanding of the computer's facility to perform arithmetic operations and the conversion of numbers from one base to another.

3-13 How many bits must a binary word contain in order to have enough possible combinations to represent 26 letters, upper and lower case, 2 punctuation symbols, and digits 0 through 9?

3-14 Devise an easy method for converting numbers from base 8 to base 2 without going through base 10.

3-15 Devise a number system for base 16.
(a) Convert to base 16 and add $397_{\boxed{10}}$ and $43_{\boxed{10}}$.
(b) Convert to base 16 and divide $720_{\boxed{10}}$ by $36_{\boxed{10}}$.

3-16 Execute the following operations with the binary numbers given and check your results by converting the numbers to their decimal equivalents.

$$\begin{array}{r}11111\\+11111\\\hline\end{array}\quad\begin{array}{r}1001\\+1100\\\hline\end{array}\quad\begin{array}{r}1111\\-1011\\\hline\end{array}\quad\begin{array}{r}10101\\-10010\\\hline\end{array}$$

3-17 Write the following numbers in the decimal system:

(a) $125.5_{\boxed{8}}$ (c) $100.101_{\boxed{2}}$ (e) $100.1_{\boxed{8}}$

(b) $130_{\boxed{8}}$ (d) $1011_{\boxed{2}}$

3-18 Convert the following decimal numbers to their binary equivalents.

(a) 47 (c) 14 (e) 3690

(b) 64 (d) 125

3-19 Convert the numbers in Exercise 3-18 to their octal equivalents.

3-20 Execute the following operations with the octal numbers given and check your results by converting the numbers to their decimal equivalents.

$$\begin{array}{r}4763\\+2342\\\hline\end{array}\quad\begin{array}{r}1234\\+7624\\\hline\end{array}\quad\begin{array}{r}6374\\-5653\\\hline\end{array}\quad\begin{array}{r}3764\\-2576\\\hline\end{array}$$

3-21 Solve the following binary multiplications and convert to decimal equivalents to verify your answer.

$$\begin{array}{r}111\\\times101\\\hline\end{array}\qquad\begin{array}{r}101\\\times101\\\hline\end{array}$$

NOTE. In the multiplication of two numbers we multiply one digit at a time from each number. We get a product and a carry digit (sometimes zero), and we work from right to left. As an example, let us multiply the binary numbers 101 and 11. The multiplication would proceed as follows:

$$\begin{array}{rl}101 & \\ \times\ 11 & \\ \hline 101 & \leftarrow (101\times 1)\\ +1010 & \leftarrow (101\times 10)\\ \hline 1111 & \end{array}$$

3-22 Solve the following octal multiplications and convert to decimal equivalents to verify your answer:

$$\begin{array}{r}26\\\times\ 5\\\hline\end{array}\qquad\begin{array}{r}62\\\times\ 5\\\hline\end{array}$$

NOTE. The multiplication of octal numbers proceeds as in Exercise 3-21. For example:

$$\begin{array}{r}7\\\times4\\\hline34\end{array}\qquad\begin{array}{r}7\\\times14\\\hline124\end{array}$$

3-23 Solve the following binary division:

$10010001 \div 101$

NOTE. The division of two numbers in the binary system is shown by dividing 101010 by 111:

```
         0110
111 ) 101010
      -111
      ------
        111
       -111
      ------
         00
```

4

PROBABILITY AND THE WILL TO DOUBT

4-1 INTRODUCTION

In Chapter 1 we discussed the will to doubt as an important attitude to problem solving. It provides the flexibility that permits us to change our ideas in the face of new evidence and to accept or reject premises or hypotheses on a tentative basis. This attitude is not to be interpreted as an attitude of no faith in anything. Not at all. It suggests that we can have faith in a premise or hypothesis before all doubt is dispelled. In fact, if we had to wait for all doubt to be dispelled before we could have faith, it would be difficult to have faith in anything.

The question of how much doubt or how much faith we have in a premise or hypothesis is a matter of degree which is subjective in nature. *The degree of our faith is related to our knowledge or the information at our disposal.*

To incorporate in a model new information that may enhance our faith in a premise (or increase our doubt), it will be necessary to *assign a quantitative value to our state of knowledge.* Such an assignment will take the form of a number between 0 and 1. The value of 1 will represent absolute *faith* in a premise; a value of 0 will represent absolute *no faith* in a premise. Note that in either case we leave no room for doubt; it takes as much complete conviction to have absolute faith as it does to have absolute no faith.* In other

*In testing the validity of a hypothesis, we continue to subscribe to it so long as we do not have sufficient evidence to reject it. We do not prove it to be right or wrong in an absolute sense. Proving it right implies exhausting all possibilities of proving it wrong. This can never be accomplished since there are always the attempts of the future. Proving it wrong on the basis of an experiment implies absolute faith in the results of the experiment, but the future may show the experiment to be inaccurate or irrelevant, or the credibility of the results may be questioned. For example, the premise that the speed of light is constant may be accepted as true because no evidence exists to the contrary. However, the future may lead to new evidence. But then the evidence may be questioned at some point in time.

words, we can always take a positive statement such as: "All men are created equal," and state it in a negative form: "Not all men are created equal." Absolute *faith* in the first statement (value 1) will require the same conviction as absolute *no faith* (value 0) in the second.

The attitude of the *will to doubt* suggests that we assign the values of 0 and 1 with great caution, or better yet, keep the options open by assigning values which are very close to 1 (0.999. . .) or to 0 (0.000. . . 1) when we have a great deal of relevant and credible information regarding a premise or hypothesis. After all, does any mortal problem-solver have perfect knowledge, i.e., *all relevant information* with no shadow of doubt regarding its *credibility*?

Man's need to know, his need for faith and will to doubt, and the problems which are a result of these needs, stem from his cognitive activity which places him between animal and the concept of God. An animal is not capable of cognitive activity as long as it is ignorant of everything, including its own ignorance and, hence, will make no effort to gain knowledge. At the other end is the concept of a God who has no reason to make an effort because he is all knowing (in advance) and knows that he knows. Only man, the intermediate being between the all-ignorant animal and the all-knowing God, must make an effort to move from ignorance to knowledge because he is endowed with the gift of being aware of his ignorance. It is the greatness of man that he knows he does not know [30], and it is this knowledge, coupled with the need to know, that creates man's problems. (Socrates attributed this attitude to himself as his only pride.)

Descartes began modern philosophy when he claimed that man can doubt everything, but in order to be in a position to doubt everything he must not doubt that he doubts. So here is the *primitive basic premise of the will to doubt: We cannot doubt that we doubt.*

One can argue that, perhaps, instead of placing the emphasis on the will to doubt, we should stress the will to believe. However, the will to believe has promoted historically an attitude to propagate dogma which stifles free thinking and suppresses questioning. The will to doubt promotes the desire to find out, to search for truth. This undogmatic attitude has proven to be of great importance in the search for truth in science. There is great hope that the same attitude of rational doubt will prove fruitful in dealing with problems of society.

4-2 PROBABILITY AND DOUBT

We shall now discuss in more detail the key concepts that were mentioned in the first section of this chapter: *information, relevance of information, credibility of information,* and *quantitative measure of state of knowledge.* We begin here with a discussion of probability that relates to all of these concepts.

We all have an intuitive concept of probability as related to the outcome of events such as flipping coins or rolling dice. But we also use the term in such sentences as, "It will probably rain tomorrow." Most authors make a distinction between the last statement and the statement "The probability of a head in the toss of a coin is 1/2." The distinction comes from the formal definition of probability as the limiting value of a relative frequency of occurrence of an event as the number of trials becomes large. Namely, if we denote by $P(H)$ the probability of the event head (H) in the toss of a coin, then

$$\frac{\text{Number of heads}}{\text{Number of tosses } n} \longrightarrow P(H)$$

as n becomes very large.

In keeping with the philosophy that unwarranted formalizing in the learning process may be stifling to innovation and creativity, we take the attitude that a little vagueness in the concept of probability may prove more useful than a formal strict description such as the one above. Let us, therefore, adopt the more broad definition proposed by Tribus [31] which concerns the assignment of probabilities:

"*A probability assignment is a numerical encoding of a state of knowledge.*"

Consider, for example, the statements:

"The probability is 0.6 that it will rain tomorrow."
"The probability is 0.8 that it will rain tomorrow."

The second statement represents a higher state of knowledge or certainty than the first. The assignment of the values of 1 or 0 to the probability signifies absolute certainty; in each case, our knowledge is complete (at least, we claim so). This is reasonable because to claim a probability of zero for rain is identical to a probability of one for no rain.

In a similar manner, consider the statements:

"The probability of a head in the toss of the coin is 0.5."
"The probability of a head in the toss of the coin is 1."

The second statement shows absolute knowledge regarding the outcome of the toss; apparently the coin has a head on each face. In the first statement we bring our experience into our assessment because we normally recall coins falling heads half the time. It is interesting to ask people for the probability of a thumbtack falling with the head flat on a table. You will be surprised at the spectrum of numbers between zero and one. Here experience is rather limited.

Note that an assignment of probability of 1/2 to a statement represents the lowest state of knowledge because a number larger than 1/2 indicates that more knowledge causes us to approach closer to certainty in our faith in the statement (value of 1); and a number smaller than 1/2 indicates that additional knowledge causes us to approach closer to certainty in *no faith* in the statement (value of 0). *The maximum doubt or uncertainty in the truth value of a statement is represented by 1/2*, namely, we have insufficient knowledge to be inclined more towards its truth than its falsehood. A change of a probability from 1/2 represents new knowledge which dispels some doubt. This is shown schematically in Fig. 4-1.

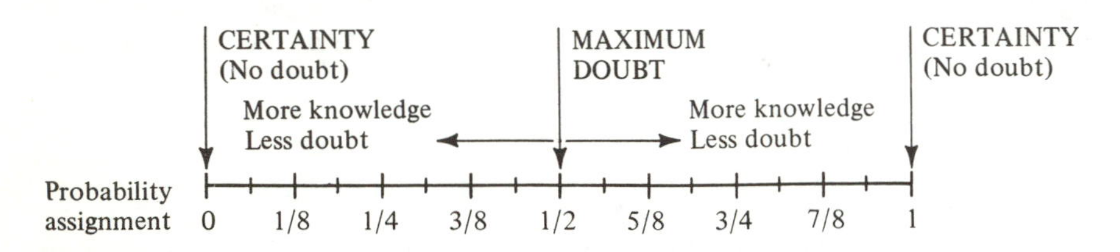

Figure 4-1 Probability and Doubt

In his book, "The Will to Doubt," Bertrand Russell [32] defines *rationality of opinion* as the habit of taking account of all relevant evidence in arriving at a belief. Where certainty is unattainable, a rational man will subscribe to the most probable opinion, but will retain the other less probable opinions because future evidence may show them to be preferable. Russell defines *rational behavior* as the habit of considering all relevant desires and not only the one which is strongest at a given moment. The probability of satisfying all desires must be weighed. This, like rationality of opinion, is subjective in nature and is a matter of degree. No claim is made here that man is capable of perfectly rational behavior in theory and practice. However, it is our belief that the will to doubt will help bring about more rational behavior. A man obsessed by a blind desire to do harm to another fails to consider other desires and, therefore, assigns no doubt to the consequences of his behavior. Such behavior may border on insanity. A rational man doubts because the consequence of his behavior can never be considered certain.

There is another important point. A man who has no will to doubt will not listen, will close his eyes to new evidence. It is this dogmatic attitude which is irrational. However, once doubt, no matter how little, is admitted, evidence will be considered and opinions will change when sufficient evidence supports change. How much doubt one has in a premise, and how much evidence he will require to change his opinion, is subjective in nature.

We shall demonstrate this point by an example and, in the process, develop the tools for measuring the relevance and credibility of information.

4-3 IS MR. X HONEST?

Consider the following question: "Is Mr. X an honest man?" Let us suppose that two people were asked this question regarding the same Mr. X and gave the following answers:

Person 1: *Mr. X is honest,* and the probability is 1 that this statement is true. I know the man and no one can tell me something that will make me change my mind.

Person 2: *Mr. X is honest,* and the probability is very close to 1, say 0.99999, that this statement is true. After all, I base my statement on what I know, and I am the first to admit that I do not know everything. Yet I have learned to have faith, knowing that one cannot wait for all doubt to be dispelled before faith is professed. *The probability of 0.99999 and not 1 is merely a display of my will to doubt, and through it I profess that I am human and, therefore, my knowledge is not complete.*

This question regarding Mr. X's honesty may appear as a strange and certainly an unorthodox way to begin a discussion of probability, but this is exactly what we shall do. Person 2 must have read the first two sections of this chapter. His lengthy, almost apologetic statement shows it. But wait, we are not through with the story. Persons 1 and 2 are now told that Mr. X was seen on weekends, standing in a glass booth in a Las Vegas casino (Fig. 4-2)

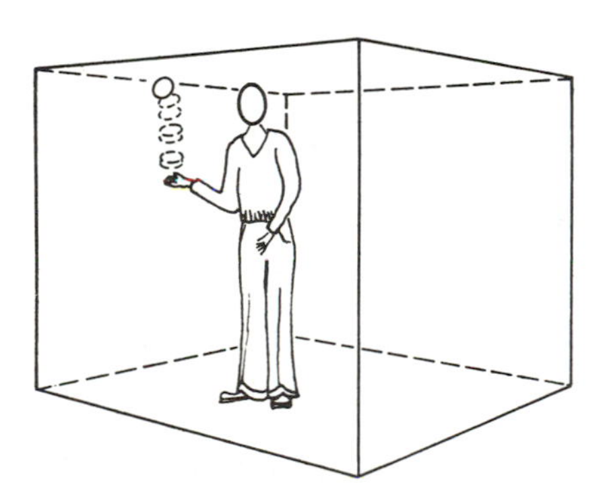

Figure 4-2 Coin Tosser in Glass Booth

and playing the following game. As customers enter they can bet one dollar against five of the house that when the man in the booth flips the dime in his hand, a tail will show. No one forces the customers to play, of course, but in the past six months, since the glass booth made its appearance, many made the bet and *all* lost. We should also mention that you cannot inspect the dime; you can only observe the flip and the outcome. You may object to this constraint, but no one forces you to play.

Person 1 reacts with the expected dogmatic attitude of one who has no will to doubt. He claims that the evidence must be false. When he is offered a free trip to observe and find out for himself, he rejects the offer, saying that even if the evidence is true, it is not relevant. He made up his mind and that is it!

Person 2 takes a different attitude. He values new evidence, but he wishes to check its credibility, in particular when such a serious matter as the reputation of a man is concerned. He goes to Las Vegas, observes the game, and verifies the story.

We now leave this story, but we shall return to it after we develop some of the tools of probability. These tools will enable us to encode by numerical values the change in attitude of the persons as they acquire new information. The procedure for translating the evidence to a change of opinion will prove rational, consistent, and in a way formal; however, it will retain the important ingredient of subjective behavior, namely, personal judgment. How much evidence—in our case, how many coin tosses—must be observed before we say that the man is dishonest because he is partner to a dishonest game, will be a matter of individual judgment.

This last point is worth emphasizing. Our education provides us with knowledge which includes a host of tools and a great deal of information in various forms. It is also supposed to provide us with mental habits and tools of learning on our own and acquiring new knowledge when we recognize the need for it, so that we can make rational and sound judgments for ourselves when we solve problems. It is the latter, i.e., the habits and tools of learning and the making of judgments for ourselves, which is the most important. But it is possible that we stifle these habits by a tradition of formalism in education that borders on conveying dogmatic patterns for problem solving.

4-4 LAWS OF PROBABILITY

Probability theory has a wide range of applications. It is used in engineering, physics, sociology, anthropology, economics, law, and medicine, to name but some areas. The theory was founded by Pascal (1623-1662) and Fermat (1601-1665) to calculate probabilities in games of chance. Since their work, many books have been written on the concept of probability.

The probability of 1/2 that an honest coin will fall heads is considered an *objective probability* because most people will subscribe to it. A *subjective* or *personal probability* is conveyed by the sentence, "The probability is 0.9 that Mr. Z will be elected president." The objective probability represents a form of knowledge which is objective in the sense that it can be verified by experiment, and the results found to agree by most experiments. A subjective probability is strictly a personal form of knowledge which cannot be verified (Mr. Z may never run), and may not be shared by others.

For our purposes we shall use the view expressed earlier that the *assignment of a probability is a numerical encoding of the state of knowledge of the assigner.* To be able to incorporate new knowledge in our judgments and assignments of probabilities we shall develop the laws of probability. The development will be axiomatic in the sense that we shall state the laws and demonstrate their significance and use. On that basis, the axioms of probability may be regarded in the same sense as the primitive undefined terms *line* and *point* in geometry. Useful theorems can be derived from the axioms.

Sample Space and Events

We refer now to our discussion of symbolic logic and sets in Chapter 2. Consider a set that contains all possible outcomes of a particular experiment, such as tossing a die. Each outcome of such an experiment is called a *sample point*, and the collection of all sample points in the set is called the *sample space*. An *event* is a subset of the sample space. An event that consists of one point in the sample space is called a *simple event*. A *compound event* consists of two or more points. For example, an outcome of an even number (2 or 4 or 6) in the toss of a die is a compound event.

Let Fig. 4-3 represent the sample space of the following experiment. We shall drop a pin within the confines of the rectangle of area C and observe

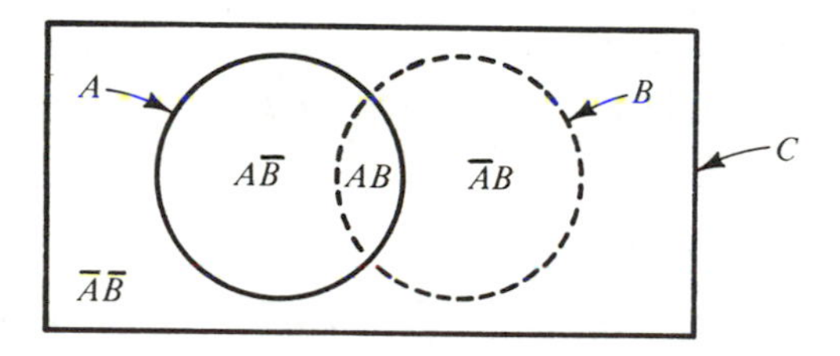

Figure 4-3

where the head lands. The area of circle A (solid line) is A, and the area of circle B (dashed line) is B. Let A be the event that the head of the pin lands in circle A, and event B that it lands in circle B. If we drop the pin in a way which makes every position equally likely to be the final rest position, then the probability of event A, namely, landing in circle A, is denoted as $P(A)$ and given by

$$P(A) = \frac{\text{area } A}{\text{area } C}$$

Similarly, for event B (landing in circle B):

$$P(B) = \frac{\text{area } B}{\text{area } C}$$

For event AB, which signifies landing in the area common to circles A and B (intersection of A and B):

$$P(AB) = \frac{\text{area } AB}{\text{area } C}$$

$P(AB)$ signifies the probability of events A *and* B; namely, the pin head lands both in circle A and circle B. Using the symbols of sets, we can write $P(A \cap B)$. However, if we are concerned with statements A and B and wish to assign a number to our confidence in the truth of both, we could write $P(A \wedge B)$.

For the events A *or* B:

$$P(A \text{ or } B) = (\text{area } A + \text{area } B - \text{area } AB)\frac{1}{\text{area } C}$$

Again, the notation $P(A \cup B)$ could be used for sets and $P(A \vee B)$ for statements.

Probability Laws—Formalization

For any event A or statement A:

$$1 \geq P(A) \geq 0 \qquad (4\text{-}1)$$

$$P(A) + P(\bar{A}) = 1 \qquad (4\text{-}2)$$

Equation 4-1 limits the values to be assigned to a probability between 0 and 1. Equation 4-2 states that the probability of event A or not A is a certainty. In Fig. 4-3 this implies that when a pin is dropped it will land with certainty in A or outside A. If A stands for the statement: "It shall rain next Monday," then $P(A) + P(\bar{A})$ is again a certainty.

Mutually Exclusive Events

Two events are *mutually exclusive* (or *disjoint*) when they cannot occur at the same time. For example, in the single toss of a die, the outcome of any two numbers cannot occur. Hence, the events 5 and 1, representing the outcomes 5 and 1, respectively, are mutually exclusive. On the other hand, the toss of a die and a coin simultaneously may result in T or H for the coin and any of the six numbers for the die. Hence, the outcomes from the coin and the die are *not mutually exclusive*. The same could be demonstrated for statements. For example,

A_1: "At three o'clock sharp, New York time, I shall be in New York."

A_2: "At three o'clock sharp, New York time, I shall be in Washington."

A_3: "At three o'clock sharp, New York time, I shall be in my car."

A_1 and A_2 are mutually exclusive or disjoint. A_1 and A_3 are not.

Independent Events

When the outcome of one event is not influenced by the outcome of another event, we say that the two events are independent. For example, in tossing a die and then a coin the outcome from the toss of the die has no influence on the outcome from the toss of the coin. Knowledge of one in no way changes our knowledge regarding the other. Therefore, we shall assign the same probability to the result of a head H in the toss of the coin irrespective of what we know about the outcome for the die.

Dependent Events

When knowledge of the outcome from one event causes a change in our state of knowledge regarding a second event, then the events are dependent. This dependence will be reflected in a change in the probability that we assign to the dependent event. Suppose we draw two balls, one after the other, from an urn containing 4 red and 2 white balls. The probability $P(R_1)$ of drawing a red ball in the first draw is 4/6. Suppose we draw a red ball. If we do not replace it, then $P(R_2) = 3/5$; namely, the probability for drawing a red ball in the second draw is different from the first. If a white ball is drawn in the first try and not replaced, $P(R_2) = 4/5$ and $P(W_2) = 1/5$; again the outcome of the first draw influences the outcome of the second.

Probability of *A* OR *B*

Let A and B be events or statements. When A and B are *not mutually exclusive*, the probability of A *or* B is given by*

$$P(A \cup B) = P(A) + P(B) - P(A \cap B) \tag{4-3}$$

When A and B are *mutually exclusive*, $P(A \cap B) = 0$ and $P(A \cup B)$ becomes

$$P(A \cup B) = P(A) + P(B) \tag{4-4}$$

These equations can be verified from Fig. 4-3. The probability of A *or* B is represented by the area confined to the two circles, divided by area C of the rectangle. This is achieved in Eq. 4-3 by adding the areas of the two circles and subtracting the area of the intersection AB because it was included twice, first in A and then in B. When A and B are disjoint, mutually exclusive, the circles A and B do not intersect, area AB is zero, and Eq. 4-3 reduces to Eq. 4-4.

Example of Equation 4-3. The probability of a head H or a 5 in the toss of a coin and die is

$$\begin{aligned} P(H \cup 5) &= P(H) + P(5) - P(H \cap 5) \\ &= \frac{1}{2} + \frac{1}{6} - \frac{1}{2} \times \frac{1}{6} \\ &= \frac{7}{12} \end{aligned}$$

*We use the notations $\cup$, $\cap$. These are, of course, applicable to sets, but the reader can replace these symbols by $\vee$ and $\wedge$, respectively, for statements. The concepts developed are identical for both.

The sample space is shown schematically in Fig. 4-4. The total area of the figure is 1, because one of the twelve outcomes is a certainty. Each of the twelve rectangles has an area of $1/2 \times 1/6 = 1/12$ and represents the probability of the number of the die to the left of the rectangle and the coin outcome H or T.

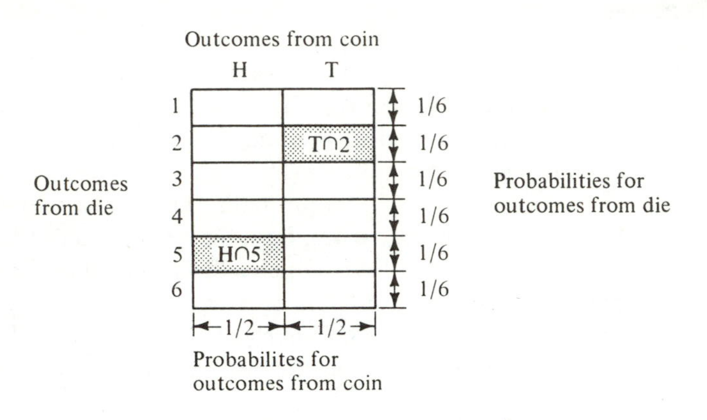

Figure 4-4 Schematic Sample Space for Toss of Coin and Die

Example of Equation 4-4. The probability of a 1 or a 4 in the toss of a die is

$$\begin{aligned} P(1 \cup 4) &= P(1) + P(4) \\ &= \frac{1}{6} + \frac{1}{6} \\ &= \frac{1}{3} \end{aligned}$$

Probability of *A* AND *B*

From Fig. 4-3 the probability of *A and B*, $A \cap B$ (also denoted as AB) is given by

$$P(A \cap B) = \frac{\text{area } AB}{\text{area } C}$$

Multiplying and dividing the right-hand side by area B or area A, the last expression can be written also in the forms

$$P(A \cap B) = \frac{\text{area } AB}{\text{area } B} \times \frac{\text{area } B}{\text{area } C}$$

and

$$P(A \cap B) = \frac{\text{area } AB}{\text{area } A} \times \frac{\text{area } A}{\text{area } C}$$

The ratio (area AB/area B) is called the *conditional probability of A, given B*, and is denoted as $P(A/B)$. Similarly, (area AB/area A) is the conditional probability of B, given A, $P(B/A)$. Using this notation, the last two equations become

$$P(A \cap B) = P(A/B)P(B) \tag{4-5a}$$

and

$$P(A \cap B) = P(B/A)P(A) \tag{4-5b}$$

Equation 4-5 is very important in our continued development. Let us, therefore, discuss it in more detail. When we first introduced Fig. 4-3, we limited the boundaries of our set to the rectangle of area C. This is a *condition* for all outcomes in the set. All probabilities such as $P(A)$ or $P(B)$ were obtained by taking the area corresponding to the event considered and dividing by area C. But we also note that no matter what the outcome was, C was always true because the pin must land within the confines of the set. We therefore should have written $(A \cap C)$ and not A, and similarly, $(B \cap C)$ and not B. The probability $P(A)$, for example, would then be written as

$$P(A) = P(A \cap C) = \frac{\text{area } AC}{\text{area } C}$$

Multiplying and dividing each side by area C:

$$P(A) = P(A \cap C) = \frac{\text{area } AC}{\text{area } C} \times \frac{\text{area } C}{\text{area } C}$$

Using the notation of conditional probability, we obtain an equation identical in form to Eq. 4-5a:

$$P(A) = P(A \cap C) = P(A/C)P(C)$$

In Fig. 4-3 area C is the complete set and, therefore, $P(C) = 1$; hence, the above equation reduces to

$$P(A) = P(A \cap C) = P(A/C)$$

Since C is the complete set, we tend to forget the condition that A must happen with C, $(A \cap C)$, and that A is conditional on C; i.e., it is limited to the boundaries of set C. We can state, therefore, that *all probabilities are conditional* and, no matter what event or statement we consider, we should put a slash followed by a symbol that represents the conditions of those events and statements. For example, Eq. 4-5a should be written

$$P(A \cap B/C) = P(A/BC)P(B/C)$$

This is not customary, but it is absolutely legitimate. Ordinarily, the conditional probability symbol is recorded when the original boundaries C are reduced to a subset, such as A and B in Fig. 4-3.

We now formalize the definition of $P(A/B)$ as follows: The conditional probability of A (actually A and B), given B, is the probability of A (actually A and B) when the confines of the sample space are reduced to subset B.

When B is given, it implies that we have acquired new knowledge; $P(A/B)$ measures our state of knowledge regarding A in the light of this new information. Information B is relevant to A only when $P(A/B)$ is different from $P(A)$. $P(A)$ is called an *a priori probability*, namely, the probability of A prior to the knowledge of B. $P(A/B)$ is called an *a posteriori probability*, namely, the probability of A after obtaining knowledge of B.

When A and B are independent, the knowledge of B does not reduce the original complete set; therefore, knowledge of B is not going to change $P(A)$,

or $$P(A/B) = P(A)$$

In this case we say that *knowledge of B is not relevant to A.*

For A and B independent, Eq. 4-5 becomes

$$P(A \cap B) = P(A)P(B)$$

or $$P(A \cap B) = P(B)P(A) \tag{4-6}$$

Example: *$P(A \cap B)$ when A and B are Dependent.* Consider an urn with 4 red and 2 white balls from which we draw two balls without replacement.* The results can be described by a *tree diagram* in a compact form that accounts for all possible outcomes, as shown in Fig. 4-5.

The probability of drawing a white ball first W_1, then a red ball R_2 without replacement (Eq. 4-5a) is

$$P(R_2 \cap W_1) = P(R_2/W_1)P(W_1)$$

Subscripts 1 and 2 refer to results of the first and second draw, respectively. It is always safer to use the form of Eq. 4-5. Equation 4-6 may be considered a special case of Eq. 4-5 when A and B are independent.

We begin from the right-most term on the right-hand side of the equation and evaluate $P(W_1) = 2/6$. Now that W_1 is given, i.e., we drew a white ball, the sample space is reduced; we have only 5 balls of which 4 are red, hence $P(R_2/W_1) = 4/5$. Thus,

$$P(R_2 \cap W_1) = \frac{4}{5} \times \frac{2}{6} = \frac{8}{30}$$

The same result can be obtained from Eq. 4-5b:

$$\begin{aligned} P(R_1 \cap W_2) &= P(W_2/R_1)P(R_1) \\ &= \frac{2}{5} \times \frac{4}{6} \\ &= \frac{8}{30} \end{aligned}$$

The result for $P(R_2 \cap W_1)$ can also be obtained by taking the ratio of the number of outcomes with $R_2 \cap W_1$ to the total number of possible out-

*With replacement, you draw a ball, observe its color, and return it to the urn before the second drawing.

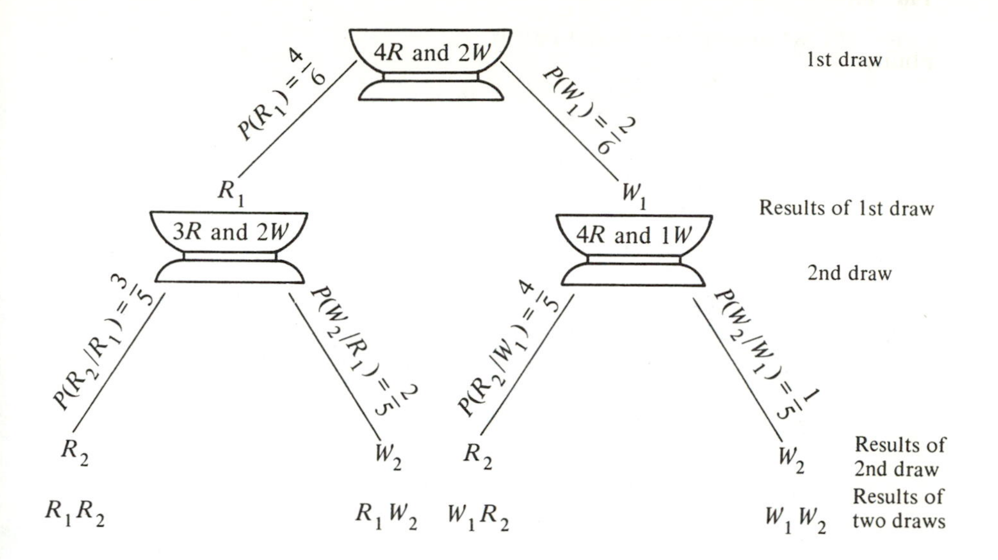

Probabilities of the results from the two draws:

$$P(R_1R_2) = \frac{4}{6} \times \frac{3}{5} = \frac{12}{30}\ ;\ P(R_1W_2) = \frac{4}{6} \times \frac{2}{5} = \frac{8}{30}\ ;\ P(W_1R_2) = \frac{2}{6} \times \frac{4}{5} = \frac{8}{30}\ ;\ P(W_1W_2) = \frac{2}{6} \times \frac{1}{5} = \frac{2}{30}$$

Figure 4-5 Tree Diagram for Drawing Two Balls from an Urn without Replacements

comes. There are 6 different balls possible in the first draw. Each of these can be combined with each of the 5 different balls for the second draw; hence, there are 30 possible ways of drawing two balls without replacement. Of these, there are 2 different white balls that can be drawn first, and each can be combined with the 4 red balls for a total of 8 ways of getting a white and then a red ball. Thus, $P(R_2 \cap W_1) = 8/30$. Use similar reasoning to obtain $P(R_1 \cap W_2)$.

The sum of all probabilities in drawing two balls must add up to one:

$$\frac{12}{30} + \frac{8}{30} + \frac{8}{30} + \frac{2}{30} = \frac{30}{30}$$

Example: $P(A \cap B)$ when A and B are Independent. Repeating the last example, but with replacement—namely, the first ball is drawn, observed, and returned to the urn—then given that W_1 is observed,

$$P(R_2/W_1) = P(R_2)$$

That is, knowledge of outcome W_1 in the first draw does not change our state of knowledge about the outcome of R_2 in the second draw. Now, the probability of a white and then a red ball is

$$\begin{aligned} P(R_2 \cap W_1) &= P(R_2/W_1)P(W_1) \\ &= P(R_2)P(W_1) \\ &= \frac{4}{6} \times \frac{2}{6} \\ &= \frac{8}{36} \end{aligned}$$

Using direct counting, we have 6 ways of drawing the first ball which, after replacement, can be combined with 6 ways for the second for a total of 36 ways. There are still 8 ways for drawing a white and then a red ball, thus:

$$P(R_2 \cap W_1) = \frac{8}{36}$$

The reader may wish to verify the result $P(R_1 \cap W_2) = 8/36$.

Suggested Problems. Consider the urn with 4 red and 2 white balls and draw 2 balls without replacement. What is the probability that a red and a white ball are drawn if the order in which they are drawn is not relevant? What is the probability of a red and a white ball, in this sequence, if the drawing is with replacement? Show a tree diagram for the events in such a drawing. What is the probability of a red and a white ball if the drawing is with replacement but the sequence red-white or white-red is irrelevant?

Summary—Laws of Probability

For any event or statement A:

I. $1 \geq P(A) \geq 0$ (4-1)

$P(A) = 1$, event A is certain to occur; statement A is true with no doubt.

$P(A) = 0$, event A is certain not to occur; statement A is false with no doubt.

II. $P(A) + P(\bar{A}) = 1$ (4-2)

III. *OR Probability Statement,* $P(A \cup B)$

For any two events or statements A and B:

(a) When A and B are not mutually exclusive,

$$P(A \cup B) = P(A) + P(B) - P(A \cap B) \quad (4\text{-}3)$$

Example: $P(H \cup 5)$ in the toss of a coin and a die.

(b) When A and B are mutually exclusive [$P(A \cap B) = 0$],

$$P(A \cup B) = P(A) + P(B) \quad (4\text{-}4)$$

Example: $P(1 \cup 4)$ in the toss of a die.

IV. *AND Probability Statement*, $P(A \cap B)$
For any two events or statements A and B:
(a) When A and B are dependent,

$$P(A \cap B) = P(A/B)P(B) \qquad \text{(a)}$$
$$\text{or} \quad P(A \cap B) = P(B/A)P(A) \qquad \text{(b)} \qquad (4\text{-}5)$$

Example: Drawing balls from an urn without replacement.
(b) When A and B are independent,

$$[P(A/B) = P(A); P(B/A) = P(B)]$$
$$P(A \cap B) = P(A)P(B) \qquad (4\text{-}6)$$

Example: Drawing balls from an urn with replacement.

4-5 BAYES' EQUATION AND RELEVANCE OF INFORMATION

Equations 4-5a and b lead to this result:

$$P(A/B)P(B) = P(B/A)P(A)$$

Dividing both sides by $P(B)$, this equation can be written in the form

$$P(A/B) = P(A)\frac{P(B/A)}{P(B)} \qquad (4\text{-}7)$$

In keeping with our earlier discussion, we could include C (the conditions or bounds of the complete set) behind a slash with each probability statement:

$$P(A/BC) = P(A/C)\frac{P(B/AC)}{P(B/C)}$$

This is rather uncommon. It is implied, however, and often forgotten.

Equation 4-7 is known as *Bayes' equation* or *Bayes' theorem*. In this equation, $P(A)$ on the right-hand side encodes our state of knowledge before we know B, *a priori probability*. $P(A/B)$ encodes the state of knowledge about A after we learn about B, *a posteriori probability*. The ratio $P(B/A)/P(B)$ which multiplies $P(A)$ is the *measure of relevance*. When this ratio is 1, information B is not relevant to A. Any other number larger or smaller than 1 indicates relevancy and is reflected by the assignment of a new value to the probability of A, i.e., $P(A/B) \neq P(A)$.

Equation 4-7 provides us with a procedure for updating our state of knowledge in the presence of new information. It also permits us to establish the relevance of new information by simply considering what our state of knowledge about A might have been if we were to get B. If the relevance ratio turns out to be 1, we derive no information from B and should expend no cost or effort to obtain it.

Before illustrating the use of Eq. 4-7, let us focus our attention on computations for $P(B)$ in the denominator of the relevancy ratio. This turns out to be important because it is a source of difficulty to many students of probability. It will also give us an opportunity to review the OR and AND statements of probability discussed in Sec. 4-4.

When probability $P(B)$ of an event or statement B is considered, we exhaust all possible mutually exclusive ways in which B can occur and include them in assigning a probability value to B. For example, in Fig. 4-3, B is true with A, namely, $B \cap A$, or without A, namely, $B \cap \bar{A}$,

$$P(B) = P(B \cap A) + P(B \cap \bar{A})$$

In a more general case shown in Fig. 4-6, B is true when

$$(B \cap A_1) \text{ or } (B \cap A_2) \dots \text{ or } (B \cap A_{16})$$

is true; namely, we go through all the alternative outcomes within the set and include all mutually exclusive ways in which B can occur with them.

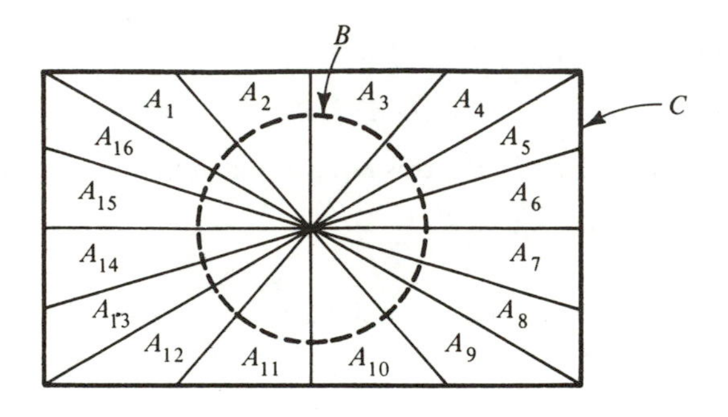

Figure 4-6

For example, in Fig. 4-6, we add the probabilities of all mutually exclusive events in which B can occur and write

$$\begin{aligned} P(B) &= P(B \cap A_1) + P(B \cap A_2) + \dots + P(B \cap A_n) \\ &= \textstyle\sum_{j=1}^{n} P(B \cap A_j) \end{aligned} \tag{4-8}$$

Using Eq. 4-5, we can write for each term on the right-hand side

$$P(B \cap A_j) = P(B/A_j)P(A_j)$$

and $P(B)$ becomes

$$P(B) = \textstyle\sum_{j=1}^{n} P(B/A_j)P(A_j) \tag{4-9}$$

Using Eq. 4-9, then Eq. 4-7 can be written

$$P(A_i/B) = P(A_i)\frac{P(B/A_i)}{\sum_{j=1}^{n} P(B/A_j)P(A_j)} \tag{4-10}$$

The right side of Eq. 4-10 is the *general form of Bayes' theorem*. It can also be written in the form

$$P(A_i/B) = \frac{P(B/A_i)P(A_i)}{\sum_{j=1}^{n} P(B/A_j)P(A_j)} \tag{4-11}$$

In this form, the conditional probability $P(A_i/B)$ can be interpreted as a probability ratio of a subset to a complete set, in which the numerator $P(B/A_i)P(A_i)$ is the probability of the subset and the denominator $\sum_j P(B/A_j)P(A_j)$ is that of the complete set. The subset is contained, of course, in the set. The set (denominator) considers all mutually exclusive and exhaustive ways in which event B can occur with all other possible events A_j ($j = 1, 2, \ldots, n$), while the subset (numerator) considers the occurrence of event B with A_i only. The denominator is the probability of event B, $P(B)$. Therefore, as we consider all subsets of events BA_j ($j = 1, 2, \ldots, n$) with associated probabilities $P(B/A_j)P(A_j)$ ($j = 1, 2, \ldots, n$) which contribute to the evaluation of $P(B)$, we can identify the numerator, $P(B/A_i)P(A_i)$, of Eq. 4-10 as one of these terms for which $j = i$. This is best illustrated by examples. The next two sections, 4-6 and 4-7, are devoted to examples for the application of Bayes' theorem as expressed by Eq. 4-10 or Eq. 4-11. In the examples we first compute the denominator in Eq. 4-10 or 4-11, i.e., the probability $P(B)$ of the complete set, then the numerator will be identified in order to determine the conditional probability on the left-hand side of Eq. 4-10 or 4-11.

4-6 APPLICATIONS OF BAYES' THEOREM

EXAMPLE 1. There are five identical boxes on a table. Each box contains 4 marbles. The first has 4 red marbles; the second, 3 red and 1 white; the third, 2 red and 2 white; the fourth, 1 red and 3 white; and the fifth, 4 white marbles.

(a) What is the probability of getting a red marble (R) if we randomly* select a box, put our hand into it, and take out one marble without looking into the box?

Event R can happen with event A_1 of selecting box 1, or event A_2 of selecting box 2, . . . , or event A_5. Using Eq. 4-9, we have (see Table 4-1 for detailed computations)

$$\begin{aligned} P(R) &= \textstyle\sum_{j=1}^{5} P(R/A_j)P(A_j) \\ &= \frac{1}{5}\left(\frac{4}{4} + \frac{3}{4} + \frac{2}{4} + \frac{1}{4} + 0\right) \\ &= \frac{1}{2} \end{aligned}$$

[Compute $P(R)$, considering that we have (1) only the first three boxes; (2) the first four boxes; (3) the first two boxes.]

*_Randomly_ means here that we have no extraneous influences (for all we know) that will cause us to favor the choice of one box over the others; therefore, each box has an equal probability of being selected.

Table 4-1

Box A_j	$P(A_j)$	$P(R/A_j)$	$P(R/A_j)P(A_j)$
A_1	1/5	4/4	4/20
A_2	1/5	3/4	3/20
A_3	1/5	2/4	2/20
A_4	1/5	1/4	1/20
A_5	1/5	0	0
			$P(R) = 10/20$

(b) Given that a red marble has been drawn (R), what is the probability it came from box 1 (A_1)?

$$P(A_1/R) = P(A_1)\frac{P(R/A_1)}{P(R)}$$
$$= \frac{1}{5} \times \frac{1}{1/2}$$
$$= \frac{2}{5}$$

The information R is relevant to A_1.

EXAMPLE 2. Draw a ball from an urn containing 3 balls, 2 red and one white. Place the ball in a bag and do not look at it. Suppose you draw a second ball now. (a) What is the probability that it is red? (b) What is the probability that it is white? (c) What is the probability of drawing a second ball (B_2)?

Let us designate by a subscript 1 or 2 the outcomes from the first and second draws, respectively. Then the probabilities we seek are $P(R_2)$, $P(W_2)$ and $P(B_2)$. Now, R_2 can occur with R_1 or with W_1. Hence,

(a) $$P(R_2) = P(R_2/R_1)P(R_1) + P(R_2/W_1)P(W_1)$$
$$= \frac{1}{2} \times \frac{2}{3} + \frac{2}{2} \times \frac{1}{3}$$
$$= \frac{2}{3}$$

Similarly, for W_2,

(b) $$P(W_2) = P(W_2/R_1)P(R_1) + P(W_2/W_1)P(W_1)$$
$$= \frac{1}{2} \times \frac{2}{3} + \frac{0}{2} \times \frac{1}{3}$$
$$= \frac{1}{3}$$

The last result is also a check of our computations for $P(R_2)$ because $P(W_2)$ is identical to $P(\bar{R}_2)$ and

$$P(R_2) + P(\bar{R}_2) = 1$$

(c) The probability of drawing a second ball (B_2), i.e., a ball of any color, is a certainty as verified by the following:

$$\begin{aligned} P(B_2) &= P(B_2/R_1)P(R_1) + P(B_2/W_1)P(W_1) \\ &= 1 \times \frac{2}{3} + 1 \times \frac{1}{3} \\ &= 1 \end{aligned}$$

(d) What is the probability that the first ball drawn was red (R_1), given that the second ball drawn was red (R_2)?

$$\begin{aligned} P(R_1/R_2) &= P(R_1)\frac{P(R_2/R_1)}{P(R_2)} \\ &= \frac{2}{3} \times \frac{1/2}{2/3} \\ &= \frac{1}{2} \end{aligned}$$

Information R_2 is relevant to R_1.

(e) What is the probability of R_1, given that the second ball was white (W_2)?

$$\begin{aligned} P(R_1/W_2) &= P(R_1)\frac{P(W_2/R_1)}{P(W_2)} \\ &= \frac{2}{3} \times \frac{1/2}{1/3} \\ &= 1 \end{aligned}$$

(f) What is the probability of R_1, given that a ball has been drawn in the second drawing (B_2)?

$$\begin{aligned} P(R_1/B_2) &= P(R_1)\frac{P(B_2/R_1)}{P(B_2)} \\ &= \frac{2}{3} \times \frac{1}{1} \\ &= \frac{2}{3} \end{aligned}$$

Information B_2 is not relevant to R_1. Is it relevant to W_1?

EXAMPLE 3. A man tosses a coin n times and gets a head (H) each time. Consider two alternatives: The coin is a good coin (G) with $P(H/G) = 1/2$, or is a two-headed dishonest coin ($\bar{G}$) with $P(H/\bar{G}) = 1$.

(a) What is the probability of getting n heads in n tosses, that is $P(nH)$, if $P(G) = 0.99$ and $P(\bar{G}) = 0.01$; namely, the tossed coin was randomly selected from 100 coins of which 99 are good (G) and 1 is dishonest $(\bar{G})$?

There are two alternatives: a good coin (G) and a bad coin $(\bar{G})$. Therefore,

$$P(nH) = P(nH/G)P(G) + P(nH/\bar{G})P(\bar{G})$$

For a good coin (G) with $P(H) = 1/2$, the probability of n heads in n throws is

$$P(nH/G) = [P(H/G)]^n = \left(\frac{1}{2}\right)^n$$

because the outcomes are independent.* For a bad coin $(\bar{G})$, $P(H) = 1$; therefore,

$$P(nH/\bar{G}) = 1$$

Using these results, we have

$$P(nH) = \left(\frac{1}{2}\right)^n P(G) + (1)P(\bar{G})$$
$$= \left(\frac{1}{2}\right)^n (0.99) + (1)(0.01)$$

NOTE. As n gets larger, the first term on the right side gets smaller, i.e.,

$$\left(\frac{1}{2}\right)^n (0.99) \longrightarrow 0, \quad \text{as } n \longrightarrow \infty$$

therefore, probability $P(nH)$ approaches the value of the second term,

$$P(nH) \longrightarrow (1)(0.01) \quad \text{as } n \longrightarrow \infty$$

(b) Given that 3 heads show in 3 tosses $(n = 3)$, what is the probability that the coin is good (G)?

$$P(G/nH) = P(G)\frac{P(nH/G)}{P(nH)}$$
$$= 0.99\frac{(1/2)^3}{(1/2)^3(0.99) + (1)(0.01)}$$

Is the information nH relevant to G?

EXAMPLE 4. In Example 3, consider that the coin may be one of five different kinds, $G_1, G_2, \ldots, G_5$. The man selects the coin in his hand randomly

*For two heads $(2H)$, $P(2H) = P(H)P(H) = (1/2)^2$; for 3 heads $(3H)$, the event $(2H)$ is independent of the third head (H), or $P(2H \cap H) = P(3H) = P(2H)P(H) = [P(H)]^3$. Similarly, $P(4H) = P(3H)P(H) = [P(H)]^4$, and $P(nH) = [P(H)]^n$.

from a large collection which contains 70% of G_1, 20% of G_2, 5% of G_3, 4% of G_4, and 1% of G_5. Each coin type is biased differently with the probability of a head as given in Table 4-2.

Table 4-2

Coin	$P(H/G_i)$	$P(G_i)$
G_1	1/2	0.70
G_2	1/4	0.20
G_3	1/8	0.05
G_4	1/16	0.04
G_5	1	0.01

(a) What is the probability of nH?

$$P(nH) = \textstyle\sum_{i=1}^{5} P(nH/G_i)P(G_i)$$
$$= \left(\frac{1}{2}\right)^n (0.70) + \left(\frac{1}{4}\right)^n (0.20) + \left(\frac{1}{8}\right)^n (0.05) + \left(\frac{1}{16}\right)^n (0.04) + 0.01$$

(b) Given that 5 heads show in 5 tosses ($n = 5$), what is the probability that coin G_4 was tossed?

$$P(G_4/5H) = P(G_4)\frac{P(5H/G_4)}{P(5H)}$$
$$= 0.04\frac{(1/16)^5}{P(5H)}$$
$$P(5H) = \left(\frac{1}{2}\right)^5 (0.70) + \left(\frac{1}{4}\right)^5 (0.20) + \left(\frac{1}{8}\right)^5 (0.05) + \left(\frac{1}{16}\right)^5 (0.04) + 0.01$$

Is information $5H$ relevant to G_4?

EXAMPLE 5. A test for detecting disease D comes out with positive and negative diagnoses as follows. For people afflicted with disease D, it shows positive (correct diagnosis) 98% of the time and negative (incorrect diagnosis) 2% of the time. For healthy people, it shows negative (correct diagnosis) 96% and positive (incorrect) 4%. For people who are not healthy but not afflicted with disease D, it shows negative (correct) 92% and positive (incorrect) 8%.

(a) Consider a population with 95% healthy people and 5% not healthy, of which 1.2% have disease D. If a person is selected at random from the population and administered the test, what is the probability that it will be positive?

Let:

POS	designate positive result of test
NEG	designate negative result
HLTY	designate healthy person
SCK	designate sick person, but not disease D
D	designate person with disease D

$$\begin{aligned} P(\text{POS}) &= P(\text{POS}/\text{HLTY})P(\text{HLTY}) + P(\text{POS}/\text{SCK})P(\text{SCK}) \\ &\quad + P(\text{POS}/D)P(D) \\ &= (0.04)(0.95) + (0.08)(0.038) + (0.98)(0.012) \end{aligned}$$

(b) Given that the result of the test is positive (POS) for a person, what is the probability that he has disease D?

$$\begin{aligned} P(D/\text{POS}) &= P(D)\frac{P(\text{POS}/D)}{P(\text{POS})} \\ &= 0.012\frac{0.98}{P(\text{POS})} \end{aligned}$$

$$P(\text{POS}) = (0.04)(0.95) + (0.08)(0.038) + (0.98)(0.012)$$

Is information POS relevant to D?

4-7 MORE APPLICATIONS OF BAYES' THEOREM

Additional examples are introduced here to enhance the experience in applying Bayes' Theorem.

EXAMPLE 6. Table 4-3 shows the relative share of the market and the percent defective upon sale of four brands of an electric appliance.

Table 4-3

Brand	% of Market	% Defective
A_1	60	18
A_2	20	12
A_3	15	4
A_4	5	2

(a) What is probability $P(D)$ of a defective appliance (D) in a home that is chosen at random from a large number of homes that have the appliance?

$$\begin{aligned} P(D) &= \textstyle\sum_{i=1}^{4} P(D/A_i)P(A_i) \\ &= (0.18)(0.60) + (0.12)(0.20) + (0.04)(0.15) \\ &\quad + (0.02)(0.05) \end{aligned}$$

(b) Given that a housewife bought a defective appliance (D), what is the probability that it is brand A_3?

$$P(A_3/D) = P(A_3)\frac{P(D/A_3)}{P(D)}$$
$$= 0.15\frac{0.04}{0.139}$$

Information D is relevant to A_3.

EXAMPLE 7. One hundred balls, 50 red and 50 white, are to be placed in two urns, U_1 and U_2. Suppose that r red balls and w white balls are placed in U_1, and the remainder in U_2.

(a) What is the probability of drawing a red ball (R) if an urn is selected at random?

$$P(R) = \sum_{i=1}^{2} P(R/U_i)P(U_i)$$
$$= \left(\frac{r}{r+w}\right)\left(\frac{1}{2}\right) + \left(\frac{50-r}{100-r-w}\right)\left(\frac{1}{2}\right)$$

(b) What should r and w be in order to maximize $P(R)$? If we place the 50 red balls in U_1 and the 50 white in U_2, then

$$P(R) = 1 \times P(U_1) + 0 \times P(U_2) = \frac{1}{2}$$

But we can achieve the contribution 1/2 to $P(R)$ from U_1 even if a single red ball is in it. So while the contribution from U_1 to $P(R)$ does not change when we remove red balls from U_1, the contribution to $P(R)$ from U_2 increases when red balls are added to it. The answer is, therefore, $r = 1$ and $w = 0$.

$$\max P(R) = 1 \times \frac{1}{2} + \frac{49}{99} \times \frac{1}{2} = \frac{148}{198} \approx 0.747$$

(c) Show that the upper bound for max $P(R)$ is 0.75 as the equal number n of red and white balls approaches infinity.

$$P(R) = 1 \times \frac{1}{2} + \frac{n-1}{2n-1} \times \frac{1}{2} = \frac{1}{2} + \frac{1-1/n}{2-1/n} \times \frac{1}{2}$$
$$\longrightarrow \frac{1}{2} + \frac{1}{4} \quad \text{as } n \longrightarrow \infty$$

(d) Given that a red ball has been drawn (R) with $r = 10$, and $w = 0$, what is the probability that it came from U_1?

$$P(U_1/R) = P(U_1)\frac{P(R/U_1)}{P(R)}$$
$$= \frac{1}{2}\,\frac{10/10}{(10/10)(1/2) + (40/90)(1/2)}$$
$$= \frac{9}{13}$$

Information R is relevant to U_1.

EXAMPLE 8. The percent of students enrolled in courses C_1, C_2, C_3, and C_4 and the percent receiving the grade A in each course are summarized in Table 4-4.

Table 4-4

Course Number	% Enrollment	% of Grade A
C_1	50	8
C_2	25	15
C_3	15	4
C_4	10	1

(a) What is the probability $P(A)$ that a student selected at random from these courses will receive the grade A (A)?

$$\begin{aligned} P(A) &= \textstyle\sum_{i=1}^{4} P(A/C_i)P(C_i) \\ &= (0.08)(0.50) + (0.15)(0.25) + (0.04)(0.15) \\ &\quad + (0.01)(0.10) \end{aligned}$$

(b) What is the probability that a student will not receive the grade A, $(\bar{A})$?

$$\begin{aligned} P(\bar{A}) &= \textstyle\sum_{i=1}^{4} P(\bar{A}/C_i)P(C_i) \\ &= (0.92)(0.50) + (0.85)(0.25) + (0.96)(0.15) \\ &\quad + (0.99)(0.10) \\ &= 1 - P(A) \end{aligned}$$

Tree Structure. The information in Table 4-4 and the results of the last two equations can be recorded in a *tree structure* as shown in Fig. 4-7 using symbols, and in Fig. 4-8 using numbers.

(c) Given that a student got an A in a course, what is the probability he got that grade in course C_2?

$$\begin{aligned} P(C_2/A) &= P(C_2)\frac{P(A/C_2)}{P(A)} \\ &= 0.25\frac{0.15}{0.0845} \end{aligned}$$

Information A is relevant to C_2.

EXAMPLE 9. Three prisoners, P_1, P_2, P_3, are in jail. One is to be freed the next morning. Prisoner P_1 approaches the jailor and, knowing the jailor will not tell who will be freed although he has the information, says to him: "You, as well as I, know that P_2 or P_3 will not be freed tomorrow. Can you tell me who?"

(a) What is the probability that the jailor will say P_2, assuming that he will give him an answer and that he tells the truth, but will not tell P_1 anything about himself (i.e., about P_1)?

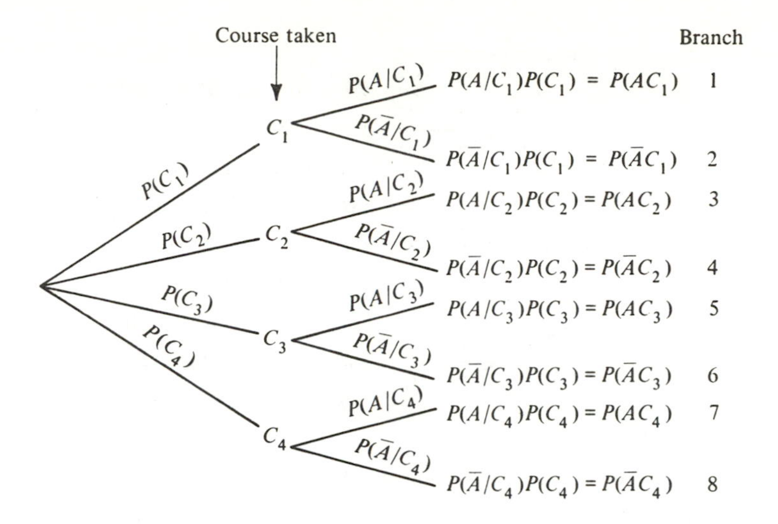

Figure 4-7

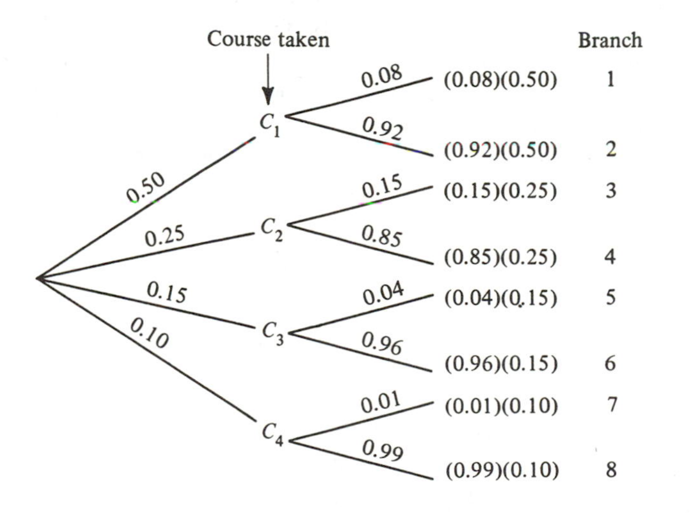

Figure 4-8

Let:

$$P_i = \text{prisoner } P_i \text{ will be freed}$$
$$JS\bar{P}_i = \text{jailor says } P_i \text{ will } \textit{not} \text{ be freed}$$

The tree for all possibilities is shown in Fig. 4-9.

$$\begin{aligned} P(JS\bar{P}_2) &= \textstyle\sum_{i=1}^{3} P(JS\bar{P}_2/P_i)P(P_i) \\ &= (1/2)(1/3) + (0)(1/3) + (1)(1/3) \\ &= 1/2 \end{aligned}$$

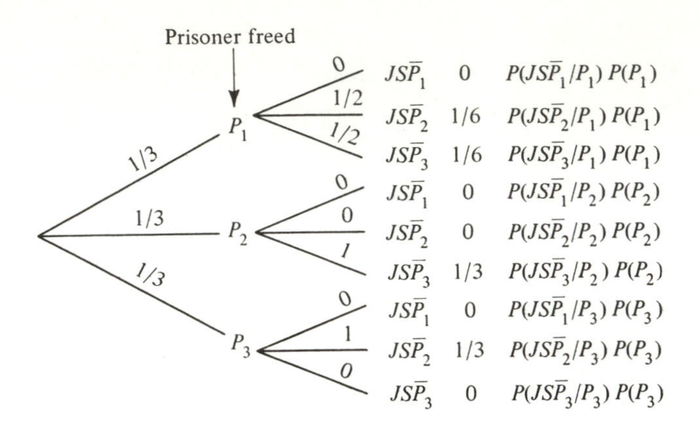

Figure 4-9

The reason for the probabilities of 1/2 for $JS\bar{P}_2$ and $JS\bar{P}_3$ at junction P_1 of the tree is based on the *principle of insufficient reason.** Since we have no knowledge of what the jailor will say given P_1 will be freed, because he may say P_2 will not be freed, or P_3 will not be freed, we assign equal probability to each statement. This is consistent with the state of maximum doubt discussed earlier (see Fig. 4-1). In general, for n possible statements or events, each will be assigned the probability of $1/n$ on the basis of this principle.

(b) Given that the jailor says P_2 will not be freed ($JS\bar{P}_2$), what is the probability P_1 will be freed (P_1)?

$$\begin{aligned} P(P_1/JS\bar{P}_2) &= P(P_1)\frac{P(JS\bar{P}_2/P_1)}{P(JS\bar{P}_2)} \\ &= \frac{1}{3} \times \frac{1/2}{1/2} \\ &= \frac{1}{3} \end{aligned}$$

Information $JS\bar{P}_2$ *is not* relevant to P_1.

Discuss the significance of this statement in the present example.

EXAMPLE 10. Consider, in Example 9, the case where prisoner P_1 finds, by chance, a note that states the name of one prisoner who will *not* be freed.

(a) What is the probability the note says P_2 will not be freed?

Let $NS\bar{P}_i$ = note says P_i will not be freed. The tree structure of Example 10 is shown in Fig. 4-10.

$$\begin{aligned} P(NS\bar{P}_2) &= \textstyle\sum_{i=1}^{3} P(NS\bar{P}_2/P_i)P(P_i) \\ &= (1/2)(1/3) + (0)1/3 + (1/2)(1/3) \\ &= 1/3 \end{aligned}$$

*Also known as the principle of maximum entropy. (See Sec. 4-10.)

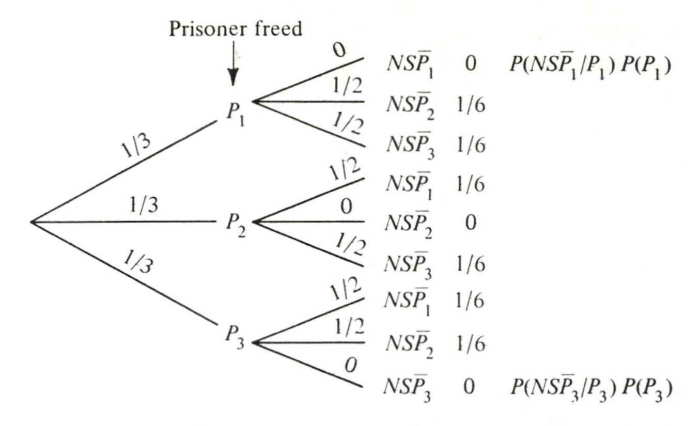

Figure 4-10

Explain the reason for the differences in the results obtained in Examples 9 and 10. Can you arrive at these results by direct reasoning?

(b) Given that the note says P_2 will not be freed ($NS\bar{P}_2$), what is the probability that P_1 will be freed?

$$P(P_1/NS\bar{P}_2) = P(P_1)\frac{P(NS\bar{P}_2/P_1)}{P(NS\bar{P}_2)}$$
$$= \frac{1}{3} \times \frac{1/2}{1/3}$$
$$= \frac{1}{2}$$

Information $NS\bar{P}_2$ is relevant to P_1. Discuss the difference between the information in Examples 9 and 10.

Tree Diagrams and Bayes' Equation

Bayes' Theorem is interpreted in Eq. 4-11 as a ratio of probabilities of a subset to a set. Considering A_i as the event of interest given B, then the subset consists of all contributions to BA_i, while the set consists of all contributions to B, i.e., $\sum_{j=1}^{n} BA_j$. The desired conditional probability is computed from Eq. 4-11, written here in the following form:

$$P(A_i/B) = \frac{P(BA_i)}{\sum_{j=1}^{n} P(BA_j)} \tag{4-12}$$

The ratio on the right-hand side of the equation can be evaluated conveniently from a tree diagram as shown by the following examples.

For instance, in Example 8 of this section we have, for part (c):

$$P(C_2/A) = \frac{P(AC_2)}{\sum_{j=1}^{4} P(AC_j)} = \frac{0.15 \times 0.25}{0.0845}$$

Symbol A here corresponds to B in Eq. 4-12, and C_j corresponds to A_j. The numerator is the probability at the end of branch 3 of the tree, shown in Figs. 4-7 and 4-8, in Example 8, and the denominator is the sum of probabilities at the ends of branches 1, 3, 5, and 7.

In Example 9, part (b):

$$P(P_1/JS\bar{P}_2) = \frac{P(JS\bar{P}_2 \cap P_1)}{\sum_{j=1}^{3} P(JS\bar{P}_2 \cap P_j)} = \frac{1/6}{1/6 + 0 + 1/3} = \frac{1}{3}$$

The numerator comes from the second branch of the tree shown in Fig. 4-9, in Example 9, and the denominator from branches 2, 5, and 8.

In Example 10, part (b), we have

$$P(P_1/NS\bar{P}_2) = \frac{P(NS\bar{P}_2 \cap P_1)}{\sum_{j=1}^{3} P(NS\bar{P}_2 \cap P_j)} = \frac{1/6}{1/6 + 0 + 1/6} = \frac{1}{2}$$

The numerator comes from branch 2 of the tree shown in Fig. 4-10, in Example 10, and the denominator from branches 2, 5, and 8.

Suggested Problems. Construct tree diagrams for Examples 1, 2, 4, 5 in Sec. 4-6 and Example 6 in Sec. 4-7.

4-8 MR. X REVISITED

We return now to Sec. 4-3 and consider the attitudes of persons P_1 and P_2 to the information regarding Mr. X's activities.

Let us designate the statement "Mr. X is honest" by HNST. Then prior to the information regarding the booth at the Las Vegas Casino, person 1 writes

$$P(\text{HNST}) = 1$$

and person 2 writes

$$P(\text{HNST}) = 0.99999$$

The assignments are referred to as *prior probabilities* (or *a priori probabilities*) to indicate that they reflect the state of knowledge prior to acquisition of new information.

Suppose, now, that we have evidence that Mr. X has flipped the coin in his booth n times and the outcome was a head each time. Let us study the relevance of this information to persons 1 and 2.

To simplify the development we consider a model with two alternatives: the coin is either good (G) with $P(H) = 1/2$, or it is not good ($\bar{G}$), namely, it is a two-headed coin with $P(H) = 1$. We shall stipulate also that a dishonest two-headed coin will be interpreted as a dishonest Mr. X; namely, an honest Mr. X (HNST) would take the job in the booth with only a good coin (G). Hence, HNST $\equiv G$.

Using Bayes' theorem, we write

$$P(\text{HNST}/nH) = P(\text{HNST})\frac{P(nH/\text{HNST})}{P(nH)}$$

in which (nH) denotes that n heads result from n tosses of the coin.

From Example 3, Sec. 4-6, we write:

$$P(nH) = P(nH/\text{HNST})P(\text{HNST}) + P(nH/\overline{\text{HNST}})P(\overline{\text{HNST}})$$

Person 1:

$$P(nH) = P(nH/\text{HNST})$$

because $P(\text{HNST}) = 1$

and $P(\overline{\text{HNST}}) = 0$

Hence, $P(\text{HNST}/nH) = P(\text{HNST})\dfrac{P(nH/\text{HNST})}{P(nH/\text{HNST})} = P(\text{HNST})$

Therefore, Person 1 does not change his mind, regardless of how much evidence (nH) is supplied. This is the important consequence of no will to doubt, which is reflected by a prior probability $P(\text{HNST}) = 1$.

Person 2:

$$P(nH/\text{HNST}) = \left(\frac{1}{2}\right)^n$$

$$P(nH/\overline{\text{HNST}}) = 1$$

Using these results, and $P(\text{HNST}) = 0.99999$, yields

$$P(nH) = (1/2)^n(0.99999) + (1)(0.00001)$$

Hence, $P(\text{HNST}/nH) = P(\text{HNST})\dfrac{P(nH/\text{HNST})}{P(nH)}$

$$= 0.99999\frac{(1/2)^n}{(1/2)^n(0.99999) + 0.00001}$$

As $n \longrightarrow \infty$, $(1/2)^n \longrightarrow 0$

and $P(\text{HNST}/nH) \longrightarrow 0$

or $P(\overline{\text{HNST}}/nH) \longrightarrow 1$

Thus, as evidence increases (large n), Person 2 becomes more uncertain of Mr. X's honesty, and no matter how little doubt was reflected in the prior probability $P(\text{HNST})$ by making it very little different from 1, there can always be sufficient evidence to drive the probability very close to zero. At what point will Person 2 claim that Mr. X is dishonest? Namely, how much evidence will he require, or how small must $P(\text{HNST}/nH)$ be? This is as personal and subjective as his original claim that Mr. X is honest with probability 0.99999, because, if he states that Mr. X is dishonest when $P(\text{HNST}/nH) = 0.00001$, it is equivalent to $P(\overline{\text{HNST}}/nH) = 0.99999$. The last statement reflects the will to doubt to the same extent as the original statement prior to knowledge of evidence nH. Here the consequences of the statements may enter the considerations, because people are likely to be more cautious with adverse statements.

4-9 PROBABILITY AND CREDIBILITY

Bayes' theorem can be viewed as a rational procedure to establish whether information B relevant to A enhances or reduces the credibility of a statement or an event A. When $P(A/B) > P(A)$, then the credibility of the statement "A is true" is enhanced. When $P(A/B) < P(A)$, then the credibility of the statement "$\bar{A}$ is true" is enhanced because $P(A/B) + P(\bar{A}/B) = 1$, $P(A) + P(\bar{A}) = 1$, and, therefore, $P(\bar{A}/B) > P(\bar{A})$.

Demonstrative and Plausible Reasoning

Consider two statements, A and B, with B the consequence of A or $A \rightarrow B$ (A implies B). In the language of conditional probability, $A \rightarrow B$ is equivalent to

$$P(B/A) = 1$$

because, given A, B is a certainty when A implies B.

To attach physical significance to that statement, consider Fig. 2-11 in which light P is on as a consequence of closing switch Q, or switch R. Thus, $Q \rightarrow P$, $R \rightarrow P$, and $P(P/Q) = 1$; $P(P/R) = 1$.

In situations where the "imply" statement holds true, we distinguish two forms of reasoning: demonstrative and plausible [33].

Demonstrative Reasoning. Consider the following syllogism (see Chapter 2):

Premise 1: $A \rightarrow B$
Premise 2: B is false
Conclusion: A is false

This conclusion follows directly from Bayes' theorem. For $A \rightarrow B$:

$$P(B/A) = 1$$
$$P(\bar{B}/A) = 0$$

The conclusion regarding the truth of A, given $\bar{B}$ (B false), can be written as

$$P(A/\bar{B}) = P(A)\frac{P(\bar{B}/A)}{P(\bar{B})} = 0 \tag{4-13}$$

The conclusion is definitive, therefore, if we accept the two premises $A \rightarrow B$ and $\bar{B}$ as being true. This form of reasoning is objective, impersonal, and holds true in all fields of knowledge. It is deductive in that it leads from the general to the particular, i.e., from knowledge about set B (B false) to knowledge about subset A which is contained in B. (See Fig. 4-11.)

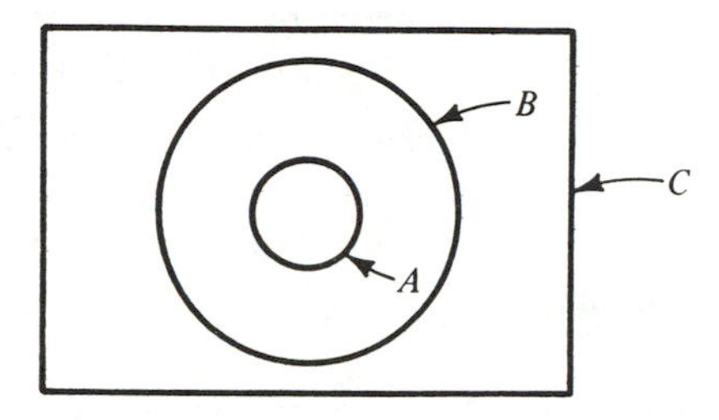

Figure 4-11 Schematic Diagram Illustrating $A \rightarrow B$, $B \rightarrow C$, $(A \cup B) \rightarrow C$

Demonstrative reasoning enhances the credibility of $\bar{A}$ from $P(\bar{A}) < 1$ to a certainty $P(\bar{A}/\bar{B}) = 1$, when $A \rightarrow B$ and B is false. That is, when consequence B of cause A is false, then the cause is false.

Plausible Reasoning. Consider now the following two premises and resulting conclusion:

Premise 1: $A \rightarrow B$
Premise 2: B is true
Conclusion: A is more credible

Here the conclusion is plausible; it indicates only direction. Credibility of A is enhanced, but it is a personal and subjective matter to establish how large $P(A/B)$ must be before we are willing to accept A as true.

Using Bayes' theorem, we can establish a numerical value for $P(A/B)$:

$$P(A/B) = P(A)\frac{P(B/A)}{P(B)}$$

Since $A \rightarrow B$, then $P(B/A) = 1$ (see Fig. 4-11), and the last equation becomes

$$P(A/B) = P(A)\frac{1}{P(B)} \tag{4-14}$$

Considering $P(A) = 0$ to signify that A is false, and $P(A) = 1$ to signify that A is true, the last equation confirms our conclusion that when B is true the credibility of A is enhanced, because $P(A) \leq P(B) \leq 1$ (see Fig. 4-11), therefore, $P(A/B) \geq P(A)$.

The smaller probability $P(B)$ is, the more the information "B is true" enhances the credibility of A. In other words, the more unexpected B is, the more information it carries in our encoding the state of knowledge regarding A. B carries maximum information regarding A when it takes on its smallest possible value, $P(B) = P(A)$, because then A becomes a certainty when B is true. (Regions A and B coincide in Figure 4-11.) B carries no information when its probability is 1, because then $P(A/B) = P(A)$. Thus, as $P(B)$ decreases from 1 to $P(A)$, $P(A/B)$ increases from $P(A)$ to 1. Stated in words: the increased confidence in the credibility of A due to the verification of B, which is a consequence of A, varies inversely as $P(B)$ (probability prior to the verification of B, i.e., prior probability of B).

Influence of New Evidence on Credibility of a Theory or a Model. For an example of how new evidence E with a small prior probability of occurrence $P(E)$ can enhance the credibility of a theory, or a postulated model, M, that is, $P(M/E) \gg P(M)$, see Chapter 5, Sec. 5-7.

4-10 THE CONCEPT OF INFORMATION AND ITS MEASUREMENT

Information and Relevance—Mutual Conveyence of Information

So far our discussion has dealt with information indirectly. Thus, we have discussed the relevance of information B to our knowledge regarding event A. This has been done through the use of Bayes' theorem. As long as $P(A/B)$ was different from $P(A)$, information B was relevant to A. While no formal statement was made on the measurement of information, a feeling was conveyed that information content is related to probability. The relevance ratio $P(B/A)/P(B)$ in Eq. 4-7 provided a means to quantify the degree of relevance by noting the deviation of this ratio from the number 1. This is the same as observing to what extent $P(B/A)$ is different from $P(B)$. In a way, there is a mutual conveyence of information between A and B. This can be demonstrated by writing Bayes' theorem in the following two ways merely by interchanging the symbols A and B:

$$P(A/B) = P(A)\frac{P(B/A)}{P(B)} \tag{4-15a}$$

$$P(B/A) = P(B)\frac{P(A/B)}{P(A)} \tag{4-15b}$$

From either one of these equations we can write

$$\frac{P(A/B)}{P(A)} = \frac{P(B/A)}{P(B)} \tag{4-16}$$

Therefore, the relevance ratios in Eqs. 4-15a and b are identical. This shows the mutual conveyence of information content in the sense that knowledge of B is as relevant to A (Eq. 4-15a) as knowledge of A is relevant to B (Eq. 4-15b).

Earlier we discussed information and credibility. Here, too, a feeling was conveyed, pointing in the direction of credibility. The degree of enhancement in the credibility of A when $A \rightarrow B$ is inversely related to the probability of B as evidenced from Eq. 4-14 (repeated here for ease of reference):

$$P(A/B) = P(A)\frac{1}{P(B)} \tag{4-14}$$

Namely, the smaller the probability of event B, the more information we obtain when we learn that B has occurred.

To clarify some of the vagueness about the concept of information as it relates to our state of knowledge, we will demonstrate in the remainder of this section that information is measurable. Information theory is based on the premise that information is measurable [34]. Our preliminary discussion of relevance and credibility of information was meant to suggest that the measurement of information is related to probability. This will be demonstrated in more detail now.

How Many Questions?

We each have an intuitive concept of information, but it is not quite the information evaluated by information theory. The information in a message

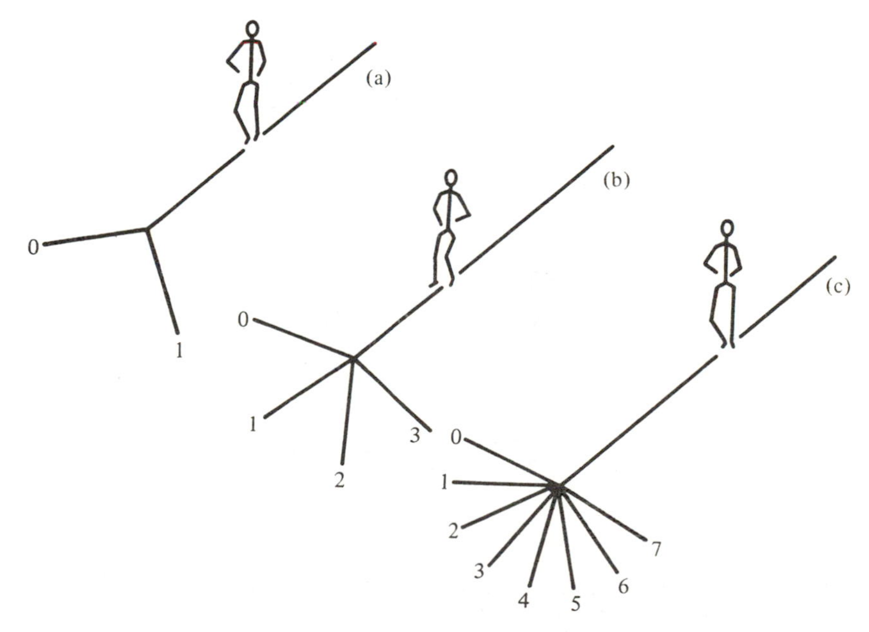

Figure 4-12 Routes at Intersection: (a) 2 Routes: 0, 1; (b) 4 Routes: 0, 1, 2, 3; (c) 8 Routes: 0, 1, 2, . . . , 7

is a measure of the amount of knowledge or intelligence that the message can convey through a symbolic form of representation, such as language.

Consider a man arriving at the intersections shown in Fig. 4-12. Only one of the routes leads to his desired destination, but he does not know which. A genie stationed at each intersection will respond with YES or NO to any question. The answer is always credible, but there is a fixed fee per question. How many questions should the man ask at each intersection? Figure 4-13 shows diagramatically that one question, "Is it road 0?" will do for the routes in Fig. 4-12(a); 2 questions for Fig. 4-12(b); and 3 for Fig. 4-12(c). In general, for 2^n routes, n questions will lead to the correct route. The first question reduces the region of search by a factor of 2. Similarly, each successive ques-

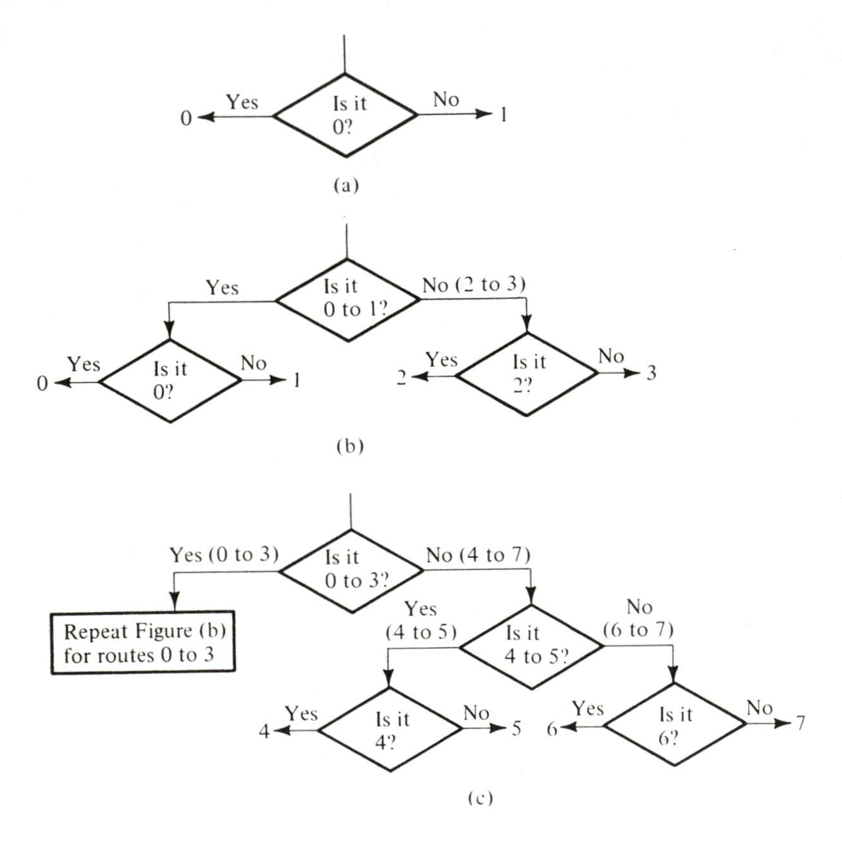

Figure 4-13 Sequence of n Questions to Find Correct Route Out of 2^n Routes: (a) 2 Routes: 0, 1; (b) 4 Routes: 0, 1, 2, 3; (c) 8 Routes: 0, 1, 2, . . . , 7

tion reduces the previous search region by a factor of 2.* This is illustrated also in the form of *decision trees* in Fig. 4-14, which is an abstraction of Fig. 4-13.

Let us agree now that the answer to each question represents the same amount of information. After all, in each case the domain of search is reduced by a factor of 2. In each case we get a response of either YES or NO. On the basis of this common unit of measurement, we conclude that n units of information are required to find the one correct route out of 2^n routes.

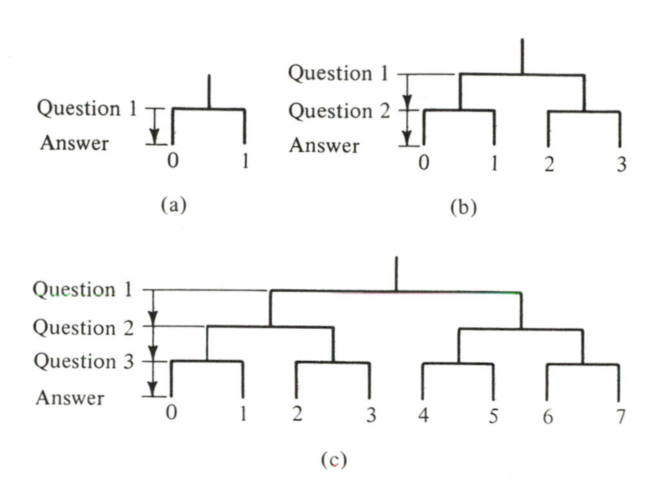

Figure 4-14 Decision Trees, Abstractions of Fig. 4-13: (a) 2^1 Routes, 1 Question; (b) 2^2 Routes, 2 Questions; (c) 2^3 Routes, 3 Questions

Probability and Measurement of Information

We now view the routes problem in terms of probability. If we consider the two routes of Fig. 4-12(a), with no knowledge as to which is more likely to be correct, we assign each a probability of 1/2 of being the one we seek. This reflects our state of maximum doubt or uncertainty (see Fig. 4-1). For four routes, we assign each an equal probability of 1/4, and for 2^n routes, $1/2^n$ each. The smaller the probability, the greater our uncertainty and the larger the amount of information required to dispel our uncertainty.

We can use a logarithmic function to transform the probability of a

*Note that it is possible to find the correct route even by a single question. For instance, considering n routes, say, $n = 128$, we may by chance receive the response YES to the first question: "Is it route r?" However, on the average, our procedure of halving the region of search will require fewer questions; in addition, our procedure *guarantees* the correct answer in n questions for 2^n routes.

route being correct to a corresponding amount of information (number of questions) required to find the correct route. The information I of event E, designated by $I(E)$, is obtained from the probability $P(E)$ of event E as follows:

$$
\begin{aligned}
I(E) &= \log_2 \frac{1}{P(E)} \\
&= -\log_2 P(E) \qquad (4\text{-}17)
\end{aligned}
$$

If we let E_n designate the event of finding the correct route out of 2^n routes, then

$$
\begin{aligned}
I(E_1) &= \log_2 \frac{1}{1/2} = \log_2 2 = 1 \\
I(E_2) &= \log_2 \frac{1}{(1/2)^2} = \log_2 2^2 = 2 \\
&\vdots \\
I(E_n) &= \log_2 \frac{1}{(1/2)^n} = \log_2 2^n = n
\end{aligned}
$$

Hence, information content $I(E_n)$ corresponds to the minimum number* of questions we have to ask on the average to find the correct route from 2^n equally likely routes. The choice of base 2 for the logarithm in the definition of $I(E)$ corresponds to our binary response to each question; however, the base of the logarithmic function amounts to a choice of a unit for measuring information. When base 2 is used, the unit of information is called a *bit*, from the contraction of binary digit, because it is information associated with an event that can happen in one of two ways with a probability 1/2 each. In a way this represents the most fundamental unit of information; that is the answer to a question by a YES or NO, or a single binary digit 1 or 0, each having probability 1/2 of being the correct response to the question.

Also note that, if we use the language of binary digits, one bit is required for each position. Hence, to specify the correct route in Fig. 4-12(a), we need a single binary digit 0 or 1. In Fig. 4-12(b), we need two bits so we can designate four different routes 00, 01, 10, 11. In Fig. 4-12(c), three bits are required to designate 8 routes 000, 001, . . . , 111; and for 2^n routes, n bit positions are required.

More on Why a Log Function

The choice of the logarithmic function in Eq. 4-17 can also be defended on other grounds. When an event has a probability of 1 (it is a certainty), its information content is zero. The log of one is indeed zero. Also, it is possible to generate the logarithmic function by requiring that the information func-

*For a proof that $I(E_n)$ corresponds to the minimum numbers of questions on the average, see [34]. Also see NOTE 2 on p. 175.

tion satisfy a number of postulates. These postulates can be viewed as constraints which reduce our choice of functions and lead to the logarithmic function as an acceptable candidate that satisfies the constraints. See NOTE 2 of this section.

Digression: Expected Value

Consider the following game. A die is tossed. You receive \$1 reward when 1 appears, \$2 reward when 2 appears, . . . , \$6 reward when 6 appears. Adding the products of each reward and the corresponding probability of receiving it, we obtain the expected value EV of the game for a single toss:

$$EV = \frac{1}{6}(\$1) + \frac{1}{6}(\$2) + \ldots + \frac{1}{6}(\$6) = \$3.5$$

Consider a new game with the same die in which you receive \$1.40 when 1 or 2 appears, and \$0.80 when any other number appears. Then, for a single toss:

$$\begin{aligned} EV &= P(1 \cup 2)(\$1.40) + P(3 \cup 4 \cup 5 \cup 6)(\$0.80) \\ &= \frac{2}{6}(\$1.40) + \frac{4}{6}(\$0.80) \\ &= \$1.0 \end{aligned}$$

The first game is called a fair game if you pay \$3.50 each time the die is tossed. The second game is fair when you pay \$1.0 each toss. In a fair game, the expected value of gain is zero for each player.

In general, when we have n possible events E_i, each with probability $P(E_i)$, and with each outcome E_i we receive a reward $R(E_i)$, then the expected value of the reward is

$$EV = \sum_{i=1}^{n} P(E_i)R(E_i) \tag{4-18}$$

Expected Value of Information—Entropy

When the rewards are $I(E_i)$ units of information associated with events E_i and probabilities $P(E_i)$, the expected value of information is given by Eq. 4-18 in which we merely replace reward R by information I. The resulting quantity is the *average information* and is also known as *entropy*, and is denoted by the letter H. Hence,

$$H = \sum_{i=1}^{n} P(E_i)I(E_i) \tag{4-19}$$

Substituting from Eq. 4-17,

$$H = -\sum_{i=1}^{n} P(E_i) \log_2 P(E_i) \tag{4-20}$$

or

$$H = \sum_{i=1}^{n} P(E_i) \log_2 \frac{1}{P(E_i)} \tag{4-21}$$

H in this equation is expressed in *bits* of information.

Let us consider the concept of entropy in the light of our earlier example of "How many questions." Suppose we wish to use a communication system which can transmit binary digits 0, 1 as shown in Fig. 2-20. The messages we wish to transmit are the symbols E_1, E_2, E_3, and E_4, each representing a different event, say, rain, snow, etc. Using two binary digits, we can encode these *source symbols* E_i in the form:

Source Symbols	*Binary Code*
E_1	00
E_2	01
E_3	10
E_4	11

These distinct code words of equal length can be generated by using the tree diagram of Fig. 4-15.

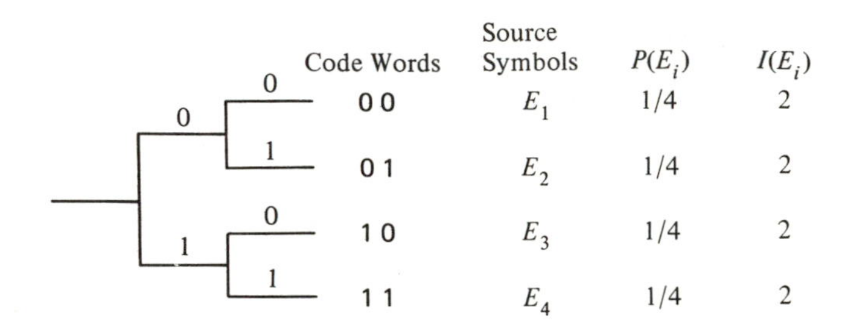

Figure 4-15 Tree Diagram for Generating Code Words of Equal Length

Now, suppose we are at the receiving end of the communication system. Knowing the code, we ask:

Question 1. Is the first bit a zero?
Question 2. Is the second bit a zero?

The answers to these two questions will establish uniquely which of the four symbols of the information source symbols was intended, provided there is no corruption due to noise in the transmission channel (such a channel is called a noiseless channel and is, of course, a theoretical utopia).

If the probability of occurrence of each of the source symbols E_i is $P(E_i) = 1/4$, then, from Eq. 4-21,

$$H = \sum_{i=1}^{4} \frac{1}{4} \log_2 4 = 4\left(\frac{1}{4} \times 2\right) = 2 \text{ bits}$$

and thus the entropy of our source of information is two bits, corresponding to the two binary questions which we must ask to establish the message.

But now suppose the symbols E_i do not have the same probability of occurrence and, instead,

$$P(E_1) = \frac{1}{2}$$

$$P(E_2) = \frac{1}{4}$$

$$P(E_3) = \frac{1}{8}$$

$$P(E_4) = \frac{1}{8}$$

If we again use the above code of two binary bits for each source symbol, then the average number of bits transmitted per symbol and the corresponding number of binary questions (Is it a zero?) remains two. But the concept of information and its definition in terms of Eq. 4-17 suggests that

$$I(E_1) = 1$$
$$I(E_2) = 2$$
$$I(E_3) = 3$$
$$I(E_4) = 3$$

Namely, the information associated with each source symbol is inversely related to its probability of occurrence and, therefore, E_1 requires one binary question while E_4 requires three questions, with the number of questions corresponding to the number of bits in the code of the source symbol. Using a tree as shown in Fig. 4-16, we can generate the following code for source

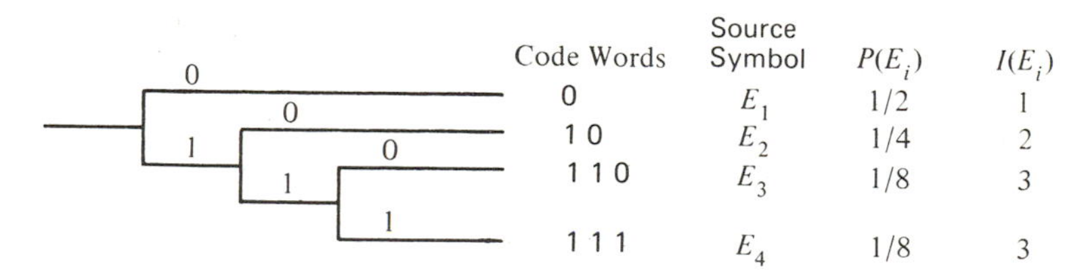

Figure 4-16 Tree Diagram for Generating Code Words of Variable Length

symbols E_i ($i = 1, 2, 3, 4$) with the number of bits in the code equal to $I(E_i)$ to the nearest integer. With this code we may ask only one question if the response to the first question leads to zero, i.e., E_1, or as many as three questions if the source symbol intended is E_3 or E_4. The sequence of questions is:

Question 1. Is the first bit a zero?

Answer: Yes, $\frac{1}{2}$ of the time, stop, message is 0. P(stop here after one question) $= \frac{1}{2}$

No, $\frac{1}{2}$ of the time, continue with question 2.

Question 2. Is the second bit a zero?

Answer: Yes, $\frac{1}{2}$ of the time $\left(\frac{1/4}{1/4 + 1/8 + 1/8}\right)$, stop, message is 10. P(stop here after two questions) $= \frac{1}{4}$

No, $\frac{1}{2}$ of the time $\left(\frac{1/8 + 1/8}{1/4 + 1/8 + 1/8}\right)$, continue with question 3.

Question 3. Is the third bit a zero?

Answer: Yes, $\frac{1}{2}$ of the time, stop, message is 110. P(stop here after three questions) $= \frac{1}{8}$

No, $\frac{1}{2}$ of the time, stop, message is 111. P(stop here after three questions) $= \frac{1}{8}$

[Draw a tree diagram of these questions and identify the probabilities for each of the code words.]

The probabilities of asking one, two, or three binary questions before the intended source symbol is identified are:

$$P(\text{one question}) = \frac{1}{2}$$

$$P(\text{two questions}) = \frac{1}{4}$$

$$P(\text{three questions}) = \frac{1}{8} + \frac{1}{8} = \frac{1}{4}$$

The expected number of questions is obtained by adding the products of the number of questions and their corresponding probabilities of being asked (Eq. 4-18, where the "reward" is the number of questions).

$$\begin{aligned} E(\text{number of questions}) &= \frac{1}{2}(1) + \frac{1}{4}(2) + \frac{1}{4}(3) \\ &= 1\tfrac{3}{4} \text{ questions} \end{aligned}$$

This result is identical with the entropy computed from Eq. 4-21:

$$\begin{aligned} H &= \frac{1}{2}\log_2 2 + \frac{1}{4}\log_2 4 + \frac{1}{8}\log_2 8 + \frac{1}{8}\log_2 8 \\ &= 1\tfrac{3}{4} \text{ bits.} \end{aligned}$$

Thus, using the measure of information $I(E_i)$, we can code the source symbols efficiently so that, on the average, 1 3/4 bits need be transmitted per source symbol, or in the transmission of 1000 source symbols, only 1750 binary bits must, on the average, be transmitted over the communication channel. This leads to a saving in cost of transmission and also reduces the average transmission time.

Entropy H of a source of symbols representing events is, therefore, equivalent to the average number of binary questions required to identify the events, or the average length of the binary code required to code the events. This is discussed further in the following notes.

It can be shown that in general for n events E_i $(i = 1, 2, \ldots, n)$, the entropy is maximum when the probabilities $p(E_i)$ are identical for all events, i.e., $P(E_i) = 1/n$ for all i. This is true, for instance, in the preceding example of four events in which the maximum entropy is 2 bits for equal probabilities, but only 1 3/4 bits for the unequal probabilities 1/2, 1/4, 1/8, 1/8. In the limit where one of the events has a probability of one, the entropy is zero.

NOTE 1—PREFIX CODES. The codes of Figs. 4-15 and 4-16 are generated at the ends of topological trees so that no code word appears as a prefix of any other code word. Such codes, known as *prefix codes* or *instantaneous codes*, permit us to decode messages uniquely, although the code words are written in one long string without separation. For example, the code

0
01
001

is not a prefix code because the code word 0 is a prefix to 01 and 001. The string 001 can be interpreted as 0 01, or 001, and therefore a separation is required which is essentially equivalent to another bit (space, no space).

NOTE 2—CODE LENGTH AND ENTROPY. The measure of information $I(E_i)$ of event E_i or of source symbol E_i (the symbol represents the event and is transmitted to signify the occurrence of the event) can be generated mathematically as the solution of the following problem:

Consider n code words for n source symbols E_i with associated probabilities of occurrence $P(E_i)$. We define length l_i of each code word as the number of bits in the word, and the average length $\bar{l}$ of the code is the expected length $E(l_i)$ of the code words:

$$\bar{l} = E(l_i) = \textstyle\sum_{i=1}^{n} l_i P(E_i) \tag{4-22}$$

Problem. Select a code for the source symbols E_i such that $\bar{l}$ is minimum, subject to the condition that the code be a prefix code.*

*The prefix condition can be stated mathematically in the form $\sum_{i=1}^{n} s^{-l_i} \leq 1$ in which s is the number of different symbols used in the code ($s = 2$ for a binary code). The inequality is known as the Kraft inequality and it indicates that we cannot have many short sequences l_i. [See (34).]

Solution (no details given):*

$$l_i = \log_s \frac{1}{P(E_i)} \tag{4-23}$$

in which s is the number of different symbols in the code, $s = 2$ for a binary code.

The optimum length of code word l_i for minimum $\bar{l}$ is, therefore, identical with the information associated with E_i. The optimum length $\bar{l}_{\text{opt}}$ (i.e., minimum length $\bar{l}$) is equal to the entropy of the source symbols E_i as can be verified by substituting Eq. 4-23 into Eq. 4-22 and comparing the result with Eq. 4-21 for $s = 2$:

$$\bar{l}_{\text{opt}} = H(E_i) = \textstyle\sum_{i=1}^{n} P(E_i) \log_2 \frac{1}{P(E_i)} \tag{4-24}$$

$\bar{l}_{\text{opt}}$ is a lower bound on average length $\bar{l}$ because the length of a code word l_i must be adjusted to the closest integer. For example, for $P(E_i) = 0.9$, $I(E_i) = l_i = 0.152$ bit, but l_i cannot be less than one bit. Hence, the average length $\bar{l}$ of actual code words will be between the following bounds in which $\bar{l}_{\text{opt}} = H$ for a binary code:†

$$\bar{l}_{\text{opt}} \leq \bar{l} < \bar{l}_{\text{opt}} + 1 \tag{4-25}$$

or

$$H(E_i) \leq \bar{l} < H(E_i) + 1$$

EXAMPLE. Show that the average word lengths $\bar{l}_1$ of code 1 and $\bar{l}_2$ of code 2 in the following codes satisfy the inequality 4-25.

Source Symbol	Code 1	Code 2	$P(E_i)$	$\log_2 \frac{1}{P(E_i)}$
E_1	1	1	0.55	0.862
E_2	01	01	0.30	1.737
E_3	000	00	0.15	2.737

Answer:
$$\begin{aligned}\bar{l}_{\text{opt}} = H &= 0.55 \times 0.862 + 0.30 \times 1.737 + 0.15 \times 2.737 \\ &= 0.474 + 0.521 + 0.410 \\ &= 1.405 \text{ bits} \qquad \text{from Eq. 4-24}\end{aligned}$$

Code 1:
$$\begin{aligned}0.55(1) &= 0.55 \\ 0.30(2) &= 0.60 \\ 0.15(3) &= 0.45 \\ \bar{l}_1 &= 1.60\end{aligned}$$

Code 2:
$$\begin{aligned}&0.55 \\ &0.60 \\ &0.30 \\ \bar{l}_2 = &1.45\end{aligned}$$

$$1.405 < \bar{l}_1 = 1.60 \leq 2.405$$
$$1.405 < \bar{l}_2 = 1.45 \leq 2.405$$

---

*See (34) for details.

†In general, $\bar{l}_{\text{opt}} = [H(\text{source symbols})/\log_2 s]$ bits, s is the number of different code symbols, say, $s = 2$ for binary code, $s = 8$ for octal code, and $s = 10$ for decimal code.

NOTE 3—HISTORICAL. The Morse code as we now know it was devised in 1838 by Samuel F. B. Morse (1791–1872) for use in telegraphy. The code consists of dots (electric current of short, one-unit duration) and dashes (current of three units duration) as Table 4-5 shows. The code was most generally used until the mid-1920s.

Table 4-5 Morse Code

Source Symbol	Code	Source Symbol	Code
A	• —	S	• • •
B	— • • •	T	—
C	— • — •	U	• • —
D	— • •	V	• • • —
E	•	W	• — —
F	• • — •	X	— • • —
G	— — •	Y	— • — —
H	• • • •	Z	— — • •
I	• •	1	• — — — —
J	• — — —	2	• • — — —
K	— • —	3	• • • — —
L	• — • •	4	• • • • —
M	— —	5	• • • • •
N	— •	6	— • • • •
O	— — —	7	— — • • •
P	• — — •	8	— — — • •
Q	— — • —	9	— — — — •
R	• — •	0	— — — — —

Note that the shortest code goes with the letter E, which has the highest probability of occurrence in the English language. Morse estimated the various frequencies of occurrence of the letters by counting the number of type letters in the bins of a printing shop [35]. Thus, less frequently used letters (smaller probability of occurrence) have longer representations in the Morse code.

Note also that Morse code is not a prefix code. For example,

• • • • • — •

could mean *site*, *sin*, *hen*, *sue*, or *5n*. For this reason the code uses spaces. One space between symbols (dash, dot), three spaces between letters, and six spaces between words.

NOTE 4—ENTROPY AND THE ENGLISH LANGUAGE. The probabilities of occurrence of symbols in the English language are given in Table 4-6.

Table 4-6 Probability of Symbols in the English Language[34]

Symbol	Probability	Symbol	Probability
Space	.1859	N	.0574
A	.0642	O	.0632
B	.0127	P	.0152
C	.0218	Q	.0008
D	.0317	R	.0484
E	.1031	S	.0514
F	.0208	T	.0796
G	.0152	U	.0228
H	.0467	V	.0083
I	.0575	W	.0175
J	.0008	X	.0013
K	.0049	Y	.0164
L	.0321	Z	.0005
M	.0198		

The entropy H of the source symbols, or the average number of information bits per symbol, is $H = 4.03$ (see Eq. 4-24). When the symbol immediately preceding an unknown symbol is given, and conditional probabilities for the following symbol are used, $H = 3.32$ bits per symbol. If we give the conditional probabilities for occurrence of symbols on the basis of eight given preceding symbols, H becomes 2.4 bits per symbol.

NOTE 5—EFFICIENCY IN AGGREGATION OF EVENTS, CODE EXTENSION. Consider two source symbols E_1, E_2 with $P(E_1) = 0.8$ and $P(E_2) = 0.2$. The shortest code we can devise is one binary bit per source symbol,. say,

$$E_1 \quad 0$$
$$E_2 \quad 1$$

Thus, the average number of bits per source symbol is 1, no different from the case of maximum entropy in which $P(E_1) = P(E_2) = 1/2$, although the entropy H and. therefore, the optimum average length $\bar{l}$ of code word, or number of bits per source symbol on the average, should be $\bar{l}_{\text{opt}} = H(E_i) = 0.8 \log_2 (1/0.8) + 0.2 \log_2 (1/0.2) = 0.7219$ bit/source symbol.

If we expect continuous strings of symbols E_1 and E_2 to be transmitted, we can improve the efficiency of transmission (fewer bits per source symbol) by an extension of source symbols in the following way. Consider strings of two symbols of E_1 and E_2 combinations (called second extension, because there are two original source symbols per string), then for independent events E_1E_2 we have

String	$P(E_iE_j)$	Code
E_1E_1	0.64	0
E_1E_2	0.16	10
E_2E_1	0.16	110
E_2E_2	0.04	111

Now, the average number $\bar{l}_2$ of bits per new source symbol (string), each consisting of two original source symbols, is:

$$\bar{l}_2 = 0.64(1) + 0.16(2) + 0.16(3) + 0.04(3)$$
$$= 1.56 \text{ bits/string}$$

or $$\frac{1.56}{2} = 0.78 \text{ bit/original source symbol}$$

compared with $\bar{l}_1 = 1$ originally.

An extension to strings of three source symbols yields:

String	Probability	Code
$E_1E_1E_1$	.512	0
$E_1E_1E_2$	.128	100
$E_1E_2E_1$	.128	101
$E_2E_1E_1$	.128	110
$E_1E_2E_2$	.032	11100
$E_2E_1E_2$	.032	11101
$E_2E_2E_1$	.032	11110
$E_2E_2E_2$	.008	11111

$$\bar{l}_3 = 0.512(1) + \ldots + 0.008(5) = 2.184 \text{ bits/string}$$

and $$\frac{2.184}{3} = 0.728 \text{ bit/original source symbol.}$$

In the limit, as string lengths go to infinity, H approaches 0.7219, which is the theoretical lower bound value of $\bar{l}$ or H in the original source, i.e.,

$$H = 0.8 \log_2 \frac{1}{0.8} + 0.2 \log_2 \frac{1}{0.2}$$
$$= 0.7219 \text{ bit/source symbol.}$$

In general, the closer we wish to approach the minimum average number of bits per original source symbol (event), the larger the sequences or extensions that must be coded. The improved efficiency and reduced cost of transmission are attended by penalties: (1) there is a delay in waiting for a group of events E_i to occur so that a string can be transmitted, (2) the encoding and decoding is more elaborate.

NOTE 6—The concepts of information, for which this section is merely an exposure, are useful in many fields. Biology, with its genetic codes, is an exciting contemporary addition to a long list of areas.

The concept of entropy is interesting and fundamental, but not easy to grasp. We shall revisit *entropy* in Chapter 10 and discuss it from another point of view.

4-11 SUMMARY

The *will to doubt* provides the flexibility that permits us to change our ideas in the face of new evidence and accept or reject premises on a tentative basis. It suggests that we can have faith in a premise before *all* doubt is dispelled, because otherwise it would be difficult to have faith in anything. The strength of our faith in a premise is related to our state of knowledge or the information at our disposal.

To incorporate information regarding a premise, we assign a quantitative value to our state of knowledge in the form of a number between 0 and 1. One represents *absolute knowledge* and corresponding *faith* in a premise, while the value zero represents absolute knowledge and a corresponding *no faith* in a premise. The value of 1/2 represents maximum doubt or the lowest state of knowledge. The will to doubt suggests that the values of 0 and 1 should be assigned with caution.

A probability assignment is a numerical encoding of a state of knowledge of the assignee. The value of 1/2 represents maximum doubt or uncertainty.

An *objective probability* represents an impersonal form of knowledge which can be verified by experiment and the results shared by most people. A *subjective probability* represents a personal form of knowledge which cannot be verified and may not be shared by others.

An outcome of an experiment, real or conceptual, is called a *sample point*, and the collection of all sample points for the outcomes of interest is the *sample space. An event* is a subset of the sample space. A *simple event* consists of one sample point, and a *compound event* consists of two or more sample points.

Laws of Probability

For any event A:

I. $0 \leq P(A) \leq 1$

II. $P(A) + P(\bar{A}) = 1$

III. OR probability statement for events A and B:

(a) When A and B are *not mutually exclusive*

$$P(A \cup B) = P(A) + P(B) - P(A \cap B)$$

(b) When A and B are *mutually exclusive*

$$P(A \cap B) = 0$$

and $$P(A \cup B) = P(A) + P(B)$$

IV. AND probability statement for events A and B:
(a) When A and B are *dependent*

$$P(A \cap B) = P(A/B)P(B)$$

or $$P(A \cap B) = P(B/A)P(A)$$

(b) When A and B are *independent*

$$P(A/B) = P(A)$$
$$P(B/A) = P(B)$$

and $$P(A \cap B) = P(A)P(B)$$

V. *Bayes' equation* for conditional probability $P(A/B)$ is:

$$P(A/B) = P(A)\frac{P(B/A)}{P(B)}$$

$P(A)$ in this equation is called an *a priori probability*, and $P(A/B)$ is called an *a posteriori probability*. The ratio $P(B/A)/P(B)$, which multiplies $P(A)$ on the right of the equation, is a *measure of relevance*. Information B is relevant to A when this ratio is different from one.

If event B can occur with many different events A_j $(j = 1, 2, \ldots, n)$ and the events $B \cap A_1, B \cap A_2, B \cap A_3, \ldots, B \cap A_n$ are mutually exclusive, then

$$P(B) = \sum_{j=1}^{n} P(B \cap A_j)$$
$$= \sum_{j=1}^{n} P(B/A_j)P(A_j)$$

and Bayes' equation for the conditional probability of any event A_i, given B, becomes

$$P(A_i/B) = P(A_i)\frac{P(B/A_i)}{\sum_{j=1}^{n} P(B/A_j)P(A_j)}$$

When an event B is a consequence of A, i.e., $A \rightarrow B$, then

$$P(B/A) = 1$$

and Bayes' equation becomes

$$P(A/B) = P(A)\frac{1}{P(B)}$$

Therefore, the *credibility* of event A is enhanced (gets closer to 1) when B occurs, provided $P(B) \neq 1$, because $P(A/B) > P(A)$. The smaller $P(B)$ is, the more B contributes to the credibility of A.

Demonstrative reasoning is objective, impersonal, and holds true in all fields of knowledge. It is a deductive process that leads from the general to the particular. For example, for events A and B,

Premise 1	$A \rightarrow B$
Premise 2	B is false
Conclusion	A is false

The conclusion is completely objective and impersonal.

Plausible reasoning is subjective and personal in character. It is an inductive process in which we attempt to go from the particular to the general. For example, for two events A and B,

Premise 1	$A \rightarrow B$
Premise 2	B is true
Conclusion	A is more credible

The conclusion is plausible, but personal. It only indicates direction (more), and the quantification of degree of enhancement in credibility is personal.

Bayes' equation can be used in many applications in which information is to be incorporated in the assessment of relevance and credibility.

Information theory is based on the premise that information is measurable. The information $I(E)$ of event E is related to probability $P(E)$ of the event by

$$I(E) = \log_2 \frac{1}{P(E)}$$

The *units of information* in this equation are *bits*. The number of units represents the number of binary questions (to which the answer is "Yes" or "No") which must be asked on the average to establish the occurrence of the event in question.

The expected value of information for a system of n events $E_1, E_2, \ldots, E_n$ which can occur with probabilities $P(E_1), P(E_2), \ldots, P(E_n)$ is called the *entropy*, H, of the system, and is given by

$$\begin{aligned} H &= \textstyle\sum_{i=1}^{n} P(E_i) I(E_i) \\ &= \textstyle\sum_{i=1}^{n} P(E_i) \log_2 \frac{1}{P(E_i)} \end{aligned}$$

The entropy H is also equivalent to the minimum average number of binary questions that must be asked to identify an event from possible events E_i $(i = 1, 2, \ldots, n)$. This is identical with the minimum average length of a binary code for the events E_i. Since H can take on noninteger values, but a binary code consists of an integer number of bits (or an integer number of binary questions must be asked), the entropy H is a *lower bound* for the average length $\bar{l}$ of a binary code for events E_i, namely,

$$H \leq \bar{l} < H + 1$$

That is, $\bar{l}$ can be no smaller than entropy H of the information source of events E_i. At best, $\bar{l} = H$ when H is an integer; otherwise, $\bar{l}$ is adjusted to the closest integer larger than H.

Aggregation of events E_i in strings can improve the efficiency of coding, approaching the theoretical lower bound on average bits per symbol established by the entropy of the source when the aggregated strings of events get very long. However, this leads to more elaborate coding and requires a time delay for events to occur before a string can be transmitted.

EXERCISES*

4-1 Given the situation described in Exercise 2-21 of Chapter 2, suppose a team of geologists has found that the probability of an earthquake of magnitude M or greater is 0.3. What is the probability that, in a 4-month period, there will be exactly *one* earthquake of magnitude M or greater?

4-2 In the game of Keno, as played in Las Vegas, 20 numbers out of 80 are picked at random by the house. The rules of the game allow the player to pick from 1 to 15 of the 80 numbers. He wins varying amounts, depending on how many numbers he has chosen and how many of these numbers are among the 20 picked by the house.

Suppose a player decides to select only one number out of the 80. He receives $1.80 if his number is among the 20 picked by the house and receives nothing if his number is not among the 20.

(a) What is the expected value of this player's strategy?

(b) If it costs $0.60 to play this game, how much should the player expect to win (lose) per game in the long run?

4-3 One hundred people have tickets for a certain raffle. Twenty tickets are selected at random. Ten win $7 each, and the other ten win free tickets for a second raffle identical to the first. The remaining 80 ticket holders get nothing. What is the expected value of this raffle?

4-4 Consider the following game. Toss a fair coin. The game will be over when you get a head for the first time. If you get a head on the first toss, you win $2; if you get a head for the first time on the second toss, you win $(\$2)^2 = \$4$; . . .; and if you get the first head on the nth toss, you win $\$2^n$.

(a) What is the expected value of the game?

(b) Would you be willing to bet an amount equal to the expected value of this game? Explain your answer.

---

*The reader who needs a stronger foundation in the concepts of probability should first attempt Exercises 4-29 through 4-39.

4-5 In a quiz game that consists of two stages, the contestant has three alternatives at each stage. These alternatives and their consequences are:

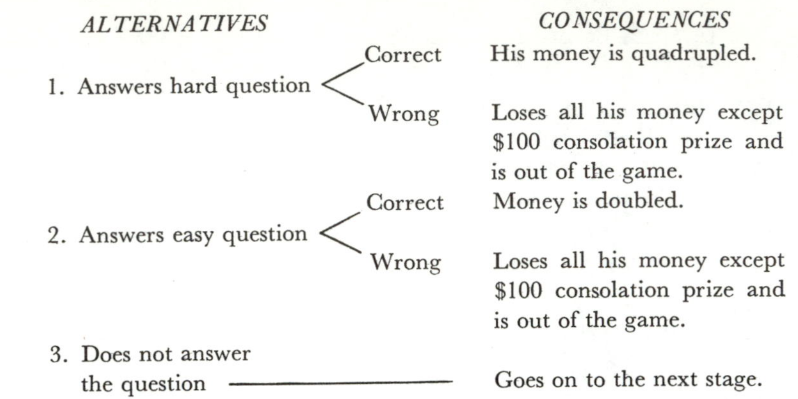

Suppose a contestant starts with \$100 and has probabilities of 0.35 and 0.15 of answering correctly an easy and hard question, respectively. Assuming he is equally likely to select a hard question, an easy question, or abstain, find the probability that he will end up with (a) \$100, (b) \$400, or (c) \$1600.

4-6 Assume that there is a diagnostic test for cancer with the following properties:

Given that a person has cancer, the test is positive (i.e., it indicates that the person has cancer) with a probability of 0.95.

Given that a person does not have cancer, the test is negative (i.e., it indicates that the person does not have cancer) with a probability of 0.9.

Let C represent the event that a person has cancer. Let T represent the event that the test is positive, i.e., that it indicates a person has cancer. Assume that 5 out of every 1000 persons have cancer. Find:

(a) $P(C)$, the probability that a person selected at random has cancer.
(b) $P(\bar{C})$, the probability that a person selected at random does not have cancer.
(c) $P(T/C)$
(d) $P(T/\bar{C})$
(e) $P(T)$, the probability that the test indicates that a person has cancer.
(f) Using Bayes' theorem, find the probability that a person, who according to the test has cancer, actually does have cancer.
(g) Show how the result obtained in (f) can be found by using a tree diagram.

4-7 In Fig. 4-17, switches A_1 and A_2 are not perfect. When A_1 is on, light B is on 80% of the time and off 20% of the time. When A_2 is on, B is on 40% of the time and off 60% of the time. B is always off whenever A_1 and A_2 are both off.

One switch (either A_1 or A_2) is turned on each evening on the basis of the result from the toss of two coins. If two heads show, A_1 is turned on; otherwise, A_2 is turned on. The switch is turned off every morning.

(a) Given that light B went on in the evening, what is the probability that switch A_1 was turned on?

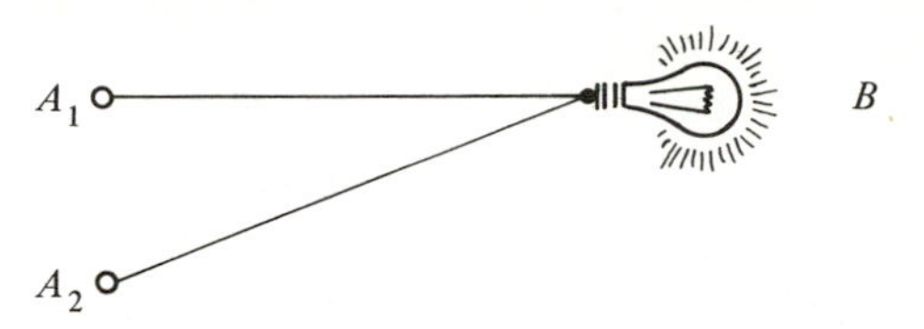

Figure 4-17

(b) Given that light B was off all night, what is the probability that switch A_2 was turned on?

4-8 In Fig. 4-18, switches A and Q are not perfect. When A is pressed, light B goes on with probability 0.1 and R goes on with probability 0.9. When Q is pressed, light B goes on with a probability of 0.8 and R goes on with a probability of 0.2.

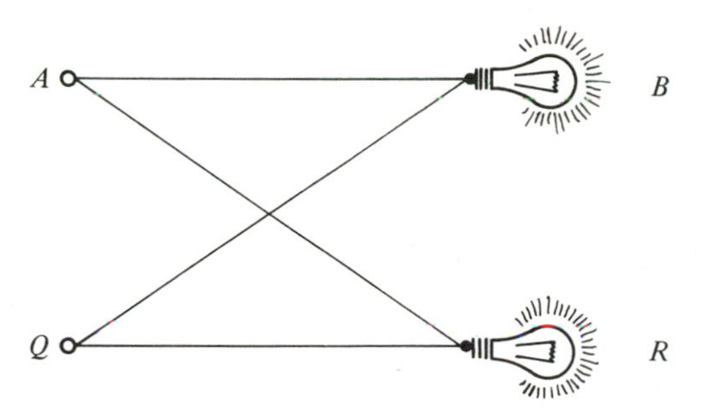

Figure 4-18

The switches are pressed on the basis of the outcome of the simultaneous toss of a die and a coin. If a head appears on the coin, switch A is pressed; if a 6 appears on the die, switch Q is pressed; and if *both* a head and a 6 appear, both switches A and Q are pressed.

It is observed that the light B is on. What is the probability that switch A was pressed?

4-9 In a community of 1000 voters, there are 600 Democrats and 400 Republicans. It is known that in an election in which all 1000 people voted, 20% of the Democrats voted Republican. Also, the probability that a Republican voted Republican was 0.9.

(a) If we speak to a citizen who voted Republican, what is the probability that this citizen is a Democrat?

(b) Who won the election?

4-10 A man is faced with the decision of driving home from work by one of two routes. If he takes the first route, the probability that he will get home before 6 o'clock is 0.7. If he takes the second route, the probability that he will get home before 6 o'clock is 0.8. Whenever it is raining he uses the first route, and the remainder of the time he uses the second route. Given that it rains 1/3 of the time and the man got home after 6 o'clock on a given day, what is the probability that he took the first route home that day?

4-11 Create one problem of interest to you which requires the use of Bayes' theorem and solve it.

4-12 (a) How is information related to probability?

(b) In Exercise 4-6, it was stated that 5 out of every 1000 persons have cancer. Assuming that we have a perfect test for predicting cancer in a person, state in which case the information received from the test is greater:

(i) The test indicates the person has cancer.

(ii) The test indicates the person does not have cancer.

4-13 The average information associated with an experiment or event which has n possible outcomes, each with probability $P(E_i)$, is given by

$$\text{Average information } H = \sum_{i=1}^{n} P(E_i) I(E_i)$$

where $I(E_i) = \log_2 \dfrac{1}{P(E_i)}$

This average information represents the number of binary questions (i.e., yes-no questions) that a person must ask in order to identify the correct outcome of the experiment.

(a) Find the average information associated with the selection of the correct route when there are four routes, and it is equally likely that any one of the four routes is the correct one.

(b) On the average, how many questions must a person ask when faced with 10 routes? Draw a tree diagram to illustrate.

4-14 In Exercise 4-13(a), a person is faced with routes with each equally likely of being the correct one. Now, assume that the four routes are not equally likely. The probabilities associated with each route being the correct one are as shown in Fig. 4-19.

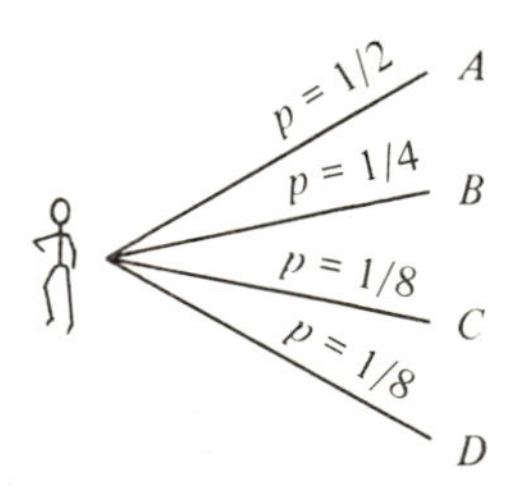

Figure 4-19

(a) Compute the average numb.r of questions that must be asked in repeated experiences with the situation, to find the correct route.

(b) Identify a strategy which minimizes the average number of questions that must be asked to find the correct route.

4-15 Consider two situations, A and B, as shown in Fig. 4-20, in which a person is faced with either 4 or 8 equally likely routes to a solution, respectively.

For example, A may be 4 possible paths to solving a certain problem in a laboratory, and B may be 8 paths to solving the same problem on paper. Because of limited availability of the laboratory, a person encounters situation A with a probability $p_A = 1/3$ and situation B with a probability $p_B = 2/3$.

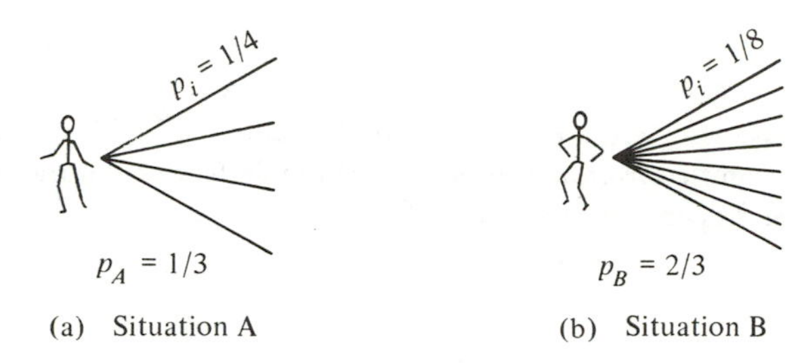

Figure 4-20

What is the average number of binary questions the person needs to ask in order to find the correct path to the solution?

4-16 Devise a binary prefix code for the following five source symbols such that the average length of coded word received (and transmitted) will be less than two bits.

Source Symbol	Probability of Occurrence
E_1	1/2
E_2	1/4
E_3	1/8
E_4	1/16
E_5	1/16

4-17 Repeat Exercise 4-16 for the following seven source symbols:

Source Symbol	Probability of Occurrence
E_1	1/2
E_2	1/4
E_3	1/8
E_4	1/16
E_5	1/32
E_6	1/64
E_7	1/64

NOTE. Exercises 4-18, 4-19, and 4-20 deal with the concept of *reliability*. The reliability of a system is defined as the probability that it will operate successfully. For example, the reliability of the system in Fig. 4-21 depends on the reliability of switches 1 and 2; i.e., if a message is sent from A to B, the probability that a message transmitted from A will be received at B is dependent on *both* the switches operating successfully.

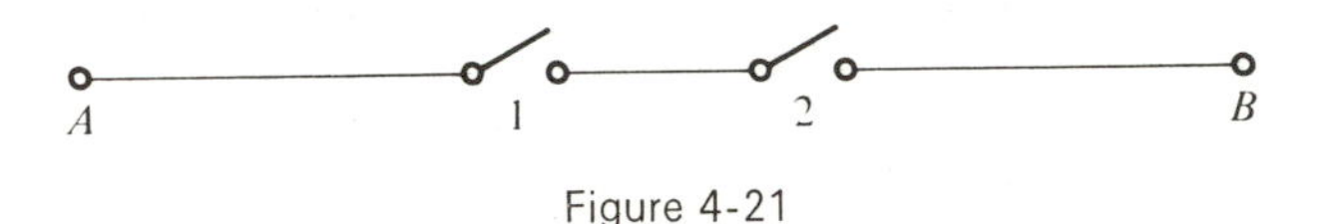

Figure 4-21

Similarly, if the switches were arranged in parallel, as in Fig. 4-22, the probability that a message transmitted from A will be received at B depends on the probability of *either* switch 1 or 2 operating successfully.

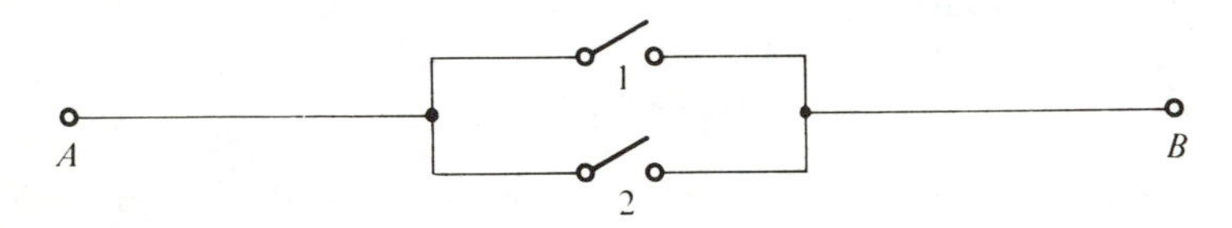

Figure 4-22

4-18 The two switches in Fig. 4-22 are not perfect. The probabilities of successful operation are 0.9 and 0.8 for switches 1 and 2, respectively.
Find:
(a) the probability that switch 1 will fail to operate; the probability that switch 2 will fail to operate.
(b) the probability that a message transmitted from A will not be received at B.
(c) the probability that a message transmitted from A will be received at B.

4-19 One can build highly reliable systems out of unreliable elements by using redundancy of components in the system. Justify this statement. Present a simple example in support of your argument.

4-20 Mr. K has to travel from Washington to Saigon via Paris. The trip from Washington to Paris takes 5 hours with a probability of 0.1 for a delay, and from Paris to Saigon it takes 12 hours with a probability of 0.3 for a delay.
Find:
(a) the probability that Mr. K will reach Saigon within 17 hours.
(b) the probability that he will not reach Saigon within 17 hours.

4-21 How can you relate the problem of routes at an intersection to the Hamming code problem? That is, why are three check bits required for a message of 4 bits? (Note that there may also be an error in the check bits.)

4-22 In a college community it is estimated that 70% of the girls have brown eyes. Let

B = event that a girl has brown eyes

C = event that a girl is not a blonde

Let us assume that $B \longrightarrow C$. Using this assumption, find the probability range for $P(C)$.

4-23 Bayes' theorem may be stated as follows:

$$P(A/B) = P(A)\frac{P(B/A)}{P(B)}$$

How is this relationship affected when $A \longrightarrow B$?

4-24 Give examples of situations in which probability is subjective. Counter these with examples in which probability is objective. Discuss how you can make a distinction between subjective and objective probability.

4-25 Refer to Sec. 4-3, "Is Mr. X honest?"

(a) If you were Person 2 and you knew that Mr. X flipped the coin only once and the player lost, would you say Mr. X is honest? Why?

(b) If Mr. X won every flip of 1000 tosses, would you say Mr. X is honest? Why?

(c) How would you go about determining at what point you would change your mind about the honesty of Mr. X?

(d) Suppose the game is changed. Now Mr. X tosses two dice. If the sum of numbers on the faces of the dice equals 2, Mr. X wins; otherwise, you win. If Mr. X rolls two 2s in the first two throws, would you consider him honest? Ten 2s in the first ten rolls? Why? (*Note:* If Mr. X is dishonest, both dice have the number 1 on each face.)

(e) How do the two games differ in the sense that the honesty of Mr. X is in question?

4-26 All people have two genes that determine their eye color, one from each parent. The chances of getting either gene from a particular parent are equal. (For example, if a person's father has one blue gene and one brown gene, that person will inherit one gene for determining eye color from his father, with a 50% chance of its being a blue gene and a 50% chance of its being a brown gene. Similarly, the person inherits one eye color gene from the mother.)

Blue eyes can occur only in individuals with two blue genes; brown eyes can occur in individuals with at least one brown gene. If my brother has blue eyes, what is the probability of my having blue eyes? Assume we consider *only* blue and brown genes, and that they are equally likely.

4-27 Suppose you are given 27 gold coins and are told that 26 of them are identical, but one is heavier, while appearing identical in all other ways.

You are given a double-pan balance and are allowed only three weighings on the balance. What is the sequence of weighings that you must perform to guarantee identifying the heavy coin? Use a tree diagram.

4-28 Use a Venn diagram to investigate the validity of the argument presented in Exercise 2-11 which is reproduced here for convenience:

Premise 1: If the sun is shining, it is not raining.
Premise 2: The sun is not shining.
Conclusion: It is raining.

NOTE. Exercises 4-29 through 4-39 may be used as drill exercises to provide practice in solving very basic probability problems. Mastery of the concepts dealt with in these exercises is necessary before many of the preceding, more difficult exercises can be attempted.

4-29 Draw a Venn diagram to illustrate each of the following:

(a) $A \cup \bar{B}$
(b) $A \cup B$
(c) $\bar{A} \cup \bar{B}$
(d) $(A \cap B) \cap \bar{C}$
(e) $\bar{A} \cup (\bar{B} \cap C)$
(f) $A \longrightarrow B$
(g) $A \longleftrightarrow B$
(h) $B \longrightarrow \bar{A}$

4-30 The sample space diagrammed in Fig. 4-23 illustrates the 52 possible outcomes of the event of drawing *one* card from a deck of cards. As you work the following

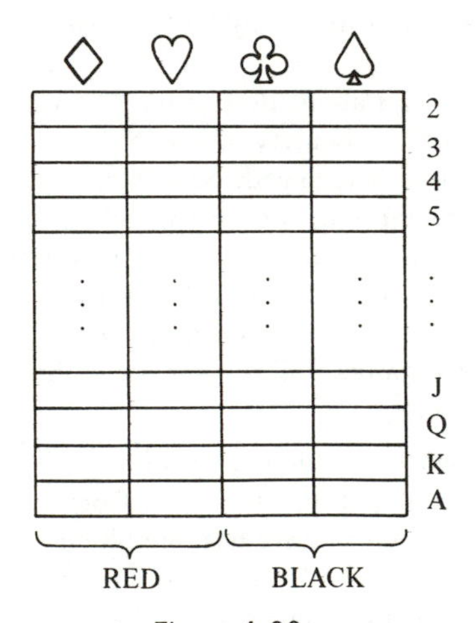

Figure 4-23

exercises, verify each answer by finding the ratio of the number of squares on the sample space that correspond to the region in question, to the total number of 52 squares.

(a) $P(\text{spade}) = ?$
(b) $P(\text{Queen}) = ?$
(c) $P(\text{spade} \cap \text{Queen}) = ?$
(d) Using part (c), show how Eq. 4-5a reduces to Eq. 4-6 when two events are independent.
(e) $P(\text{spade} \cup \text{Queen}) = ?$

(f) $P(\text{four}) = ?$ $P(\text{five}) = ?$
(g) $P(\text{four} \cap \text{five}) = ?$
(h) $P(\text{four} \cup \text{five}) = ?$
(i) Using part (h), show how Eq. 4-3 reduces to Eq. 4-4 when two events are mutually exclusive.

4-31 Consider the following game. First you roll a die. If 1 through 4 come up, you flip a coin. If 5 or 6 comes up, you pay $1. If you flip the coin and it is heads, you win $3. If it is tails, you pay $1. What is a fair price to pay for this game?

4-32 Two dice are in a drawer. One is fair, and the other has probability of 1/2 of rolling a 1 and probability of 1/10 each for rolling a 2, 3, 4, 5, or 6. A single die is selected at random and rolled twice. Both outcomes are 1. What is the probability that the die selected was the fair one?

4-33 A deck of cards contains ten cards numbered 1 to 10.
(a) What is the probability of drawing two cards whose sum is 10?
(b) What is the probability of (a) if the number 5 card is removed from the deck?

4-34 Consider the following game. If you roll a 1 or 2 on a fair die, you win $2. If you roll a 3 or 4, you lose $1. If you roll a 5 or 6, you lose $.50. What is the expected value of the game?

4-35 In a room there are 8 type I urns (4 red balls and 6 white balls inside) and 2 type II urns (8 red balls and 2 white balls inside). The urns appear identical from the outside. An urn is selected at random and a red ball is drawn from it. What is the probability that the urn selected was of type I?

4-36 Box "A" contains 30 red balls and 50 white balls. Box "B" contains 40 green balls and 60 orange balls. If one ball is selected at random from each box, what is the probability that the two include a red ball or a green ball, but not both?

4-37 In a card game in which you receive 5 cards, what is the probability you will be dealt: (a) four aces? (b) four of any kind?

4-38 Three balls are drawn successively from a box containing 6 red, 4 white, and 5 blue balls. What is the probability that one ball of each color is drawn if each ball is: (a) replaced? (b) not replaced?

4-39 The telephone company decides to raise its rates but wishes not to use pennies for pay phones. As a result they charge 15¢ for some fraction (f) of the calls, while the rest of the calls are free.
(a) If $f = .9$, what is the expected cost of a call?
(b) What must the value of f be for an average charge of 12¢ per call?

5

MODELS AND MODELING

5-1 INTRODUCTION

The concept of a *model* is so fundamental to problem solving that it is present at all stages, from problem definition to solution. It is a concept characterized by ubiquity; the words and symbols we use, the responses recorded by our senses are all models. *A model is an abstract description of the real world; it is a simple representation of more complex forms, processes, and functions of physical phenomena or ideas.*

The first four chapters of this book mention and use models repeatedly. In Chapter 1, models of problem solving are discussed. The four stages of preparation, incubation, inspiration, and verification are described as a model suggested by some psychologists. The information-processing specialist and the behaviorist offer different models. While the behaviorist's model concentrates on input-output or stimulus-response aspects of problem solving, the information-processing model emphasizes the process that intervenes between input and output. This matter of emphasis is important because, if indeed a model is a simpler representation of reality, some elements present in the real-world problem may not be present in the model. The decision to include or exclude an element will depend on its importance or the emphasis it is given in terms of its contribution to the description of the form, process, or function of the real world.

Chapter 2 introduces natural language and symbolic logic as models of real-world objects, and models of concepts and ideas. In the same chapter, the diagram of Fig. 2-19 is a model of a communication system in which each element is an abstraction of a much more complex real-world counterpart. For example, consider the number of alternative schemes and detailed descriptions that could accompany an information source, a signal, a channel, etc. Chapter 3, Sec. 2 introduces a simple computer model in which both the structure and function of the digital computer are described through a gross abstraction. The descriptions of a half-adder (Fig. 3-9), full adder (Fig. 3-11), and parallel adder (Fig. 3-13) are models because they represent abstractions of more complex real-world objects or devices. In Chapter 4, Fig. 4-1 is a model for the assessment of doubt or level of knowledge. Through the use of a scaled yardstick it represents an abstraction of the complex concepts of doubt and knowledge. Similarly, the formulas for information I and for entropy H are mathematical models of complex concepts. In fact, the table of contents is a model for the entire book.

In the light of the preceding discussion, which points out the ubiquity of models, it is appropriate to stop here and introduce the objective of this chapter on Models and Modeling. *The objective is to consolidate the experience in modeling in various fields and unify those elements of the modeling process that appear most productive in problem solving.* We discuss the purpose of models and their classification, provide guidelines for the modeling process, and give examples of models in different fields. The remainder of the book can be viewed as devoted to models and their use: probabilistic models (Chapter 6), decision models (Chapter 7), optimization models (Chapter 8), dynamic models (Chapter 9), and models of ethical behavior (Chapter 10).

5-2 THE PURPOSE OF MODELS

A model is constructed to *facilitate understanding and enhance prediction.* We understand an event or an idea when we identify it as part of a larger frame in terms of structure, functional relations, cause-effect relations, or combinations of these frames. There is a definite link between understanding and prediction. When we can identify functional relations or cause-effect relations between events, we can then construct better models to predict the occurrence of future events, and in some cases we can *cause* the occurrence of future events through the control of relevant parameters. Let us consider some examples in order to give a more concrete meaning to the concepts of understanding and prediction.

Models for Understanding

When we claim to understand a person's hostile behavior on a particular occasion, we mean that we view it as part of a more complete pattern of behavior, a larger frame, which characterizes this person or people in general. On the other hand, when a particular mode of behavior seems not to follow a pattern that we know, and we cannot place it in a larger frame comprehensible to us, we claim not to understand the behavior. When we understand a word, we mean we know how it can be used in context with other words, a larger frame. We understand what bone, leg, finger, or head signify in their relation to the larger structure of which they are a part. The fall of an apple from a tree is understood only in the larger frame of gravity. Here we pause to ask the natural question: Where does understanding stop? It stops when we cannot find a frame for the idea or event which we wish to understand. This in fact was the status of Newton's laws until the advent of the model* (theory) of relativity. Newton's model was not embedded in a larger frame, although a great amount of evidence was available to support it, as a frame for understanding many physical phenomena and as a tool for prediction.

Historical Example of a Model for Understanding

Of particular historical interest is Kant's quest for understanding Newton's inverse-square model of gravitation. In this model, each mass throughout the universe is attracted by every other mass. The force of attraction between two masses, m_1 and m_2, increases in direct proportion to their product, and decreases in proportion to the square of their distance r. In the language of mathematics, this can be stated in the form

$$F = c\,\frac{m_1 m_2}{r^2} \tag{5-1}$$

in which F is the force of attraction, and c is a constant which represents the magnitude of F for two unit masses, $m_1 = m_2 = 1$, one unit of distance, $r = 1$, apart. To Newton, the inverse-square equation was a model which explained a multitude of natural phenomena and served as the frame for understanding such diverse events as free fall of objects, tides, motion of planets, and oscillation of a pendulum. The model was accepted because of this multitude of observed evidence, but there was no further explanation why the model applied. Kant was not satisfied with this state of events, and in his quest to understand the inverse-square model (or law, since the evidence was so strong) he proposed the following model. Consider gravitational forces emanating from a source of attraction and spreading out as fluid or light in all directions. Then, by analogy to fluid and light, the intensity of the forces will decrease with distance. Viewing spherical surfaces of different radii from the source, it is known that the surface area increases as the square of the radius. If the gravitational forces are conserved, thus their density and, hence, intensity is decreased as the square of the distance. Using this model, Kant claimed to understand the inverse-square law, namely, from the spatial relation of spherical surfaces, the presence of $1/r^2$ in the equation was understood to fit, while any other factor $1/r^n$ with $n \neq 2$ would not fit. Kant's

*In the context of this chapter, we consider a theory to be a model in a more formal sense.

explanation, while appealing in form, depends on the implied assumption in his model that gravitational forces both propagate and are conserved in much the same way as momentum is conserved or energy and matter are conserved. This was the area of difficulty; the assumption was not confirmed nor rejected until seventy years later when a distinction between the concept of energy and force was introduced.* The *form* of Kant's argument is valid. In fact, it applies to the intensity of illumination produced on a surface at a distance from a source of light. However, while light is understood in the larger frame of electromagnetic theory (model) to be a form of electromagnetic energy, gravitational forces are understood to be distinguished from energy.

The digression to Kant's model should serve as an illustration of man's endeavors in his quest for understanding. It is quite possible to claim understanding when it is not so. The test is in verification through observations and measurements (which, in the case of gravitational forces, have always been difficult to perform), and in the usefulness of the model or proposed frame for further understanding and prediction.

However, a model must be rejected with caution and, to some degree, on a tentative basis. For example, one day we may find it productive to describe a model for the gravitational field in terms of quantized discrete corpuscles of gravitational energy† analogous to the "photons" of electromagnetic theory. Should this happen, then Kant's "larger frame" for understanding may apply.

Models for Prediction

Models for prediction normally include factors that are considered to contribute to the outcomes of interest. Thus, a small-scale model of an airplane which is placed in a wind tunnel, to facilitate prediction of stability for the real object in flight, will include such factors as the shape of the wings and their flexibility which influence the stability, but will not include the shape of the seats in detail or the color of the exterior.

A map is a model. It can be used to predict the distance and time of travel between various points included in the map. The amount of detail or level of abstraction depends on the objective. To find the distance from San Francisco to New York through some intermediate stops, we can use a map in which each city is marked by a small dot. However, once in New York, this model is of little value to find the Statue of Liberty.

To predict the economic well-being of a society (in terms of level of employment, level of income, etc.), models are constructed and include factors considered relevant in the sense that they influence the economic

---

*Energy is the product of force and distance.

†Dirac suggested the name "gravitons" for these discrete packets of gravitational energy. See [36].

behavior. Chapter 7 discusses a model of economic behavior which abstracts the behavior of people in terms of a measure of utility of various alternative outcomes. Man has been modeled as the *economic man* by abstracting key features of stimulus and response to describe a model that will serve to predict his behavior in our industrialized society. Man has been modeled as the *psychological man* in which the abstracted key features serve to predict his behavior in the more general environment.

Models can, of course, serve both the purposes of understanding and prediction. Historians, for example, have suggested models of history to serve as frames for understanding the past and, at the same time, improve prediction of the future course of history. The models differ both in terms of their structure and their success in affording understanding and facilitating prediction. Examples of such models are discussed later in this chapter.

5-3 THE NATURE OF MODELS

In Chapter 2 we introduced the concepts of language and metalanguage. From the language of objects we were led to metalanguage which deals with the names of the objects as distinct from the objects themselves. To deal with metalanguage, we progressed to yet a higher level of abstraction of meta-metalanguage. At each step we moved further from the world of the concrete real objects. This increased remoteness from the "real thing" as we progress from one level of abstraction to a higher one is a characteristic of the modeling process which is fundamental to its very nature. The more abstract a model, the less detail it contains and yet the more productive and useful it may be. Consider a large television screen that contains a large number of dots in various shades of grey. If we fail to identify the picture, it is sometimes possible to move far enough from the screen and observe a familiar face or form emerge. This, of course, can also be obtained by projecting the same figure on a smaller screen. What happens is a case of giving up detail for the sake of the total picture. As the spacing between dots disappears (or gets very small), we deal with a more abstract figure in terms of aggregates of points or chunks of the total figure. Our brain, which can manipulate a limited number of unrelated items, can then proceed more productively to manipulate the relations between the aggregates of the total picture and identify it.

From the preceding we conclude that, just as words in language are models of things and not the things themselves, in general *a model is not the real thing; it is an abstract representation of the real thing*. The representation may emphasize form and its relation to content, or functions and operations for serving a purpose.

Models evolve, and they change as new understanding is gained, with old models giving way to new ones that are more useful and productive in achieving the purpose for which they were constructed in the first place.

Modeling Process at Initial Stage

The modeling process at the early stage, to achieve a simple high level of abstraction, consists of these fundamental steps:

1. Establish the purpose of the model.
2. List the possible elements (observations, measurements, ideas) which may relate to the purpose, however remote.
3. Select those elements of step 2 which are relevant to the purpose in step 1.
4. Aggregate elements which can be chunked together by virtue of the strong structural, functional, or interactive connections between them. This is a process of classification, in a sense.
5. Repeat step 4 several times until a model consisting of seven, plus or minus two, chunks emerges.

The above process can be repeated by a subaggregation of each chunk in step 5 and, thus, the overall modeling process is conducted at various levels of abstraction. Here are some examples of this process.

EXAMPLE 1. *The Limits to Growth* [37]* is a book in which a team of researchers constructed a model to predict the future of the world population. A brief summary of their modeling process in terms of the preceding five steps is:

1. Purpose: Predict the relationship between actual population size and population carrying capacity of the world to the year 2100.

2,3. Possible (and relevant) elements relating to purpose: We leave it as a problem for the reader to list the possible elements which may relate to the purpose, and to select the relevant ones among them.

4,5. Aggregation steps: These steps led to the following five chunks or aggregated elements in the model:

Population
Capital
Food
Natural Resources
Pollution

The model includes, of course, much more detail. Relationships between the five elements are postulated to study the influence of, say, capital invested in factories and machines, in agriculture and farming, in food, pollution, population (i.e., fertility and mortality), or the effects of depletion of natural resources on population, etc. Each of the five major elements is

*The reader is urged to read this book. Therefore, the detail is rather sketchy here. The model will be discussed further in Chapter 9.

described in terms of elements which contribute to it. For example, population growth is influenced by the difference between birth rate and death rate (as shown schematically in Fig. 5-1); these are, in turn, influenced by other factors. For instance, birth rate per year is related to the population size and fertility rate. Fertility rate depends on effectiveness of birth control, the

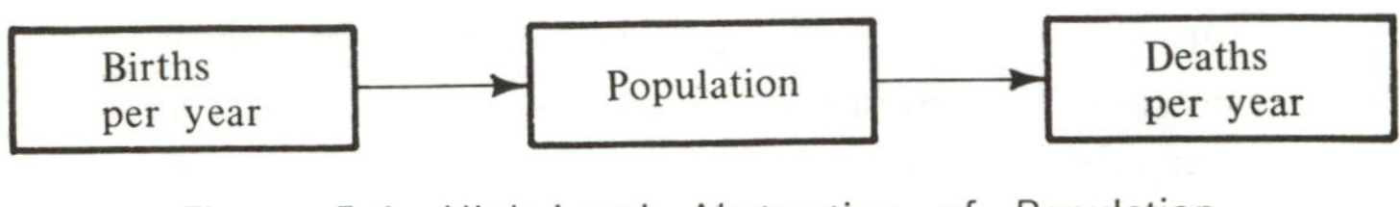

Figure 5-1 High-Level Abstraction of Population Model

desired birth rate by society, health care, food, pollution, etc. Similarly, death rate depends on the size of population and on mortality. Mortality depends on health care, food, pollution, etc.

The model of five main aggregates described here represents a high level of abstraction or aggregation. The researchers claim [37] that it was necessary for them to do so "to keep the model understandable." No questions of detail can be dealt with or answered by using the model. Only gross predictions regarding actual population and population carrying capacity compatible with the level of aggregation of the model can be considered.

The study predicts under what conditions* the trend in population size and capacity (as a function of time) is likely to be one of those shown in Fig. 5-2. Each time actual population exceeds capacity, the result is a subsequent decline in population. In Fig. 5-2(d), the sharp drop from a peak signifies a massive death rate.

Problem. Discuss any major elements that you would include in this model.

EXAMPLE 2. We now consider a more tractable example, the design of a tea kettle masterfully treated by Christopher Alexander [38]:

1. Purpose: Invent a kettle to fit the context of its use.
2. Possible elements relating to purpose: size; weight; handling: not hard to pick up when hot, not easy to let go by mistake, storage in kitchen; ease of water flow in and out, pour cleanly; maintain water temperature, i.e., water not to cool too fast; cost of material, material should withstand temperature of boiling water; not hard to clean; shape not too difficult to machine; shape compatible with material of reasonable cost;

*The conditions considered by the study include the five major aggregates: capital investment in industry, availability of food, rate of use of natural resources, resulting pollution, and population, both actual and possible.

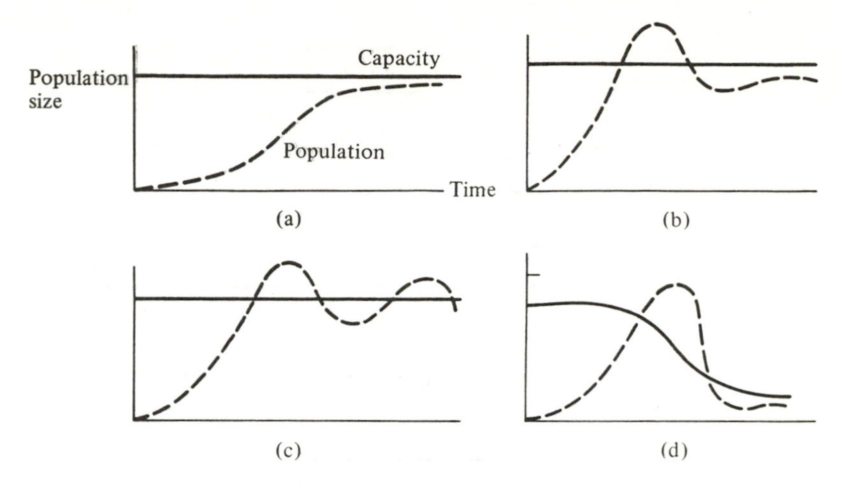

Figure 5-2 Population Size and Capacity Versus Time

cost of assembly; not corrode in steamy kitchen; inside not difficult to keep free of scale; economical to heat small amounts of water; can be used with gas or electricity; cost of use, namely, gas or electricity, not high; should have a reasonable life; should satisfy a large class of customers; safe for children, not burn out dry without warning, stable on stove when boiling; color of exterior, pleasant appearance in shape, compatible in shape and choice of colors with other common kitchenware.

3. Relevant elements: Let us consider all but the last three elements as relevant. Namely, and here comes aggregation, the aesthetic features of color and appearance are not considered relevant.
4. Aggregation: Use, safety, production, initial cost, cost of maintenance.
5. Further abstraction leads to the aggregation of the chunks in step 4 into two larger chunks of *Function* and *Economics*.

Alexander [38] shows the hierarchical model in Fig. 5-3. Note how the level of abstraction increases as we proceed from smaller to larger chunks. Fill in the items at the lowest level of abstraction in Fig. 5-3. How would you modify the model if your purpose was to purchase a kettle which will fit the context of its use by *you* only?

EXAMPLE 3. Figure 5-4 shows a model for the purpose of selection of a transportation system [39]. Discuss this model in the context of the five steps in the modeling process.

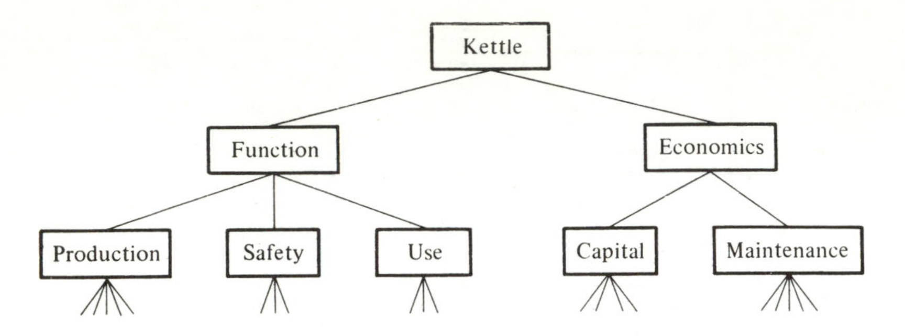

Figure 5-3 Model of Kettle Design Problem

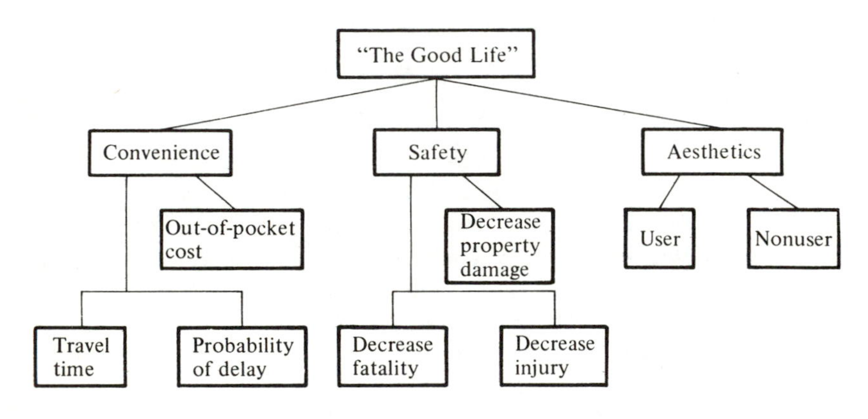

Figure 5-4 See [39].

Decomposition and Complexity

The nature of models is such that their construction requires a creative effort which is inductive. For example, if you consider the 28 elements listed in step 2 of the tea kettle model, then in the process of aggregation we may have to list 2^{28} subsets as potential elements in chunks. Namely, each of the 28 elements may be included or excluded in a subset. The creative effort avoids studying such a list which, incidently, contains more elements than the entire English language. The words of our language and associated concepts such as economics, function, and aesthetics are most helpful in the aggregation process. In principle, we attempt to simplify the real-world problem through modeling by aggregating elements that are strongly connected through structure and function, or both, and selecting the chunks so the connection between them in terms of structure and function is weaker than that inside. This is shown schematically in Fig. 5-5. This process of modeling

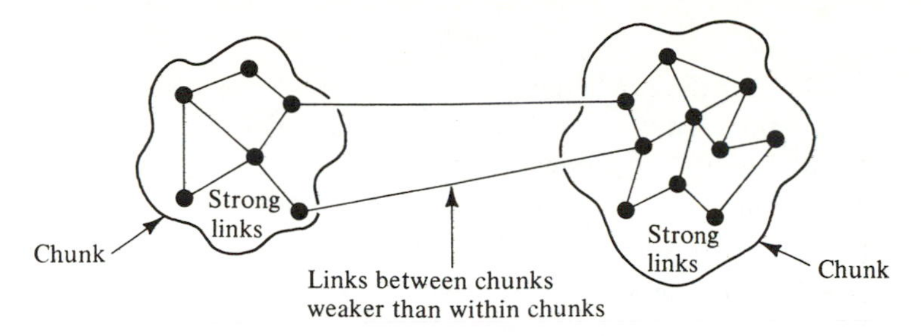

Figure 5-5

permits tearing or decomposing the model into smaller parts and enhances the ability to deal with complexity. The weakly linked parts are studied separately and then the complete model is synthesized from the constituent parts. Complexity is a function of the number of links between elements and not the number of elements. Therefore, to reduce complexity, we aggregate elements with many links into chunks that are then viewed as single elements, but in the chunking process we are also guided by the criteria of minimizing the number and strength of links between the chunks selected so they can be treated separately. This is the heart of the modeling process, the essence of its very nature which justifies its definition as an abstraction of reality.

5-4 VALIDATION OF MODELS

Errors of Omission and Commission

A model does not tell the whole "truth." At best it comes close to it. It may have much less content than the real thing by virtue of gross aggregation or because some elements of the real thing are ignored in the model. This inadequacy of models is referred to by Simon and Newell [40] as errors of omission or "Type I errors," if we borrow a concept from statistics which is discussed in Chapter 6. Addressing himself to this inadequacy, Whitehead [41] wrote: "If you only ignored everything which refuses to come into line, your powers of explanation were unlimited."

A second inadequacy results when we include in the model more content than the real thing. This is an error of commission or "Type II error" in the language of statistics (see Chapter 6).

Model Validation Process

To reduce the misfit between the model and the real world which it represents, the elements of the model are manipulated and the model is modified as a result of tests of its validity. The validation process has a recurrent pattern called the *scientific method*, and it consists of these steps:

1. Postulate a model.
2. Test the prediction or explanation of the model against measurements or observations in the real world.
3. Modify the model as a result of step 2 to reduce the misfit by detecting errors of omission or commission.

The validation process is shown schematically in Fig. 5-6. The figure may suggest that the process is never ending. In a sense, this is so; the validation always continues. However, a model that is supported continuously by evidence from observations and measurements, and requires no modifications, becomes a valid theory; the postulates of the model which represent an observed regularity then become *laws*. Kepler's laws are an example.

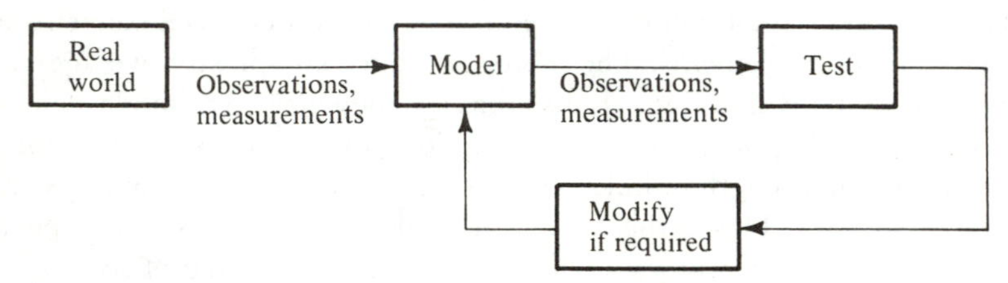

Figure 5-6 Model Validation Process

In some ways we may think of models in terms of Darwin's concept of evolution. The survival of a model in competition with rival models will depend on its advantage in terms of the degree of misfit between it and the test environment of validation. To carry the analogy further, then just as some species have a survival advantage in one environment over other species and little survival-value in a different environment, similarly a model suitable to one field may not survive the validation tests when applied to other fields, or may even fail to compete with new models in its field of greatest usefulness. The new models may result from creative discoveries, or the injection of random elements into the model as by mutation, to borrow from biology. Namely, a new feature may first be introduced in the model by chance and subsequently acquire great merit in meeting the test of validation. A classic example of such "mutation" is the concept of *atomic number*. When the chemical elements were first fit into a model, they were arranged in order of the atomic weights. The atomic weight was considered a fundamental characteristic. The form, i.e., ordering, and the use of an index number to signify the position in the ordering model was incidental, merely a matter of convenience. Letters could be used instead of numbers to serve the same purpose. Subsequent understanding gave these index numbers the status of properties which characterize the atoms in a more fundamental way than the atomic weights. They became identified with the number of electrons of negative charge orbiting the nucleus, or with the number of positive units of electric charge in the nucleus. This feature greatly enhanced the predictive capacity of the model and afforded the understanding of a much wider range of physical phenomena. The "mutation" enhanced the survival potential of the model.

Can All Models be Validated from Observations and Measurements?

Consider models that explain the creation of the universe. Such models depict the relevant elements and their interconnection which will cause, so to speak, creation. Or consider models for the creation of the solar system. We can consider such models without waiting for another creation in order to have observations which will cause a validation or modification of the models. Aspects of the models may be validated on a smaller, shorter-term basis. For example, Darwin's model for the origin of the species, with mutation and natural selection as basic features, has not been tested in total to predict new creatures as a form of model validation. However, aspects of the model have been validated. For instance, in the control of the rabbit population in Australia the model correctly predicted that a new strain of rabbits with more resistance to a particular disease than the original population will become dominant. Similar validation was observed in other small-scale tests of the natural-selection aspect of the model.

5-5 CLASSIFICATION OF MODELS

Form and Content

Two fundamental features characterize all models: *form* and *content*. It is possible to describe different contents by the same form of a model, and one content can be fitted into different model forms. The choice of form, which signifies here a form of representing the content, establishes the ease of manipulating the content and detecting errors of omission and commission. The choice of form, therefore, establishes the facility to refine and improve the model to better serve its purpose. In Chapter 2 we discussed the fact that the Hebrews and Greeks used letters to represent numbers. This mode of representation was not productive and useful for manipulation of the content which these forms represented. The same was true for Roman numerals. The invention of the positional number system provided a new form with much greater ease of manipulation. In Fig. 1-6 of Chapter 1, the magic square provided a form that facilitated manipulation for the game of number scrabble. In the chess problem of Fig. 1-5, a change in representation to the form shown in Fig. 5-7, in which the idea of the circular mobility of the knights is brought into focus, makes manipulation of content, movement of knights, so much easier that the solution becomes almost transparent. Figure 5-7(a) repeats Fig. 1-5 with the squares numbered (sequence is arbitrary). Figure 5-7(b) shows the same initial state in terms of the position of each

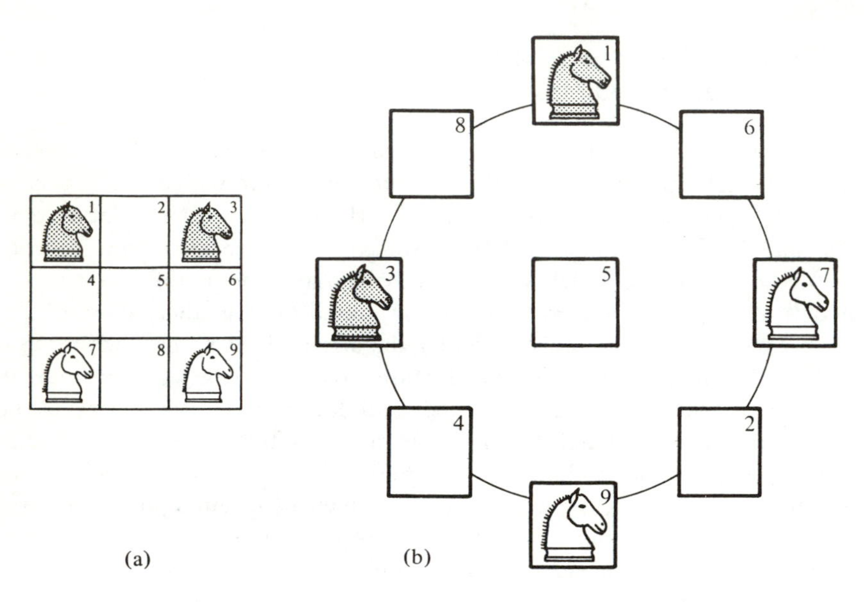

Figure 5-7 Change in Form of Representation of a Model

knight and those squares in the array into which each can move from the initial state. To change positions, each knight must move four steps in either a clockwise or counterclockwise direction.

The fundamental forms of models may be classified as:

Verbal
Mathematical
Analog

Verbal Models. Consider the statement: *The volume of yearly sales increases at a constant rate with increase in the number of salesmen. But the increase is less than proportional. There is a volume* a *of sales with no salesman (automated self-service, as sales through vending machines)*. The statement is a verbal model that can be used to predict sales. However, the predictions will be qualitative. For example, if we triple the number of salesmen, then sales should less than triple for $a = 0$. To be more explicit, we need a more accurate definition of the words "less than proportional."

Verbal models have advantages and disadvantages. Natural language can introduce ambiguity, more meaning than intended, or different meaning. As such, it may make the content obscure, but, on the other hand, it provides fertile ground for new ideas or new directions which enrich the field of possible manipulation. Consider the word *simultaneity* discussed in Chapter 1 and the role it had, by virtue of its ambiguity, to enrich the search for understanding and better models of the real world.

Verbal models can appear in conjunction with block diagrams, trees (as in Chapter 2), and other modes of geometric representations. The basic elements of the verbal model, *words*, constitute a third level of abstraction in terms of remoteness from the real thing they symbolize. The first level is a *sign* of the real thing; a visible head of a bull behind a barn is a sign of a bull. The second level is a *resemblance* or likeness in the form of a figure with features similar to the real thing, say, a drawing or picture of a bull. Words are symbols which are not signs, not resemblances (at least not after what we did to the bull in Chapter 2), but *symbols* which depend on a code as discussed in Chapter 2. It is this level of abstraction and associated remoteness which gives the symbols their power of consolidation, but also opens the door to ambiguity and richness. Remember "3" and "sheet" on the one hand, and "simultaneity" on the other.

Mathematical Models. The equation

$$Y = a + bX \qquad a > 0, \quad 0 < b < 1$$

is the mathematical form of the verbal model discussed above. Y stands for volume of yearly sales, X for number of salesmen, a is the level of sales when $X = 0$, and b is the constant of proportionality which is less than 1 but larger than zero. The mathematical model becomes more powerful when a and b are assigned explicit values. The above equation is a form called a linear equation in which Y is represented as a function of X. X is the *independent variable* which can be controlled, and Y is the *dependent variable* because its value depends on the choice of a value for X. a and b are *parameters of* the model. a is the value of Y for $X = 0$, and b represents the incremental change in Y for a unit change in X. The equation may represent many different contents such as the length Y of a spring as a function of a weight X suspended from its end, as shown in Fig. 5-8. Here a represents the length when $X = 0$, i.e., with no weight, and b is the change in length caused by one unit of X. b is referred to as a spring constant. a, b are then *parameters* of the model. The *parameters* characterize the spring in terms of two abstract elements that serve to predict the length of the spring from the model of the linear equations. The *parameters* can be changed, of course, to describe entire families of springs. The equation is limited in its modeling power. For example, Y cannot have negative values and, in fact, it has an upper and lower bound for which the

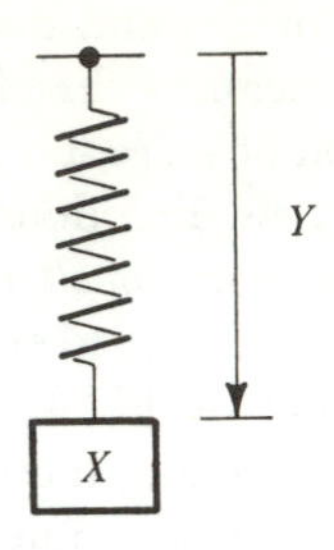

Figure 5-8

linear behavior holds. To describe more completely the behavior of the spring, a modified model will be required.

While verbal models tend to be qualitative, mathematical models are normally quantitative. For example, in studying the acceptance by society of new technologies, a verbal model may suggest that the acceptance is a function of the risk associated with the introduction of a new technology. A quantitative model studied by Starr [42] is described by a nonlinear equation of the form $R = f(B^3)$ in which R stands for risk in terms of fatalities per hour of exposure. B stands for benefit in terms of average annual income of dollars per person in the society. f signifies the word function, namely, risk (R) is a function (f) of benefit (B) raised to the power of 3. Starr's model proceeds to suggest that acceptable risk is inversely related to the number of people participating in an activity, and validates this hypothesis by studying the risk level and participating population in motor vehicles, commercial aviation, and general aviation. The model can be expressed mathematically and serve the purpose of predicting future trends in acceptance of new technologies.

It is of interest to note that in the early 1900s elevator accidents were very common, and it was not unusual for as many as thirty people to be killed in elevator accidents in a large city in one year. It was more risky to ride an elevator then than to fly in an airplane today. Starr indicates in his model that the risks associated with commercial aviation and motor vehicles at present are very close to the risk of disease for the entire U.S. population in terms of fatalities, when we use the product of people times hours of exposure as a basis in each case, i.e., fatalities/(number of people $\times$ hours of exposure).

We can consider a wide range of other mathematical models in which the central feature is an effort to eliminate the ambiguity of natural language in verbal models, and inject the opportunity for quantitative rather than qualitative assessment of elements in terms of their characteristic parameters and interconnections.

Analog Models [43]. The word *analog* is derived from the Greek word *analogia* which means proportion. This explains the idea behind the concept of an *analog model*. When we note a similarity in relations, we identify an analog model. For example, in the earlier discussion of this section, *sales relate to salesmen* is an analog model to *spring length relates to weight*. Namely, volume of sales is to length of spring as number of salesmen is to units of weight.

Many of the greatest creative efforts in modeling in history have had their origin in analog models of known systems. The clock was the analog to Kepler's model; the solar system, to Bohr's model of the atom; the water pump, to circulation of the heart (an earlier model was the analog of an irrigation system in agriculture with the blood absorbed in the tissues, no circulation); the steam engine, to models in thermodynamics; machine, to mechanical model of the universe; gambling, to probability models; games, to game theory models of decision under conflict; and currently the computer is beginning to serve as an analog model of the brain [44]. Analog models can be physical models such as electric circuits as analogs to mechanical systems, to hydraulic systems, to economic systems and others, or they can be conceptual models that take the form of mental pictures rather than material physical models. We can view geometric models as mechanical analogs of mathematical models. This analog which fused geometry and algebra is due to Descartes (see Chapter 1).

Analog models can be very powerful and productive; however, they may have the inadequacies that models in general have, namely, errors of omission and commission. We should normally consider only those features of the analog model which serve to describe the real-world problem of concern and not fall into a trap in an effort to make an analog model *complete*. Let us explain this point further through the following example from history. In the search for understanding the phenomenon of light, the analog models of water and sound were used. Water and sound were known to propagate in waves. The wave propagation takes place in a medium; water is the medium for water waves and air the medium for sound waves. This led to a search for a medium for the propagation of light waves. The medium was even given a name: ether. For years radio waves were referred to as waves of ether in many languages in the world, to the point where it became almost a tangible real thing "out there" in much the same sense that water and air are real things. The quest for completing the analogy led to a search for the ether, but its existence was never detected. We now know that light, and electromagnetic waves in general, need no medium of propagation.

Analog computers use electronic components to model automobiles, power distribution systems, and economic systems, to name but a few. Here the analog is represented by physical material elements. The digital computer can also be used as an analog model of other systems. But here the analog is more abstract. It is represented by the symbols in the program and not by the physical make-up of the digital computer.

Other Classifications

There are other classifications of models, such as physical versus abstract [45], in which the physical model is a material model; descriptive versus functional models, in which the functional model is operational as is a small model train (in which case the descriptive is a set of drawings of the train); causal versus correlative models, in which the causal model represents a cause-effect relation, while the correlative model merely indicates a statistical correlation between variables [46]. The causal model permits prediction and possible control of outcomes, the correlative model may be used only to forecast a trend. For example [46], the extent of damage in a fire is known to be correlated with the number of fire trucks at the site of the fire; the larger the number of trucks, the larger the damage:

$$\text{Fire damage} = f(\text{number of fire trucks})$$

This equation can be used as a model to forecast future trends: given the number of fire trucks at a site of a fire, we can forecast the damage. However, note that the fire damage *is not caused* by the fire trucks, but rather by the size of the fire. The size of the fire is the cause of the number of trucks on the site. The forecast model as presented could not be used to control future outcomes. In fact, if the above correlative equation is to be used as it stands, then to reduce the fire damage we should reduce the number of fire trucks dispatched to the site. This, of course, will achieve the opposite result. The causal model should be

$$\text{Fire damage} = f\left(\frac{1}{\text{number of fire trucks}}\right)$$

Namely, the more trucks that are dispatched to the site of a fire, the smaller the damage. The correlative model is based on observation only. The causal model is based on understanding the connection between the elements (variables) of the model to the point where it can be used for control and adequate prediction of outcomes. The choices of criteria by which a distinction between a mere correlation and causality is made is not a simple matter [47]. This can be understood in the context of conditional probabilities discussed in Chapter 4. Consider two events, A and B. If $A \rightarrow B$, then

$$P(B/A) = 1$$

and we may claim that A causes B. But suppose

$$P(B/A) = 0.70$$

Here we cannot make a claim of causality with certainty.

Models can also be classified as deterministic versus probabilistic, in which the probabilistic models derive their structure from probability theory and statistics, and deal with uncertainty. Probabilistic models are discussed in Chapter 6. More specific classes of models are decision models of Chapter 7, optimization models of Chapter 8, and dynamic models of Chapter 9. *Simulation models* are another class in which a sequence of events in the real world is acted out on the model. A rehearsal of a play is in this category, or flying a model airplane, or going through a sequence of computations to predict outcomes as a consequence of various inputs such as in the world model in the limit to growth described in Sec. 5-3. An example of a simulation model is given in Chapter 6.

Models can also be classified as isomorphic or homomorphic. In an isomorphic model, each element corresponds to only one element of the original system, and vice versa, i.e., each element of the original system corresponds to one element in the model. Figure 5-7(b) is an isomorphic model of the system in Fig. 5-7(a). In a homomorphic model, there does not exist a unique two-way correspondence between the elements of the original system and the model. The process of aggregation leads to homomorphic models. Figure 5-9 shows schematically the distinction between isomorphic and homomorphic modeling for seven elements of a system.

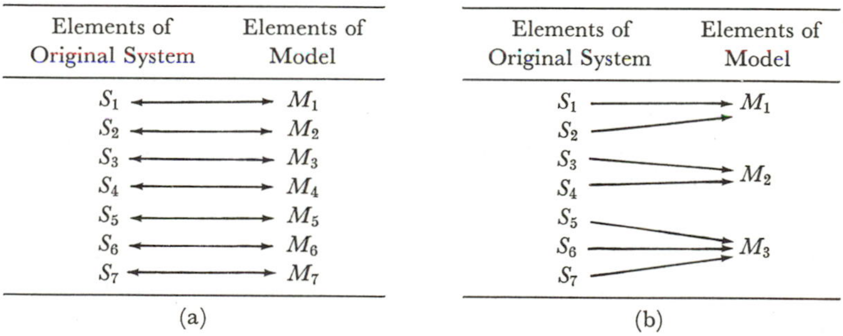

Elements of Original System		Elements of Model
S_1	←——→	M_1
S_2	←——→	M_2
S_3	←——→	M_3
S_4	←——→	M_4
S_5	←——→	M_5
S_6	←——→	M_6
S_7	←——→	M_7

(a)

Elements of Original System		Elements of Model
S_1, S_2	——→	M_1
S_3, S_4	——→	M_2
S_5, S_6, S_7	——→	M_3

(b)

Figure 5-9 (a) Isomorphic Modeling; (b) Homomorphic Modeling

The classifications of models discussed in this section are certainly not exhaustive. The main features which may be extracted are those of *form* of representation; *content* of real-world problem of interest; *level of abstraction* in terms of remoteness from real thing; *qualitative* and *quantitative* models. Most important of all, models have a *purpose*. It is the purpose which serves as a criterion for the choice of model from the wide range of model types. One type of model will replace another in the modeling process as intermediate objectives of modeling are stipulated. Thus, a change in representation may serve the intermediate objective of ease in manipulation or ease of computation on a digital computer.

Although there are many common features to the various classes of models, and although the modeling process can be guided by the fundamental concepts of aggregation and ease of manipulation discussed in this section, there exists no general theory (model) for the modeling process. The process is so ubiquitous and so creative in nature that any efforts to construct a general rigid and formal prescriptive model of the modeling process may prove counterproductive and stifling to the ingenuity of man as a model builder in dealing with his inner and outer environment.

The following sections give examples of models in various areas to serve as inspiration to the reader and give further emphasis to the truism that models are ubiquitous.

5-6 MODELS OF HISTORY

Models of history, as models in general, should be constructed *to understand* past events and possibly *predict* future events of historical character. We are quite familiar with the kind of history taught in some schools in which the model does not contain a foundation for possible understanding and prediction. This model consists of an orderly arrangement of events aggregated in terms of successions of dynasties, battles, strong and weak leaderships, strong and weak social justice, strong and weak morality, etc. The succession is orderly but sterile. It is usually not productive as a framework for prediction of the future.

There are, however, more sophisticated models of history. The *geographic model of history* originated with the Greeks. This model claims that the attributes of the geographic terrain, in terms of topography, soil, water, and general climate, are the fundamental elements that contribute to the formation of character of people. Man's social institutions, his individual morality, and social justice are explained in terms of these elements of the physical environment. Events of history are then explained in terms of these parameters. For those who consider this model useful, the predictions for future events may be based on the fundamental elements of the physical environment of future societies, whether natural or man-made. It is also possible that subscription to the model is limited to a society dominated by scarcity and characterized by a primitive technology which is in no position to mold the physical environment and create a man-made world. If this is the case, then the model may serve to explain events of the distant past (or present) where scarcity and primitive technology were (or are) the order of the day, but it cannot be used to predict the future of an affluent, highly industrialized society.

Marx proposed an *economic model of history* in which its course is explained by the way goods are produced and distributed. According to this model, the way goods are produced shapes a people's cultural and social institutions, their religion, morals, ethics, system of social justice, values, and total pattern of behavior. If the geographic model was predicated on the concept of the "*environmental man*," then the economic model focuses on the concept of the "*economic man*" as the most relevant element of characterization to understand history. The economic model may also have its limitations and may apply more to societies of the past than of the future. In societies where the basic needs of food and shelter are almost guaranteed by a welfare state, other factors may become more dominant in shaping the course of history. The drifting of the capitalist and communist societies closer to each other, despite fundamental differences in the production and distribution of goods, is a sign that the economic model needs modification.

Freud suggested a *psychoanalytic model of history* on the basis of what we might call the "*psychological man*." The model focuses on man's unconscious hostilities and his ability to repress them as the most relevant links to the course of history. Only when man has learned to control his impulses to murder, incest, sadism, unrestrained sexual gratification, and all forms of violence, can he release energy for creative endeavors along channels leading to culture, art forms, social institutions, and a civilization that leaves its mark on history. The course of history is determined by the kind of unconscious drives which are repressed or controlled, and by the pattern used to achieve this control.

Analog models of history were suggested by Spengler [48] and Toynbee [49]. These are philosophical models that view history as an analog to a living thing. Each civilization has a predictable pattern in terms of the analog. Spengler (1880–1936) predicted in *The Decline of the West* that the western civilization was in the winter of its life cycle and would die by the twenty-third century. Other civilizations in the spring of their life cycle (such as Slavic or Sinic) would supersede it. Spengler's model of history is referred to as a *cyclic model* because it considers the life cycle of civilization as an analog to annual flowers which emerge in the spring, mature in the summer, grow into the autumn, then decline into winter, and finally die. Or the analog may refer to man's life cycle with the milestones from cradle to death. First, the spring of life, maturity in the summer marked by peak physical strength, autumn marked by a high point in intellectual achievement, and decline into winter with progressive deterioration and decline in physical strength and intellectual capacity to ultimate death. Spengler studied a number of civilizations to test the validity of his model: Egyptian, Babylonian, Chinese, Indian, Arabian, and Western. He found much uniformity in patterns in the civilizations which he studied. The spring was characterized by a rural economy and receptive attitude to religion. The summer was marked by urbanization, rational attitude in philosophy and religion; science and art flourished. In the winter technology expanded and the giant cities with their enormous architectural creations made their appearance. It was the phase of the megalopolis, crowding; the age of large-scale wars of destruction and annihilation; a period of emphasis on the material and practical, and decline in creative productivity in science, art, and literature. No great men and no heroes characterized the winter season; these were treasures of earlier seasons in the past. At the last phase the civilization collapsed under attacks from within and from outside amidst attempts to return to the religion of the spring.

The validity of the model with respect to the Western civilization is controversial. The prediction of megalopolis, wars of large scale, and expanded technology are quite accurate, considering the fact that Spengler's book was first published in 1918, but the model was conceived by him before World War I which started in 1914. But the creativity in science, in particular since 1914, has been marked by achievements not predicted by Spengler's model.

Toynbee [49] proposes a linear evolutionary model of history. A civilization progresses by evolution from primitive to more sophisticated forms. A civilization can continue to survive if it adapts to new challenges by changing its form and content and responds in new and novel ways to match the new challenges. A civilization dies when it fails to change in form and content to meet new challenges.

Other models of history are the great personality models which claim that history is shaped by the moving forces of great men. These men create the events of history. There are also religious models of history, which claim that history is shaped by what happens between God and man. In some models God interferes directly in shaping history; in others, such as the existential theologian's model, God does not interfere directly in shaping history—it is the relation which man thinks exists between him and God that shapes history.

5-7 MODELS OF THE UNIVERSE

The quest for understanding the origin of the universe has its roots early in recorded history. The biblical story of creation is one of many models constructed by man to understand where it all came from in terms of a larger frame, a more complete context, and perhaps *The Cause* of all causes. In Chapter 1 we discussed Newton's search for absolute space as a search for *The Cause* of all causes.

Observations made since the early part of this century, and measurements of the light coming from distant galaxies, made it possible to establish their distances from our galaxy. Further understanding of the spectral characteristics of the elements led to the concept of the expanding universe. The distribution of colors of light emitted by an element, such as an excited hydrogen atom, characterizes the element in terms of frequencies of the various components of the spectrum. It was observed that the more distant the galaxy, the more red the light emitted by hydrogen gas tends to be. In other words, the frequency of the spectral components was decreasing. This could be explained in terms of the *Doppler effect*. When a source of a cyclic phenomenon (light, sound), emitted at a particular frequency, is moving toward an observer, the observer notes an increase in the frequency of the phenomenon; similarly, when the source moves away from the observer, the frequency decreases, i.e., the time spacing between cycles of the phenomenon gets longer. The more reddish colors of the hydrogen spectrum of the more distant galaxies, called the *red shift*, can be explained by the Doppler effect. The more distant a galaxy, the faster it moves away, i.e., the universe is expanding in all directions.*

The red shift and the associated interpretation of the expanding universe led Gamow [50] to his *big bang model* of creation. In this model, the universe had its origin in a giant explosion of closely packed matter several billion years ago. All the galaxies are fragments of this explosion. The faster a galaxy, the further it is, and the faster it continues to travel away in the expanding universe. Gamow contrived a simple analog model in the form of a balloon marked by dots. As the balloon is inflated the dots move with respect to any reference dot, and the further a dot is from the reference dot, the faster it seems to move away or recede. A simpler linear model may be viewed as many balls all lined up at the beginning of an experiment along parallel grooves. At a given instant the "big bang" takes place and each ball is given a different initial velocity. As time goes on, each ball maintains its velocity, but each observes all its neighbors receding from it. The ones that are faster continue to get further away ahead, and the slower ones continue to stay further behind. The Gamow model of the big bang assumes that the total amount of matter in the universe is conserved and remains constant.

An alternate model was proposed by Hoyle [51] in which the amount of matter in the universe increases steadily. New galaxies are formed from new matter, and although the galaxies recede, moving away from each other, the average separation of the galaxies remains the same and the universe of the past looked the same as it appears today. No tests have yet been devised to prove the validity of either rival model and thus provide further understanding to make a choice between the two.

Models of the motion of the stars and planets have been of great concern to man throughout the ages. Man's ego and his preoccupation with his

*Other explanations are also possible, such as a change in frequency with distance of travel or perhaps the atoms emitting the light were different when the light was emitted.

importance caused him to place Earth in the center of the universe for many generations in the past. Thus, an early Greek model considered the stars as sources of light fastened to a revolving shell of a hollow sphere with Earth at its center. When some of the brightest stars were observed not to fit the model because they wandered among the constellations, the single shell was supplemented by additional spherical hollow shells of different radii, one for each of the wandering stars. These wandering stars were called the planets, which is Greek for "wanderers." The model still failed to predict certain phenomena. Some of the planets were observed to move backwards with respect to the pattern of the distant fixed stars. Copernicus (1473–1543) proposed a new model in which Earth and the other planets moved in circular orbits around the sun. When Earth passed a slower moving planet, the planet appeared to be moving backward. It was quite difficult to reject the fundamental notion of man and his earth as the center of the universe, but in time the massive evidence of new observation caused even persistent skeptics to reject the deeply rooted notion of *man in the center of the universe*. Kepler (1571–1630) formulated his celebrated model in the form of three main elements that became known as Kepler's laws because of the great deal of evidence which validated the model continuously since its inception. Kepler's model, for the solar system, contains the following laws:

1. Each planet revolves in an elliptical orbit with one focus in the sun.
2. The line from the sun to a planet sweeps out equal areas in equal times (thus, a planet moves fastest when it is closest to the sun).
3. The square of the period of a planet (time to complete its orbit around the sun) is proportional to the cube of its greatest distance from the sun.

The predictions of Kepler's model have been validated with so much accuracy that no modifications have been introduced in it. The model was later *understood* in the context of a larger frame and a more general model when Newton (1642–1727) introduced his celebrated laws of motion and the law of gravitation mentioned in Chapter 1, and repeated here for completeness:

Newton's Three Laws of Motion:

1. A particle not acted on by an outside force moves in a straight line with constant velocity.
2. When a force acts on a particle, the particle moves in the direction of the force, and the force equals the time rate of change of momentum of the particle.
3. For every action (force) there is a precisely equal opposite reaction.

Newton's Law of Gravitation: Every particle attracts every other particle with a force that is proportional to the product of their masses and the inverse square of their separation.

Validation of a Model—An Example [33]

Kepler knew of six planets revolving around the sun. In 1781, about 150 years after Kepler's death, the astronomer Herschel observed a seventh* planet, Uranus, revolving beyond Saturn. The observed orbit of Uranus did not agree with the predictions of Newton's model.† Astronomers hypothesized that the deviations of Uranus from the orbit predicted by the model must be attributed to attraction from a planet beyond Uranus. Leverrier, a French astronomer, investigated this hypothesis and predicted the existence of an ultra-Uranian planet. He mapped the orbit of this planet on the basis of the orbit of Uranus, using Newton's model, and assigned its position in the skies. He wrote about his findings to a colleague at a well-equipped observatory and asked that the planet be observed on the night of September 23, 1846. The letter arrived on that day, and the same evening the new planet was found within 1 degree of the location predicted by Leverrier. The mass and orbit were later determined to agree also with Leverrier's prediction. Using the tools of Chapter 4, we can quantify the enhancement of credibility for Newton's model on the basis of this new evidence of validation. Let

$P(M)$ = Credibility of Newton's model before evidence
$P(E)$ = Probability of evidence occurring
$P(M/E)$ = Credibility of Newton's model given evidence
$P(E/M)$ = Probability of evidence given Newton's model to be true.

From Bayes' theorem

$$P(M/E) = P(M)\frac{P(E/M)}{P(E)}$$

$P(E/M)$ can be taken as close to 1. Then

$$P(M/E) = P(M)\frac{1}{P(E)}$$

The enhancement in credibility of the model M is inversely related to $P(E)$. This probability can be considered equivalent to the probability of a random meteor appearing in a spherical cap of 2 degrees (1 degree all around the center) on the shell of a hollow sphere. With all 2-degree spherical caps on the shell being equally likely to be the location, i.e., symmetry, the probabil-

*Hundreds of minor planets have been observed since then to revolve between the orbits of the major planets.

†Newton's model is referred to here as a more general model which contains Kepler's model for the solar system.

ity is computed as the ratio of the area of the 2-degree cap to the surface area of the entire shell, and yields 0.00007615. The credibility is enhanced by more than 10,000-fold. That is to say, even the great skeptic who would assign $P(M)$ the value of 0.00001 would be shaken from his skepticism, and the one who would assign $P(M) = 0.00007$ would become almost a "fanatic" believer in the model.

The ultra-Uranian planet predicted by Leverrier is the planet Neptune. It is interesting to note that, after the discovery of Neptune, irregularities were observed in its orbit. These were also explained by Newton's model and were attributed to the existence of a small planet beyond Neptune. This small planet is Pluto, which was discovered in 1930.

Leverrier also observed that the planet Mercury, which is closest to the sun, displayed irregularities in its orbit. Specifically, the perihelion, which is the point of its orbit closest to the sun, was precessing around the sun describing a rosette path. Astronomers were overwhelmed by the mounting evidence that was continually enhancing the credibility of Newton's model. The model already explained the irregularities in the orbit of Uranus by predicting the existence of Neptune. Then it explained the irregularities in the orbit of Neptune (after its discovery) by predicting the existence of Pluto. It is not surprising, therefore, to learn that astronomers predicted the existence of another planet in the vicinity of Mercury to explain the irregularities observed in its orbit. So strong was their conviction that they even named this planet Vulcanus. To the present day no trace of Vulcanus has ever been found.

This was one of a number of observations that started casting some doubt on the domain of validity of Newton's model. It took the general theory of relativity to explain the irregularities in the orbit of Mercury. These irregularities were attributed to the curvature of space-time in the neighborhood of a large mass, such as the mass of the sun. (See Fig. 5-10 for the proximity

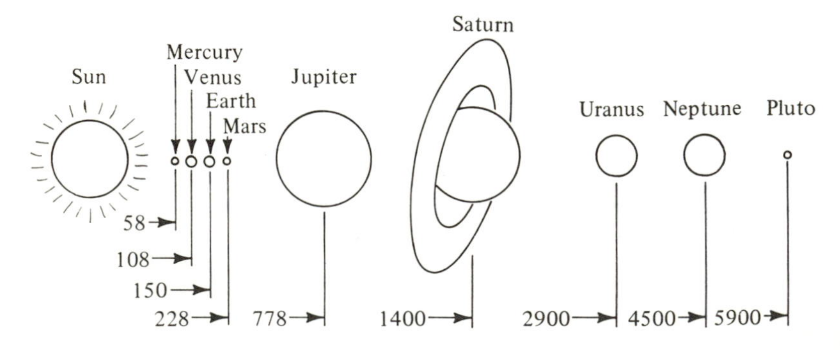

Figure 5-10 Planets and Their Distances from the Sun in Millions of Kilometers

of Mercury to the sun.) The planet Mars also exhibits a precession of its perihelion, but this precession is smaller than that exhibited by the planet Mercury. The model of general relativity took over at the point where Newton's model reached the limit in its ability to explain and predict natural phenomena. Newton's model became, then, a subset of the more general and more powerful model of relativity.

Just to refresh your memory, Table 5-1 lists the planets and their distances in millions of kilometers from the sun (see also Fig. 5-10).

Table 5-1

Planet	*Distance from the Sun in Millions of Kilometers*
Mercury	58
Venus	108
Earth	150
Mars	228
Jupiter	778
Saturn	1,400
Uranus	2,900
Neptune	4,500
Pluto	5,900

*5-8 MODELS OF THE ATOM

From our study of descriptive, functional analog models of small scale, by means of which we try to comprehend the enormous size and extent of the universe, we proceed now to the study of the domain of matter in its most minute (microlevel) dimensions, and attempt thereby to comprehend the elementary structure of materials by analog models that are of much larger scale than their counterparts in nature. The solar system has been considered as an analog to the atom, and vice versa.

In 1913 Niels Bohr proposed a model for the hydrogen atom:

- A negatively charged electron revolves in a circular orbit around a massive, positively charged nucleus.
- The orbit may be computed from Newton's model by replacing gravitational attraction by attraction of electrical charge.
- The electron can occupy any of a number of orbits. The possible orbits are those for which the angular momentum* of the electron is an integral multiple of Planck's constant h divided by 2π. (The significance of h is discussed below.)

*Angular momentum is the product of mass m and angular velocity $\omega = v/r$ in which v is the linear velocity and r the radius of the circular orbit.

- The electron radiates a photon (see below) of light energy when it makes a transition from a larger to a smaller orbit. The energy of the photon depends on the frequency of the transition, and it establishes the color of the light. The energy emitted is equal to the decrease of the electron energy in going to the smaller orbit. To excite an electron from a smaller to a larger orbit, energy must be supplied to the electron. The electron is normally in the smallest orbit representing the lowest energy level.

The constant h, known as Planck's constant, was postulated by Max Planck (1858–1947) as a model to understand the observation that all solids, when heated in a cavity to a glow, emitted light which changed from a pale orange and yellow to a bluish white in all cases. Planck postulated that radiant energy comes in discrete chunks, not continuously. The amount of energy is equal to the frequency of the light multiplied by the constant number h. The frequency of the light determines its color. Blue light has about twice the frequency of red light. Therefore, the discrete amounts of energy, which Planck called *quanta* and are now called *photons*, differ for different colors of light. A photon of blue light has about twice the energy of red light. Einstein validated this concept of discrete quanta of light energy (photons) by studying the photoelectric effect, namely, the behavior of electrons ejected from a metal surface by incident radiation of a light source on it. Einstein received the Nobel prize for this contribution.

Bohr's model served to explain why hydrogen gas, the lightest and simplest of the elements (one electron), gave off a spectrum of light colored from a bright red to blue. The Bohr model predicted the behavior of hydrogen and helium (with two electrons) very well, but for lithium (three electrons) and heavier elements, predictions by the model became less and less accurate, and the model required modifications.

Planck's model led to the idea that light behaves both as a particle and a wave. In 1924 Louis de Broglie suggested that a moving particle might have wave properties, and he proposed a method of calculating the wave length λ of an electron by dividing Planck's constant h by the momentum, p, of the electron, i.e., $\lambda = h/p$. Note that until de Broglie proposed his model the electron was thought of as a particle, although no one had seen it as such, or has to the present day; *the characterization of the electron is merely a model of its manifested behavior.* Later experiments by Davisson and Germer in 1927 validated the possible wave properties of the electron.

The wave aspect of the electron also fit Bohr's model. The various possible orbits have circumferences that fit an integer number of electron wavelength. Namely, the smallest orbit is 1 wavelength in circumference, the next orbit is 2 wavelengths in circumference, and so on. The mental picture of the model is not a particle in orbit, but a wave spread around the entire orbit. To establish the location of an electron, it must be observed. The observation may take place by impinging light on it and assessing its position from the reflected energy of light. But this process disturbs the original position of the electron which we wished to measure. Therefore, a certain degree of uncertainty in the measurement of position will always exist. The same will hold true for measurements of momentum, energy, and time. This led Werner Heisenberg to postulate in 1927 his *Uncertainty Principle* which states that the inaccuracy Δx in the measurement of position x and the inaccuracy Δp in the measurement of momentum p of an electron are limited by the inequality

$$\Delta x \cdot \Delta p \geq h/2\pi \tag{5-2}$$

($h = 6.626 \times 10^{-34}$ joule-sec; 1 joule $= 10^7$ erg
1 erg $=$ 1 dyne $\times$ 1 cm
1 dyne $=$ force which will produce an acceleration of 1 cm/sec^2 on a mass of 1 gram

The inequality can be restated in the form

$$\Delta E \cdot \Delta t \geq h/2\pi \qquad \text{or} \qquad \sigma_E \cdot \sigma_t \geq h/2\pi$$

where E is kinetic energy and t is time. These results are derived from Eq. 5-2 as follows:

Multiplying the left side of Eq. 5-2 by the identity $(\Delta x/\Delta t) \cdot (\Delta t/\Delta x)$ and recalling that $p = m(\Delta x/\Delta t)$, then

$$\frac{\Delta t}{\Delta x} \Delta x \cdot \Delta m\left(\frac{\Delta x}{\Delta t}\right)^2 \geq h/2\pi$$

$$\Delta t \cdot \Delta E \geq h/2\pi$$

where E is kinetic energy and t is time.

Δx can be thought of as the standard deviation σ_x (see Chapter 6) of position x; and, similarly, Δp as σ_p, ΔE as σ_E, and Δt as σ_t.

The models described so far in this section, including Bohr's model and the modifications introduced by de Broglie and Heisenberg, led to the development of the *quantum mechanics model* of energy and matter with prime contributions by Heisenberg and Schrödinger. In quantum mechanics the position of small particles is specified by a probability distribution as a function of space and time. For large objects, the model reduced to the *classical mechanics model* of Newton in which position is exactly specified (zero variance or $\sigma^2 = 0$). In quantum mechanics the wave character of the electron is representative of the probability distribution of its position around the orbit described by the Bohr model.

I still recall the days when the nucleus of the atom was defined as the smallest fundamental element of matter which is characterized by the fact that it cannot be further subdivided. This model has been discarded. Efforts are still going on to model the nucleus in terms of the multitude of more fundamental particles that comprise it.

5-9 A MODEL OF THE BRAIN

In an article entitled "What Kind of Computer Is Man?" Earl Hunt [44] proposes the model shown in Fig. 5-11 which he calls a *Distributed Memory Model*. The main element of the model is the long-term memory

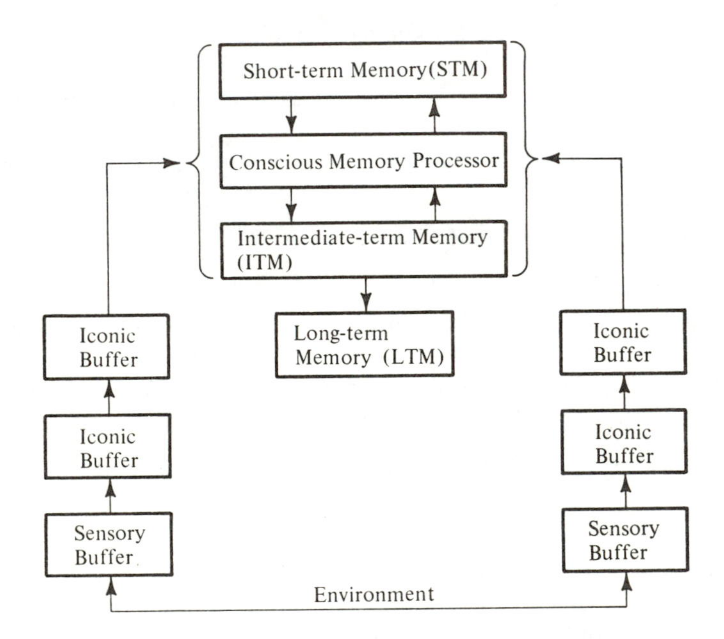

Figure 5-11 Model of Distributed Memory of Man [44]

LTM in which information is stored permanently. The LTM is surrounded by buffers, temporary memory units, which have associated computing devices in the form of neural circuits to examine information. The *sensory buffers* examine raw information from the environment and code it in a fixed manner. First, an analog signal (light, sound, pressure) is detected by a transducer mechanism that transforms it to a digital code. Namely, messages of light, sound, and pressure are first transduced in a digital electric signal. The digital code is then examined in the sensory buffer by a feature detector looking for patterns of zeros and ones, i.e., off/on elements. The sensory buffer transmits a feature which classifies the raw data entering the brain in a reduced form by matching it with a feature which is stored in the brain. The coded features, such as curves, straight-line segments, etc., pass through intermediate or iconic buffers. The intermediate or iconic buffers code the received information under the control of programs and data in the LTM, using such abstractions as transition from collections of lines to letters, from letters to words, and words to sentences.

The parallel tracks of peripheral buffers in the model of Fig. 5-11 indicate that the brain can process several sensory paths simultaneously. The tracks terminate at the conscious memory which contains a *processing unit*, and two memory units, *short-term*, STM, and *intermediate-term*, ITM. The conscious memory can receive information from several sources and can selectively block information from sources. All input to the concious memory is through the STM which is smaller and faster than the ITM. The ITM is the only unit that transmits coded information to the LTM. The ITM stores a general picture while the STM holds the most recent picture of the received input.

The Model of the Brain and Problem Solving

The influence of the digital computer is evidenced in Hunt's model. Hunt uses his model to explain learning, verbal comprehension, concept identification, and problem solving. The problem-solving process follows the state-space searching in terms of a means-ends pattern [52] discussed in Chapter 1. In problem solving, Hunt [44] considers the data describing the states and operators being considered at the moment to be stored in the STM. The ITM contains the list of states visited and information about frequently used operators. The LTM stores rules which define operators and states. Computer problem-solving programs focus attention on generating a process that minimizes the number of states visited before a solution is reached.

Models of the brain will continue to be modified, with the computer serving as a prominent analog model. The computer can be programmed to modify its own program and, thus, the computer can be programmed to learn. The details of the learning process will depend on the experiences, such as games of chess the computer may play with human opponents. These details will not be known in advance to the programmer, nor will he be aware of them when they occur, any more than he is aware of the minute elements of his own thought and learning process. Such a learning computer program represents a higher level of abstraction and aggregation than the details of the neural activity in the human thinking process. The program could be designed, however, to retain all the minute details of its learning process, something human beings find difficult to do because it requires a process of reliable and comprehensive introspection. This, in turn, could lead to a better understanding of what is going on in the human mind in learning and problem solving.

A Pragmatic Plan for Remembering

Man has been concerned with the practical art of improving the efficiency of the memory (mnemonics) for many centuries. Various schemes were devised which normally contain two main features: learn to pay attention and learn to organize raw information. Patterns form the framework for organization which may prove most useful in efforts to memorize raw unrelated information. The pattern may take the form of a rhyme or a picture associating unrelated items. All mnemonic techniques try to relate the information to be memorized to a pattern that is already stored in our long-term memory. The following is an example taken from Miller, Glanter, and Pribram [53] which illustrates the power of patterns in memorizing. To appreciate the power of this, memorize the following list:

one is a bun,
two is a shoe,
three is a tree,
four is a door,
five is a hive,
six are sticks,
seven is heaven,
eight is a gate,
nine is a line,
ten is a hen.

Now consider a list of ten unrelated words which are given one at a time. As each word is given, form an association with the corresponding words in the above list no matter how bizzare the association. For example, given the first word *hand*, say to yourself, "One is a *bun* in the *hand*." For a second word *house*, say, for instance, "Two is a *shoe* in the *house*." For a third word *cat*, say, "The *cat* is on the *tree*," and so on. Try to do this with the following list of ten words: book, radio, lamp, pencil, boy, floor, box, horse, cup, pot. Take about five seconds to form an association for each word, then stop and relax for a few minutes. Then ask yourself what the sixth, the fourth, ninth, etc., words were in any random order. You will be amazed how much you can remember when you use a pattern, or a plan, for remembering.

5-10 MODELS IN ENGINEERING

A fundamental distinction between models in science and engineering has its roots in the tradition that science addresses itself to the natural world while engineering is primarily concerned with the artificial man-made world. While science generates models to understand the way things are in the natural world, so we can predict more accurately how they will be, engineering deals with how things *ought to be* and generates models for man-made systems to achieve that which ought to be, i.e., the stipulated goals. Engineering uses the validated models of science to manipulate parameters and establish bounds on what models are possible for artificial creations in terms of such principles as conservation, continuity, laws of mechanics, thermodynamics, and others. The heart of engineering modeling can best be described as a design synthesis process in which an objective end goal is stated and a system is to be synthesized to achieve the goal. A preliminary model is examined via a means-ends analysis by subjecting it to a model of the environment which is most representative of that in which the real system will reside, and observing how close the system behavior fits the goal behavior. The detection of misfits leads to modifications in the parameters of the model and possibly to complete changes in the conceived system. Once an acceptable model is synthesized, alternate acceptable models are conceived and from all these acceptable or feasible models, which meet the constraints on the behavior of the system, one is selected as best in terms of some criterion such as least cost, maximum benefit-to-cost ratio, maximum reliability of performance, minimum risk to users, maximum expected utility, or other possible criteria. The search for the best choice among alternatives is facilitated by the use of mathematical programming models that are discussed in Chapter 8. However, it must be emphasized that the "best" is selected from the alternative models conceived. In most problems of reasonable complexity there is no assurance that *THE BEST* alternative has been included in the list of feasible alternatives, namely, the list is not exhaustive.

This preoccupation of engineering with what ought to be, and a search to achieve it, demands models that can be manipulated with relative ease. Engineering models must, as a first requirement, be *working models*. This may lead to gross simplifications in the form of aggregations of variables and properties of a system, but no sight is lost of the fact that a model must serve to illuminate, i.e., serve to better understand and predict behavior of the ultimate real thing, *but it is not the real thing*. Therefore, a primitive model is first generated for ease in manipulation and is progressively modified and refined until confidence is gained in its usefulness to understand and predict the behavior characteristics of the real system.

In this section we introduce some concepts and terminology which are useful in a wide range of engineering models, because they serve as guides in classifying relevant factors in the synthesis process to facilitate the construction of models useful for manipulation.

In a broad sense, engineering problems can be classified as problems of analysis and problems of synthesis or design. Analysis is normally a link in the design problem. This classification was mentioned in connection with problems in general in Chapter 1. To clarify the distinction, consider the *system* of Fig. 5-12. The word system is used here to identify abstractly a

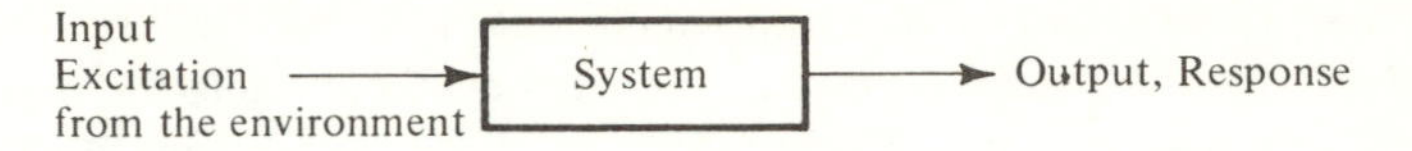

Figure 5-12 Abstract Representation of a Physical System

relatively complex assembly of physical elements characterized by parameters which can be measured. The boundaries of the system are well defined and its behavior in terms of its response to excitations (disturbances) from the environment is predictable in terms of the parameters.

Analysis. The analysis problem requires the prediction of a response of a prescribed model of a system to a given input. The system model may be given in terms of a set of parameters and associated relationships which link the excitation to the resulting response. The analysis problem has usually a unique solution, namely, to each input excitation there corresponds one, and only one, output response. The analysis involves one of a number of prescribed methods to generate the solution. The method may consist of inverting a matrix of coefficients, using an iterative method, solving a differential equation, etc.

Synthesis. The synthesis problem begins with a statement of what ought to be the response to a prescribed excitation. It is required that a system be synthesized so that the desired goal be achieved. The parameters of the synthesized system are not a unique set; there may be a number of different alternative systems that will be acceptable. To check the acceptability of any proposed model, an analysis is performed to predict the response of the model to the prescribed excitation and compare it with the desired response. Thus, analysis is a link in the synthesis process. To select one system from the possible alternative candidate systems, a criterion of choice must be introduced. This could be minimum cost, minimum weight, maximum safety, or others.

Control. A problem somewhat related to analysis is the *control problem.* In the control problem the system and outputs are specified, and it is required to find the inputs which will cause the desired output.

Identification. Identification is related to the synthesis problem. Given measured input and output of an existing system, it is required to identify the system in terms of the parameters which will link the measured input and output.

Variables. The variables of engineering systems can be classified as *across* and *through* variables. An across variable (also called *a potential*) relates conditions at one point to another reference point in or out of the system. Displacement, velocity, and voltage are across variables. A through variable (also known as *flux*) is defined at a single point in the system. Force and current are through variables. The system elements are characterized by parameters which relate input and output variables in the system. The input may consist of through or across variables, or a mixture of the two. The same holds true for the response.

Parameters. The system parameters may be classified in terms of their energy transfer mechanism as energy dissipators or energy storing devices for potential and flux. (See Sec. 9-6 for examples of energy dissipation and energy storage devices.)

System Behavior. Two fundamental principles of conservation and continuity from physics govern the behavior of physical engineering systems. The conservation principle states that when a certain quantity is applied to a system it must be conserved. In mechanical systems momentum (product of mass and velocity) is conserved; in electrical systems electrical charge is conserved. The continuity principle requires that the through variable must emanate from a source and return there or to another source.

The abstraction of classes of variables, the classification of parameters in terms of energy transfer characteristics, and the principles of conservation and continuity describing system behavior serve as an aggregation of concepts useful in a number of areas. Models of systems in these areas result in similar form, inasmuch as the mathematical equations representing the model are concerned. Figure 5-13 shows a dynamic system and an electric system and the associated equations. Note the similarity and analogy between the two equations.

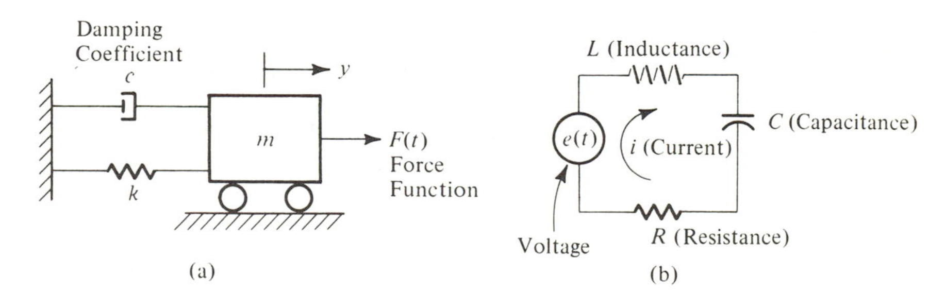

Figure 5-13

Equation for Figure (a):

$$m\frac{d\dot{y}}{dt} + c\dot{y} + ky = F(t)$$

Equation for Figure (b):

$$L\frac{di}{dt} + Ri + \frac{1}{C}\int_0^t i\,dt = e(t)$$

Force $F(t)$	analogous to	Voltage $e(t)$
Velocity $\dot{y}$	" "	Current i
Displacement $y = \int_0^t \dot{y}\,dt$	" "	Charge $q = \int_0^t i\,dt$
Mass m	" "	Inductance L
Compliance $1/k$	" "	Capacitance C
Damping c	" "	Resistance R

EXAMPLE. To illustrate the difference between analysis and synthesis through a simple example, consider the two-story building of Fig. 5-14. The forces F_i and static displacements u_i at the two floors are related through the system parameters in a stiffness matrix $[k]$:

$$\begin{Bmatrix} F_1 \\ F_2 \end{Bmatrix} = \begin{bmatrix} k_{11} & k_{12} \\ k_{21} & k_{22} \end{bmatrix} \begin{Bmatrix} u_1 \\ u_2 \end{Bmatrix} \tag{5-3}$$

or, in general, for n floors:

$$\{F\} = [k]\{u\} \tag{5-4}$$

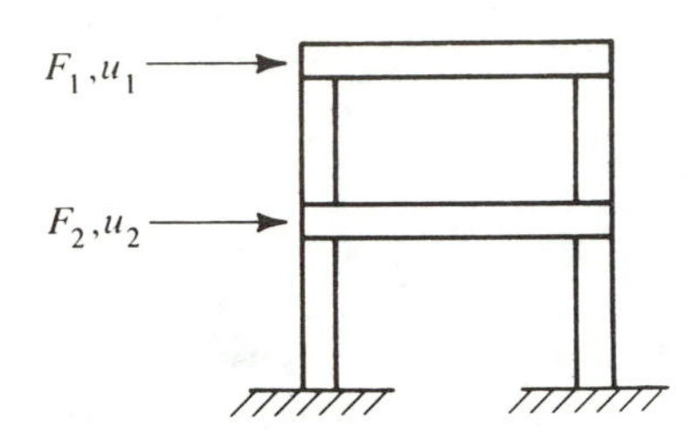

Figure 5-14

The analysis of the system consists of computing $\{F\}$ for given $\{u\}$, or vice versa. As long as the system is specified through the elements (system parameters) of matrix $[k]$, the analysis can be performed through direct multiplication when $\{u\}$ is given, or through the solution of two equations in two unknowns (n equations and n unknowns in Eq. 5-4) when $\{F\}$ is given. However, should $\{F\}$ be specified together with upper bounds for u_1 and u_2, then we must solve in Eq. 5-3 for parameters k_{ij} from two equations. Although $[k]$ is, in general, a symmetrical matrix with $k_{ij} = k_{ji}$, we still have no unique solution for three unknowns: k_{11}, k_{12}, k_{22} from two equations. The choice of a system from all feasible candidates requires, therefore, an additional criterion such as minimum weight, minimum cost, or others.

Some problems of current history, such as urban transportation, housing, mass communication, and urban development, pose engineering problems which are much more complex than the physical systems described so far in this section. Such problems are of a sociotechnical character. They require a much broader modeling space which includes the physical system as a subset. Goals must be stated in terms of the need of a community of people with varying values. The subgoals of safety, economy, and others must be weighed, and alternate systems compared in terms of criteria such as cost/benefit ratios, acceptability by the users in terms of risk levels, potential undesirable side effects on the environment, etc. The modeling for such complex problems with their attending multitudes of interconnections to politics, culture, and ecology, requires far-reaching insight, understanding, and heuristic sociotechnical-system-modeling power of the highest degree. To decompose such a problem into logically aggregated chunks with strong ties (dependence) within the elements of a chunk, but weak links between chunks to the point where each chunk can be studied and modeled independently, is a monumental modeling task. Yet it must be done more often and more urgently now then ever in the history of man. The wisdom of man, his ingenuity to cope with the environment, his creative genius which has contrived a man-made world, will ultimately shine in this new task, too.

5-11 MODELS IN PHYSICAL SCIENCE AND IN HUMAN AFFAIRS

There is an important difference between models in physical science, which deal with objects, and models which deal with sociotechnical systems. The objects of interest in physical science models are not affected by our models. For example, Kepler's model exerts no influence on the motion of the planets. However, a model that attributes a low learning capacity to a class of students may become self-fulfilling when a teacher treats a class in accordance with the statement of the model. Predictive models in human affairs are thus prone to self-fulfillment. Those who take the prediction seriously will adopt a mode of behavior consistent with it, and thus contribute to the potential of its realization.

When Kant said that a Kepler or a Newton was needed to find laws for the "movement of civilization" in analogy with the laws of planetary movements, he expressed the frustrating difficulty in attempts to model the "movement of civilization." Yet some efforts have been made to construct such models, as was discussed earlier in this chapter in the brief review of models of history. The models of Marx, Freud, Spengler, and Toynbee were efforts that attempted to aggregate vast observations into a small number of key parameters to explain *what is* and predict *what is likely to be.* The validation of these models is much more difficult, if possible at all. It is quite easy to understand why an aggregation in the form of a standardized pattern of behavior is helpful in economics. But standardization is contrary to man's biological make up. If we consider ten attributes of human beings, each at ten different levels, then we have 10 billion distinctly different possible human beings. This number is three times the population of the world. The unity sought by models to explain diverse events ignores unique features and differences and concentrates on what they all have in common. The imposed or discovered order which results is of great value.

In human affairs, the uniqueness of people has a high value in certain cultures, and in such cultures the imposition of order and standardization which masks these unique features may not be acceptable. Natural language demonstrates how much we strive for distinction instead of unity in human behavior. There are thousands of words to express different emotions in the English language. Models of human behavior seek to identify more uniquely each individual, while models of physical sciences attempt to order and unify.

To reach consensus on models in human affairs, more emphasis must be placed on what we have in common and what we agree on, rather than what we disagree on. As Dalkey points out [133], perhaps a new language must be developed for this purpose just as the language of mathematics was developed to deal with physical models. But central to all this is the need to cultivate a realization that to reach consensus is in itself a high value. This could form the basis for collective or group rationality as a model of ethical behavior in human affairs.

5-12 MORE ON MATHEMATICAL MODELS

Mathematical models are often very powerful forms for representing problems. We consider here a number of such models.

Classification Model

Consider three attributes P, Q, and R. We wish to classify a collection of objects according to their share in these attributes. For example, P may represent purple color; $\bar{P}$, not purple; Q, heavy; $\bar{Q}$, not heavy; R, round; $\bar{R}$, not round. Each class of objects will be described by a sequence of the letters P, Q, R, with or without overbars. There are eight distinct classes that can be generated by a tree diagram, as discussed in Chapter 2 and shown in Fig. 5-15. The eight distinct sequences of symbols PQR at the end branches

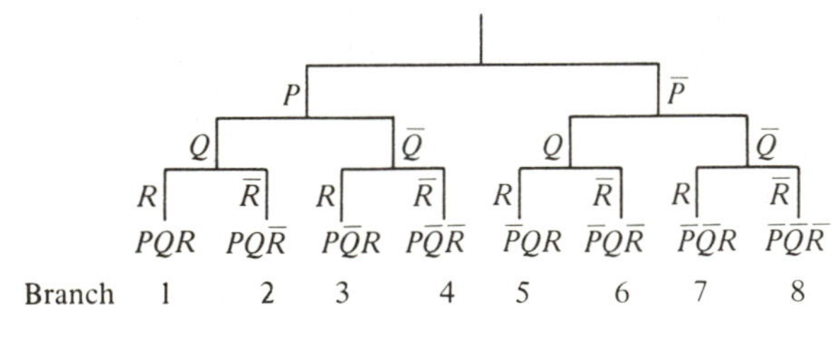

Figure 5-15

represent the eight classifications of the objects. If we employ the plus symbol (+) to represent the OR operation, and use multiplication to represent the AND operation, we can generate the classifications from the following operations (see Sec. 2-13 on Boolean algebra):

$$(P + \bar{P})(Q + \bar{Q})(R + \bar{R}) \tag{5-5}$$

Namely, objects are classified as (P or $\bar{P}$) AND in addition as (Q or $\bar{Q}$) AND in addition (R or $\bar{R}$). The multiplication yields the same results as those obtained by the multiplication down the branches of Fig. 5-15. P multiplies Q or $\bar{Q}$, and the result multiplies R or $\bar{R}$. The same is true for $\bar{P}$. Simply view each "fork" in Fig. 5-15 as a parenthesis in the following products:

$$\begin{aligned}&(P + \bar{P})(Q + \bar{Q})(R + \bar{R}) \\ &= PQR + PQ\bar{R} + P\bar{Q}R + P\bar{Q}\bar{R} + \bar{P}QR + \bar{P}Q\bar{R} + \bar{P}\bar{Q}R \\ &\quad + \bar{P}\bar{Q}\bar{R}\end{aligned} \tag{5-6}$$

Now note the power of the mathematical model for representation when we consider n attributes $A_1, A_2, \ldots, A_n$ with n large. Instead of generating the entire tree, we can write

$$(A_1 + \bar{A}_1)(A_2 + \bar{A}_2) \ldots (A_n + \bar{A}_n) \tag{5-7}$$

In Eqs. 5-5 and 5-6 we can use T and F to designate, respectively, the presence and absence of an attribute. The nature of the attribute can be identified by the position of T or F in the sequence of three letters. Thus TFF is equivalent to $P\bar{Q}\bar{R}$, i.e., first position signifies attribute P, second $\bar{Q}$, third $\bar{R}$. The classifications in Fig. 5-15 or Eq. 5-6 then become:

$$\begin{aligned}(T + F)(T + F)(T + F) = {} & TTT + TTF + TFT + TFF \\ & + FTT + FTF + FFT + FFF\end{aligned} \tag{5-8}$$

Binomial Model

In case we wish to identify in Fig. 5-15 *the number* of distinct sets that have none of the attributes, the number that share one attribute only, the number with two attributes, and the number with three attributes, we can use the binomial model and write Eq. 5-8 as

$$(T + F)^3 = T^3 + 3T^2F + 3TF^2 + F^3 \tag{5-9}$$

Namely, there is one set with all attributes (T^3), three sets with two (T^2F), three with one (TF^2), and one with none (F^3). For n attributes we can write

$$\begin{aligned}(T + F)^n = {} & T^n + nT^{n-1}F + \ldots + \frac{n!}{r!(n-r)!}T^{n-r}F^r \\ & + \ldots + nTF^{n-1} + F^n\end{aligned} \tag{5-10}$$

The binomial model groups the classification in larger chunks. Thus, for three attributes we have eight distinct classifications when the nature of the attributes is significant; however, when we are concerned only with the number of attributes, and not their nature, we have four classes as shown by Eq. 5-9. The notation $3T^2F$ signifies that there are three ways to have two attributes present and one absent, namely, the first and second present and third not, the first and third present and second not, or the second and third present and first not.

The binomial model can be used to represent other phenomena. Consider, for example, the man in Fig. 5-16 who wishes to move down from level 0. He can travel from any level to the one immediately beneath it by moving along a single link in an easterly (E) or a southerly direction (S).

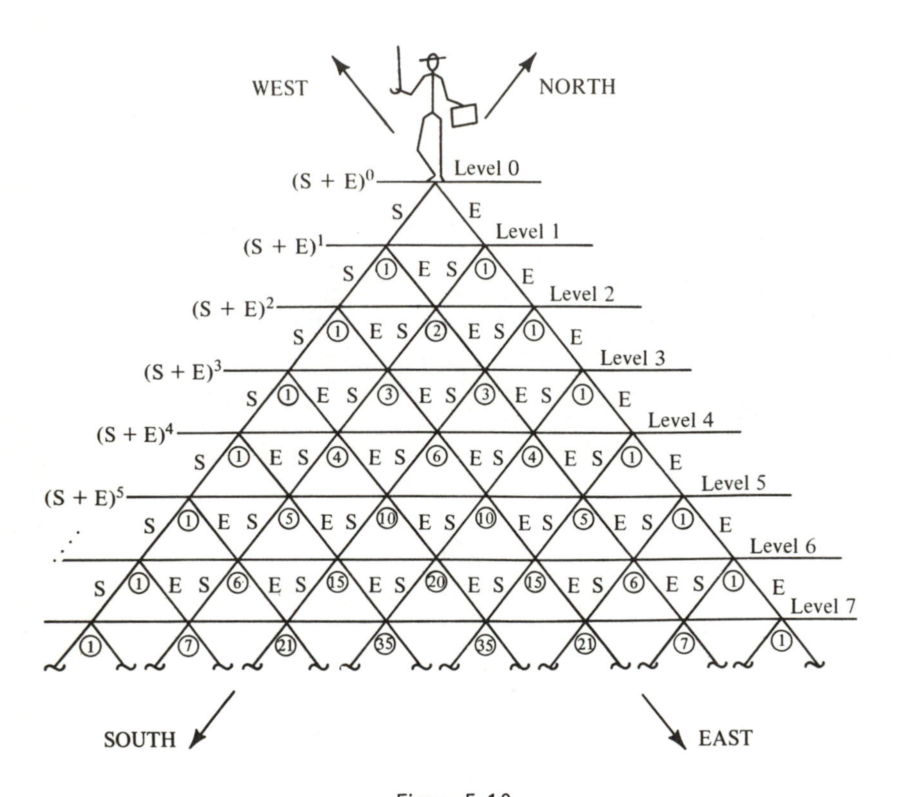

Figure 5-16

Now, let us classify the number of distinct routes that he can use to reach level 3 from level 0. There are exactly eight such routes. These are, respectively,

$$SSS\ SSE\ SES\ SEE\ ESS\ ESE\ EES\ EEE$$

This sequence is identical with that of Eq. 5-8 when S replaces T and E replaces F. The structure of Fig. 5-16 is a tree structure similar to that of Fig. 2-10 in which each level of Fig. 5-16 is represented by a row in Fig. 2-10. There is, however, one difference. In Fig. 5-16 the routes are clustered in terms of the number of S or E links in them. Thus, level 3 can be reached by one route with no E links, three routes with one E link, three routes with two

E links, and one route with three E links. These numbers 1, 3, 3, 1 are marked at the appropriate nodes of level 3 in the figure, and can be generated from the binomial model $(S + E)^3$. To reach the five nodes of level 4 from level 0, we can use the model $(S + E)^4$, and for any level n, $(S + E)^n$. The number of distinct routes leading to any node can also be generated by adding the number of distinct routes which can be used to reach it from the two nodes linked to it from the level above. For example, at level 5, starting from the left, $1 = 1$, $5 = 1 + 4$, $10 = 4 + 6$, $10 = 6 + 4$, $5 = 4 + 1$, $1 = 1$. This expansion of the number of routes by adding the numbers on the left and the right at the level above results in what is known as Pascal's triangle as shown below:

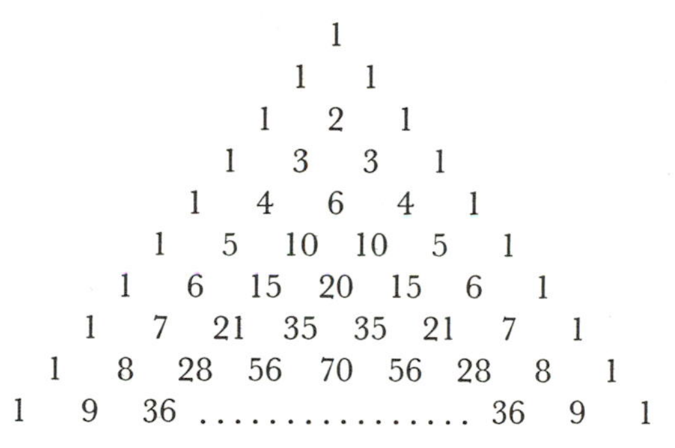

1
1 1
1 2 1
1 3 3 1
1 4 6 4 1
1 5 10 10 5 1
1 6 15 20 15 6 1
1 7 21 35 35 21 7 1
1 8 28 56 70 56 28 8 1
1 9 36 .... 36 9 1

The same binomial model (or Pascal triangle) can be used to classify the results of tossing n coins. Namely, $(H + T)^n$ will yield the number of patterns with no heads (H designates head and T tail), one head, two heads, and finally n heads.

Problem. In how many distinctly different ways in Fig. 5-17 can a man reach point B, starting from A and walking only on links between nodes which are heading south or east? There are eight links in the east direction and eight in the south direction. What is the number of distinct routes from A to B if the array of links is $n \times n$ instead of 8×8?

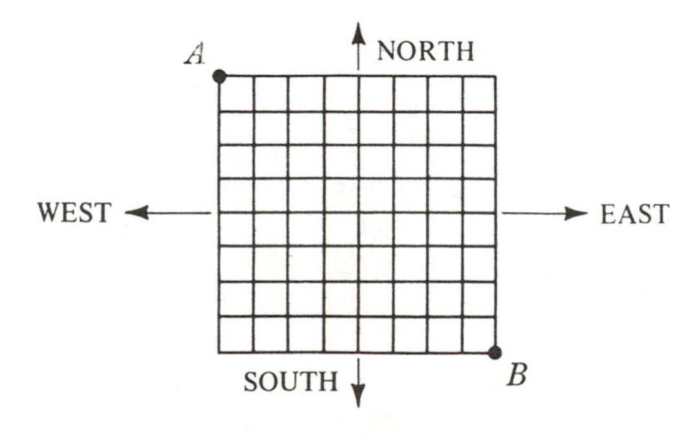

Figure 5-17

* The Sine and Cosine Model

Many phenomena are cyclic in nature. They have a pattern of ups and downs, as shown schematically in Fig. 5-18, in which dependent variable y varies as a function of independent variable t. y may represent the distance traveled by a mass suspended from a spring when disturbed from a rest position, and time is t. The peak value Y of y is known as the *amplitude* (3 inches in Fig. 5-18), and the time T to complete a cycle is known as the *period* (8

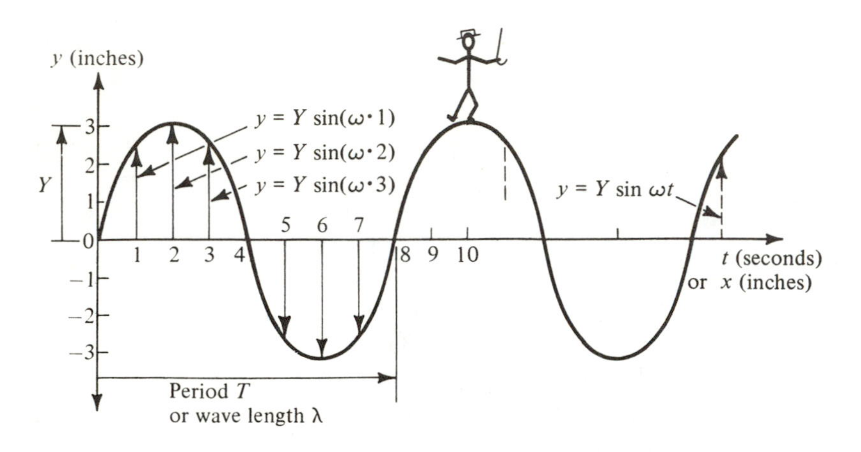

Figure 5-18 Sine Function $y = Y \sin \omega t$ or $y = Y \sin \omega x$

seconds in the figure). The *frequency*, f_T, of the function is defined as the number of cycles completed per unit of time. In the present example, $f_T = 1/8$. The period T and frequency f_T are the reciprocal of each other:

$$T = \frac{1}{f_T}, \quad f_T = \frac{1}{T}; \qquad \text{or} \qquad f_T T = 1 \tag{5-11}$$

The function of Fig. 5-18 can be obtained from an isomorphic transformation, or a one-to-one correspondence between the states in Figs. 5-19 and 5-18. Figure 5-19 shows an arrow, or pointer, of length Y corresponding to the amplitude of the cyclic function in Fig. 5-18. The pointer rotates at constant angular velocity of ω radians per second in a counterclockwise direction. We now inspect the vertical projection of the pointer, or the height of the arrowhead from its initial horizontal position at $t = 0$, which corresponds to $t = 0$ and $y = 0$ in Fig. 5-18. At $t = 1$ the angle swept by the pointer is $\omega \cdot 1$ and the vertical projection is $Y \sin(\omega \cdot 1)$ corresponding to y at time $t = 1$ in Fig. 5-18. At $t = 2$, $y = Y \sin(\omega \cdot 2)$; at $t = 3$, $y = Y \sin(\omega \cdot 3)$; and, in general, at time t, $y = Y \sin \omega t$. The pointer in Fig. 5-19 returns to its initial position after T units of time, i.e., the period, and completes one circular revolution which corresponds to one cycle in Fig. 5-18. Since the pointer

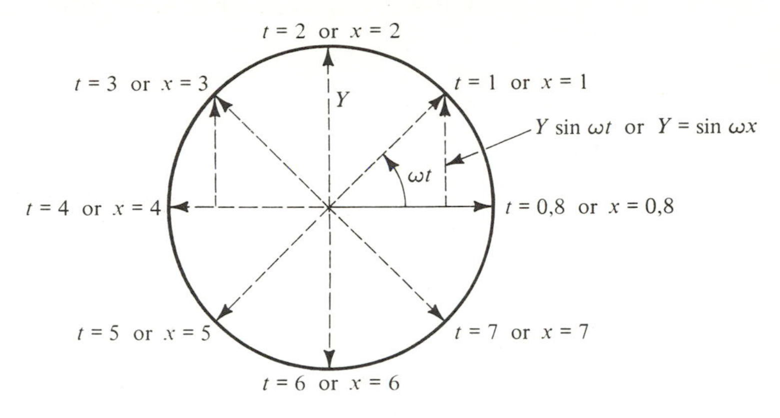

Figure 5-19

moves at a constant angular velocity of ω radians per second, and there are 2π radians in a complete revolution, the time period T required to complete 2π radians is

$$T = \frac{2\pi}{\omega} \tag{5-12}$$

or $$\omega T = 2\pi$$

Since $f_T T = 1$, it follows that

$$f_T = \frac{\omega}{2\pi} \tag{5-13}$$

Spatial Independent Variable. If time t in Figs. 5-18 and 5-19 is replaced by distance x in inches, we may consider the following situation. In Fig. 5-18 a man is walking on the wave form which was constructed by elevating and excavating successive portions of an initial zero elevation terrain $y = 0$. As the man travels he reports his elevation y for successive values of the horizontal projection of his position x. For each unit of x, the pointer in Fig. 5-19 is rotated ω radians and the vertical projections of Y correspond to the elevations y in Fig. 5-18. Hence, at $x = 1$, $y = Y \sin(\omega \cdot 1)$; at $x = 2$, $y = Y \sin(\omega \cdot 2)$; and in general for x, $y = Y \sin \omega x$. Now, each ω radians in Fig. 5-19 correspond to one unit distance x in the horizontal direction in Fig. 5-18. Therefore, when the pointer completes one revolution in Fig. 5-19, our man has completed one cycle or one wave of his travel and has progressed horizontally by one *wavelength* λ. Since the pointer moves ω radians for each unit of x, and there are 2π radians in a complete revolution, the wavelength λ (or number of units of x) required to complete 2π radians is

$$\lambda = \frac{2\pi}{\omega} \tag{5-14}$$

or $$\omega\lambda = 2\pi$$

The frequency f_λ of cycles per unit length is the reciprocal of λ; hence,

$$f_\lambda \lambda = 1$$

and
$$f_\lambda = \frac{\omega}{2\pi} \tag{5-15}$$

f_λ is also known as the *wave number*.

Phase. The phase establishes the value of the independent variable t (or x) when (or where) the sinusoidal response starts with respect to the origin of the coordinates. For example, let the sine wave of Fig. 5-18 be written in the form

$$y = Y \sin(\omega t + \phi) \tag{5-16}$$

ϕ is the phase angle in radians and has the following interpretation in terms of Fig. 5-19. For $y = 0$, $(\omega t + \phi) = 0$; therefore, $t = -\phi/\omega$. Namely, the sinusoidal response begins ϕ/ω units of time earlier than shown in Fig. 5-18. For instance, if $\phi/\omega = 1.5$, then the sine wave is shifted (phased) 1.5 units of time to the left. For the equation

$$y = Y \sin(\omega x + \phi) \tag{5-17}$$

$y = 0$ when $(\omega x + \phi) = 0$ or $x = -\phi/\omega$. Namely, the sinusoidal response begins ϕ/ω units of distance to the left of the origin of coordinate x in Fig. 5-18. The phase is of great significance, because two waves of identical frequency but different phases may be combined to achieve various results. If the phases are the same, the waves reinforce each other; if they differ by π radians and are of the same amplitude, they nullify each other. The interference or combination of many waves of various frequencies, amplitudes, and phases can be made to produce any desired arbitrary disturbance as a function of time and space, and is the basis for radar detection and a host of other applications.

Cosine. When the origin of coordinates in Fig. 5-18 is shifted one-fourth of a period or wavelength (two units in the figure) to the right, the function becomes $y = Y \cos \omega t$ or $y = Y \cos \omega x$. The value of y in Fig. 5-19 becomes, then, the horizontal projection of pointer Y. Verify this.

* Exponential Function Model

Consider a population of size N at time t which grows at a constant rate of λ percent per unit of time. Stated in mathematical symbols, we write for the rate of change of population N per unit time,

$$\frac{dN}{dt} = \lambda N \tag{5-18}$$

or
$$\frac{dN}{N} = \lambda\, dt$$

Integrating between the limits of $t = 0$ and t:

$$\int_{N_0}^{N_t} \frac{dN}{N} = \lambda \int_0^t dt$$

$$\ln N_t - \ln N_0 = \lambda t$$

$$\ln \frac{N_t}{N_0} = \lambda t$$

$$\frac{N_t}{N_0} = e^{\lambda t}$$

$$N_t = N_0 e^{\lambda t} \tag{5-19}$$

In the above, N_0 is N at time $t = 0$, N_t is N at time t, and ln stands for the natural logarithm, i.e., log to base e.

To generate the same result in an alternate way, consider time interval t to be divided into n equal increments of duration $\Delta t = t/n$ each. Starting with population N_0, which increases at a rate of λ percent per unit time, then proceeding from increment to increment, we have

$$\begin{aligned}
N_1 &= N_0 + \lambda N_0 \frac{t}{n} = N_0\left(1 + \lambda \frac{t}{n}\right) \\
N_2 &= N_1 + \lambda N_1 \frac{t}{n} = N_1\left(1 + \lambda \frac{t}{n}\right) \\
N_3 &= N_2 + \lambda N_2 \frac{t}{n} = N_2\left(1 + \lambda \frac{t}{n}\right) \\
&\vdots \\
N_{n-1} &= N_{n-2} + \lambda N_{n-2} \frac{t}{n} = N_{n-2}\left(1 + \lambda \frac{t}{n}\right) \\
N_n &= N_{n-1} + \lambda N_{n-1} \frac{t}{n} = N_{n-1}\left(1 + \lambda \frac{t}{n}\right)
\end{aligned} \tag{5-20}$$

Substituting backwards for N_{n-1} in the last equation from the equation above it, then for N_{n-2}, etc., until N_1 is replaced finally by the right-hand side of the first equation, we have

$$N_n = N_0\left(1 + \lambda \frac{t}{n}\right)^n \tag{5-21}$$

Let $$\lambda \frac{t}{n} = \frac{1}{m}, \text{ or } n = m \lambda t \tag{5-22}$$

Then $$N_n = N_0\left[\left(1 + \frac{1}{m}\right)^m\right]^{\lambda t} \tag{5-23}$$

For λ and t finite, the only way to make the time intervals $\Delta t = t/n$ infinitesimally small, $dt \longrightarrow 0$, is to increase n to ∞ which is the same as increasing m to ∞. For $m \longrightarrow \infty$,

$$\left(1 + \frac{1}{m}\right)^m \longrightarrow e$$

and the last equation becomes

$$N_n = N_0 e^{\lambda t} \tag{5-24}$$

in which $N_n \equiv N_t$. Equations 5-24 and 5-19 are identical.

Doubling Time. Let us denote by letters DT the doubling time t of an original population N_0. In terms of Eq. 5-19 we can write

$$N_{DT} = 2N_0 = N_0 e^{\lambda DT}$$

or

$$e^{\lambda DT} = 2$$

and

$$\left.\begin{aligned} \lambda DT &= \ln 2 \\ DT &= \frac{1}{\lambda} \ln 2 \\ DT &\approx \frac{1}{\lambda} 0.693 \end{aligned}\right\} \quad (\ln 2 \approx 0.693) \tag{5-25}$$

For example, if $\lambda = 0.02$, i.e., 2 percent growth per year, then $DT = 34.65$ years, or about 35-year doubling time. For $\lambda = 0.03$, $DT \approx 23$ years; and for $\lambda = 0.05$, $DT \approx 14$ years.

Using the definition of DT from Eq. 5-25, we have $\lambda = \ln 2/DT$. Substituting this expression for λ in Eq. 5-19, we can write

$$\begin{aligned} N_t &= N_0 e^{\ln 2 (t/DT)} \\ &= N_0 (e^{\ln 2})^{t/DT} \end{aligned}$$

But $\quad e^{\ln 2} = 2$

Therefore, $N_t = N_0 (2)^{t/DT}$ (5-26)

An equation similar to Eq. 5-26 can be generated if we are interested in tripling time, TT, or other measures of exponential growth. Show that for tripling time TT we obtain

$$N_t = N_0 (3)^{t/TT}$$

* Model for Exponential Decay

When λ is a constant rate of decrease in percent per unit time, then an initial population N_0 at time $t = 0$ decays or diminishes in size to N_t at time t:

$$N_t = N_0 e^{-\lambda t} \tag{5-27}$$

The *half-life* of an exponentially decaying population is defined as the time *HL* in which the population N_0 is reduced to half, i.e., to $(1/2)N_0$. From

Eq. 5-27 we can write

$$N_{HL} = \frac{1}{2} N_0 = N_0 e^{-\lambda HL}$$

or

$$e^{-\lambda HL} = \frac{1}{2}$$

and

$$\left.\begin{aligned} -\lambda HL &= \ln \frac{1}{2} = -\ln 2 \\ HL &= \frac{1}{\lambda} \ln 2 \\ HL &\approx \frac{1}{\lambda} 0.693 \end{aligned}\right\} \quad (\ln 2 \approx 0.693) \qquad (5\text{-}28)$$

Substituting for $\lambda[\lambda = -(1/HL) \ln 1/2]$ from Eq. 5-28 into Eq. 5-27 gives

$$\begin{aligned} N_t &= N_0 e^{(\ln 1/2)(t/HL)} \\ &= N_0 (e^{\ln 1/2})^{(t/HL)} \end{aligned}$$

But

$$e^{\ln 1/2} = \frac{1}{2}$$

Therefore, $N_t = N_0 \left(\frac{1}{2}\right)^{t/HL}$ (5-29)

Carbon Dating and Exponential Decay

The radioactive decay of carbon is used to establish the date at which an object lived in the past. This involves the measurement of the amount of the carbon 14 isotope contained in remnants of plants and animals. C_{14} is a rare variety of carbon that enters with ordinary carbon C_{12} into the structure of living matter. C_{14} is unstable and decays exponentially with a half-life HL = 5568 years, while C_{12} is not radioactive. C_{14} atoms are constantly produced at the top of the atmosphere when neutrons produced by cosmic rays hit the nuclei of nitrogen atoms. These new carbon isotopes spread through the atmosphere and combine with oxygen to form CO_2. Every plant absorbs some C_{14} atoms. C_{14} enters all living things directly, and indirectly through plants consumed by animals. The amount of C_{14} is small in living things, and its ratio to ordinary carbon C_{12} remains constant because of continued exchange through the life of the plant or animal. Upon death, when the remains of the living thing are buried in the ground and are isolated from the air, the C_{14} atoms diminish in number through decay, and the ratio C_{14}/C_{12} gets smaller, the older the remains. From this information the age at which the remains ceased to live can be computed.

Combination of Oscillation and Exponential Growth or Decay

The sine and cosine functions can be combined with the exponential function to yield a model in which the response y oscillates with progressively increasing or decreasing amplitude. Figure 5-20 shows an exponentially increasing and exponentially decreasing oscillation which are obtained from products of functions treated in this section.

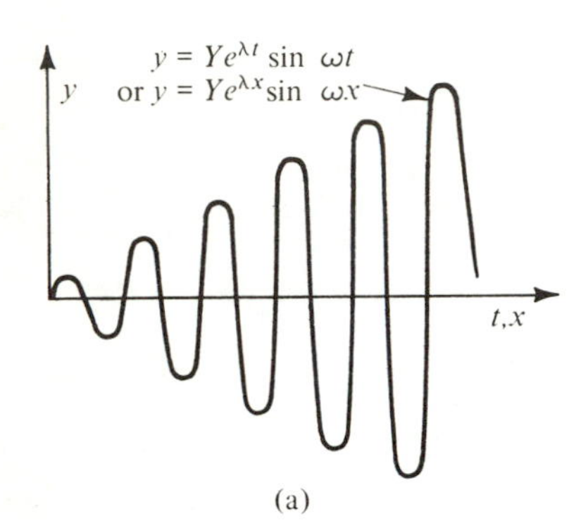

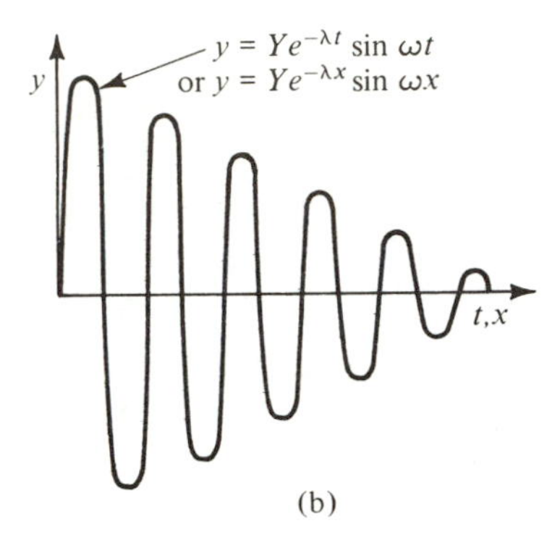

Figure 5-20

5-13 SUMMARY

Chapter 5 explores:

1. The structure of a model
2. The purpose of a model
3. The modeling process
4. Model validation
5. Model classification
6. Examples of models in history and the sciences
7. Examples of models in a man-made world

A model is an abstract description of the real world. It is a simple representation of more complex forms, processes, and functions of physical phenomena or ideas.

A model is constructed to facilitate understanding of relationships between elements, forms, processes, and functions, and to enhance the capacity to predict outcomes in the natural and man-made world.

Understanding a phenomenon or an idea implies an ability to place it in a larger context or framework. Understanding may improve prediction and provide means for control of outcomes when cause-effect relations are identified.

Models Evolve. The modeling process at the early stage to achieve a simple high-level abstraction consists of five main steps:

1. Establish purpose.
2. List elements related to purpose.
3. Select elements of Step 2 relevant to purpose.
4. Chunk.
5. Continue chunking until 7 ± 2 chunks result at the highest level of aggregation.

To cope with complexity, decomposition is used to subdivide a large and complex model into aggregates or submodels that are virtually independent and can be studied separately. The decomposition is performed so that the links among elements within submodels are strong while the links between submodels are weak.

Models may fail to contain some elements of the real world or contain elements not present in the real world. These are called, respectively, errors of omission and commission.

Validity of a model is tested against measurements and observations in the real world when possible. Not all models can be validated.

The two fundamental features which characterize all models are *form* and *content.* The form of representation has great influence on the ease of manipulation of content and on the richness or sterility of possible refinements and changes in the modeling process. The form interacts strongly with the content and contributes to it.

The fundamental forms of models may be classified as: *verbal*, *mathematical*, and *analog.*

Other classifications of models are possible:

physical and abstract
descriptive and functional
causal and correlative
deterministic and probabilistic
simulation models
isomorphic and homomorphic
qualitative and quantitative

The following models of history are described:

The Greeks' geographic model
Marx's economic model
Freud's psychoanalytic model
Spengler's cyclic analog model
Toynbee's linear evolutionary model

Models of creation are discussed in terms of the expanding universe. Both Gamow's big bang model and Hoyle's steadily increasing universe model are treated. Kepler's model of the solar system is described and an example is given of an observation which served to enhance the credibility of Newton's model.

Bohr's model of the atom is described and the history is related briefly on how understanding and observations led to the present-day quantum mechanics model of the atom.

As a third main example, a model of the brain suggested by Hunt is described. The analog of the model to a computer and its explanation of the problem-solving process are highlighted.

Models in science help us understand the natural world as it is and predict its behavior in the future. Models in engineering deal with what *ought to be*. Namely, engineering addresses itself to a man-made world and creates models of systems conceived to achieve stipulated goals (what ought to be).

Engineering models are models of systems dealing with the input from the environment, with system response and system characteristics. Two main classes of problems can be identified for the modeling process: analysis, synthesis.

The behavior of physical systems can be characterized by parameters in terms of their energy transfer mechanism. The input-output variables are classified as across and through variables, and can be related by using the principles of conservation and continuity.

There is an important difference between models of physical science and models of human affairs. The objects of interest in physical science are not affected by our models, whereas models of human affairs are prone to self-fulfillment, and thus affect the behavior that is being modeled.

The following mathematical models are developed:

To list all possible mutually exclusive regions for n attributes or sets $A_1, A_2, \ldots, A_n$, we can use the following classification model:

$$(A_1 + \bar{A}_1)(A_2 + \bar{A}_2) \ldots (A_n + \bar{A}_n)$$

The binomial model can be used to generate the number of elements in aggregated subsets in classification problems:

$$(T + F)^n = T^n + nT^{n-1}F + \ldots + \frac{n!}{r!(n-r)!}T^{n-r}F^r + \ldots + nTF^{n-1} + F^n$$

The sine function with both time t and distance x as the independent variable are developed:

$$y = Y \sin \omega t$$

and $$y = Y \sin \omega x$$

The following terms are defined and physically interpreted:

Amplitude Y is the maximum response y
Frequency is cycles per second f_T, or cycles per unit distance f_λ (f_λ is also called wave number)
Period T is the time to complete one cycle
Wave length λ is the distance which contains one wave

Phase ϕ is defined from the equation

$$y = Y \sin (\omega t + \phi)$$

or

$$y = Y \sin (\omega x + \phi)$$

$$f_T = \frac{\omega}{2\pi} \qquad f_\lambda = \frac{\omega}{2\pi}$$

$$f_T T = 1 \qquad f_\lambda \lambda = 1$$

Exponential growth and decay models are developed. For an initial population N_0 growing at a constant rate of λ percent per unit time, the population N_t after t units of time is given by

$$N_t = N_0 e^{\lambda t}$$

The doubling time DT is the time required for an initial population N_0 to double and is computed from λ:

$$DT = \frac{1}{\lambda} \ln 2 \approx \frac{1}{\lambda} 0.693$$

When λ is a constant rate of decay, then

$$N_t = N_0 e^{-\lambda t}$$

and the half-life HL, i.e., the time required for N_0 to decay to half its value, is computed from

$$HL = \frac{1}{\lambda} \ln 2 \approx \frac{1}{\lambda} 0.693$$

Carbon dating is described as an application of exponential decay and half-life observations.

EXERCISES

Structure of a Model

5-1 Give an example of a model in your field of interest.

Purpose of a Model

5-2 What is the purpose of the model you discussed in Exercise 5-1? Why or why not was this purpose achieved?

5-3 Discuss the difficulties involved in making predictions by means of a model. What difficulties are involved in making predictions without a model?

Modeling process

5-4 Using the five steps of the modeling process at the initial stage, devise a model for the selection of a candidate for President of the United States.

5-5 Go through the five steps of the modeling process that are needed to construct the model you selected in Exercise 5-1.

5-6 Go through the five steps of the modeling process in order to construct a model of the project you chose in Exercise 1-1.

5-7 Your mission is to develop a plan to launch a manned space vehicle to Mars. Limiting yourself to relevant information, use the five steps of the modeling process to create such a plan.

5-8 In manufacturing, there are certain overhead costs that remain constant no matter how many units (within reason) of goods are produced. Also, there are costs associated with each unit produced. That is, for *each* unit produced there is a constant cost, independent of the overhead costs. Therefore, the total cost of production may be represented by a model which is the equation of the straight line:

$$T = K + cx$$

where $T =$ total cost, $K =$ overhead cost (a constant), $c =$ unit cost (a constant), and $x =$ number of units produced. Using the five steps of the modeling process, show how you would derive this model equation for total production cost.

Validation of Models

5-9 How would you validate the model you constructed in Exercise 5-6?

5-10 List any errors of omission or commission you feel are included in your model of Exercise 5-6.

Classification of Models

5-11 How would you classify the model you constructed in Exercise 5-6? Justify your classification in terms of *form* and *content*.

5-12 Discuss how the model of your project (Exercise 5-6) may be simulated. What can you say about any results you derive from simulation in general?

Models of Natural Science

5-13 *Break-even Point.* The break-even point is the quantity of units sold at which point revenue equals cost. It can be expressed as units of output or as the minimum amount of revenue, in dollars, that must be received to cover cost. If the cost curve and the revenue curve are modeled on a graph, the break-even point is at their intersection.

If $F =$ fixed cost (independent of the number of units)
$P =$ price (of a unit of output)
$V =$ variable cost (varies with number of units of output)

then the break-even quantity B_Q may be expressed as

$$B_Q = \frac{F}{P - V}$$

Namely,

$$\text{Revenue} = PB_Q$$
$$\text{Cost} = F + VB_Q$$

At the break-even point:

$$F + VB_Q = PB_Q$$
$$\therefore \quad B_Q = \frac{F}{P - V}$$

Problem. Assume that \$50,000 is required to cover the research and development cost of a product. Let the variable cost of the product be \$200 per unit for direct labor, material, and sales commission. If the price of the product is \$300, what is the break-even quantity? Graph the two curves and indicate the break-even point.

5-14 *Present Value.* Present value is the value of future income expressed in current dollars. The concept is based on the fact that a dollar today is worth more than a dollar one year hence. This is not because of inflation, but because a dollar today can be invested to earn more money during the time one would be waiting to receive the future dollar. A mathematical model to evaluate the present worth of future income is

$$PW = \frac{S}{(1 + i)^n}, \quad \text{or} \quad S = PW(1 + i)^n$$

PW = present worth
S = sum received n years in the future
i = discount rate (commonly known as interest)

Problem. The marketing division for GM estimates gross profits at the end of a 10-year period of 100 million, 350 million, or 300 million for the car divisions A, B, or C, respectively. The investments required for these mutually exclusive alternatives are $\$80 \times 10^6$, $\$200 \times 10^6$, $\$150 \times 10^6$. Which is the best investment if the discount factor is 5%?

Engineering Models

5-15 A bacterium (10^{-9} grams) under ideal conditions will undergo binary fission every 20 minutes. If a single bacterium falls into a very large vat of ideal medium, how long will it be before the vat contains 100 kilograms of bacteria?

5-16 Explain why the equations for doubling time and half-life as given in this chapter are identical.

$$DT = \frac{0.693}{\lambda} \quad \text{and} \quad HL = \frac{0.693}{\lambda}$$

5-17 The half-life of carbon 14 is approximately 5000 years. If a fossil has only 12.5% of the carbon 14 that one would expect in a bone of a living creature, (a) How old is the fossil? (b) What is the annual decay rate?

5-18 An island has 52 square miles and a carrying capacity of 1 person per square mile. At present there are 13 people there, but the population is increasing at a rate of 10% per year. How long will it be before the population reaches the capacity of the island?

5-19 The mayor of a small town must decide whether or not to annex a large amount of land. He knows that his town presently has enough area for 18,000 people, more than sufficient for the 5,000 people now living there. On the other hand, if this opportunity is passed up, no new land will be available for 20 years. Knowing that the town's population is predicted to grow at a steady rate of 7%/year over that period, must the mayor annex the land now, or can he wait the additional 20 years?

5-20 Suppose that the average quantity of a particular food needed to sustain life is 300 lbs per person per year. Suppose also that all of this food provided in a year is either consumed by the population or the surplus discarded. In the initial year, 1,000,000 pounds of the food are produced and the population is 2000. The population increases at a rate of 7%/year and the food production increases linearly at a rate of 10,000 lbs every year.

(a) Write an equation to solve for t', the time it takes the average food consumption to become less than 300 pounds per person per year.

(b) Approximate t' by using a graphical approach.

5-21 Is a "3-dimensional Pascal triangle" possible? What can it represent and what complications can arise in constructing it and identifying its points?

5-22 Extend your discussion in Exercise 5-21 to include an "n-dimensional Pascal triangle."

NOTE. The following problems illustrate a situation in which it is helpful to model a system in terms of Boolean algebra components, and then perform Boolean operations on them. For example, suppose we want to know all the possible ways we can pick two cards from a deck and have first a red card, then a black card. Intuitively, we know there are four ways: (H, S), (H, C), (D, S), and (D, C), where $H =$ hearts, $D =$ diamonds, $S =$ spades, $C =$ clubs.

Using Boolean algebra, we say a red is a heart *or* a diamond: $R = (H + D)$. Similarly, a black is a spade *or* a club: $B = (S + C)$. Finally, a red *and* a black $= R \cdot B = (H + D) \cdot (S + C)$. "Multiplying," we get the four combinations (H, S), (H, C), (D, S), (D, C), which checks with the result we obtained by intuition.

5-23 A company with stores in four cities A, B, C, D wishes to locate a warehouse in one or more of those cities such that every store can be reached in one day from at least one of the warehouses. Fig. 5-21 represents all the possible connections that can be made in one day, due to existing roads and prevailing

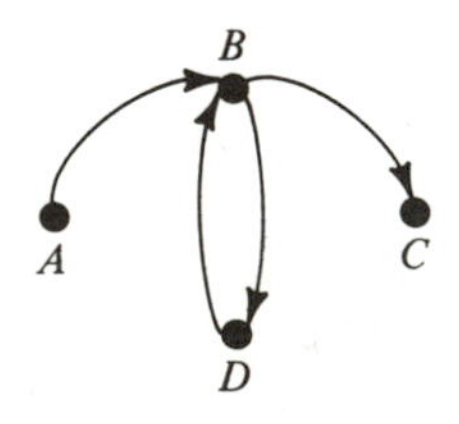

Figure 5-21

traffic patterns. Each arrow describes a possible flow and direction of flow in one day from a warehouse in one city to a store in another.

Where should the company locate the warehouse(s) in order to have one-day coverage for each store while minimizing the number of warehouses?

5-24 Refer to Exercise 5-23. Assume that there are now six cities, with the possible connections as shown in Fig. 5-22.

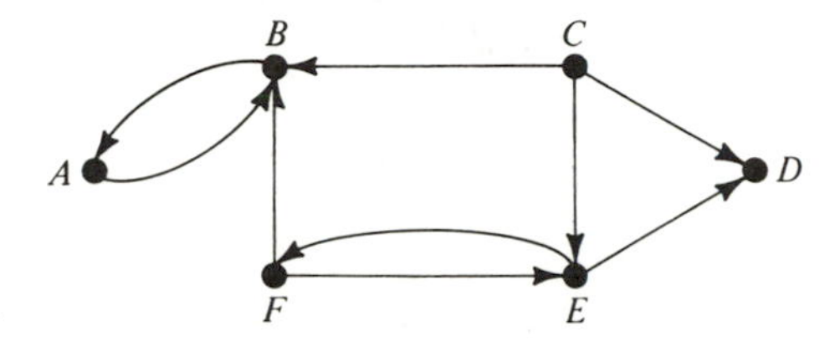

Figure 5-22

(a) Set up the equations for the solution to the problem as stated in Exercise 5-23.

(b) Can you see some possible solutions without multiplying out the set-up in (a)?

6

PROBABILISTIC MODELS

6-1 PROBABILISTIC MODELS—PRELIMINARY CONCEPTS

Before proceeding with a more formal discussion of probabilistic models, let us consider some preliminary concepts through an example. These concepts will be discussed in more detail later in this chapter. The treatment here is in the nature of an exposure.

Let us suppose that we are producing necklaces. These necklaces have some objects of varying weights suspended from them. We wish to establish a quantitative measure for the load- or weight-carrying capacity of a typical representative necklace. Each necklace is constructed from the same material and has the same dimensions. However, imperfections in the production and assembly process cause the load-carrying capacity C (in pounds) to vary. To establish a measure of C, we design an experiment in which each necklace in a sample of necklaces is subjected to a progressively increasing load until it breaks. The breaking load is designated as C. C may, of course, vary from necklace to necklace. We can plot the results of our experiment as shown in Fig. 6-1. Each time a necklace fails at loads between the limits $[C + (1/2)\Delta C]$ and $[C - (1/2)\Delta C]$ we draw one square above C. In Fig. 6-1 we have such a plot for a sample of 50 necklaces, with $\Delta C = 2$ pounds. The total number of necklaces that failed for loads smaller than 41 pounds, for example, is given by the number of squares to the left of $C = 41$ pounds, i.e., 30. We can use this number to predict the probability that a necklace will fail (break) under a load smaller than 41 pounds by taking the ratio of the total number of necklaces that failed below 41 pounds divided by the total number of necklaces in the sample tested, i.e., 30/50, or, using compact notation, $Pr(C < 41) = 30/50 = 0.6$. If, instead of the *number* of squares in Fig. 6-1, we consider the

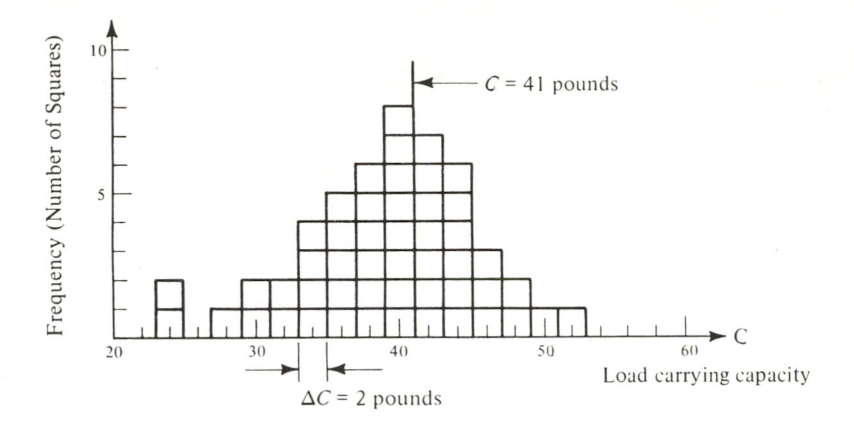

Figure 6-1 A Plot of Frequency Versus Load-Carrying Capacity for 50 Necklaces

area represented by the squares, then the same result will be obtained for $Pr(C < 41)$ by taking the area to the left of $C = 41$ and dividing it by the total area of the squares. We can also adjust the area of each square so that the total area of all squares will be unity. In this case, $Pr(C < 41)$ will simply be equal to the area to the left of $C = 41$.

Extending the experiment to a large number N of necklaces and taking ΔC vanishing small, the plot of Fig. 6-1 will approach a continuous curve (Fig. 6-2) as $N \longrightarrow \infty$ and $\Delta C \longrightarrow 0$. Again, we can make the total area under the curve in Fig. 6-2 equal to unity so that $Pr(C < C_1)$ will be equal to the area under the curve $p(C)$ to the left of C_1. This is expressed mathematically as

$$Pr(C < C_1) = \int_0^{C_1} p(C)\, dC \tag{6-1}$$

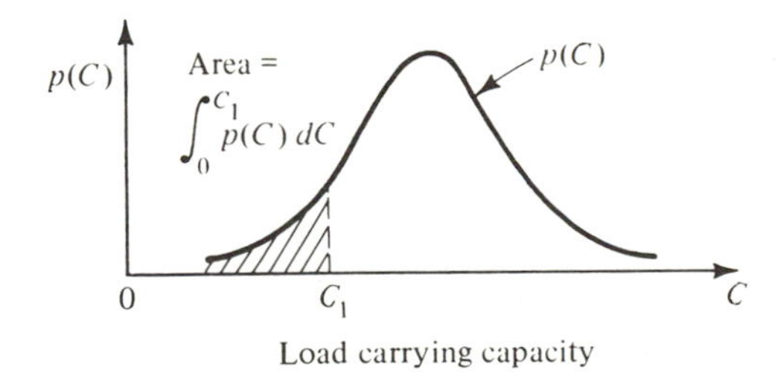

Figure 6-2 Probability Density of Load-Carrying Capacity C for Necklaces

The function $p(C)$ of Fig. 6-2 is called a probability density function or distribution function of random variable* C, where in the present example C represents the load-carrying capacity of a necklace. To evaluate the integral of Eq. 6-1, the shape of the curve $p(C)$ must be established. This is not a simple task in most cases. For the purpose of this presentation we will assume that $p(C)$ is the *normal distribution function* which is given by

$$p(C) = \frac{1}{\sigma_C\sqrt{2\pi}} e^{-1/2[(C-\mu_C)/\sigma_C]^2} \tag{6-2}$$

in which μ_C and σ_C^2 are, respectively, called the *mean* and *variance* of C. The mean μ_C is a measure of the location of the distribution (analogous to center of gravity of an area). The variance σ_C^2 is a measure of the dispersion (spread) of the distribution $p(C)$ (analogous to the moment of inertia of an area with respect to a vertical line through its center of gravity). The larger the variance, the more spread is the area under $p(C)$ about the mean. The normal distribution is completely characterized by the two parameters μ_C and σ_C^2, as seen from Eq. 6-2.

The normal distribution is very important and has great utility. This is due to one of the most important theorems in mathematics, called the Central Limit Theorem. To point out the significance of this theorem in terms of our present example, let us consider Fig. 6-2. Suppose that in this figure each value of C represents the average load-carrying capacity of n necklaces; then, invoking the central limit theorem, it can be shown that as n gets large the distribution $p(C)$ of these average capacities C will approach the normal distribution (also referred to as the Gaussian distribution) irrespective of what shape the distribution $p(C)$ has when $n = 1$.

So far, we have considered the load-carrying capacity C of the necklaces. The load intensity L to which a necklace may be subjected in actual use is also not deterministic and has a probability distribution function $p(L)$. For simplicity, let us assume that L is also normally distributed with mean μ_L and variance σ_L^2. The functions $p(C)$ and $p(L)$ are plotted on the same load scale in Fig. 6-3. The condition

$$C - L \leq 0 \tag{6-3}$$

determines failure; i.e., whenever the load-carrying capacity C is smaller than the applied load L, the necklace will fail. Let us define a new variable R as

$$R = C - L \tag{6-4}$$

R is a linear combination of C and L. It can be shown by probability theory [54] that if C and L are normally distributed with means μ_C, μ_L and variances

*C is a random variable when all known extraneous effects contributing to the variability of C have been removed from the experiment.

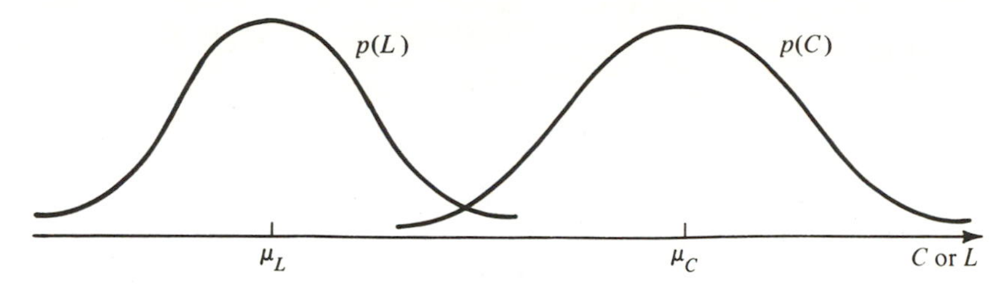

Figure 6-3 Probability Density Functions for Load Intensity and Load-Carrying Capacity

σ_C^2, σ_L^2, then R is also normally distributed* with mean

$$\mu_R = \mu_C - \mu_L \tag{6-5}$$

and variance

$$\sigma_R^2 = \sigma_C^2 + \sigma_L^2 \tag{6-6}$$

so that

$$p(R) = \frac{1}{\sigma_R\sqrt{2\pi}} e^{-1/2[(R-\mu_R)/\sigma_R]^2} \tag{6-7}$$

Plots of Eq. 6-7 are shown in Fig. 6-4. From Eqs. 6-3 and 6-4, the condition of failure is determined by $R \leq 0$, and the probability of failure is given by the area under $p(R)$ to the left of $R = 0$. Mathematically, this can be written as

$$Pr(R \leq 0) = \int_{-\infty}^{0} p(R)\, dR \tag{6-8}$$

* The probability of failure [namely, $Pr(R \leq 0)$] can be made smaller in a number of ways by reducing the region of overlap of $p(C)$ and $p(L)$ in Fig. 6-4 as follows:

1. μ_R is made larger with σ_R^2 unchanged (μ_R^* in Fig. 6-4). This means that either μ_C is made larger or μ_L is made smaller, or both, so that the function $p(R)$ will be shifted to the right, causing the area to the left of $R = 0$ to decrease (see Fig. 6-4).
2. σ_R^2 is made smaller (σ_R^{*2} in Fig. 6-4). This can be accomplished by making σ_C^2, σ_L^2, or both, smaller, thus concentrating more of the area in the vicinity of μ_R and reducing the area to the left of $R = 0$.
3. A combination of 1 and 2. The decision as to which measures should be taken to decrease the probability of failure will depend on the degree to which the parameters μ_C, μ_L, σ_C^2,

*Provided that C and L are statistically independent. (See [54] or Sec. 6-4.)

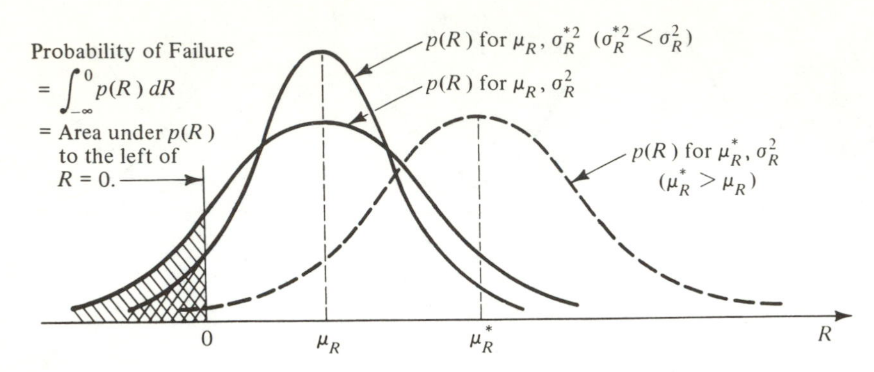

Figure 6-4 Probability Density Functions for $R = C - L$

and σ_L^2 can be controlled, and on the costs associated with their control.

6-2 POPULATIONS AND SAMPLES

Population

A population is defined as a set which contains all the possible observations (real and hypothetical) of a particular attribute (phenomenon or measurement). For example, we may speak of the population of heights of USA citizens, the population of yearly incomes of USA citizens, the population of all possible measurements of length of a pencil, the population of outcomes resulting from the flip of a coin n times, etc. A population can be finite or infinite in terms of the number of observations it contains. The concept of a population is similar to the concept of the *universal set* discussed in Sec. 2-14. The universal set is the original complete set which contains *all the elements* in the subject of discourse. In a similar way, *population* refers to the total of all measurements or observations of interest or concern.

Sample

A sample is a subset of a population. For example, we may speak of a sample of five measurements of height of five different USA citizens. Similarly, we may obtain a sample of yearly incomes, a sample of measurements of length of a pencil, and a sample of outcomes in the flip of a coin. Schematically, the relationship between a population and a sample from the population is shown in Fig. 6-5. In the language of sets (Sec. 2-14) we have then *Sample* $\subset$ *Population*, i.e., the sample is contained in the population. The population is the *sample space*.

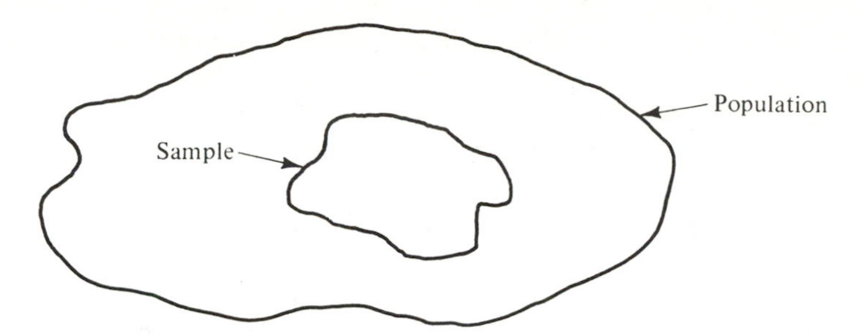

Figure 6-5 Relationship Between Population and Sample (The *Population* is the *Sample Space*.)

Aggregation of Measurements

Let us suppose that the President of the United States desires to determine the economic well-being of the country. To do so, he requests that the income of each family in the country in the last year be presented to him. The president soon discovers that the rule of "seven plus or minus two" is presenting him with a real constraint, and the only way he can get the total picture is to give up detail and aggregate the millions of numbers, or income figures, which represent the elements of the population.

Suppose we have N families and m different incomes x_i $(i = 1, 2, \ldots, m)$. $m \leq N$ because some families may have the same income, so that x_i is the income of each of n_i families and x_j the income of each of n_j families; $\sum_{i=1}^{m} n_i = N$ and $n_i \leq N$. The President aggregates the measurements x_i into parameters to describe the population. One parameter is a measure of *central location* or the *mean* of the x_i, also referred to as the *expected value* μ of x_i:

$$\mu = \frac{1}{N} \sum_{i=1}^{m} n_i x_i = \sum_{i=1}^{m} f_i x_i \tag{6-9}$$

in which

$$f_i = \frac{n_i}{N}, \quad \sum_i f_i = 1$$

f_i represents a measure of "density," i.e., the fraction of families that have yearly income x_i. The mean μ gives an idea of income per "average family." However, it is quite possible that very few families have an income close to the mean, while many have a smaller income and very few have a much larger income. Yet it is also possible that most families have, indeed, an income close to μ. To get a better idea as to which of those situations is more representative of reality, we aggregate the data in the form of a *parameter of dispersion* called the *population variance* σ^2, which is computed as follows:

$$\sigma^2 = \sum_i f_i (x_i - \mu)^2 \tag{6-10}$$

σ^2 is the sum of the squares of the deviations of individual family incomes x_i from the mean income μ. The smaller σ^2, the more representative is μ as a model of *family income* in the country. For the extreme case of $\sigma^2 = 0$, μ is the income of each family. The larger σ^2, the larger the dispersion in incomes and the less representative μ is as a model of family income.

Population Parameters and Sample Statistics

μ and σ^2 are population parameters descriptive of location and dispersion. In many cases these quantities are not known, cannot be computed, or require too much computational effort. In such cases, *estimates* for the population parameters μ and σ^2 are computed from a sample of the population.* Such estimates are called *statistics*. Consider a sample of size n with measurements $x_i, \ldots, x_n$. The estimate $\hat{\mu}$ for μ is the *sample mean* $\bar{x}$:

$$\hat{\mu} = \bar{x} = \frac{1}{n} \sum_i x_i \tag{6-11}$$

and the estimate $\hat{\sigma}^2$ for σ^2 is the *sample variance* s^2:

$$\hat{\sigma}^2 = s^2 = \frac{1}{n-1} \sum_i (x_i - \bar{x})^2 \tag{6-12}$$

For large sample sizes, i.e., n large, $n - 1$ can be replaced by n with little influence on the results.†

Deductive and Inductive Inference

When a population is known, we can deduce certain conclusions regarding the attributes of a sample from the population. This is a process of deductive inference in which we go from the general to the limited, i.e., from the universal set to a subset or from the population to a sample. The reverse process takes place in inductive inference, when we use the measurements from a sample, i.e., the sample statistics, to infer the population parameters. Here we attempt to go from the limited to the general, from a subset to the universal set. Inductive inference involves states of partial knowledge and, therefore, requires that the degree of uncertainty in the inferences be assessed.

Deductive inference can be likened to a process in which conclusions are derived from a population with no new knowledge generated. Inductive inference is a process of generalization in which new knowledge may be generated.

---

*μ and σ^2 are not the only parameters although very commonly used. Additional parameters can be generated for a more complete description of the population.

†The reason for $n - 1$ in Eq. 6-12 derives from the fact that s^2 has the property of an unbiased estimate, namely, in the limit for many samples, $s^2 \longrightarrow \sigma^2$. If n is used instead of $n - 1$, then in the limit $s^2 \longrightarrow [(n-1)/n]\sigma^2$. This is treated in Sec. 6-5.

6-3 PROBABILITY DISTRIBUTION MODELS

Consider the function $p(x)$ (x may refer to the strength of a necklace) of Fig. 6-6(a) which has the following properties:

1. $p(x)$ is equal to or larger than zero.
2. The area under the curve for $p(x)$ is one unit.
3. The probability that a randomly selected necklace will have a strength x in pounds larger than a and smaller than b is given by the area under $p(x)$ between the values of $x = a$ and $x = b$.

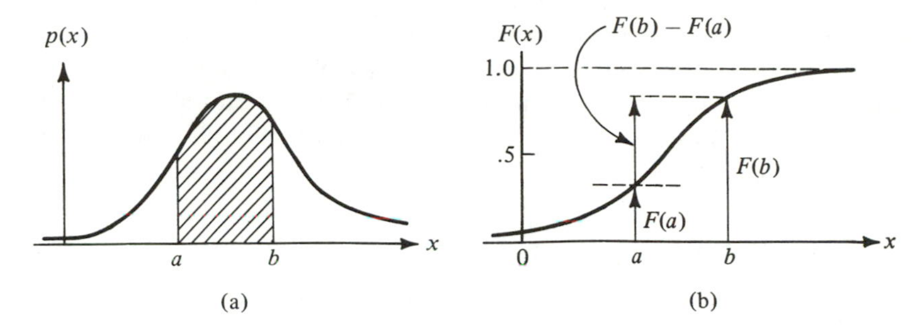

Figure 6-6 (a) Probability Density Function or Frequency Function; (b) Cumulative Distribution Function

Mathematically, these three properties can be written in the form:

$$\left.\begin{aligned} & p(x) \geq 0 \qquad \text{for } -\infty < x < \infty \\ & \int_{-\infty}^{\infty} p(x)\, dx = 1 \\ & Pr(a < x < b) = \int_a^b p(x)\, dx \end{aligned}\right\} \tag{6-13}$$

A function $p(x)$ with the above properties is called a *probability density function* or *frequency function* of random variable x. The variable x is random when, for all we know, all extraneous influences which may affect outcomes have been removed from the model.

The *cumulative distribution function** $F(x)$ is a monotonically increasing function derived from $p(x)$ [Fig. 6-6(b)]. For example, $F(a)$ is the probability of a necklace having a strength less than a pounds, and $F(b) - F(a)$ is the probability of a necklace having a strength smaller than b pounds but larger than a pounds. Mathematically, this can be written as:

$$\begin{aligned} F(a) &= Pr(x \leq a) \\ &= \int_{-\infty}^{a} p(x)\, dx \\ F(b) - F(a) &= Pr(a < x < b) = \int_a^b p(x)\, dx \end{aligned} \tag{6-14}$$

*The probability density function is sometimes referred to as a distribution function. The density function and the cumulative distribution are related by $p(x) = dF(x)/dx$.

In Fig. 6-6(b), the ordinate $F(a)$ for $x = a$ has a numerical value equal to the area under $p(x)$ in Fig. 6-6(a) to the left of $x = a$. Similarly, $F(b)$ has a numerical value equal to the area under $p(x)$ to the left of $x = b$. $F(b) - F(a)$ is then equal to the area under $p(x)$ between $x = a$ and $x = b$.

There are a number of well-known distribution functions which can be characterized by a small number of parameters and can serve as models to represent various phenomena of interest. Probability density functions fall into two main classes: discrete and continuous. In a *discrete function*, the random variable takes on discrete values, such as the number of heads in the toss of 50 coins. In a *continuous function*, the random variable is continuous, such as the length of pencils in a pencil factory. A number of discrete distribution models are discussed in the remainder of this section. Section 6-4 discusses the most important of the continuous distribution models, the normal distribution. Equations 6-13 and 6-14 are written for continuous distribution functions.

Discrete Probability Density Functions

Binomial Distribution. Consider the man in Fig. 5-16 in Chapter 5. Let us suppose that at each node there is a probability p that he will take link S to the next level and a probability q that he will take link E; $p + q = 1$. The event taking place at any level is *independent* of the events that have taken place at preceding levels. Namely, whether our man heads S or E from level k will in no way affect what he does at level $k + 1$ in terms of heading along S or E. Such independent events with two possible outcomes at each level (or state) are referred to as Bernoulli events or *Bernoulli trials.* For example, trials in repeated tosses of a coin with probability p for head and q for tail are Bernoulli trials.

Suppose we focus attention on the *number of links S* our man is taking in going from level 0 to level n as the attribute of interest or the subject of discourse. In doing so, we identify the *sample space*, i.e., the population which contains all possible samples obtained with repeated experiments in which our man in Fig. 5-16 walks down from level 0 to level n. We let the random variable x represent the discrete number of links S in a route and, therefore, x can take on the values $0, 1, \ldots, n$. For $x = 0$, the route consists of a sequence of n E links, i.e., the man goes down along a straight line heading east. Similarly, for $x = n$, the route is a straight line heading south. There are n different routes with only a single link S or with only a single link E. In general, there are $\binom{n}{r}$ different routes with r links S, and $(n - r)$ links E. The probability of our man following *a particular* route with r links S and $(n - r)$ links E is given by $p^r q^{n-r}$ because the choices of links at each level are independent. However, since we are concerned only with the number of links S in a route, and not with the pattern or order in which they appear, we can cluster all routes with exactly x links S. The probability, $p(x)$, of a route (any) with x links S out of a total of n links, is the sum of the probabilities of all routes from level 0 to level n which have x links S because these routes are mutually exclusive. There are $\binom{n}{x}$ or $n!/x!(n-x)!$ such routes, and each has a probability $p^x q^{n-x}$ of being followed by our man. Therefore,

$$p(x) = \binom{n}{x} p^x q^{n-x} = \frac{n!}{x!(n-x)!} p^x q^{n-x}, \quad x = 0, 1, \ldots, n \tag{6-15}$$

Equation 6-15 is the *binomial distribution function*. It is a discrete function (random variable x takes on discrete integer values) and has the following properties analogous to those of Eq. 6-13 for a continuous distribution:

$$\left.\begin{array}{l} p(x) \geq 0 \quad \text{for all } x \\ \sum_{\text{for all } x} p(x) = 1 \\ Pr(a < x < b) = \sum_{x=a+1}^{b-1} p(x) \\ \quad (a \text{ and } b \text{ are integers}) \end{array}\right\} \tag{6-16}$$

* *Poisson Distribution*. For Bernoulli trials in which n is large and p small, say, $n \geq 50$, $p \leq 0.1$, and $np = \lambda$ is a constant of moderate size in the order of 5, the binomial distribution can be approximated by the Poisson distribution:

$$p(x) = \frac{\lambda^x}{x!} e^{-\lambda}, \quad x = 0, 1, \ldots, n \tag{6-17}$$

in which $\lambda = np$.

The Poisson distribution is useful as a model for queueing problems such as incoming calls at a telephone exchange, computer time sharing, customer service at markets and banks, and many other service systems in which arrival of customers is random, with successive arrivals being independent during small time intervals. For a time interval t, the probability $p(x)$ that x customers will arrive is given by*

$$p(x) = \frac{(\lambda t)^x}{x!} e^{-\lambda t} \qquad x = 0, 1, \ldots, n \tag{6-18}$$

λ is the arrival rate, i.e., average number of arrivals per unit time. If the average number of customers served per unit time, i.e., *service rate*, is designated by μ, then $h = \lambda/\mu$ is the *utilization factor*. h is the proportion of the time the service is in use, or probability that a customer will have to wait; $h < 1$, for otherwise the waiting line will grow indefinitely long. The following expressions can be derived from the Poisson distribution model of a waiting-

*The customers can be replaced by machines to be serviced, trucks to be loaded or unloaded, airlines arriving at airport, etc.

line service system:

average number of customers in system at any time	$\dfrac{h}{1-h}$
average time between arrival and departure for a customer	$\dfrac{1}{\mu-\lambda}$
average length of waiting line	$\dfrac{h^2}{1-h}$
average time a customer waits to be served	$\dfrac{h}{\mu-\lambda}$

* *Hypergeometric Distribution.* When a set contains a total of $a+b$ elements in which a are of one kind and b of another, then a random selection of $n \leq a+b$ elements without replacements will yield x elements of type a with probability $p(x)$ given by this ratio:

$$p(x) = \frac{\binom{a}{x}\binom{b}{n-x}}{\binom{a+b}{n}}, \quad x = 0, 1, \ldots, n \tag{6-19}$$

$$x \leq a, \quad n - x \leq b$$

When n is small compared to $a+b$, the binomial distribution is a good approximation to the hypergeometric distribution.

6-4 NORMAL DISTRIBUTION MODEL

One of the most important and useful continuous distributions is the bell-shaped function of Fig. 6-7, called the *normal distribution.* It is completely described from Eq. 6-20 by specifying two parameters, the mean μ and the variance σ^2:

$$p(x) = \frac{1}{\sigma\sqrt{2\pi}} e^{-1/2[(x-\mu)/\sigma]^2} \tag{6-20}$$

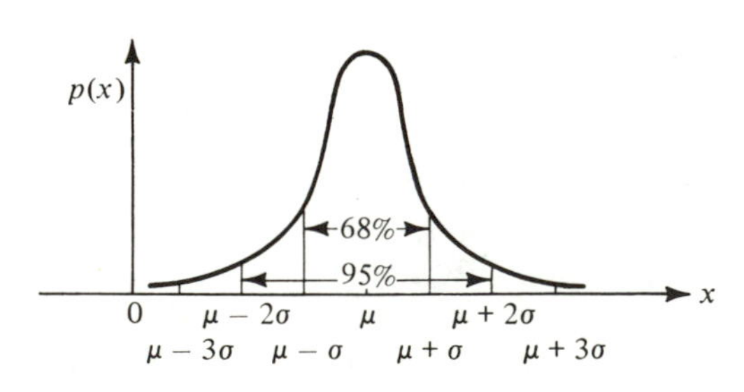

Figure 6-7

The cumulative normal distribution is given by (see Eq. 6-14):

$$F(a) = \int_{-\infty}^{a} p(x)\, dx \tag{6-21}$$

The parameters μ and σ^2 are defined from:

$$\mu = \int_{-\infty}^{\infty} x\, p(x)\, dx \tag{6-22}$$

and

$$\sigma^2 = \int_{-\infty}^{\infty} (x - \mu)^2\, p(x)\, dx \tag{6-23}$$

As seen in Fig. 6-8, the larger σ^2, the more spread is the distribution. To use an analog from mechanics, the location μ can be viewed as the center of gravity of the area under the probability density function along the x axis, and σ^2 is the moment of inertia of the area with respect to a vertical axis going through μ. Namely, each strip $p(x)\, dx$ is multiplied by the square of its distance from μ, $(x - \mu)^2$; the products are added for all strips and in the limit, as $dx \longrightarrow 0$, the sum becomes the integral of Eq. 6-23.

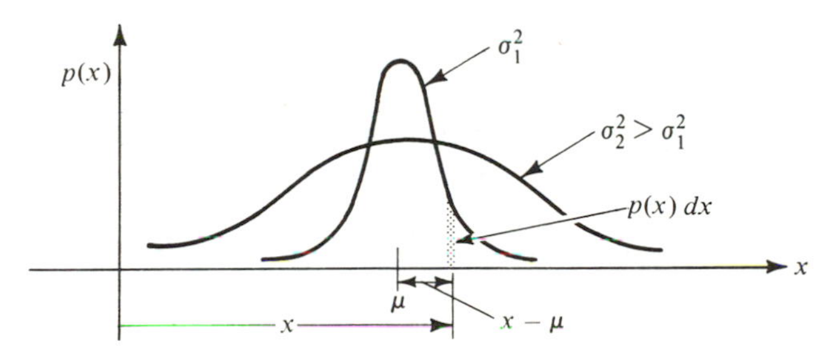

Figure 6-8

The square root of the variance is called the *standard deviation* σ. The standard deviation provides some standard measures regarding the normal distribution as follows. The probability of obtaining an observation x between the values of $\mu - \sigma$ and $\mu + \sigma$ is 0.68. Namely, the probability that x will not differ from the mean μ by more than σ units is 0.68. (See Fig. 6-7.) Similarly, the probabilities are 0.95 and 0.998 that x will not differ from μ by more than 2σ and 3σ, respectively. For example, consider the scores in an examination to be represented by a model of the normal distribution function with mean $\mu = 65$ points and variance $\sigma^2 = 49$, i.e., $\sigma = 7$ points. On the basis of this model we can make the following probabilistic predictions. The probabilities are 0.68 that a student will score between $65 - 7$ and $65 + 7$ points, 0.95 for a score in the range 51 to 79, and 0.998 for a score in the range 44 to 86. Since the distribution is symmetrical with respect to the mean μ with one-half the area to the right and one-half to the left, we can also answer the following questions. What is the probability of a student scoring above $\mu + \sigma$, or above $\mu + 2\sigma$, or $\mu + 3\sigma$? The probabilities are 0.16, 0.025, and 0.001, respectively, and are generated as follows. The probability of $\mu - \sigma < x < \mu + \sigma$ is 0.68. Therefore, the probability of x outside this range is $1 - 0.68 = 0.32$; but since we are interested in only the high range ($x > \mu + \sigma$), the answer is $1/2(0.32) = 0.16$ because of symmetry of the normal distribution. Similar arguments lead to the values of 0.025 and 0.001. Thus, the probability of a student scoring more than $\mu + 2\sigma = 65 + 14 = 79$ is 0.025, and 0.001 for more than 86.

We can easily commit to memory the numbers 0.68, 0.95, and 0.998. However, suppose a student wishes to calculate the probability of a grade B or better in the examination, when the teacher announces that a grade of B or better requires a score above 75. This score is $\mu + (10/7)\sigma$. We know that the desired probability is smaller than 0.16 which is the probability for $x > \mu + \sigma = 72$, but larger than 0.025 which is the probability for $x > \mu + 2\sigma = 79$. To get the answer, we will need to compute the area $F(x)$ under the normal distribution curve for $\mu = 65$ and $\sigma = 7$ to the right of $x = 75$. We could calculate such areas and record the values of the cumulative distribution for values of x spaced at very close intervals such as 0.1 apart. However, for a different problem in which the normal distribution model is used, μ and σ^2 may be different and the table of $F(x)$ values for $\mu = 65$ and $\sigma = 7$ will require modification. It is, therefore, customary to resort to a *Standard Normal Distribution* table which can be used regardless of what the values of μ and σ^2 are in the model considered, as long as the variable is considered to be normally distributed.

Standard Normal Distribution and Its Use. The standard normal distribution is characterized as distinct from all other normal distributions by the fact that it has a mean $\mu = 0$ and variance $\sigma^2 = 1$ and, hence, standard deviation $\sigma = 1$. Table 6-1 gives the value of the cumulative distribution function $F(z)$ for various values of the standard normal variable which is designated by the letter z. In general, it is customary to denote the fact that variable x is normally distributed with mean μ_x and variance σ_x^2 (the subscript x identifies the variable of the distribution) in the form $N(\mu_x, \sigma_x^2)$. Using this notation, variable z has a distribution $N(0, 1)$, i.e., a normal distribution with mean $\mu_z = 0$ and variance $\sigma_z^2 = 1$. Equation 6-20 then becomes

$$p(z) = \frac{1}{\sqrt{2\pi}} e^{(-1/2)z^2} \tag{6-24}$$

Values of $p(z)$ are given in Table 6-2.

Using Table 6-1, we can find for example $Pr(z < 1.66)$, designated $F(1.66)$, by going to the value recorded in the row to the right of $z = 1.6$ under the column for .06, i.e., $Pr(z < 1.66) = .9515$. Using this result, $Pr(z > 1.66) = 1 - .9515 = .0485$. The values of $F(z)$ for negative values of z are obtained from the symmetry of the distribution, $F(-z) =$

$1 - F(z)$. For example, $F(-0.82) = 1 - 0.7939 \doteq 0.2061$. As another example,

$$\begin{aligned} Pr(-0.44 < z < 1.53) &= F(1.53) - F(-0.44) \\ &= F(1.53) - [1 - F(0.44)] \\ &= 0.9370 - (1 - 0.6700) \\ &= 0.6070 \end{aligned}$$

Table 6-1* Values of Cumulative Distribution $F(z)$ for the Standard Normal Distribution $N(0, 1)$

$$F(z) = \int_{-\infty}^{z} p(z)\, dz \qquad p(z) = \frac{1}{\sqrt{2\pi}} e^{(-1/2)z^2}$$

$F(z)$ for negative z are obtained from symmetry
$F(-z) = 1 - F(z)$. $F(1.27) = 0.8980$; $F(-0.82) = 1 - 0.7939 = 0.2061$

z	.00	.01	.02	.03	.04	.05	.06	.07	.08	.09
.0	.5000	.5040	.5080	.5120	.5160	.5199	.5239	.5279	.5319	.5359
.1	.5398	.5438	.5478	.5517	.5557	.5596	.5636	.5675	.5714	.5753
.2	.5793	.5832	.5871	.5910	.5948	.5987	.6026	.6064	.6103	.6141
.3	.6179	.6217	.6255	.6293	.6331	.6368	.6406	.6443	.6480	.6517
.4	.6554	.6591	.6628	.6664	.6700	.6736	.6772	.6808	.6844	.6879
.5	.6915	.6950	.6985	.7019	.7054	.7088	.7123	.7157	.7190	.7224
.6	.7257	.7291	.7324	.7357	.7389	.7422	.7454	.7486	.7517	.7549
.7	.7580	.7611	.7642	.7673	.7704	.7734	.7764	.7794	.7823	.7852
.8	.7881	.7910	.7939	.7967	.7995	.8023	.8051	.8078	.8106	.8133
.9	.8159	.8186	.8212	.8238	.8264	.8289	.8315	.8340	.8365	.8389
1.0	.8413	.8438	.8461	.8485	.8508	.8531	.8554	.8577	.8599	.8621
1.1	.8643	.8665	.8686	.8708	.8729	.8749	.8770	.8790	.8810	.8830
1.2	.8849	.8869	.8888	.8907	.8925	.8944	.8962	.8980	.8997	.9015
1.3	.9032	.9049	.9066	.9082	.9099	.9115	.9131	.9147	.9162	.9177
1.4	.9192	.9207	.9222	.9236	.9251	.9265	.9279	.9292	.9306	.9319
1.5	.9332	.9345	.9357	.9370	.9382	.9394	.9406	.9418	.9429	.9441
1.6	.9452	.9463	.9474	.9484	.9495	.9505	.9515	.9525	.9535	.9545
1.7	.9554	.9564	.9573	.9582	.9591	.9599	.9608	.9616	.9625	.9633
1.8	.9641	.9649	.9656	.9664	.9671	.9678	.9686	.9693	.9699	.9706
1.9	.9713	.9719	.9726	.9732	.9738	.9744	.9750	.9756	.9761	.9767
2.0	.9772	.9778	.9783	.9788	.9793	.9798	.9803	.9808	.9812	.9817
2.1	.9821	.9826	.9830	.9834	.9838	.9842	.9846	.9850	.9854	.9857
2.2	.9861	.9864	.9868	.9871	.9875	.9878	.9881	.9884	.9887	.9890
2.3	.9893	.9896	.9898	.9901	.9904	.9906	.9909	.9911	.9913	.9916
2.4	.9918	.9920	.9922	.9925	.9927	.9929	.9931	.9932	.9934	.9936
2.5	.9938	.9940	.9941	.9943	.9945	.9946	.9948	.9949	.9951	.9952
2.6	.9953	.9955	.9956	.9957	.9959	.9960	.9961	.9962	.9963	.9964
2.7	.9965	.9966	.9967	.9968	.9969	.9970	.9971	.9972	.9973	.9974
2.8	.9974	.9975	.9976	.9977	.9977	.9978	.9979	.9979	.9980	.9981
2.9	.9981	.9982	.9982	.9983	.9984	.9984	.9985	.9985	.9986	.9986
3.0	.9987	.9987	.9987	.9988	.9988	.9989	.9989	.9989	.9990	.9990

*Reprinted by permission of John Wiley and Sons, Inc., from *Statistical Tables and Formulas*, A. Hald, John Wiley and Sons Inc., New York, 1952.

Table 6-2* Ordinates of the Standard Normal Density Function $N(0, 1)$

$$p(z) = \frac{1}{\sqrt{2\pi}} e^{-z^2/2}$$

z	.00	.01	.02	.03	.04	.05	.06	.07	.08	.09
.0	.3989	.3989	.3989	.3988	.3986	.3984	.3982	.3980	.3977	.3973
.1	.3970	.3965	.3961	.3956	.3951	.3945	.3939	.3932	.3925	.3918
.2	.3910	.3902	.3894	.3885	.3876	.3867	.3857	.3847	.3836	.3825
.3	.3814	.3802	.3790	.3778	.3765	.3752	.3739	.3725	.3712	.3697
.4	.3683	.3668	.3653	.3637	.3621	.3605	.3589	.3572	.3555	.3538
.5	.3521	.3503	.3485	.3467	.3448	.3429	.3410	.3391	.3372	.3352
.6	.3332	.3312	.3292	.3271	.3251	.3230	.3209	.3187	.3166	.3144
.7	.3123	.3101	.3079	.3056	.3034	.3011	.2989	.2966	.2943	.2920
.8	.2897	.2874	.2850	.2827	.2803	.2780	.2756	.2732	.2709	.2685
.9	.2661	.2637	.2613	.2589	.2565	.2541	.2516	.2492	.2468	.2444
1.0	.2420	.2396	.2371	.2347	.2323	.2299	.2275	.2251	.2227	.2203
1.1	.2179	.2155	.2131	.2107	.2083	.2059	.2036	.2012	.1989	.1965
1.2	.1942	.1919	.1895	.1872	.1849	.1826	.1804	.1781	.1758	.1736
1.3	.1714	.1691	.1669	.1647	.1626	.1604	.1582	.1561	.1539	.1518
1.4	.1497	.1476	.1456	.1435	.1415	.1394	.1374	.1354	.1334	.1315
1.5	.1295	.1276	.1257	.1238	.1219	.1200	.1182	.1163	.1145	.1127
1.6	.1109	.1092	.1074	.1057	.1040	.1023	.1006	.0989	.0973	.0957
1.7	.0940	.0925	.0909	.0893	.0878	.0863	.0848	.0833	.0818	.0804
1.8	.0790	.0775	.0761	.0748	.0734	.0721	.0707	.0694	.0681	.0669
1.9	.0656	.0644	.0632	.0620	.0608	.0596	.0584	.0573	.0562	.0551
2.0	.0540	.0529	.0519	.0508	.0498	.0488	.0478	.0468	.0459	.0449
2.1	.0440	.0431	.0422	.0413	.0404	.0396	.0387	.0379	.0371	.0363
2.2	.0355	.0347	.0339	.0332	.0325	.0317	.0310	.0303	.0297	.0290
2.3	.0283	.0277	.0270	.0264	.0258	.0252	.0246	.0241	.0235	.0229
2.4	.0224	.0219	.0213	.0208	.0203	.0198	.0194	.0189	.0184	.0180
2.5	.0175	.0171	.0167	.0163	.0158	.0154	.0151	.0147	.0143	.0139
2.6	.0136	.0132	.0129	.0126	.0122	.0119	.0116	.0113	.0110	.0107
2.7	.0104	.0101	.0099	.0096	.0093	.0091	.0088	.0086	.0084	.0081
2.8	.0079	.0077	.0075	.0073	.0071	.0069	.0067	.0065	.0063	.0061
2.9	.0060	.0058	.0056	.0055	.0053	.0051	.0050	.0048	.0047	.0046
3.0	.0044	.0043	.0042	.0040	.0039	.0038	.0037	.0036	.0035	.0034

*Reprinted by permission of John Wiley and Sons, Inc., from *Statistical Tables and Formulas*, A. Hald, John Wiley and Sons Inc., New York, 1952.

Table 6-1 can be used as a model for any normal distribution $N(\mu_x, \sigma_x^2)$. This is achieved by transforming distribution $N(\mu_x, \sigma_x^2)$ of random variable x to a standard normal distribution $N(0, 1)$ of random variable z. Figure 6-9 shows on the same abscissa (horizontal coordinate) a normal distribution $N(19,9)$ of variable x, i.e., $\mu_x = 19$, $\sigma_x^2 = 9$, and the standard normal distribution $N(0, 1)$ of variable z. If we now subtract μ_x from variable x, the distribution $p(x)$ of x will be shifted to the left, as shown by the dashed line curve $p(x - \mu_x)$ in Fig. 6-9. The mean of the shifted distribution is zero; namely, it was transformed from $N(\mu_x, \sigma_x^2)$ to $N(0, \sigma_x^2)$. Next, we squeeze in

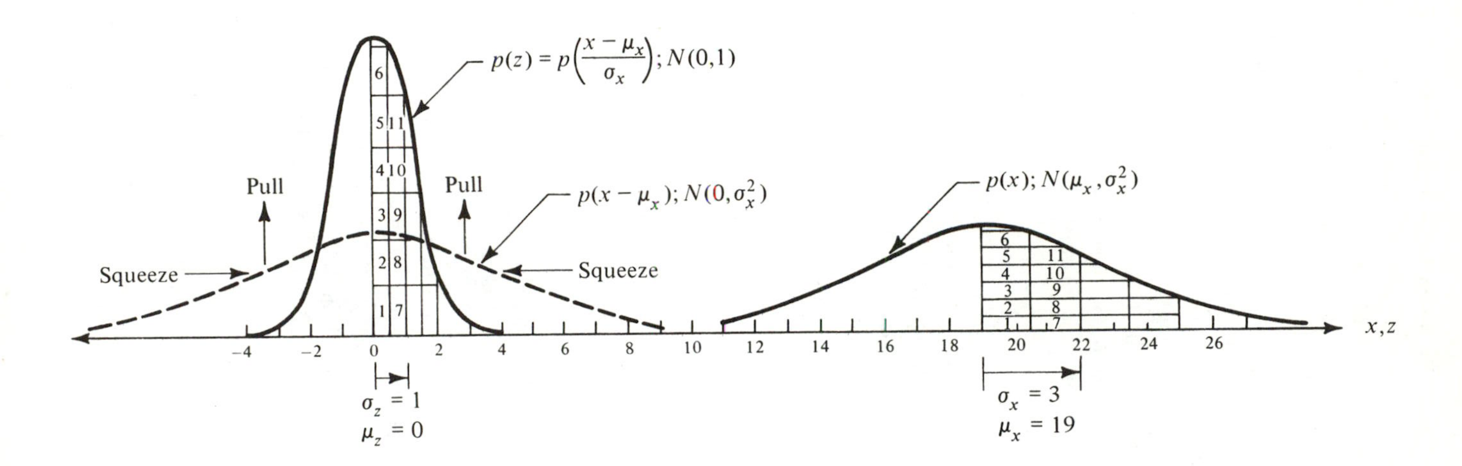

Figure 6-9 Standard Normal Distribution $N(0, 1)$, Distribution $N(\mu_x, \sigma_x^2)$, and Transformation of $N(\mu_x, \sigma_x^2)$ to $N(0, 1)$

the distribution symmetrically toward the mean and pull it up at the same time, so that the area under the curve remains one, until the dashed curve $p(x - \mu_x)$ becomes identical to $p(z)$ in Fig. 6-9. The squeezing in will result in reducing the horizontal scale of $p(x - \mu_x)$ by a factor of $1/\sigma_x$ and increasing the vertical scale by a factor of σ_x.

Figure 6-9 shows 11 numbered rectangles, one standard deviation from the mean for $p(x)$ and $p(z)$. The areas of the rectangles are identical as is the area one standard deviation σ to the right of the mean in any normal distribution, i.e., 0.34 or 34% of the total area. The squeezing operation is merely a change in the dimensions of the rectangles in $p(x)$ to those in $p(z)$ without change in area. The two-step operation described is a transformation of variable x to variable z from this formula:

$$z = \frac{x - \mu_x}{\sigma_x} \tag{6-25}$$

$x - \mu_x$ is the shifting of the distribution to make the mean zero, and factor $1/\sigma_x$ is the squeezing*, or change in scale, to make the variance one.

Transformation Eq. 6-25 reduces $N(\mu_x, \sigma_x^2)$ to $N(0, 1)$ and now Table 6-1 for standard normal variable z with $N(0,1)$ can be used to make predictions regarding x in $N(\mu_x, \sigma_x^2)$. For any value of x and corresponding z from Eq. 6-25, we have

$$F(x) = F(z)$$

and, in general,

$$Pr(x_1 < x < x_2) = Pr(z_1 < z < z_2)$$

in which

$$z_1 = \frac{x_1 - \mu_x}{\sigma_x}$$

and $$z_2 = \frac{x_2 - \mu_x}{\sigma_x}$$

EXAMPLE 1. Suppose we wish to compute $Pr(x < \mu_x + \sigma_x)$ in Fig. 6-9. This is equivalent to the area to the left of $x = 22$ under $p(x)$, but this area is also the same as the area to the left of $z = 1$ under $p(z)$, so we can read this value from Table 6-1, i.e., 0.8413. In this example, $x = 22$ is transformed to $z = 1$. This transformation is obtained directly from Eq. 6-25:

$$z = \frac{22 - 19}{3} = 1$$

Thus, $Pr(x < 22) = Pr(z < 1) = 0.8413$.

*When $\sigma_x < 1$, the squeezing becomes a stretching away from the mean accompanied by a downward push of $p(x)$ to keep the area the same.

EXAMPLE 2. Consider a random variable x with distribution $N(65, 144)$. What is the probability of x larger than 80, i.e., $Pr(x > x_1)$, $x_1 = 80$?

$$Pr(x > x_1) = Pr(z > z_1)$$

$$z_1 = \frac{x_1 - \mu_x}{\sigma_x} = \frac{80 - 65}{12} = 1.25$$

$$Pr(z > 1.25) = 0.1056, \quad \text{thus } Pr(x > 80) = 0.1056$$

EXAMPLE 3. What is the value x_1 of x in Example 2, for which the probability is 0.1788, that a random observation of x will be smaller than x_1? Namely, $Pr(x < x_1) = 0.1788$, $x_1 = ?$

Now, $Pr(x < x_1) = Pr(z < z_1) = 0.1788$ from Table 6-1; z_1 must be negative because $Pr(z < 0) = 0.5$. It is equal to -0.92. But

$$z_1 = \frac{x_1 - \mu_x}{\sigma_x}$$

$$-0.92 = \frac{x_1 - 65}{12}$$

$$x_1 = 65 - 12 \times 0.92 = 53.96 \approx 54$$

EXAMPLE 4. What is the variance σ_x^2 of a distribution $N(\mu_x, \sigma_x^2)$ if $\mu_x = 70$ and $Pr(x > 80) = 0.0062$?

$$Pr(x > x_1) = Pr(z > z_1) = 0.0062$$

Using Table 6-1,

$$z_1 = 2.50$$

Using Eq. 6-25,

$$z_1 = \frac{x_1 - \mu_x}{\sigma_x}$$

Since $\quad z_1 = 2.50,\ x_1 = 80,\ \mu_x = 70$

then $\quad 2.50 = \dfrac{80 - 70}{\sigma_x}$

and $\quad \sigma_x = \dfrac{10}{2.5} = 4$

$$\sigma_x^2 = 16$$

EXAMPLE 5. Find the mean μ_x of $N(\mu_x, 36)$, given that $Pr(x < 75) = 0.8315$

$$Pr(x < x_1) = Pr(z < z_1) = 0.8315, \quad \text{thus } z_1 = 0.96$$

$$z_1 = \frac{x_1 - \mu_x}{\sigma_x}$$

$$z_1 = 0.96,\ x_1 = 75,\ \sigma_x = 6$$

$$0.96 = \frac{75 - \mu_x}{6}$$

$$\mu_x = 75 - 5.76 = 69.24$$

6-5 EXPECTED VALUES OF RANDOM VARIABLES AND THEIR AGGREGATES

Expected Value*

The mean μ which describes location of a population, variance σ^2 which describes population dispersion, and other measures of population distribution are derived from an *operator* called *expected value*. The expected value of any function $f(x)$ of random variable x is a *weighted average* of the $f(x)$ values given by

$$E[f(x)] = \int_{-\infty}^{\infty} f(x)\, p(x)\, dx \tag{6-26}$$

for a continuous probability density function $p(x)$. For a discrete distribution $p(x_i)$,

$$E[f(x_i)] = \textstyle\sum_{\text{all } x_i} f(x_i)\, p(x_i) \tag{6-27}$$

The expected value operator E is a *linear operator* and has the following properties:

- The expected value of the sum of two functions $f(x)$ and $g(x)$ of random variable x is equal to the sum of the expected values. Namely,

  $$E[f(x) + g(x)] = E[f(x)] + E[g(x)] \tag{6-28}$$

  This follows from the definition of expected value in Eq. 6-26. (We can show the same for Eq. 6-27.)

  $$\begin{aligned} E[f(x) + g(x)] &= \int_{-\infty}^{\infty} [f(x) + g(x)]\, p(x)\, dx \\ &= \int_{-\infty}^{\infty} f(x)\, p(x)\, dx + \int_{-\infty}^{\infty} g(x)\, p(x)\, dx \\ &= E[f(x)] + E[g(x)] \end{aligned} \tag{6-29}$$

- The expected value of a constant a is a, i.e.,

  $$E(a) = a \tag{6-30}$$

  $$E(a) = \int_{-\infty}^{\infty} a\, p(x)\, dx = a \int_{-\infty}^{\infty} p(x)\, dx = a$$

  because $\int_{-\infty}^{\infty} p(x)\, dx = 1$

- The expected value of a constant a multiplying $f(x)$ is equal to a times the expected value of $f(x)$:

  $$\begin{aligned} E[af(x)] &= \int_{-\infty}^{\infty} af(x)\, p(x)\, dx = a \int_{-\infty}^{\infty} f(x)\, p(x)\, dx \\ &= aE[f(x)] \end{aligned} \tag{6-31}$$

---

*Read the note preceding Exercise 6-25 and work Exercises 6-25 through 6-27 on expected value before proceeding with the reading of this section.

Some Results Using the Above Properties. For a random variable x with mean μ_x and variance σ_x^2 and probability density function $p(x)$:

1. $E(x) = \int_{-\infty}^{\infty} x\,p(x)\,dx = \mu_x$ (6-32)
   (See Eq. 6-22.)

2. $E[(x-\mu_x)^2] = \int_{-\infty}^{\infty} (x-\mu_x)^2\,p(x)\,dx = \sigma_x^2$ (6-33)
   (See Eq. 6-23.)

3. $$\begin{aligned} E[(x-\mu_x)^2] &= E(x^2 - 2\mu_x x + \mu_x^2) \\ &= E(x^2) - 2\mu_x E(x) + \mu_x^2 \\ &= E(x^2) - 2\mu_x^2 + \mu_x^2 \\ &= E(x^2) - \mu_x^2 \end{aligned}$$
   Hence, $\sigma_x^2 = E(x^2) - [E(x)]^2$ (6-34)

4. The mean of transformed random variable
   $$z = \frac{x-\mu_x}{\sigma_x}$$
   takes the following form by using expected values (Eq. 6-32):
   $$\begin{aligned} \mu_z = E(z) &= E\left(\frac{x-\mu_x}{\sigma_x}\right) \\ &= \frac{1}{\sigma_x} E(x-\mu_x) \\ &= \frac{1}{\sigma_x}[E(x) - E(\mu_x)] \\ &= \frac{1}{\sigma_x}(\mu_x - \mu_x) \\ &= 0 \end{aligned}$$

5. The variance of z becomes (Eq. 6-33):
   $$\begin{aligned} \sigma_z^2 = E[(z-\mu_z)^2] &= E\left(\frac{x-\mu_x}{\sigma_x}\right)^2 \\ &= \frac{1}{\sigma_x^2} E(x-\mu_x)^2 \\ &= \frac{1}{\sigma_x^2}(\sigma_x^2) \\ &= 1 \end{aligned}$$

The last two results confirm the transformation in Sec. 6-4 in which random variable x with μ_x and σ_x^2 is transformed to variable $z = (x-\mu_x)/\sigma_x$ and results in mean $\mu_z = 0$ and variance $\sigma_z^2 = 1$.

6. $E(\bar{x}) = E\left[\frac{1}{n}\sum_i x_i\right] = \frac{1}{n}\sum_i E(x_i) = \frac{1}{n}(n\mu_x) = \mu_x$

* 7. Why $n - 1$ in Eq. 6-12? Recall that Eq. 6-12 has the form

$$\hat{\sigma}^2 = s^2 = \frac{1}{n-1} \sum_i (x_i - \bar{x})^2$$

The answer is that the expected value of s^2, as defined by Eq. 6-12, is equal to σ^2, i.e., $E(s^2) = \sigma^2$ as shown in the following derivation:

$$\begin{aligned}\sum_i (x_i - \bar{x})^2 &= \sum_i [(x_i - \mu) - (\bar{x} - \mu)]^2 \\ &= \sum_i (x_i - \mu)^2 - 2\sum_i (x_i - \mu)(\bar{x} - \mu) \\ &\quad + \sum_i (\bar{x} - \mu)^2 \\ &= \sum_i (x_i - \mu)^2 - n(\bar{x} - \mu)^2\end{aligned}$$

The last result is due to the fact that

$$\begin{aligned}2\sum_i (x_i - \mu)(\bar{x} - \mu) &= 2\sum_i n\left(\frac{x_i}{n} - \frac{\mu}{n}\right)(\bar{x} - \mu) \\ &= 2n(\bar{x} - \mu)\left(\sum_i \frac{x_i}{n} - \sum_i \frac{\mu}{n}\right) \\ &= 2n(\bar{x} - \mu)(\bar{x} - \mu) \\ &= 2n(\bar{x} - \mu)^2\end{aligned}$$

and $\quad \sum_i (\bar{x} - \mu)^2 = n(\bar{x} - \mu)^2$

We can, therefore, write

$$\begin{aligned}E[\sum_i (x_i - \bar{x})^2] &= E[\sum_i (x_i - \mu)^2] - E[n(\bar{x} - \mu)^2] \\ &= n\sigma^2 - n\frac{\sigma^2}{n} \\ &= (n-1)\sigma^2\end{aligned}$$

Hence, $\quad E(s^2) = E\left[\frac{1}{n-1}\sum_i (x_i - \bar{x})^2\right] = \sigma^2$

The above result shows that the statistic s^2, as expressed by Eq. 6-12, is an *unbiased estimate of the parameter* σ^2, because its expected value is equal to σ^2.

Mean and Variance of an Aggregate

Consider a single coin with probability of heads 1/2. The distribution function for the number of heads x in a single toss is $p(0) = 1/2$ and $p(1) = 1/2$ as shown in Fig. 6-10 (the sample space consists of $x = 0$ and $x = 1$).

The mean μ_x and variance σ_x^2 are given by

$$\mu_x = E(x) = \sum_{\text{all } x} xp(x) = 0 \cdot \frac{1}{2} + 1 \cdot \frac{1}{2} = \frac{1}{2}$$

$$\begin{aligned}\sigma_x^2 &= E[(x - \mu_x)^2] = \sum_{\text{all } x} (x - \mu_x)^2 p(x) \\ &= \left(0 - \frac{1}{2}\right)^2 \cdot \frac{1}{2} + \left(1 - \frac{1}{2}\right)^2 \cdot \frac{1}{2} = \frac{1}{4}\end{aligned}$$

$\mu_x = 1/2$ and $\sigma_x^2 = 1/4$ computed above are, respectively, the *population*

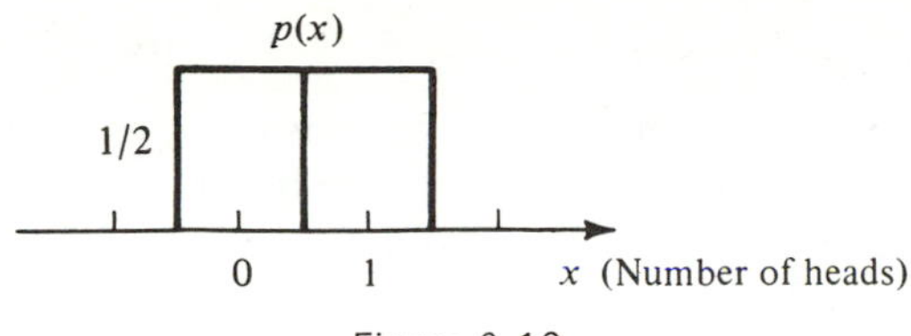

Figure 6-10

mean and *variance* of random variable x which represents the number of heads in the toss of single coin which has a probability of 1/2 of falling heads.

Let us define a *new population* by using the above coin. *The new population consists of the aggregated number of heads w from n tosses of the coin.* The sample space consists now of all integer values of w from 0 to n. w is the sum of the x's in each of the n tosses, i.e.,

$$w = \sum_{i=1}^{n} x_i$$

x_i stands for the number of heads in toss i ($i = 1, 2, \ldots, n$), and it can take on the values of 0 or 1. Now let us define a variable $\bar{x}$ in this new population which is the average number of heads in n tosses of the coin, i.e.,

$$\bar{x} = \frac{1}{n} w = \frac{1}{n} \sum_{i=1}^{n} x_i$$

$\bar{x}$ is the *aggregated proportion* of heads in n tosses. The mean $\mu_{\bar{x}}$ of the population of random variables $\bar{x}$ is:

$$\mu_{\bar{x}} = E(\bar{x}) = E\left[\frac{1}{n} \sum_{i=1}^{n} x_i\right] = \frac{1}{n} \sum_i E(x_i)$$

$$= \frac{1}{n} [E(x_1) + E(x_2) + \ldots + E(x_i) + \ldots + E(x_n)]$$

in which $E(x_i)$ is the expected value in the ith toss. $E(x_i) = \mu_x$ for all i. Hence,

$$\mu_{\bar{x}} = E(\bar{x}) = \frac{1}{n} n\mu_x = \mu_x = \frac{1}{2} \tag{6-35}$$

The variance $\sigma_{\bar{x}}^2$ of the population of random variables $\bar{x}$ is computed as follows. (The reader who is endowed with a strong *will to believe* may skip the following derivation and accept the result given by Eq. 6-36.)

$$\begin{aligned}
\sigma_{\bar{x}}^2 &= E(\bar{x} - \mu_{\bar{x}})^2 \\
&= E\left[\frac{1}{n} \sum_i x_i - \mu_x\right]^2 \\
&= E\left[\frac{1}{n} \sum_i (x_i - \mu_x)\right]^2 \\
&= E\left\{\frac{1}{n^2}\left[\sum_i (x_i - \mu_x)^2 + \sum_i \sum_j (x_i - \mu_x)(x_j - \mu_x)\right]\right\} \\
&= E\left[\frac{1}{n^2} \sum_i (x_i - \mu_x)^2\right] + E\left[\sum_{\substack{i \\ i \neq j}} \sum_j (x_i - \mu_x)(x_j - \mu_x)\right]
\end{aligned}$$

In the last expression, for each i, j with $i \neq j$,

$$E(x_i - \mu_x)(x_j - \mu_x) = 0$$

when x_i and x_j are independent. Let us demonstrate this point. x_i is the number of heads in toss i, and x_j the number of heads in toss j. $(x_i - \mu_x)$ can assume the values $0 - 1/2 = -1/2$ and $1 - 1/2 = 1/2$, each with probability of 1/2. Similarly, $(x_j - \mu_x)$ can assume the values $-1/2$ and $1/2$ each with a probability of 1/2. The expected value of the cross products $(x_i - \mu_x)(x_j - \mu_x)$ is the sum of the products of these values when multiplied by their corresponding probabilities $[p(0) = p(1) = 1/2]$:

$$\left(0 - \frac{1}{2}\right)\left(0 - \frac{1}{2}\right)p(0)\,p(0) + \left(0 - \frac{1}{2}\right)\left(1 - \frac{1}{2}\right)p(0)\,p(1)$$
$$+ \left(1 - \frac{1}{2}\right)\left(0 - \frac{1}{2}\right)p(1)\,p(0) + \left(1 - \frac{1}{2}\right)\left(1 - \frac{1}{2}\right)p(1)\,p(1)$$
$$= \left(\frac{1}{2}\right)^4 - \left(\frac{1}{2}\right)^4 - \left(\frac{1}{2}\right)^4 + \left(\frac{1}{2}\right)^4 = 0$$

Therefore, the expression for $\sigma_{\bar{x}}^2$ reduces to

$$\begin{aligned}\sigma_{\bar{x}}^2 &= E\left[\frac{1}{n^2}\sum_{i=1}^n (x_i - \mu_x)^2\right] \\ &= \frac{1}{n^2}\sum_{i=1}^n E(x_i - \mu_x)^2 \\ &= \frac{1}{n^2}(n\sigma_x^2) \\ &= \frac{1}{n}\sigma_x^2 \qquad (6\text{-}36)\end{aligned}$$

The fact that aggregation leads to a smaller variance should not be surprising because in the limit, when we aggregate the entire population, the variance is zero. An analogy is the computation of moments of inertia of mass. The moment of inertia is analogous to the variance, and it gets smaller as we aggregate more and more mass as chunks concentrated at their respective centers of gravity. The center of gravity of the mass is analogous to the mean.

Generalization. In general it can be shown that for *independent* random variables x_i, each with mean μ_i and variance σ_i^2, a linear combination (a_i are constants)

$$w = \sum_i a_i x_i \qquad (6\text{-}37)$$

has a mean

$$\mu_w = \sum_i a_i \mu_i \qquad (6\text{-}38)$$

and variance

$$\sigma_w^2 = \sum_i a_i^2 \sigma_i^2 \qquad (6\text{-}39)$$

Equations 6-37, 6-38, and 6-39 are consistent with the preceding derivation for the mean and variance of $\bar{x}$. Because $\bar{x}$ is obtained from x_i,

$$\bar{x} = \frac{1}{n} \sum_i x_i \tag{6-40}$$

$1/n$ in this equation is a_i in Eq. 6-37, $a_i = 1/n$ for all i, and hence $\bar{x}$ is a linear combination of random variables x_i, each with $\mu_i = \mu_x$ and $\sigma_i^2 = \sigma_x^2$. Therefore, from Eqs. 6-38 and 6-39,

$$\mu_{\bar{x}} = \sum_{i=1}^{n} \frac{1}{n} \mu_i = \mu_x \tag{6-41}$$

$$\sigma_{\bar{x}}^2 = \sum_{i=1}^{n} \left(\frac{1}{n}\right)^2 \sigma_i^2 = \frac{1}{n^2} n\sigma_x^2 = \frac{1}{n}\sigma_x^2 \tag{6-42}$$

6-6 CENTRAL LIMIT THEOREM AND ITS APPLICATION

Central Limit Theorem

If a random variable x is distributed with mean μ and variance σ^2, then an aggregation of n independent observations x_i, i.e., a linear combination $w = \sum_i x_i$ which is obtained from random samples of size n, will be approximately distributed in a normal distribution with mean $n\mu$ and variance $n\sigma^2$. The approximation will become progressively better as the sample size n increases.

The sample mean $\bar{x} = (1/n) \sum_i x_i$ will approach a normal distribution with mean μ and variance $(1/n)\,\sigma^2$ as the sample size n increases (see Sec. 6-5 for mean and variance of $\bar{x}$).

Note that the theorem makes no mention of the distribution of random variable x. All it requires is a mean μ and variance σ^2 for x and then, regardless of the distribution of x, the distribution of $\bar{x}$ approaches the normal distribution $N(\mu, \sigma^2/n)$ as n increases. This theorem is the reason for the great importance attached to the normal distribution. Whenever variables are aggregated, the normal distribution becomes a good approximation, i.e., a reasonable model for the distribution of the aggregated variable.

As an example, let us consider again our man in Fig. 5-16 of Chapter 5. Let x represent the number of S links the man is taking from a node at any level in order to reach the level below it; then x can take on the value 0 if he moves along a link E (0 links S), or 1 if he moves along a link S. Now suppose we have only levels 0 and 1 in Fig. 5-16 and we flip a coin to decide on S or E. Then $P(x = 0) = P(x = 1) = 1/2$, and the probability distribution of x is as shown in Fig. 6-11(a). The distribution is a uniform distribution (same probability for each outcome 0 or 1) and is far from the bell-shaped normal distribution. Next, we consider the distribution of the number of links S in going from level 0 to level 2, i.e., two levels; then $w = \sum_{i=1}^{2} x_i$ can assume

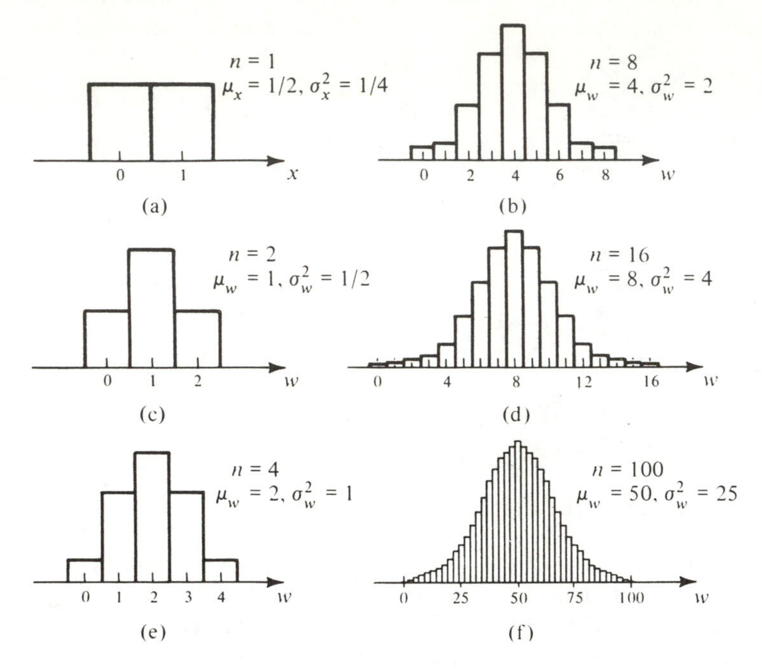

Figure 6-11 Binomial Distribution for $p = \frac{1}{2}$ for Values of $n = 1$, Variable $x = 0, 1$; and for $n = 2, 4, 8, 16, 100$, Variable $w = \sum x_i$

the values 0, 1, 2, and the distribution is shown in Fig. 6-11(b). For five levels $w = \sum_{i=1}^{5} x_i = 0, 1, \ldots, 5$, and in general for n levels $w = \sum_{i=1}^{n} x_i = 0, 1, \ldots, n$ and the distribution of w is calculated from the binomial distribution

$$p(w) = \binom{n}{w} p^w q^{n-w} \tag{6-43}$$

Figure 6-11 shows the probability distributions of w for $n = 1, 2, 4, 8, 16, 100$. The distribution appears progressively closer to the bell-shaped normal distribution as n increases. Thus, the normal distribution curve with mean μ_w and variance $\sigma_w{}^2$ can be used as a model to calculate the probabilities associated with random variable w instead of using Eq. 6-43.

EXAMPLE 1. Suppose the man in Fig. 5-16 proceeds from level 0 to 100, i.e., $n = 100$. What is the probability that he will follow a route with at least 60 links S, i.e., $w \geq 60$? Using Eq. 6-43, we must calculate the following:

$$\begin{aligned} Pr(w \geq 60) &= Pr(w = 60) + Pr(w = 61) + \ldots + Pr(w = 100) \\ &= \sum_{w=60}^{100} \binom{n}{w} p^w q^{n-w} = 0.028 \end{aligned}$$

The above requires a great deal of computation.

Instead, let us use the normal distribution as a model for the binomial distribution.

For a random variable x which can assume the value of 0 or 1 with $Pr(x = 1) = p$ and $Pr(x = 0) = q = 1 - p$, $\mu_x = p$ and $\sigma_x^2 = pq$ as seen from:

$$\mu_x = E(x) = (0)Pr(x = 0) + (1)Pr(x = 1) = 0 \cdot q + 1 \cdot p = p \tag{6-44}$$

$$\begin{aligned}\sigma_x^2 = E(x - \mu_x)^2 = E(x - p)^2 &= (0 - p)^2 q + (1 - p)^2 p \\ &= p^2(1 - p) + (1 - p)^2 p \\ &= p(1 - p)[p + 1 - p] \\ &= p(1 - p) \\ &= pq\end{aligned} \tag{6-45}$$

Eqs. 6-44 and 6-45 apply to each random variable x_i at each level i (x_i is 0 or 1 in proceeding to each level i).

The aggregated random variable w is the summation of variables x_i:

$$w = \textstyle\sum_i x_i$$

Hence, from Eqs. 6-37, 6-38, and 6-39, with $a_i = 1$,

$$\mu_w = n\mu_x = np$$
$$\sigma_w^2 = n\,\sigma_x^2 = npq$$

For $w = 60$, $n = 100$, $\mu_x = p = 1/2$, and $\sigma_x^2 = pq = 1/4$, we have $\mu_w = 100 \times 1/2$ and $\sigma_w^2 = 100 \times 1/4$. We now consider the normal distribution $N(\mu_w, \sigma_w^2)$ as a model for the distribution of w. To compute $P(w \geq 60)$ we transform w to the standardized normal variable z. Since $P(w \geq 60)$ implies that we take the area under the distribution of $p(w)$ to the right of $w = 59.5$ as shown in Fig. 6-12, z is computed as follows:

$$z = \frac{59.5 - 100 \cdot 1/2}{\sqrt{100 \cdot 1/4}} = \frac{9.5}{5} = 1.9$$

From Table 6-1,

$$P(w \geq 60) \approx P(z > 1.9) = 0.0287$$

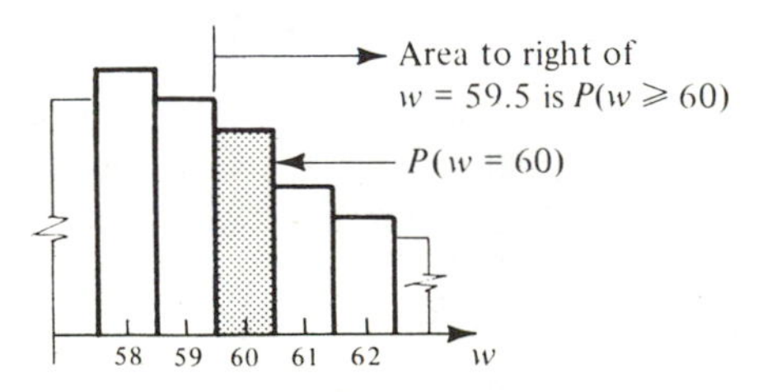

Figure 6-12

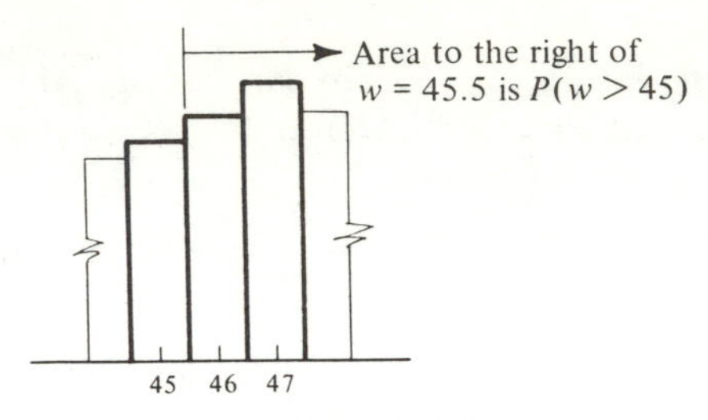

Figure 6-13

which is very close to 0.028, the result obtained from the binomial distribution.

EXAMPLE 2. For $P(w > 45)$ (Fig. 6-13),

$$z = \frac{45.5 - 50}{5} = -0.9$$

$$P(w > 45) \approx P(z > -0.9) = 0.8159$$

The binomial distribution yields 0.816.

EXAMPLE 3. Using the normal distribution model for the binomial distribution is a good approximation, even when $p \neq 1/2$ (but not close to 0 or 1) and for small values of n. As a practical guide, whenever $np > 5$ and $nq > 5$, the approximation is good.

Suppose $p = 1/5$ and $n = 40$. Then $P(w \geq 8) = 0.563$. Using the normal approximation,

$$z = \frac{7.5 - 40 \cdot \frac{1}{5}}{\sqrt{40 \cdot \frac{1}{5} \cdot \frac{4}{5}}}$$

$$= -0.197$$

$$P(w \geq 8) \approx P(z > -0.197) = 0.578$$

EXAMPLE 4. Verify for $n = 4$ and $p = 1/2$ that we obtain the following results from the binomial distribution and its normal approximation.

$P(w \geq w_i)$

w_i	*binomial distribution*	*normal approximation*
0	1.000	.9938
1	.938	.9332
2	.688	.6915
3	.312	.3085
4	.062	.0668

NOTE. The Central Limit Theorem can be applied to aggregates of any independent random variable. The examples in this section can be considered in general as follows:

Given variable x with mean μ_x and variance σ_x^2 and *unknown* distribution $p(x)$, then $w = \sum_{i=1}^{n} x_i$, in which x_i is the ith observation x in a random sample of n x's, can be studied by a model of the normal distribution $N(n\mu_x, n\sigma_x^2)$. The sample mean $\bar{x} = (1/n) \sum_i x_i$ can be studied by a model $N(\mu_x, \sigma_x^2/n)$. The model gets closer to describing the real-world phenomena as the level of aggregation (n) increases.

*6-7 RANDOM WALK

As another application of the central limit theorem we consider the problem of a *random walk*. Consider the man, shown in Fig. 6-14, who can walk along a straight line in either direction, taking one step at a time. If our man takes a walk, selecting randomly at the end of each step the direction for the next one (left or right), how far from the original position is he likely to be if he takes n steps, or what is the probability that he will be within a distance of 20 steps from the origin after a sequence of 100 steps?

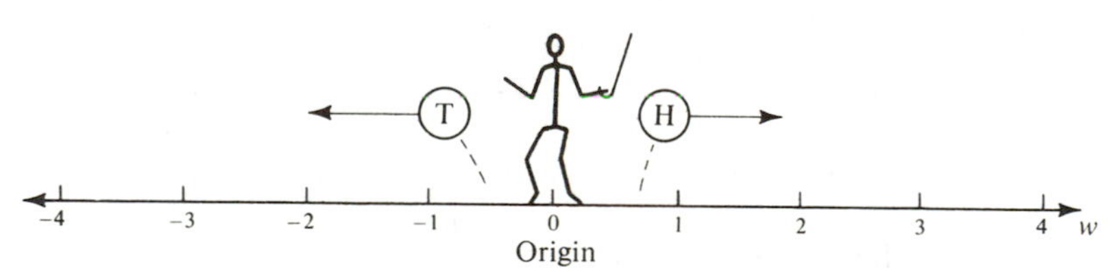

Figure 6-14 Random Walk

Digression. An attempt to answer the question may appear as an impossible task. However, it is the very essence of the statistical method and probability theory to put order into what is seemingly a problem consisting of total disorder. Probability suggests a kind of regularity. If a coin falls heads with probability $p = 2/5$, then we have good reason to believe that in 1000 ($n = 1000$) random independent tosses of the coin the number of heads is going to be quite close to $np = 1000 \times 2/5 = 400$. As the number of tosses increases, the probability that the proportion of heads will differ from the probability p of a head in a single toss by an amount α gets smaller and smaller. This illustrates the significance of the *law of large numbers* which asserts that as the number of trials (tosses) increases, the proportion of successes (heads) approaches the probability of success on an individual trial. This is the regularity we extract from probability.

The *Tchebysheff inequality* [54, p. 135] gives a precise formulation of the law of large numbers. Stated in words, this inequality reads: *The probability that a random variable w deviates from its population mean μ_w by more than k standard deviations σ_w is smaller than $1/k^2$, where k is an arbitrary positive constant.* Using our earlier notation, we can write the Tchebysheff inequality in the form

$$P(|w - \mu_w| > k\sigma_w) < 1/k^2$$

or

$$P(\mu_w - k\sigma_w < w < \mu_w + k\sigma_w) > 1 - 1/k^2 \tag{6-46}$$

The bound $1/k^2$ provided by the Tchebysheff inequality is too broad for most practical applications. The probability statement 6-46 can be improved when we use the central limit theorem to approximate the distribution of w by the normal distribution $N(\mu_w, \sigma_w^2)$. For example, to qualify our belief that in 1000 tosses the number of heads will be quite close to np, we can calculate the probability of the aggregate number of heads w in 1000 tosses being, for instance, between 370 and 430, i.e., $P(370 \leq w \leq 430)$. This can be achieved by using the central limit theorem to approximate the distribution of w by the normal distribution $N(\mu_w, \sigma_w^2)$ in which in the present example, with $p = 2/5$,

$$\mu_w = np = 1000 \times \frac{2}{5} = 400$$

$$\sigma_w^2 = n\sigma^2 = 1000\left(\frac{2}{5} \times \frac{3}{5}\right) = 240$$

$$\sigma_w = 15.5$$

$$z_1 = \frac{369.5 - 400}{15.5} = -1.97$$

$$z_2 = \frac{430.5 - 400}{15.5} = 1.97$$

$$P(370 \leq w \leq 430) = P(-1.97 < z < 1.97) = 0.9522$$

Using the Tchebysheff inequality, we write for deviation $k\sigma_w$

$$\begin{aligned} k\sigma_w &= 430 - 400 \\ &= 400 - 370 \\ &= 30 \end{aligned}$$

$$k = \frac{30}{15.5}$$

Using Eq. 6-46,

$$P(400 - 30 \leq w \leq 400 + 30) > 1 - \left(\frac{15.5}{30}\right)^2$$

or

$$P(370 \leq w \leq 430) > 1 - 0.267 = 0.733$$

This statement is quite broad compared to the more definitive statement obtained by using the central limit theorem.

Back to Random Walk. Suppose our man starts on a straight line from the origin 0 in Fig. 6-14 and takes one step to the left or the right, depending on the outcome from the toss of a coin. For H, he goes one step to the right and $x = 1$; for T, one to the left and $x = -1$. Let $P(T) = P(H) = 1/2$. At each station the process is repeated and the results T or H are independent of preceding results. Thus, μ_x and σ_x^2 have the following values at *each* station:

$$\begin{aligned}\mu_x &= E(x) = (-1)P(-1) + (1)P(1)\\ &= (-1)\left(\frac{1}{2}\right) + (1)\left(\frac{1}{2}\right)\\ &= 0\\ \sigma_x^2 &= E(x - \mu_x)^2\\ &= E(x^2)\\ &= (-1)^2\,P(-1) + (1)\,P(1)\\ &= 1 \times \frac{1}{2} + 1 \times \frac{1}{2}\\ &= 1\end{aligned}$$

Now, the distance from the origin, following n steps, is given by $w = \sum_{i=1}^{n} x_i$ if we consider each step to be one unit of distance. Negative w indicates distance to the left of the origin in Fig. 6-14. The mean distance μ_w from the origin in n steps and the variance* σ_w^2 of the distance are

*But since here

$$\sigma_w^2 = E(w - \mu_w)^2 = E(w^2) = E\left(\textstyle\sum_i x_i\right)^2$$

σ_w^2 represents the expected value of the square of the distance from the origin. For $n = 1$, one step,

$$E(w^2) = (-1)^2P(-1) + (1)^2P(1) = 1 = \sigma_x^2$$

position (w)	-1	0	1
probability	$\frac{1}{2}$		$\frac{1}{2}$

For $n = 2$, two steps,

$$E(w^2) = (-2)^2(\tfrac{1}{4}) + (0)(\tfrac{1}{2}) + (2)^2(\tfrac{1}{4}) = 2$$

position (w)	-2	-1	0	1	2
probability	$\frac{1}{4}$		$\frac{1}{2}$		$\frac{1}{4}$

For $n = 3$, three steps,

$$E(w^2) = (-3)^2\tfrac{1}{8} + (-1)^2\tfrac{3}{8} + (1)^2\tfrac{3}{8} + (3)^2\tfrac{1}{8} = 3$$

position (w)	-3	-2	-1	0	1	2	3
probability	$\frac{1}{8}$	0	$\frac{3}{8}$	0	$\frac{3}{8}$	0	$\frac{1}{8}$

For $n = 4$, four steps, $E(w^2) = 4$

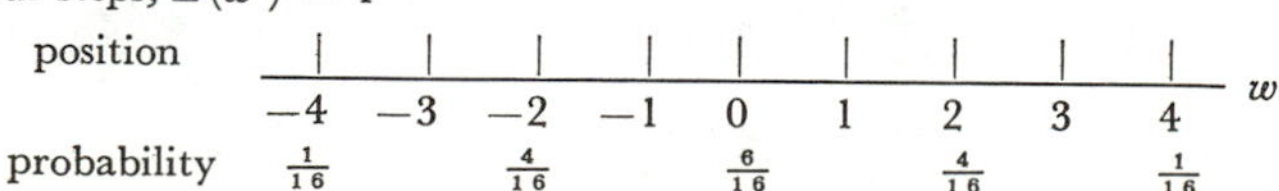

position (w)	-4	-3	-2	-1	0	1	2	3	4
probability	$\frac{1}{16}$		$\frac{4}{16}$		$\frac{6}{16}$		$\frac{4}{16}$		$\frac{1}{16}$

For n steps, $E(w^2) = n$

given by

$$\mu_w = n\mu_x = 0$$

and $$\sigma_w^2 = n\sigma_x^2 = n$$

For n large we can use the normal distribution model for $w = \sum_{i=1}^{n} x_i$ with $\mu_w = 0$ and $\sigma_w^2 = n$. Thus, the probability that our man in Fig. 6-14 will be within a distance of $\sqrt{n}$ steps from the origin (one standard deviation σ_w) is 0.68. For any distance d from the origin (left or right), the probability is computed as follows:

$$z_d = \frac{d - 0}{\sqrt{n}}$$

$$P(\text{distance} \leq d) = P(-z_d < z < z_d)$$

EXAMPLE 1. The probability of being a distance equal to or larger than 120 steps from the origin in 6400 steps is:

$$z = \frac{119.5 - 0}{\sqrt{6400}} = 1.49$$

$$P(|w| \geq 120) \approx 1 - P(-z_d < z < z_d)$$

$$P(-120 < w < 120) \approx P(-1.49 < z < 1.49) = 0.8638$$

and therefore $P(|w| > 120) \approx 1 - 0.8638 = 0.1362$

EXAMPLE 2. Suppose $P(H) = 1/5$ and $P(T) = 4/5$ in Fig. 6-14. What is the probability that in 2500 steps our man will be 1560 steps or more to the left of the origin?

Solution: $P(H) = P(x = 1) = p; \qquad P(T) = P(x = -1) = q = 1 - p$

$$\mu_x = 1p + (-1)q = p - (1 - p) = 2p - 1$$

$$\begin{aligned}\sigma_x^2 &= (1 - 2p + 1)^2 p + (-1 - 2p + 1)^2(1 - p)\\ &= (2 - 2p)^2 p + (-2p)^2(1 - p)\\ &= 4(1 - p)^2 p + 4p^2(1 - p) = 4p(1 - p)[1 - p + p]\\ &= 4pq\end{aligned}$$

$$w = \sum_{i=1}^{n} x_i \qquad x_i = -1, 1$$

$$\mu_w = \sum_{i=1}^{n} \mu_x = n(2p - 1)$$

$$\sigma_w^2 = \sum_{i=1}^{n} \sigma_x^2 = n4pq = 4npq \tag{6-47}$$

NOTE. For the special case $p = q = 1/2$, the above reduces to $\mu_w = 0$, $\sigma_w^2 = n$ as it should, confirming our results earlier in this section.

Answer:

$$P(w \leq -1560) \approx P(z < z_1)$$

$$z_1 = \frac{-1559.5 - \mu_w}{\sigma_w}$$

$$\mu_w = n(2p - 1) = 2500\left(-\frac{3}{5}\right) = -1500$$

$$\sigma_w^2 = 4 \times 2500 \times \frac{1}{5} \times \frac{4}{5}$$

$$\sigma_w = \frac{4}{5} \times 50 = 40$$

$$z_1 = \frac{-59.5}{40} = -1.49$$

$$P(w \leq -1560) \approx P(z < -1.49) = 0.0681$$

Problems

1. In Example 2, what is the probability of a distance between 1500 and 1420 steps to the left of the origin?
2. In Example 2, what is the probability of a distance between 10 and 90 steps to the right of the origin?
3. Explain the reason for the large difference in results in problems 1 and 2 although each considers a region of 80 steps: 1420 to 1500, and 10 to 90.
4. Repeat Example 2 for $p = 9/10$ and $q = 1/10$.

*6-8 FROM SAMPLE TO POPULATION—ESTIMATION OF PARAMETERS

Probability theory and the central limit theorem can be applied to problems of inference in which we attempt to generate knowledge about a population by using the limited information that can be extracted from a sample of the population. The uncertainty which will attend our inference in going from the limited to the general will be qualified by a probability statement. An example will clarify these points.

Consider a manufacturer of light bulbs who wishes to establish the probability of failure p of a bulb in less than 100 hours of service. p is a parameter of the population of bulbs manufactured in the plant in much the same sense that $p = 1/6$ is the probability of the number 1 in the toss of a perfectly balanced die. To estimate p we take a *random sample* of n bulbs and test each bulb. By *random sample* we imply that each bulb in a large warehouse of bulbs has equal probability of being included in the sample. Suppose that a number w of the sample of n bulbs fails in less than 100 hours. Then the fraction of failures is w/n.

Shall we use this ratio as an estimate for the population parameter p? The entire warehouse of bulbs may consist of N bulbs of which W will fail in less than 100 hours; then $p = W/N$, and the probability of randomly selecting a bulb which will fail in less than 100 hours is then $p = W/N$. But we do not know W unless we test all N bulbs. This could be quite costly and often not practical or possible. We therefore attempt to generate knowledge about p from the sample of size $n \ll N$. Let us call the ratio w/n an *estimate* $\hat{p}$ or $\bar{p}$ of p.

The estimate $\bar{p}$ is also called a *statistic* as discussed in Sec. 6-2. The statistic varies from sample to sample and, therefore, has a probability distribution function. The variance $\sigma_{\bar{p}}^2$ (dispersion) of the distribution gets smaller as the sample size n increases, until $\sigma_{\bar{p}}^2 = 0$ when $n = N$. The statistic $\bar{p}$ appears to be a plausible estimate of p. But can we qualify the degree of plausibility or the degree of confidence in its function as a representative of the population parameter p when the sample $n \ll N$? Also, how much does our confidence increase with increase in n? To answer these questions we proceed as follows.

Mean and Variance of $\bar{p}$

Let us designate by $x = 1$ and $x = 0$ failure and nonfailure in the test of any bulb. In a sample of n bulbs, let $x_1 (= 0 \text{ or } 1)$ designate the result with the first bulb, x_2 with the second, x_i with the ith bulb and x_n with the last of the n bulbs. Then the total number w of bulbs that fail is given by our old friend

$$w = \textstyle\sum_{i=1}^{n} x_i \tag{6-48}$$

Since the x_i are independent,

$$\mu_w = \textstyle\sum_i \mu_i$$

and

$$\sigma_w^2 = \textstyle\sum_i \sigma_i^2$$

μ_i is the expected value of x_i for the ith bulb in the sample, namely (see also Eq. 6-44),

$$\begin{aligned} \mu_i &= (0)P(x_i = 0) + (1)P(x_i = 1) \\ &= (0)(1 - p) + (1)p \\ &= p \end{aligned}$$

Similarly, we get the familiar result for the variance σ_i^2 of x_i for the ith bulb (see also Eq. 6-45):

$$\begin{aligned} \sigma_i^2 &= (0 - \mu_i)^2 P(0) + (1 - \mu_i)^2 P(1) \\ &= (-p)^2(1 - p) + (1 - p)^2 p \\ &= p(1 - p) \end{aligned}$$

Since $\mu_i = p$ and $\sigma_i^2 = p(1 - p)$ for all i, then

$$\mu_w = np \tag{6-49}$$

$$\sigma_w^2 = np(1 - p) \tag{6-50}$$

If we now define the *statistic* $\bar{p}$ as the fraction of failures in a sample of size n, then from Eq. 6-48,

$$\bar{p} = \frac{w}{n} = \sum_{i=1}^{n} \frac{1}{n} x_i \tag{6-51}$$

$$0 \leq \bar{p} \leq 1$$

From Eqs. 6-37, 6-38, and 6-39 we have $(a_i = 1/n)$:

$$\mu_{\bar{p}} = \sum_{i=1}^{n} \frac{1}{n} \mu_i = \frac{1}{n} \sum_{i=1}^{n} \mu_i = \frac{1}{n}(np) = p \tag{6-52}$$

$$\sigma_{\bar{p}}^2 = \sum_{i=1}^{n} \left(\frac{1}{n}\right)^2 \sigma_i^2 = \frac{1}{n^2} \sum_i \sigma_i^2 = \frac{1}{n^2} npq = \frac{1}{n} pq \tag{6-53}$$

in which $q = 1 - p$.

Distribution of $\bar{p}$

As the sample size n increases, the statistic $\bar{p}$ approaches the normal distribution $N(\mu_{\bar{p}}, \sigma_{\bar{p}}^2)$ in which $\mu_{\bar{p}} = p$ and $\sigma_{\bar{p}}^2 = (1/n)pq$. Using the normal distribution $N(p, (1/n)pq)$ as a model for the distribution of statistic $\bar{p}$, we can write the following probability statements regarding the deviation of the observed statistic $\bar{p}$ from the population parameter p (see Fig. 6-15):

1. The probability that a computed value $\bar{p}$ from a random sample of size n will differ from the population parameter p by more than one standard deviation $\sqrt{pq/n}$ is 0.32.
2. The probability that $\bar{p}$ *will not* differ from p by more then $\sqrt{pq/n}$ is 0.68.
3. For two standard deviations $2\sqrt{pq/n}$, statements 1 and 2 above become 0.05 and 0.95, respectively.
4. Using a transformation to standardized normal variable z, we can use Table 6-1 to make statements 1 and 2 for any deviation d of $\bar{p}$ from p. The transformation will also enable us to give a more plausible interpretation to the inference regarding p and our confidence in it.

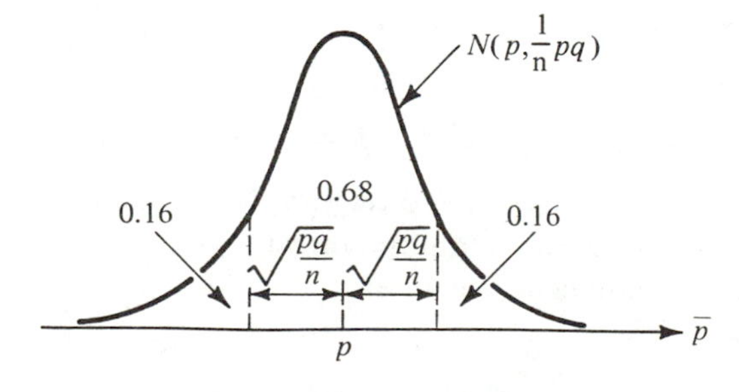

Figure 6-15 Normal Distribution Model $N(p, pq/n)$ for Statistic $\bar{p}$

Transformation to Standard Normal Variable z and Confidence Intervals

Let $z_{\alpha/2}$ designate a value for standard normal variable z with an area $\alpha/2$ to its right under the distribution curve $N(0, 1)$, namely,

$$P(z > z_{\alpha/2}) = \alpha/2 \tag{6-54}$$

as shown in Fig. 6-16.

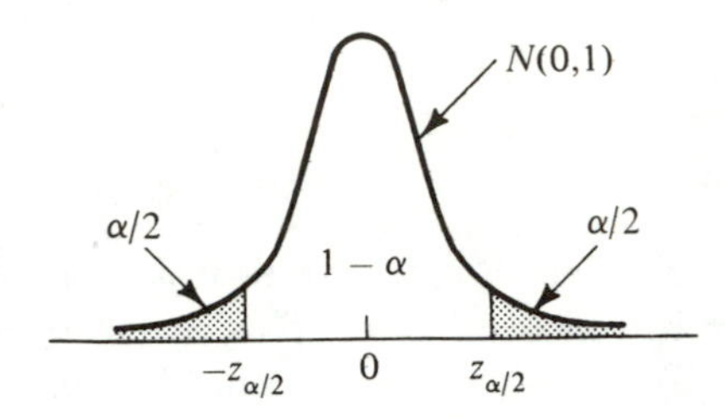

Figure 6-16 $P(z > z_{\alpha/2})$

Let this designation hold only for positive values of z so that $-z_{\alpha/2}$ is merely the negative value of $z_{\alpha/2}$ and, therefore (see Fig. 6-16),

$$P(-z_{\alpha/2} < z < z_{\alpha/2}) = 1 - \alpha \tag{6-55}$$

For variable $\bar{p}$ with $\mu_p = p$ and $\sigma_{\bar{p}}^2 = (1/n)pq$:

$$z = \frac{\bar{p} - p}{\sigma_{\bar{p}}} \tag{6-56}$$

Thus, Eq. 6-55 becomes

$$P\left(-z_{\alpha/2} < \frac{\bar{p} - p}{\sigma_{\bar{p}}} < z_{\alpha/2}\right) = 1 - \alpha \tag{6-57}$$

Multiplying the inside of the parentheses by $\sigma_{\bar{p}}$ does not alter the statement and yields

$$P(-z_{\alpha/2}\sigma_{\bar{p}} < \bar{p} - p < z_{\alpha/2}\sigma_{\bar{p}}) = 1 - \alpha \tag{6-58}$$

Subtracting $\bar{p}$ gives

$$P(-\bar{p} - z_{\alpha/2}\sigma_{\bar{p}} < -p < -\bar{p} + z_{\alpha/2}\sigma_{\bar{p}}) = 1 - \alpha$$

Multiplying by -1 changes the direction of inequalities; for example, $5 > -3 > -6$ becomes (after multiplying by -1) $-5 < 3 < 6$. Hence, the last probability statement becomes

$$P(\bar{p} + z_{\alpha/2}\sigma_{\bar{p}} > p > \bar{p} - z_{\alpha/2}\sigma_{\bar{p}}) = 1 - \alpha \tag{6-59}$$

or

$$P(\bar{p} - z_{\alpha/2}\sigma_{\bar{p}} < p < \bar{p} + z_{\alpha/2}\sigma_{\bar{p}}) = 1 - \alpha \tag{6-60}$$

In this equation, p is a constant population parameter unknown to us.

$\bar{p}$ is an observed statistic from a random sample of size n. $z_{\alpha/2}$ is obtained from Table 6-1, depending on our choice of α.

The meaning of Eq. 6-60 is:

Suppose we compute $\bar{p}$ from a sample of size n. Then we can say that *p is a number covered or contained in the range between $\bar{p} - \delta$ to $\bar{p} + \delta$* ($\delta = z_{\alpha/2}\sigma_{\bar{p}}$). In the spirit of Chapter 4, we qualify this statement by professing incomplete knowledge and adding that *the probability of the statement being true is* $1 - \alpha$.

The interval of 2δ between $\bar{p} - \delta$ and $\bar{p} + \delta$ is called the *confidence interval*, and $1 - \alpha$ represents our *level of confidence*. The boundaries $\bar{p} - \delta$ and $\bar{p} + \delta$ of the confidence interval are called *confidence limits*.

Of course, we wish to make the confidence interval as narrow as possible and α as small as possible. For $n = N$, i.e., the entire population, $\delta = 0$ (because $\sigma_{\bar{p}} = 0$), $\alpha = 0$ and $\bar{p} = p$. But, in general, we can make the confidence interval 2δ smaller for a fixed α, or make α smaller for a fixed interval by increasing the sample size n.

The Variance $\sigma_{\bar{p}}^2$. What about $\sigma_{\bar{p}}^2$? Since $\sigma_{\bar{p}} = \sqrt{pq/n}$ and we do not know p, how are we to evaluate $\sigma_{\bar{p}}$? We can proceed with a conservative confidence interval by selecting p for which $\sigma_{\bar{p}}$ and, therefore, the interval will be a maximum for a given n. This occurs when $p = q = 1/2$ because

$$\sigma_{\bar{p}}^2 = \frac{p(1-p)}{n} = \frac{p - p^2}{n}$$

$$n\sigma_{\bar{p}}^2 = p - p^2$$

and $\sigma_{\bar{p}}^2$ is a maximum when $p - p^2$ is maximum:

$$\frac{\partial}{\partial p}(p - p^2) = 0$$

$$1 - 2p = 0$$

$$p = \frac{1}{2}$$

This would be a reasonable approach if indeed p is expected to be close to 1/2, but for a case such as ours, where $p \ll 1/2$, we can take $\bar{p}$ as p for the computation of $\sigma_{\bar{p}}$:

$$\sigma_{\bar{p}} = \sqrt{\bar{p}(1 - \bar{p})/n} \tag{6-61}$$

EXAMPLE 1. Suppose $w = 36$ in a sample $n = 100$. Then $\bar{p} = 0.36$ and

$$\sigma_{\bar{p}} \approx \sqrt{\frac{0.36 \times 0.64}{100}} = 0.048$$

(a) For a 0.80 level of confidence, $\alpha = 0.20$, $\alpha/2 = 0.10$, and $z_{\alpha/2} = 1.28$:

$$\bar{p} - \delta = 0.36 - 1.28 \times 0.048 = 0.299$$
$$\bar{p} + \delta = 0.36 + 1.28 \times 0.048 = 0.422$$

Thus, $P(0.299 < p < 0.422) = 0.80$. Namely, the probability is 0.80 that the population parameter p is covered by the confidence interval 0.299 to 0.422. The values 0.299 and 0.422 are the confidence limits.

(b) If we are satisfied with a 0.68 level of confidence, then the confidence interval decreases (verify):

$$P(0.312 < p < 0.408) = 0.68$$

(c) If we are not satisfied with a 0.80 level of confidence and wish to make it 0.90, then the confidence interval increases (verify):

$$P(0.281 < p < 0.439) = 0.90$$

EXAMPLE 2. Suppose we obtain $w = 3600$ in a sample $n = 10{,}000$. Then $\bar{p} = 0.36$, as in Example 1. However, because of the larger sample size n, our levels of confidence will be greatly enhanced for the same confidence intervals as in Example 1, or the confidence intervals will be much narrower for the same levels of confidence as shown by the following computations. This is expected because the variance $\sigma_{\bar{p}}^2$ is smaller, the larger the sample size and, therefore, $\bar{p}$ is less dispersed about its expected value p:

$$\sigma_{\bar{p}} \approx = \sqrt{\frac{0.36 \times 0.64}{10{,}000}} = 0.0048$$

(a) For $1 - \alpha = 0.80$:

$$\bar{p} - \delta = 0.36 - 1.28 \times 0.0048 = 0.354$$
$$\bar{p} + \delta = 0.36 + 1.28 \times 0.0048 = 0.366$$
$$P(0.354 < p < 0.366) = 0.80$$

(b) For $1 - \alpha = 0.68$:

$$\bar{p} - \delta = 0.36 - 0.0048 = 0.355$$
$$\bar{p} + \delta = 0.36 + 0.0048 = 0.365$$
$$P(0.355 < p < 0.365) = 0.68$$

(c) For $1 - \alpha = 0.90$:

$$\bar{p} - \delta = 0.36 - 1.645 \times 0.0048 = 0.352$$
$$\bar{p} + \delta = 0.36 + 1.645 \times 0.0048 = 0.368$$
$$P(0.352 < p < 0.368) = 0.90$$

EXAMPLE 3. Consider the following claim by a pollster in a very close political race for the Senate between two candidates, R and D: The proportion p of votes for D will not differ from the value $\bar{p}$ (which the pollster specifies) by more than 0.02. If the pollster desires a 99% confidence level in the claim, how large a sample n must he take? For $1 - \alpha = 0.99$, $\alpha/2 =$

0.005, and $z_{\alpha/2} = 2.575$. The problem statement requires that (see Eq. 6-58):

$$P(-0.02 < \bar{p} - p < 0.02) = 0.99$$

or

$$z_{\alpha/2}\sigma_{\bar{p}} = 0.02$$

$$\sigma_{\bar{p}} = \sqrt{\frac{p(1-p)}{n}}$$

For a close political race, $p \approx 1/2$ (otherwise, use $\bar{p}$ instead of p in computing $\sigma_{\bar{p}}$). Then

$$\sigma_{\bar{p}} = \sqrt{\frac{1}{4n}}$$

$$z_{\alpha/2} = 2.575 \qquad \sigma_{\bar{p}} = \frac{1}{2\sqrt{n}}$$

$$2.575\frac{1}{2\sqrt{n}} = 0.02$$

$$\sqrt{n} = 2.575 \times \frac{1}{2 \times 0.02}$$

$$n = 4140$$

EXAMPLE 4. In Example 3, suppose $\bar{p}$ is computed from a sample of 4200 and comes out 0.52. Is the claim of the pollster in Example 3 still valid (we now use $\bar{p}$ to compute $\sigma_{\bar{p}}$)?

$$\sigma_{\bar{p}} = \sqrt{\frac{0.52 \times 0.48}{4200}} = 0.00771$$

$$z_{\alpha/2}\sigma_{\bar{p}} = 2.575 \times 0.00771 = 0.0198 < 0.2$$

Claim is valid.

EXAMPLE 5. Consider a machine that produces pills. Suppose the diameter x is of interest. From past experience it is known that the variance is $\sigma_x^2 = 0.160$ mm, but the mean μ_x is not known. A random sample of 25 pills is taken and $\bar{x}$ is computed to be 12.000 mm.

(a) What are the confidence limits for a confidence level $1 - \alpha = 0.80$?

(b) Repeat (a) for $1 - \alpha = 0.68$.

(c) Repeat (a) for $1 - \alpha = 0.90$.

(d) How large a sample is required if the value of μ_x is not to exceed the computed mean $\bar{x}$ by more than 0.015 mm with a 0.99 confidence level?

All these questions can be answered by using Eqs. 6-58, 6-59, and 6-60, with $\bar{x}$ replacing $\bar{p}$:

(a) $z_{\alpha/2} = 1.28 \qquad \sigma_{\bar{x}}^2 = \dfrac{\sigma_x^2}{n} = \dfrac{0.16}{25} = 0.0064$

$$\sigma_{\bar{x}} = 0.08 \text{ mm}$$

$$\bar{x} = 12 \text{ mm}$$

Confidence limits:

$$\bar{x} - \delta = 12.000 - 1.28 \times 0.08 = 11.898 \text{ mm}$$
$$\bar{x} + \delta = 12.000 + 1.28 \times 0.08 = 12.102 \text{ mm}$$
$$P(11.898 < \mu_x < 12.102) = 0.80$$

(b) $z_{\alpha/2} = 1.0$

$$\bar{x} - \delta = 12.000 - 0.08 = 11.920$$
$$\bar{x} + \delta = 12.000 + 0.08 = 12.080$$
$$P(11.920 < \mu_x < 12.080) = 0.68$$

(c) $z_{\alpha/2} = 1.645$

$$\bar{x} - \delta = 12.000 - 1.645 \times 0.08 = 11.868$$
$$\bar{x} + \delta = 12.000 + 1.645 \times 0.08 = 12.132$$
$$P(11.868 < \mu_x < 12.132) = 0.90$$

(d) For $1 - \alpha = 0.99$: $\alpha/2 = 0.005$ $z_{\alpha/2} = 2.575$

The problem statement requires (see Eq. 6-58):

$$P(-0.015 < \bar{x} - \mu_x < 0.015) = 0.99$$

or

$$z_{\alpha/2}\sigma_{\bar{x}} = 0.015$$

$$2.575 \frac{\sigma_x}{\sqrt{n}} = 0.015$$

$$2.575 \frac{\sqrt{0.16}}{\sqrt{n}} = 0.015$$

$$n = \left(\frac{2.575}{0.015}\right)^2 0.16$$
$$= 4715$$

NOTE. In most applications, σ_x^2 is not known. In such cases it is estimated by using s^2 from the sample, i.e. (see Eq. 6-12),

$$\sigma^2 \approx \hat{\sigma}^2 = s^2 = \frac{1}{n-1} \sum_i (x_i - \bar{x})^2$$

This, of course, is an approximation—a rather good one for large samples. For small samples, it is possible to use a different model for the distribution of $\bar{x}$ when σ_x^2 is not known without resorting to this approximation. This is the *Student t distribution* [55], but it requires that variable x be normally distributed.

The variable t of this model is given by

$$t = \frac{\bar{x} - \mu_x}{s_x} \sqrt{n}$$

in place of

$$z = \frac{\bar{x} - \mu_x}{\sigma_x} \sqrt{n}$$

when σ_x is known or assumed known.

We can also estimate the variance and establish a confidence interval on σ^2 with a given confidence level by using the statistic s^2 computed from a random sample of size n. For this purpose, *the* χ^2 *distribution model* is used [55]. The variable χ^2 of this distribution has the following form in terms of unknown parameter σ^2, computed statistic s^2, and sample size n:

$$\begin{aligned}\chi^2 &= \frac{(n-1)s^2}{\sigma^2} \\ &= \frac{(n-1)}{\sigma^2}\sum_i \frac{1}{n-1}(x_i - \bar{x})^2 \\ &= \frac{1}{\sigma^2}\sum_i (x_i - \bar{x})^2\end{aligned}$$

6-9 TESTING HYPOTHESES—ERRORS OF OMISSION AND COMMISSION

In Chapter 4, Sec. 4-1, in our discussion of the will to doubt, we indicated that our faith in a hypothesis is maintained on a tentative basis until "sufficient evidence" can be introduced as cause for rejecting it. The hypothesis may be our subscription to a particular population as a model for a problem of interest. In this section we shall attempt to qualify the significance of "sufficient evidence" through the use of information extracted from a statistical sample from which we can make an inference regarding the population that is stipulated (hypothesized) as our model. Such a sample contains incomplete information and, therefore, we may commit one of two errors in our decision regarding the hypothesis. We may reject it when it is true, or continue to subscribe to it when it is false. These two types of errors were discussed in Sec. 5-4 as errors of omission and commission in the modeling process in general.

As is often the case with general introductions of this type, the sooner we get to a concrete example the less likely we are to lose the reader. So here is Example 1.

EXAMPLE 1. A young physician is stationed on a remote island when he discovers that an epidemic is spreading among the inhabitants. In his office he finds a large supply of little white pills which, for a moment, he thinks are the kind prescribed for this epidemic. However, on reflection he recalls that a pill very similar in appearance, taste, and size is used for a different purpose and is detrimental if taken for the present epidemic. There are no facilities for testing the composition, and the doctor is not certain about it in the first place. All the doctor can find is a statement to the effect that the pills which would help stop the epidemic have a mean diameter of 10.00 mm and a standard deviation of 0.60 mm. The pills detrimental to the epidemic have a mean diameter of 10.50 mm and the same standard deviation of 0.60 mm. Our doctor is inclined by his original thought to make the claim or state the hypothesis referred to as the *null hypothesis* and designated as H_0:

The pills come from the population with $\mu_0 = 10.00$ mm, or
H_0: $\mu_0 = 10.00$ mm

However, there is the *alternate hypothesis* H_1 which is possible:

The pills come from the population with $\mu_1 = 10.50$ mm, or
H_1: $\mu_1 = 10.50$ mm

The doctor takes a sample of the pills, measures the sample diameter $\bar{x}$, and now must make a decision. There are two possibilities here: accept H_0 or reject H_0 (accept H_1). The truth is, of course, H_0 or H_1. Before proceeding with the guides to a decision rule, let us study the consequences of our decisions, something we must always do in problem solving. Table 6-3 is a model

Table 6-3

	H_0 True	H_1 True
Accept H_0	correct	type II error
Reject H_0 (Accept H_1)	type I error	correct

of the possible consequences. There are two situations of correct decisions and two errors from wrong decisions. *Type I error* is rejecting H_0 when true, namely, the doctor will not use the pill when he should. *Type II error* is accepting H_0 when false, i.e., H_1 is true, namely, the doctor will use the pills when he should not. As we shall discover, for a given sample size, we can make the probability of one of the errors smaller, but at the same time increase the other. The decision rule will, therefore, depend on which error is more costly or more damaging. In the language of utility theory, we may wish to make the expected utility a maximum.* Thus, if the utility of the type I error, u(error I), is larger than the utility of the type II error, u(error II), because it is better not to give the correct medicine than to give the wrong one, and if we designate by α and β the probabilities of type I and type II errors, respectively, then we write the following equations for the expected utility $E(u)$ associated with each decision (see Table 6-4):

Decision to accept H_0:

$$E(u) = (1 - \alpha)u(\text{correct decision}) + \beta u(\text{error II})$$

---

*Utility theory is discussed in Chapter 7. For the present example, consider utility as a quantitative measure, on a scale from zero to one, of benefit associated with each of the four outcomes in Table 6-3. Thus, say, for example, $u(\text{error II}) = 0$, $u(\text{error I}) = 0.3$, $u(\text{correct decision}) = 1$.

Decision to reject H_0:

$$E(u) = \alpha u(\text{error I}) + (1 - \beta)u(\text{correct decision})$$

In the present example we may wish to make β smaller than α as seen from Table 6-4. Error I will cause a larger spread of the epidemic that may lead to some deaths and a prolonged suffering on the part of the infected people who will survive. Error II can be detrimental to survival by leading to complications in both the infected and uninfected people.

Table 6-4

	Pills are good	Pills not good
Give pills	correct $P(\text{correct}) = 1 - \alpha$	$P(\text{error II}) = \beta$ (can be detrimental to survival)
Do not give pills	$P(\text{error I}) = \alpha$ (more people infected, suffering of infected prolonged)	correct $P(\text{correct}) = 1 - \beta$

Decision Rule. Let us suppose that the physician takes a sample of $n = 25$ pills and computes $\bar{x}$ to be 10.15 mm. Using the central limit theorem, $\bar{x}$ is approximately distributed as $N(\mu_0, \sigma^2/n)$ if the pills are from the population with mean μ_0, or $N(\mu_1, \sigma^2/n)$ if from the alternate population with mean μ_1. These two distributions are shown in Fig. 6-17, $\mu_0 = 10.00$ mm, $\mu_1 =$ 10.50 mm; $\sigma^2/n = 0.36/25$.

It is plausible to decide qualitatively that for values of $\bar{x}$ close to 10.00 mm we will be inclined to consider H_0 to be true, and for values closer to 10.50 mm we will be more inclined to consider H_1 to be true. Let us suppose that we wish to be more explicit and select a value $\bar{x}$(decision) such that, if $\bar{x} > \bar{x}$(decision), we accept H_1 and if $\bar{x} < \bar{x}$(decision), we accept H_0.

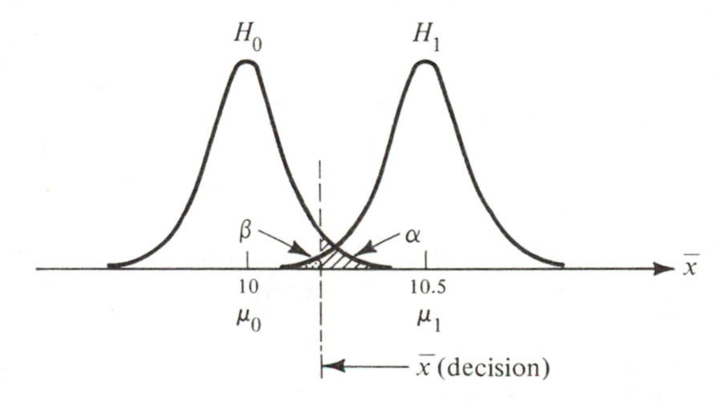

Figure 6-17 Distribution of $\bar{x}$ for $H_0: \mu_0 = 10.00$ mm and $H_1: \mu_1 = 10.50$ mm

Figure 6-17 shows the probability α for error type I (hatched area under H_0 to the right of $\bar{x}$(decision)), i.e., accepting H_1 when sample comes from H_0; and the probability β of type II error (dark area under H_1 to the left of $\bar{x}$ (decision)), i.e., accepting H_0 when sample comes from H_1. As we move the decision value $\bar{x}$(decision) to the right, β gets larger and α smaller. The reverse is true when $\bar{x}$(decision) is moved to the left. We can make both α and β smaller for a given $\bar{x}$(decision) by increasing the sample size n because then the variance $\sigma_{\bar{x}}^2 = \sigma_x^2/n$ gets smaller and each distribution is more concentrated near its mean. However, for a fixed sample size n, our choice of $\bar{x}$(decision) will depend on our choice of α or β. If we elect to choose β, then α is established by this choice; conversely if we elect to choose α, then β is established, as shown by the following:

(a) Suppose β is selected as 0.001 in the present example. Then, using the distribution to which β belongs (H_1), we write:

$$z_1 = \frac{\bar{x}(\text{decision}) - \mu_1}{\sigma_{\bar{x}}}$$

$$\bar{x}(\text{decision}) = \mu_1 + z_1\sigma_{\bar{x}}$$

For $\beta = 0.001$, $z_1 = -3.09$; since $\mu_1 = 10.50$ mm

and $\sigma_{\bar{x}} = \sqrt{\dfrac{0.36}{25}} = 0.12$ mm

$$\bar{x}(\text{decision}) = 10.5 - 3.09 \times 0.12 = 10.13 \text{ mm}$$

The decision rule is:

For $\bar{x} > \bar{x}(\text{decision}) = 10.13$ mm accept H_1

For $\bar{x} < \bar{x}(\text{decision}) = 10.13$ mm accept H_0

In the present example $\bar{x} = 10.15$ mm; therefore, reject H_0.

The probability α of error type I in the present decision rule for $\beta = 0.001$ is computed by using the distribution to which α belongs, namely, $N(\mu_0, \sigma^2/n)$.

$$z_0 = \frac{\bar{x}(\text{decision}) - \mu_0}{\sigma_{\bar{x}}}$$

$\bar{x}(\text{decision}) = 10.13$ mm, $\mu_0 = 10.00$ mm, $\sigma_{\bar{x}} = 0.12$ mm

$$z_0 = \frac{10.13 - 10.00}{0.12} = 1.08$$

$$\alpha = 0.140$$

(b) Suppose $\alpha = 0.140$ is considered too high and we wish to make it 0.05. Then $\bar{x}$(decision) is computed by using the distribution to which α belongs:

$$z_0 = \frac{\bar{x}(\text{decision}) - \mu_0}{\sigma_{\bar{x}}}$$

$$\bar{x}(\text{decision}) = \mu_0 + z_0\sigma_{\bar{x}}$$

For $\alpha = 0.05$, $z_0 = 1.645$; $\mu_0 = 10.00$ mm, $\sigma_{\bar{x}} = 0.12$ mm

$$\bar{x}(\text{decision}) = 10.1975 \approx 10.20 \text{ mm}.$$

Then for $\bar{x} = 10.15$ mm in the present example, we accept H_0.

With α fixed and the corresponding $\bar{x}$(decision) established, we are faced with the following β:

$$z_1 = \frac{\bar{x}(\text{decision}) - \mu_1}{\sigma_{\bar{x}}} = \frac{10.20 - 10.50}{0.12} = -2.5$$

$$\beta = 0.0062$$

In general, the alternatives to the null hypothesis (for instance, $H_0: \mu = \mu_0$) may be of different forms. For example, in addition to $H_1: \mu = \mu_1$, which we used, we can have

(a) $H_1: \mu_1 \neq \mu_0$
(b) $H_1: \mu_1 > \mu_0$
(c) $H_1: \mu_1 < \mu_0$

In $\mu_1 \neq \mu_0$ we have a two-sided test and all we can compute is α at the two tails of the distribution under H_0 (Fig. 6-18a). $\mu_1 > \mu_0$ and $\mu_1 < \mu_0$ are one-sided tests for which, again, only α can be established by using the right or left tail of the distribution under H_0. See Fig. 6-18b and c.

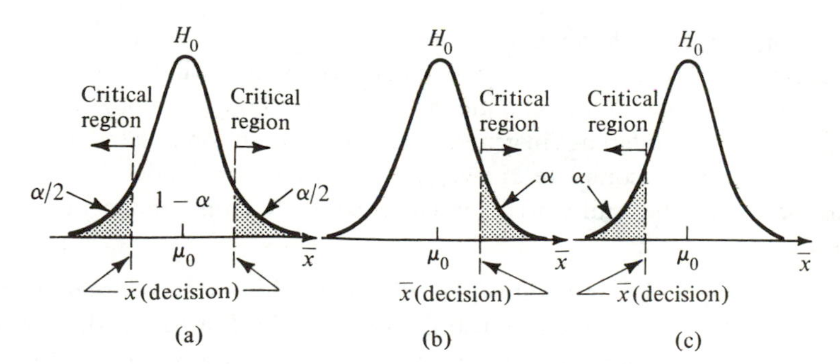

Figure 6-18 (a) Critical Region and α for Testing $H_0: \mu = \mu_0$ Against Alternative $H_1: \mu \neq \mu_0$; (b) Critical Region and α for $H_0: \mu = \mu_0$, $H_1: \mu > \mu_0$; (c) Critical Region and α for $H_0: \mu = \mu_0$, $H_1: \mu < \mu_0$

Terminology. The region to the right of $\bar{x}$(decision), is called the *critical region* of the test of hypothesis because for values of $\bar{x}$ in this region the null hypothesis H_0 is rejected.

When a sample yields $\bar{x} > \bar{x}$(decision), the result is said to be *significant;* when $\bar{x} < \bar{x}$(decision), the result is said to be *not significant.* The probability α of type I error is called the *significance level* of the test.

NOTES. Using the Student t distribution model, it is possible to test hypotheses regarding the mean of a population when σ^2 is not known.

Using the χ^2 distribution, it is possible to test hypotheses regarding the variance of a population.

EXAMPLE 2. Let us consider, model 2 of Sec. 7-2, Chapter 7, in which the null hypothesis is H_0: The man is innocent, I, and the alternative hypothesis is H_1: guilty, G. Table 6-5 describes the consequences of the possible decisions. Which error should we strive to make smaller? Which error is smaller in our Western society? Is the attitude of our culture shared by other cultures? Can you describe situations in which our attitude toward the balance between type I and type II errors will change?

Table 6-5

	Innocent, H_0 true	Guilty, H_1 true
Claim Innocent, H_0	correct $P(\text{correct}) = 1 - \alpha$	$P(\text{error II}) = \beta$ (accused may hurt someone)
Claim Guilty, H_1	$P(\text{error I}) = \alpha$ (punish innocent man)	correct $P(\text{correct}) = 1 - \beta$

Conceptually, the model of Table 6-5 is similar to that of Table 6-4. What is the difference with regard to information which will help in establishing a decision rule? This example is to serve as food for thought and discussion regarding inference in general, which by definition is based on limited information and incomplete knowledge, and can be attended by errors of omission (send the guilty free) or commission (commit innocent man to prison).

EXAMPLE 3. In automatic pattern recognition, a straightness ratio x [138] is used as a parameter to distinguish the letter B from the digit 8. The value of x is established by dividing the height h of the symbol by the arc length a of the left portion between the high and low point of the symbol, as in Fig. 6-19.

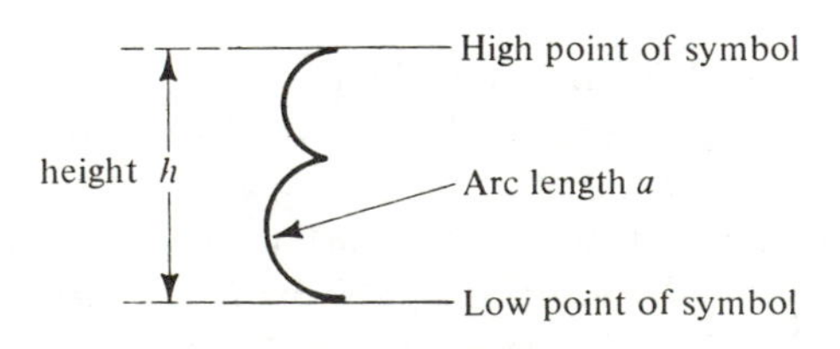

Figure 6-19

The straightness ratio is closer to 1 for the symbol B, $x = h/a \leq 1$. Experiments with various shapes of B and 8 symbols yield the conditional probability density functions for $p(x/B)$ and $p(x/8)$, as shown in Fig. 6-20. Suppose we use the decision rule:

It is B for $x > 0.90$.
It is 8 otherwise.

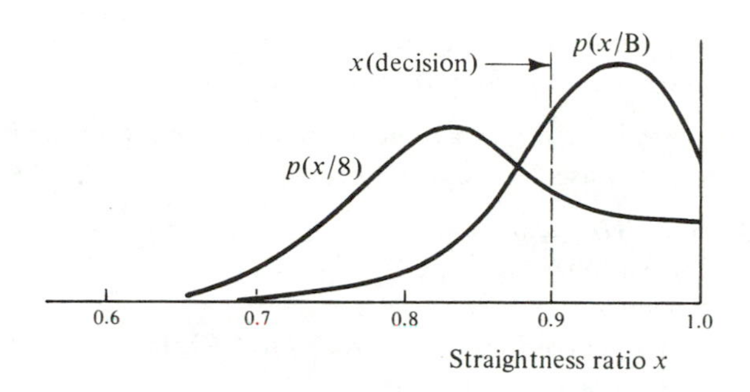

Figure 6-20 Probability Distributions for Straightness Ratios of Symbols 8 and B

Suggested Problem 1. Identify the type I and II errors in Fig. 6-20, and evaluate the corresponding probabilities of their occurrence by approximating the appropriate areas under the curves.

Consider a special case in which symbols 8 and B only are transmitted and must be identified, and suppose experience indicates that these symbols occur with probabilities $P(8) = 0.15$ and $P(B) = 0.85$, respectively. What is the probability that a symbol is B, given that the straightness ratio x is larger than 0.9 ($x > 0.9$)? Using Bayes' theorem, we can write:

$$P(B/x > 0.9) = P(B)\frac{P(x > 0.9/B)}{P(x > 0.9)}$$

Similarly,

$$P(8/x > 0.9) = P(8)\frac{P(x > 0.9/8)}{P(x > 0.9)}$$

in which

$$P(x > 0.9) = P(x > 0.9/B)P(B) + P(x > 0.9/8)P(8)$$

Suggested Problem 2. (a) Identify α and $(1 - \beta)$ in the above equations. (b) What is the expected number of wrong identifications in processing one thousand symbols (8 and B) if the decision rule remains: B for $x > 0.9$ and 8 otherwise? Give approximate answers working with the appropriate areas in Fig. 6-20.

The choice of x(decision), such as 0.90 above, depends on the cost of errors. Let C_{8B} be the cost of claiming a symbol is 8 when it is B, and let C_{B8} be the cost of claiming a symbol is B when it is 8. Then the expected cost, $E(C)$, of errors is [x_D designates x(decision)]:

$$\begin{aligned} E(C) &= C_{8B}P(\text{claim } 8 \cap B \text{ true}) \\ &\quad + C_{B8}P(\text{claim } B \cap 8 \text{ true}) \\ &= C_{8B}P(x \leq x_D / B \text{ true})P(B \text{ true}) \\ &\quad + C_{B8}P(x > x_D / 8 \text{ true})P(8 \text{ true}) \\ &= C_{8B}\beta P(B) + C_{B8}\alpha P(8) \end{aligned}$$

α and β are the probabilities of type I and type II errors, respectively. x_D is adjusted, changing α and β to make $E(C)$ minimum.

Summary of steps in testing hypothesis regarding mean μ when σ^2 is known:

(a) Procedure based on choice of α:

1. State H_0: $\mu = \mu_0$.
2. State alternative, say, H_1: $\mu = \mu_1 > \mu_0$.
3. Select α.
4. Get z_0 corresponding to α from $N(0, 1)$.
5. Compute $\bar{x}$(decision) from

$$z_0 = \frac{\bar{x}(\text{decision}) - \mu_0}{\sigma_{\bar{x}}}$$

$$\bar{x}(\text{decision}) = \mu_0 + z_0\sigma_{\bar{x}}, \ \sigma_{\bar{x}} = \frac{\sigma_x}{\sqrt{n}}$$

6. Take sample of size n and compute $\bar{x}$.

$$\bar{x} = \frac{1}{n} \sum_i x_i$$

7. Make decision.

$\bar{x} < \bar{x}$(decision), accept H_0
$\bar{x} > \bar{x}$(decision), reject H_0

(b) Procedure based on choice of β;

1. State H_0: $\mu = \mu_0$.
2. State alternative, say, H_1: $\mu = \mu_1 > \mu_0$.
3. Select β.
4. Get z_1 corresponding to β from $N(0, 1)$.
5. Compute $\bar{x}$(decision) from

$$z_1 = \frac{\bar{x}(\text{decision}) - \mu_1}{\sigma_{\bar{x}}}$$

$$\bar{x}(\text{decision}) = \mu_1 + z_1\sigma_{\bar{x}}, \ \sigma_{\bar{x}} = \frac{\sigma_x}{\sqrt{n}}$$

6. Take sample of size n and compute $\bar{x}$.

$$\bar{x} = \frac{1}{n} \sum_i x_i$$

7. Make decision.

$\bar{x} < \bar{x}(\text{decision})$, accept H_0

$\bar{x} > \bar{x}(\text{decision})$, reject H_0

Problem. How does Step 7 change in the above procedures if $H_1: \mu = \mu_1 < \mu_0$?

6-10 SIMULATION OF PROBABILISTIC MODELS—MONTE CARLO METHOD

A simulation model is a model in which a sequence of events in the real world is acted out on the model. Most simulation models are probabilistic. The Monte Carlo method generates the samples for the simulation in a way which is independent of the generating mechanism, and which permits each event to happen with a probability that, in the long run, will approach its true probability of occurrence. We use simulation whenever we wish to gain a better understanding of the contributions of certain parameters to outcomes or events of interest. By varying the values of various parameters in the simulation model, we can answer the question: "What if. . . ?" This may improve our understanding of the relationships between elements of the model and enhance our ability to predict and possibly control outcomes.

EXAMPLE 1. Suppose we wish to play a game based on the outcomes from a toss of two dice. Our outcome of interest is the sum of the numbers that will show on the faces of the dice. The sample space of outcomes and the corresponding frequencies (i.e., points in the sample space of 36 points) and the probabilities are shown in Table 6-6. The probability of any number on each die is taken to be the same, 1/6.

Table 6-6 Frequencies and Probabilities of Scores in Toss of Two Dice

Score	Frequency (Points in Sample Space of 36 Outcomes)	Probability of Score
2	1	1/36
3	2	2/36
4	3	3/36
5	4	4/36
6	5	5/36
7	6	6/36
8	5	5/36
9	4	4/36
10	3	3/36
11	2	2/36
12	1	1/36

Table 6-7 Scores and Probabilities Corresponding to Random Numbers 1 to 36

Score	Random Number	Frequency (Points in Sample Space of Random Numbers	Probability of Score
2	1	1	1/36
3	2–3	2	2/36
4	4–6	3	3/36
5	7–10	4	4/36
6	11–15	5	5/36
7	16–21	6	6/36
8	22–26	5	5/36
9	27–30	4	4/36
10	31–33	3	3/36
11	34–35	2	2/36
12	36	1	1/36

The game can be *simulated* by taking 36 pieces of paper, writing on each a number from 1 to 36, and selecting a piece of paper at random. The scores of the dice corresponding to the random numbers on paper are shown in Table 6-7. The probabilities are the same as those of Table 6-6. Thus using the pieces of paper, we employ a Monte Carlo method for a simulation model of the outcomes from the toss of two dice.

EXAMPLE 2. Random Walk Simulation. The story on the discovery of the Monte Carlo method goes back to a legend of a mathematician who observed a drunk man moving from a lamp post, as shown in Fig. 6-21. The

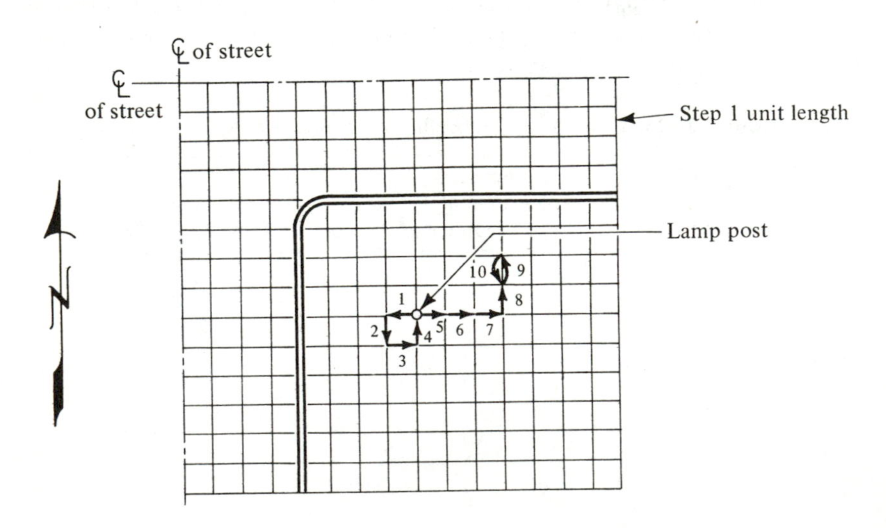

Figure 6-21 Random Walk

drunkard can move one step at a time in any of the four directions N, S, E, or W in the grid shown in the figure. The probability is 1/4 that he moves in any of these directions from any station. The question is: What is the most probable distance the drunkard will be from the lamp post after n steps? The simulation of this problem may tell us how likely the drunkard is to end up in the street traffic after n steps. The simulation may consist of tossing a coin twice at each station. Each of the four outcomes from the two tosses will lead to a move in a different direction (subscript indicates result of first or second toss):

for H_1H_2: move North, N
for H_1T_2: move East, E
for T_1H_2: move West, W
for T_1T_2: move South, S

A sample of ten trials gives the directions W S E N E E E N N S. The distance is

$$\sqrt{3^2 + 1} = \sqrt{10}$$

This happens to agree with the theoretical result for the square root of the expected value of the square of the distance d:

$$\sqrt{E(d)^2} = a\sqrt{n} \tag{6-62}$$

in which a is the length of each step (taken as 1 in the present example).

Derivation of Equation 6-62. Let $x_i = -1$ or 1 indicate, respectively, a move W or E in step i; and let $y_i = -1$ or 1 indicate, respectively, a move N or S in step i. Then in n steps the square of the distance d is given by

$$\begin{aligned} d^2 &= (x_1 + x_2 + \dots + x_n)^2 + (y_1 + y_2 + \dots + y_n)^2 \\ &= (\textstyle\sum_{i=1}^n x_i)^2 + (\sum_{i=1}^n y_i)^2 \end{aligned}$$

The expected value of d^2 is computed from

$$E(d^2) = E(\textstyle\sum_i x_i)^2 + E(\sum_i y_i)^2$$

Since $E(x_ix_j) = E(y_iy_j) = 0$ for $i \neq j$ because of independence, it follows that

$$\begin{aligned} E(d^2) &= E(\textstyle\sum_i x_i^2) + E(\sum_i y_i^2) = n \\ \sqrt{E(d^2)} &= \sqrt{n} \end{aligned}$$

EXAMPLE 3. The Monte Carlo method can be used to compute the area A in Fig. 6-22. We divide the height of the rectangle and its base into, say, 100 equal intervals, and number the dividing lines as coordinates from 0 to 100. We now select at random a number between 0 and 100 for the horizontal coordinate and make a similar choice for the vertical coordinate. Each such pair of numbers establishes a point in the rectangle of the figure. For example, the random pair 10, 80 lies in area A, and pair 55, 30 is outside A.

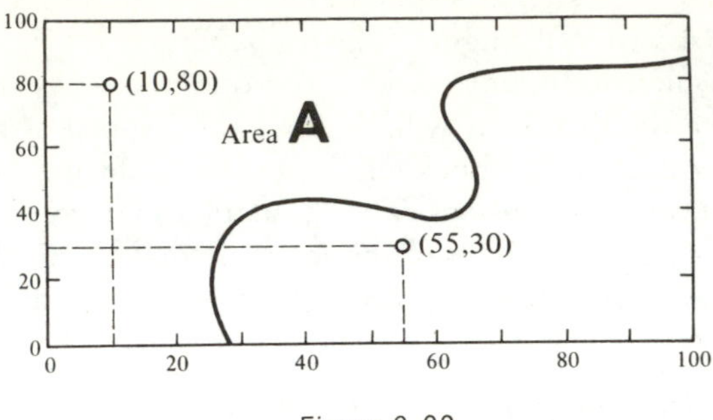

Figure 6-22

The area A is related to the area of the rectangle approximately by the fraction of points falling in the area A, i.e.,

$$\frac{\text{area } A}{\text{area of rectangle}} = \frac{\text{number of points in } A}{\text{total number of points tried}}$$

EXAMPLE 4. Suppose we are in the business of forming necklaces by linking two kinds of chains. The two kinds differ in their strength. Tensile strength, S_1 (breaking force in pulling), for chain type 1 has a normal distribution with $\mu_1 = 100$ pounds and $\sigma_1 = 8$ pounds; tensile strength, S_2, of chain type 2 is normally distributed with $\mu_2 = 92$ pounds and $\sigma_2 = 15$ pounds. We select randomly a chain from each of the two populations and link them. We wish to conduct a simulation before we proceed with an actual linking and testing of the necklaces so that we may have a better feeling about the strength of the necklaces. We now invoke the time-honored principle, "the necklace is as strong as its weakest chain." In our case the weaker chain may be of type 1 or type 2 because the distributions for S_1 and S_2 overlap, as shown in Fig 6-23.

The simulation will consist of a random selection of numbers from normal distributions $N(\mu_1, \sigma_1^2)$ and $N(\mu_2, \sigma_2^2)$ for the values S_1 and S_2, respectively. Of any pair S_1, S_2 we shall take the smaller and this will represent the strength of the necklace in the simulation. To do this we must first

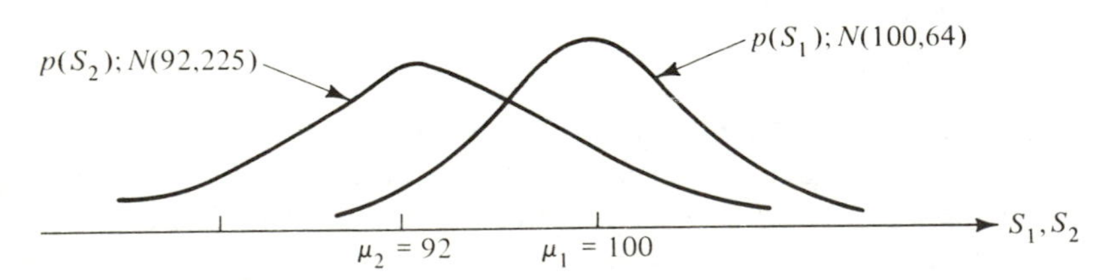

Figure 6-23 Distributions $p(S_1)$ and $p(S_2)$ for Tensile Strength of Chains Type 1 and 2

show how a random selection of numbers from a normal population can be simulated.

Simulated Sample from a Normal Distribution

Figure 6-24 shows the standard normal distribution and its corresponding cumulative curve. The ordinate in the cumulative curve for any value of standard normal variable z_α is equal to the probability

$$P(z < z_\alpha) = 1 - \alpha \tag{6-63}$$

For example, for $z_\alpha = 1$, $P(z < z_\alpha) = 0.84$. The ordinates of the cumulative curve go from 0 to 1. Now, a sample of one observation from the normal distribution $N(0, 1)$ in Fig. 6-24(a) should result in a value between -1 and 1 68% of the time, a value between -2 and 2 95% of the time, and so on, according to the probability of occurrence of an observation in the range of values stipulated.

The procedure for generating such results from the sampled numbers can be obtained by using Fig. 6-24(b) as follows. We divide the vertical scale

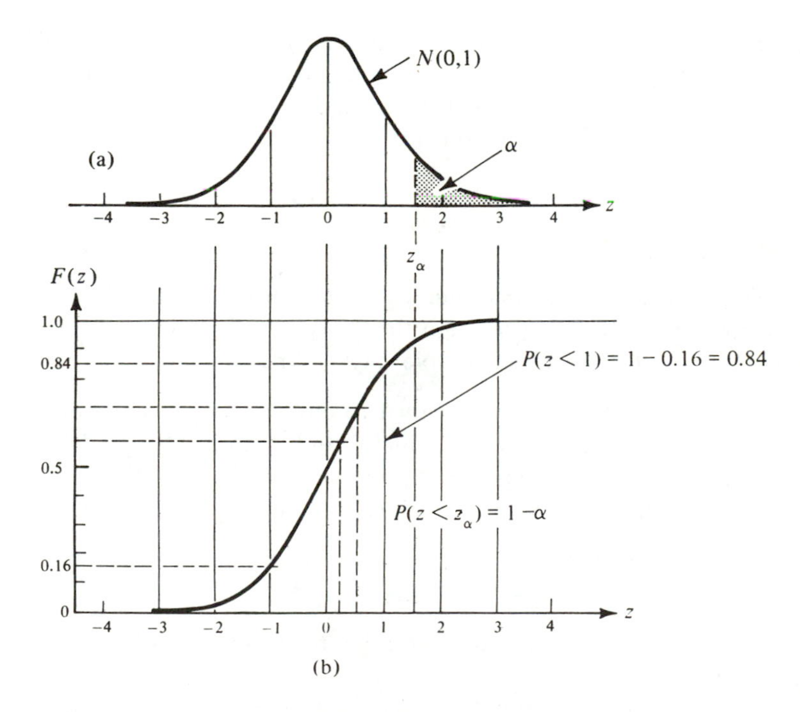

Figure 6-24 (a) Standard Normal Distribution $N(0, 1)$; (b) Cumulative Distribution for $N(0, 1)$

in Fig. 6-24(b) into, say, 100 (could be 1000, or 10,000, etc.) equally spaced intervals and then select at random a marble from a container with 100 marbles numbered from 01 to 100 to direct us to an interval. Thus, if we select the random number 60 (marble number 60), we put a decimal point in front of the number, reading .60, and draw a horizontal line from ordinate 0.60 in Fig. 6-24(b) to the cumulative curve. From the intersection of this horizontal line and the cumulative curve we drop a perpendicular line and read the value of z from the abscissa which, for ordinate 0.60, is 0.25. Similarly, for 70, i.e., 0.70, we read $z = 0.50$. For 16 (0.16), we read $z = -1$; for 84 (0.84), we read $z = 1$. Now, note that the probability of values between 16 and 84 in our random selection of marbles, which have equal probability of being drawn, is the ratio of the vertical distance on the ordinate scale from 0.16 to 0.84 divided by the total length of the vertical which is 1. Hence, the probability is 0.68. The numbers between 16 and 84 lead then to z values between -1 and 1 with the same probability of 0.68 just as we desired.

Random Sample from $N(\mu, \sigma^2)$

To simulate a random drawing from $N(\mu_x, \sigma_x^2)$, we use Fig. 6-24(b), as discussed above, to generate values for standard variable z. Then we transform each value of z so generated to the corresponding x in $N(\mu_x, \sigma_x^2)$ by using the following transformation (Eq. 6-64):

$$z = \frac{x - \mu_x}{\sigma_x}$$

$$x = \mu_x + z\sigma_x \tag{6-64}$$

Tables of random normal numbers z have been generated and can be used directly without going through the procedure described here by using Fig. 6-24(b). A portion of such a table is given in Table 6-8.

Back to Example 4. We now return to our normal variables S_1 and S_2. Using Table 6-8, let us read values for z_1 and z_2 from arbitrarily selected columns 3 and 4, respectively, taking five samples. These values are transformed to corresponding S_1 and S_2 values by using Eq. 6-64, i.e.,

$$S_1 = \mu_1 + z_1\sigma_1 = 100 + 8z_1$$

$$S_2 = \mu_2 + z_2\sigma_2 = 92 + 15z_2$$

These results are recorded together with the smaller of S_1 and S_2, i.e., strength of necklace, in Table 6-9.

An extension of this simulation as given by Table 6-9 could give sufficient data for an approximate probability distribution of the strength of a necklace in pounds.

Table 6-8 Random Numbers from $N(0, 1)$
(Part of RAND Corp. Tables "A Million Random Digits with 100,000 Normal Deviates." Reproduced with permission.)

Column 1	2	3	4	5
−0.513	−0.525	0.595	0.881	−0.934
−1.055	0.007	0.769	0.971	0.712
−0.488	−0.162	−0.136	1.033	0.203
0.756	−1.618	−0.445	−0.511	−2.051
0.225	0.378	0.761	0.181	−0.736
1.677	−0.057	−1.229	−0.486	0.856
−0.150	1.356	−0.561	−0.256	0.212
0.598	−0.918	1.598	0.065	0.415
−0.899	0.012	−0.725	1.147	−0.121
−1.163	−0.911	1.231	−0.199	−0.246
−0.261	1.237	1.046	−0.508	−1.630
−0.357	−1.384	0.360	−0.992	−0.116
1.827	−0.959	0.424	0.969	−1.141
0.535	0.731	1.377	0.983	−1.330
−2.056	0.717	−0.873	−1.096	−1.396
−2.008	−1.633	0.542	0.250	0.166
1.180	1.114	0.882	1.265	−0.202
−1.141	1.151	−1.210	−0.927	0.425
0.358	−1.939	0.891	−0.227	0.602
−0.230	0.385	−0.649	−0.577	0.237
−0.208	−1.083	−0.219	−0.291	1.221
0.272	−0.313	0.084	−2.828	−0.439
0.606	0.606	−0.747	0.247	1.291
−0.307	0.121	0.790	−0.584	0.541
−2.098	0.921	0.145	0.446	−2.661
0.079	−1.473	0.034	−2.127	0.665
−1.658	−0.851	0.234	−0.656	0.340
−0.344	0.210	−0.736	1.041	0.008
−0.521	1.266	−1.206	−0.899	0.110
2.990	−0.574	−0.491	−1.114	1.297

Table 6-9

z_1	S_1 (pounds)	z_2	S_2 (pounds)	Strength of necklace (pounds)
0.595	104.75*	0.881	105.22	104.75
0.769	106.15*	0.971	106.55	106.15
−0.136	98.99*	1.033	107.50	98.99
−0.445	96.44	−0.511	84.33*	84.33
0.761	106.10	0.181	94.72*	94.72

*Indicates smaller of S_1, S_2.

6-11 SUMMARY

A population is a set which contains all possible observations of a particular attribute.

A sample is a subset of a population; *Sample* $\subset$ *Population.*

The *mean,* μ, of a population is a measure of *central location* of the distribution. For population size N with m different values $x_i (i = 1, 2, \ldots, m)$ for random variable x,

$$\mu = \frac{1}{N} \sum_{i=1}^{m} n_i x_i = \sum_{i=1}^{m} f_i x_i \tag{6-9}$$

in which $f_i = n_i/N$ and n_i is the number of elements in the population with attribute x_i.

The *variance,* σ^2, of a population is a *measure of dispersion*

$$\sigma^2 = \sum_i f_i (x_i - \mu)^2 \tag{6-10}$$

The square root of the variance is called the *standard deviation,* σ.

The mean μ and variance σ^2 are *population parameters.* A sample of a population can be used to generate *estimates* of population parameters. Such estimates are called *sample statistics.*

The estimate $\hat{\mu}$ for μ is given by the *sample mean,* $\bar{x}$, (n is the sample size):

$$\hat{\mu} = \bar{x} = \frac{1}{n} \sum_i x_i \tag{6-11}$$

The estimate $\hat{\sigma}^2$ for σ^2 is given by sample variance s^2

$$\hat{\sigma}^2 = s^2 = \frac{1}{n-1} \sum_i (x_i - \bar{x})^2 \tag{6-12}$$

Probability density function $p(x)$ can be *discrete or continuous.* The *binomial distribution* is discrete:

$$p(x) = \binom{n}{x} p^x q^{n-x} \qquad x = 0, 1, \ldots, n \tag{6-15}$$

The *normal distribution* is continuous:

$$p(x) = \frac{1}{\sigma\sqrt{2\pi}} e^{-1/2[(x-\mu)/\sigma]^2} \tag{6-20}$$

A discrete probability distribution $p(x)$ has the following properties:

$$\left.\begin{aligned} &p(x) \geq 0 \quad \text{for all } x \\ &\textstyle\sum_{\text{for all } x} p(x) = 1 \\ &Pr(a < x < b) = \textstyle\sum_{x=a+1}^{b-1} p(x) \end{aligned}\right\} \tag{6-16}$$

The mean μ and variance σ^2 of a discrete distribution are given by Eqs. 6-9 and 6-10, respectively.

A continuous probability distribution $p(x)$ has the following properties:

$$\left.\begin{aligned} &p(x) \geq 0 \text{ for } -\infty < x < \infty \\ &\int_{-\infty}^{\infty} p(x)\,dx = 1 \\ &Pr(a < x < b) = \int_a^b p(x)\,dx \end{aligned}\right\} \tag{6-13}$$

For a continuous distribution $p(x)$

$$\mu = \int_{-\infty}^{\infty} x\,p(x)\,dx \tag{6-22}$$

and

$$\sigma^2 = \int_{-\infty}^{\infty} (x - \mu)^2\,p(x)\,dx \tag{6-23}$$

The normal distribution for random variable x with mean μ_x and variance $\sigma_x{}^2$ is designated by $N(\mu_x, \sigma_x{}^2)$. Variable x can be transformed to a *standard normal variable*, z, with $N(0, 1)$ by using the transformation

$$z = \frac{x - \mu_x}{\sigma_x} \tag{6-25}$$

Expected value E of any function $f(x)$ of random variable x with distribution $p(x)$ is given by

$$E[f(x)] = \int_{-\infty}^{\infty} f(x)p(x)\,dx \tag{6-26}$$

for a continuous distribution, and by

$$E[f(x_i)] = \textstyle\sum_{\text{all } x_i} f(x_i)p(x_i) \tag{6-27}$$

for a discrete distribution. The expected value E is a *linear operator*, and has the properties of a linear operator.

Examples of expected values:

$$\begin{aligned} &E(x) = \mu_x \\ &E(x - \mu)^2 = \sigma_x{}^2 \\ &E(z) = \mu_z = 0 \\ &E(z - \mu_z)^2 = \sigma_z{}^2 = 1 \\ &E(\bar{x}) = \mu_x \\ &E(s^2) = \sigma_x{}^2 \end{aligned}$$

A linear combination w of independent random variables x_i with means μ_i and variance $\sigma_i{}^2$, namely,

$$w = \textstyle\sum_i a_i x_i \tag{6-37}$$

has mean μ_w:

$$\mu_w = \textstyle\sum_i a_i \mu_i \tag{6-38}$$

and variance σ_w^2:

$$\sigma_w^2 = \sum_i a_i^2 \sigma_i^2 \tag{6-39}$$

When w is the sample mean $\bar{x}$ of a sample of size n, then

$$w = \bar{x} = \frac{1}{n} \sum_i x_i \tag{6-40}$$

$$\mu_{\bar{x}} = \mu_x \tag{6-41}$$

$$\sigma_{\bar{x}}^2 = \frac{1}{n} \sigma_x^2 \tag{6-42}$$

CENTRAL LIMIT THEOREM: If a random variable x is distributed with mean μ and variance σ^2, then an aggregation of n independent observations x_i, i.e., a linear combination $w = \sum_i x_i$, which is obtained from random samples of size n, will be approximately distributed in a normal distribution with mean $n\mu$ and variance $n\sigma^2$. The approximation will become progressively better as the sample size n increases.

The sample mean $\bar{x} = (1/n) \sum_i x_i$ will approach a normal distribution with mean μ and variance $(1/n)\sigma^2$ as the sample size n increases (see Sec. 6-5 for mean and variance of $\bar{x}$).

The binomial distribution can be approximated by the normal distribution. For random variable $x_i = 0$ or 1, with $P(0) = 1 - p = q$ and $P(1) = p$, we can write

$$w = \sum_i x_i$$

$$\mu_w = np$$

$$\sigma_w^2 = npq$$

As n increases, the distribution of w approaches the normal distribution $N(\mu_w, \sigma_w^2)$.

Distribution of proportion $\bar{p}$. The variable $\bar{p}$ is derived from

$$\bar{p} = \frac{w}{n} = \sum_{i=1}^{n} \frac{1}{n} x_i \tag{6-51}$$

in which $x_i = 0, 1$; $P(0) = 1 - p = q$, $P(1) = p$:

$$u_{\bar{P}} = p \tag{6-52}$$

$$\sigma_{\bar{P}}^2 = \frac{1}{n} pq \tag{6-53}$$

For large samples n, $\bar{p}$ is approximately distributed as $N(\mu_{\bar{p}}, \sigma_{\bar{p}}^2)$.

Distribution of sample mean, $\bar{x} = (1/n) \sum_i x_i$. For large samples n, $\bar{x}$ is approximately distributed as $N(\mu_{\bar{x}}, \sigma_{\bar{x}}^2)$ with $\mu_{\bar{x}} = \mu_x$ and $\sigma_{\bar{x}}^2 = (1/n)\sigma_x^2$.

In *inference* regarding parameter p, using statistic $\bar{p}$, we have

$$P(-z_{\alpha/2}\sigma_{\bar{p}} < \bar{p} - p < z_{\alpha/2}\sigma_{\bar{p}}) = 1 - \alpha \tag{6-58}$$

$$P(\bar{p} + z_{\alpha/2}\sigma_{\bar{p}} > p > \bar{p} - z_{\alpha/2}\sigma_{\bar{p}}) = 1 - \alpha \tag{6-59}$$

or

$$P(\bar{p} - z_{\alpha/2}\sigma_{\bar{p}} < p < \bar{p} + z_{\alpha/2}\sigma_{\bar{p}}) = 1 - \alpha \tag{6-60}$$

The interval $\bar{p} + z_{\alpha/2}\sigma_{\bar{p}} - (\bar{p} - z_{\alpha/2}\sigma_{\bar{p}}) = 2z_{\alpha/2}\sigma_{\bar{p}}$ is called the *confidence interval*. The boundaries of the interval $\bar{p} + z_{\alpha/2}\sigma_{\bar{p}}$ and $\bar{p} - z_{\alpha/2}\sigma_{\bar{p}}$ are called the *confidence limits*. The probability $1 - \alpha$ is the *confidence level*. Equation 6-60 states that the confidence interval contains the population parameter p with probability $1 - \alpha$.

In *inference* regarding parameter μ of a population, using statistic $\bar{x}$, we have

$$P(-z_{\alpha/2}\sigma_{\bar{x}} < \bar{x} - \mu < z_{\alpha/2}\sigma_{\bar{x}}) = 1 - \alpha$$

$$P(\bar{x} + z_{\alpha/2}\sigma_{\bar{x}} > \mu > \bar{x} - z_{\alpha/2}\sigma_{\bar{x}}) = 1 - \alpha$$

or

$$P(\bar{x} - z_{\alpha/2}\sigma_{\bar{x}} < \mu < \bar{x} + z_{\alpha/2}\sigma_{\bar{x}}) = 1 - \alpha$$

In *testing hypotheses*, two types of error can be introduced. *Error type I* is rejecting the null hypothesis H_0 when it is true. *Error type II* is accepting H_0 when it is false. The probabilities of type I and type II errors are designated by α and β, respectively.

The *test of hypotheses* consists of these steps:

(a) Procedure based on choice of α:

1. State H_0: $\mu = \mu_0$.
2. State alternative, say, H_1: $\mu = \mu_1 > \mu_0$.
3. Select α.
4. Get z_0 corresponding to α from $N(0, 1)$.
5. Compute $\bar{x}$(decision) from

$$z_0 = \frac{\bar{x}(\text{decision}) - \mu_0}{\sigma_{\bar{x}}}$$

$$\bar{x}(\text{decision}) = \mu_0 + z_0\sigma_{\bar{x}}, \quad \sigma_{\bar{x}} = \frac{\sigma_x}{\sqrt{n}}$$

6. Take sample of size n and compute $\bar{x}$.

$$\bar{x} = \frac{1}{n}\sum_i x_i$$

7. Make decision.

$\bar{x} < \bar{x}$(decision), accept H_0

$\bar{x} > \bar{x}$(decision), reject H_0

(b) Procedure based on choice of β:

1. State H_0: $\mu = \mu_0$.
2. State alternative, say, H_1: $\mu = \mu_1 > \mu_0$.
3. Select β.
4. Get z_1 corresponding to β from $N(0,1)$.
5. Compute $\bar{x}$(decision) from

$$z_1 = \frac{\bar{x}(\text{decision}) - \mu_1}{\sigma_{\bar{x}}}$$

$$\bar{x}(\text{decision}) = \mu_1 + z_1\sigma_{\bar{x}}, \quad \sigma_{\bar{x}} = \frac{\sigma_x}{\sqrt{n}}$$

.6. Take sample of size n and compute $\bar{x}$.

$$\bar{x} = \frac{1}{n} \sum_i x_i$$

7. Make decision.

$\bar{x} < \bar{x}(\text{decision})$, accept H_0
$\bar{x} > \bar{x}(\text{decision})$, reject H_0

Probabilistic models can be simulated by using the *Monte Carlo method.* This method generates samples for the simulation in a way that permits each event to happen with a probability equal to or approaching its true probability of occurrence.

EXERCISES

NOTE. *Summation Notation:*

The capital Greek letter sigma ($\sum$) is the symbol used in mathematics to denote summation. If we have a set of numbers to add, for example (3, 7, 123), we can label each number as follows: the first number (3) as x_1, the second (7) as x_2, etc. In general, we label the ith number as x_i. Of course, if there are n numbers in the set, we have x_is running from x_1 to x_n.

As an abbreviation for the operation of adding all n numbers, we write: $\sum_{i=1}^{n} x_i$. Or, in words, sum all the x_is for i going from 1 to n. That is:

$$\sum_{i=1}^{n} x_i = x_1 + x_2 + \ldots + x_n$$

Suppose we have two sets of numbers that we wish to combine in the following way. Label the elements in the first set a_1, a_2, a_3, etc., and those in the second x_1, x_2, etc., as above. We may be interested in $a_1x_1 + a_2x_2 + a_3x_3 + \ldots + a_nx_n$. We can write this as: $\sum_{i=1}^{n} a_ix_i$.

For example, suppose we know that three people make \$5000, two make \$6000, and five make \$4000, and we wish to know how much money is made by all these people. Denoting the number of people by a_i(i.e., $a_1 = 3$, $a_2 = 2$, $a_3 = 5$) and the income by x_i, we can write the problem as:

$$\sum_{i=1}^{3} a_ix_i = ?$$

The solution is
$$\begin{aligned}\sum_{i=1}^{3} a_ix_i &= a_1x_1 + a_2x_2 + a_3x_3 \\ &= (3)(5000) + (2)(6000) + (5)(4000) \\ &= \$47{,}000\end{aligned}$$

6-1 (a) We are interested in finding out the number of passenger-miles flown by a small airline in one week. Passenger-miles is the summation of all the miles flown by all the passengers. It was found that 1000 passengers flew the airline in one week. Of these, 20% flew 500 miles each, 30% flew 1000 miles each, and 50% flew 1500 miles each. Write the problem out in summation notation and solve for the number of passenger-miles.

(b) What is the average number of miles flown per passenger? Solve by using Eq. 6-9, first using the middle expression, and again using the right-hand side expression.

6-2 In a class of 10 students the scores for the final examination are distributed as follows: three grades of 60, six grades of 65, and one grade of 70. The mean grade, μ, for this population is 64.

(a) Verify that μ is 64.

(b) Using Eq. 6-10, find the variance, σ^2.

NOTE. We can consider the labels x_i in the above note as if they were addresses in a column of compartments. Compartment x_i contains the number we label x_i:

x_1
x_2
x_3
x_4

Let us suppose that the above column represents students in a history class. x_1, x_2, x_3, and x_4 represent the number of freshmen, sophomores, juniors, and seniors in the class, respectively. Then $\sum_{i=1}^{4} x_i = x_1 + x_2 + x_3 + x_4$ represents the total number of students in the class.

Now, suppose we have a second column representing students in a biology class, and a third column representing students in an art class. In order to identify which class a student is in, let us give the x's a *second* subscript, j, in which $j = 1$ represents history class, $j = 2$ biology class, and $j = 3$ art class. Thus, x_{23}, for example, represents the number of sophomores ($i = 2$) in the art class ($j = 3$). In general, we now have an array of x_{ij}' s which we can arrange as follows:

	$j = 1$ History	$j = 2$ Biology	$j = 3$ Art
$i = 1$ Freshmen	x_{11}	x_{12}	x_{13}
$i = 2$ Sophomores	x_{21}	x_{22}	x_{23}
$i = 3$ Juniors	x_{31}	x_{32}	x_{33}
$i = 4$ Seniors	x_{41}	x_{42}	x_{43}

If we want to know the total number of students enrolled in the three classes, we add the 12 x_{ij}'s above. In summation notation this is written as:

$$\sum_{i=1}^{4} \sum_{j=1}^{3} x_{ij} = x_{11} + x_{12} + x_{13} + x_{21} + x_{22} + x_{23} + x_{31} + x_{32} + x_{33} + x_{41} + x_{42} + x_{43}$$

Notice that we keep the i constant at $i = 1$ while we add the x_{1j}'s from $j = 1$ to $j = 3$, and then we make $i = 2$ and add the x_{2j}'s from $j = 1$ to $j = 3$, etc.; namely, proceeding from row 1 to row 2 to row 3. We could, of course, proceed from column to column with the addition by writing $\sum_{j=1}^{3} \sum_{i=1}^{4} x_{ij}$.

6-3 Referring to the above note, write in summation notation and expand the following:

(a) Total number of freshmen in the three classes.

(b) Total number of students in the art class.

(c) Total number of juniors and seniors in the three classes.

6-4 We wish to extend Exercise 6-1 as follows: Assume that, in addition to the airline described in 6-1, there is another airline in which in one week 600 passengers flew 500 miles, 1200 flew 1000 miles, and 1300 flew 1500 miles. Write in summation notation and solve the following:

(a) How many total passenger-miles were flown by the two airlines in one week?

(b) How many passenger-miles were flown by passengers flying less than 1500 miles?

6-5 Returning to the note preceding Exercise 6-3, suppose the three classes are each taught at three different universities. We then have data on how many students of each group are enrolled in each class in each university. We can represent these data by symbols x_{ijk} where subscript k takes on the values 1, 2, 3 to identify the three universities.

(a) Write in summation notation and expand the following:

(i) Total number of students in the three history classes.

(ii) Total number of freshmen in all classes in the three universities.

(b) Analogous to the column for x_i's, and the array for x_{ij}'s, devise a pictorial method for displaying the x_{ijk}'s by using compartments.

6-6 Referring to Exercise 6-5, how would you represent the situation if, in addition, the three classes are each taught in the three universities in each of two semesters? Represent it both pictorially and by using symbols.

6-7 Expand the following summation (write out the terms):

$$\sum_{i=1}^{2} \sum_{j=1}^{3} \sum_{k=5}^{6} x_{ijk}$$

BINOMIAL DISTRIBUTION

6-8 Solve Exercise 4-1 by using the binomial distribution, and compare your answer with the answer you found by using a tree diagram.

NOTE. *The Number of Distinct Routes Problem:*

A person wants to go from point A to point B, and at each point he can go only North or East. How many distinct paths can he take? (A point is an intersection on the grid in Fig. 6-25.) We have seen this type of problem before. In Chapter 2, Exercise 2-22, the answer was found by drawing a tree diagram in order to enumerate the possible paths.

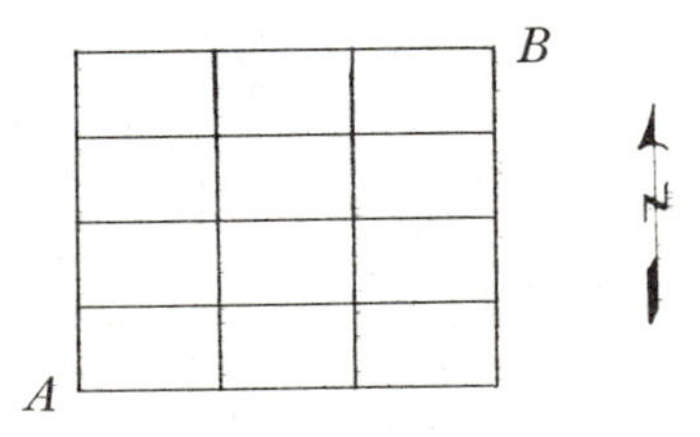

Figure 6-25

In Chapter 5, Sec. 5-12, it was shown how the enumeration of paths is described in general by the binomial model, and the answer to the question, "How many distinct paths?" could be found by using the Pascal triangle expansion method. In this chapter we see that the number of distinct paths from one point to another is one of the elements of the binomial distribution equation (Eq. 6-15). It is the expression $\binom{n}{x}$ commonly referred to as "n choose x."

6-9 In Fig. 6-25, a person walking from A to B would have to take a total of 7 steps in all cases, 3 in the East direction and 4 in the North direction.
(a) What would n and x be in this situation?
(b) How many distinct paths can be taken from A to B?
(c) Convince yourself that $\binom{7}{4} = \binom{7}{3}$. Explain the significance of this.

6-10 Find the probability of hitting a target at least once when five torpedoes are fired, each with the probability 1/3 of scoring a hit.

6-11 Three candidates run for different offices in different states. Each has 1 chance in 3 of being elected in his state. What is the chance that at least one of them is elected?

6-12 A quiz has 6 multiple-choice questions, each with three alternatives. What is the probability of 5 or more right answers by sheer guessing?

6-13 In a binomial experiment consisting of three trials, the probability of exactly 2 successes is 12 times as great as that for 3 successes. Find p.

6-14 An owner of five overnight cabins is considering buying television sets to rent to cabin occupants. He estimates that about half of his customers would be willing to rent sets. Finally, he buys three sets. Assuming 100% occupancy at all times:
(a) What fraction of the evenings will there be more requests than TV sets?
(b) What is the probability that a customer who requests a television set will receive one?

NORMAL DISTRIBUTION

6-15 Given a normal distribution with mean 150 and standard deviation 10 (i.e., $\mu = 150$ and $\sigma = 10$). Find:
(a) $P(x < 130)$
(b) $P(x < 190)$
(c) $P(x > 130)$
(d) $P(130 < x < 190)$

6-16 Suppose the curve in Fig. 6-26 describes the distribution of a normal random variable x with $\mu = 10$ and $\sigma = 5$.
(a) How many standard deviations away from the mean is the point $x = 20$? $x = 0$?
(b) What is the probability that x lies between 0 and 20?
(c) If you were asked to find the probability that x will not differ from the mean by more than one standard deviation, then
(i) what would this probability be?

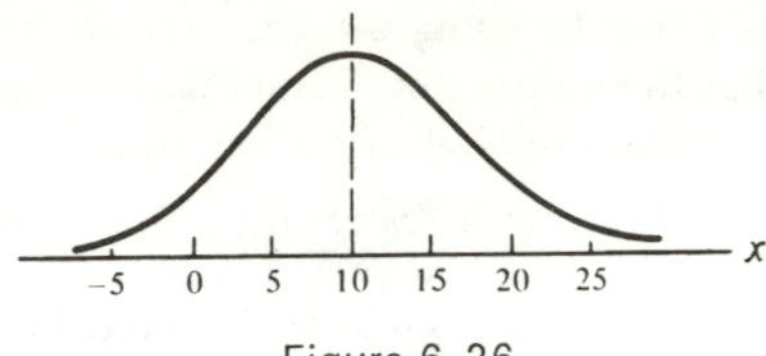

Figure 6-26

(ii) between what two numbers would x lie, on the Fig. 6-26 curve, with this probability?

6-17 (a) If x is a standard normal random variable, find

(i) $P(x > 0.3)$

(ii) $P(-1 < x < 2)$

(b) If y is a normal random variable with $\mu = 8$, $\sigma = 2$, find

(i) $P(y > 8.6)$

(ii) $P(6 < y < 12)$

(c) Compare your answers to (a) and (b) and explain.

6-18 What is the probability that a standard normal random variable takes on a value between -3 and $+3$? Between -2 and $+2$? Between -1 and $+1$?

6-19* Mr. Commuter has a statistician friend who believes that Mr. Commuter's time of arrival, measured in minutes after 8: 55, is approximately a normal random variable T, with mean 0 and standard deviation 2.5. If this approximation is valid, what is the probability that Mr. Commuter will arrive between 8: 50 and 9: 00?

6-20* The College Entrance Examination Board test scores are scaled to approximate a normal distribution with mean $\mu = 500$ and standard deviation $\sigma = 100$.

(a) What is the probability that a randomly selected student will score

(i) 700 or more?

(ii) 580 or less?

(b) What is the probability that three randomly selected students will score 700 or more?

(c) What is the probability that at least two of three randomly selected students will score less than 700?

6-21 The Highway Patrol radar checks show that the average speed on the San Diego Freeway between Wilshire Boulevard and Jefferson Boulevard is 58.3 miles per hour. The standard deviation is 5.2 mph. What is the percentage of cars exceeding 65 mph if the speeds of the cars obey a normal distribution?

6-22 Results from a survey follow a normal distribution with a mean $\mu = 100$ and standard deviation $\sigma = 10$. However, the researcher considers only the middle 95% of the values to be relevant. Between which numbers should values of the results be included in the survey?

---

*Taken from Mosteller, Rourke, and Thomas [139], which is an excellent reference for a lucid exposition on probability and statistics.

6-23 For a given population, the distribution of IQs is found to be normal with mean of 120 and a standard deviation of 15.

(a) Draw the curve of such a distribution.

(b) What is the probability that when five people are selected at random, two of them will each have an IQ over 140?

6-24 An archaeologist on a dig in the Middle East discovered several clay figurines near a dry river bed. He saw that the figurines were similar to those produced by inhabitants of two ancient cities, *Aleph* and *Beth*, and hypothesized that flood waters had carried the figurines to the river bed. He knows that the figurines from Aleph and Beth are indistinguishable except for weight. The weights of figurines Aleph and Beth have normal distributions with means of 7 ozs, and 8 ozs, respectively, and an identical standard deviation of 0.4 oz.

Suppose one of the figurines that the archaeologist found in the river bed weighed 7.5 oz. He would like to be sure that no more than 10% of the time will he identify a figurine as from Beth, when it really came from Aleph (i.e., $\alpha = 0.1$).

(a) What is the weight below which he should claim identity Aleph, and above which, Beth?

(b) From which town should he claim that the 7.5-oz figurine came?

NOTE. *Expected Value:*

As expressed in Sec. 6-5, the *expected value* of any function, $f(x)$, of a random variable x is the *weighted average* of the $f(x)$. The values of $f(x)$ for various x are weighted by their probabilities of occurrence.

From Eq. 6-27, we know that

$$E[f(x_i)] = \sum_{\text{all } x_i} f(x_i)p(x_i) = f(x_1)p(x_1) + f(x_2)p(x_2) + \ldots$$

Equation 6-26 is for the continuous case and is identical to Eq. 6-27, except the summation sign $\sum_{\text{all } x_i}$ is replaced by the integration symbol $\int_{-\infty}^{\infty}$.

For those without background in calculus, it suffices to say that the symbol $\int$ (integral sign) performs the continuous addition equivalent to $\sum$ in the discrete case.

In Chapter 4, Sec. 4-10, Eq. 4-18 is used to find the expected value of a game in which a die is tossed and the reward in dollars is equal to the number that appears on the die. Let us put this in the form of Eq. 6-27.

Let x_1 = the event that a 1 appears, x_2 = the event that a 2 appears, etc. Then, $p(x_1) = 1/6$, $p(x_2) = 1/6$, and so on for $p(x_3)$ through $p(x_6)$. Now, let $f(x_i)$ equal the reward we receive for the occurrence of event x_i. Thus, $f(x_1) = \$1$, $f(x_2) = \$2$, up through $f(x_6) = \$6$.

From Eq. 6-27, then:

$$E[f(x_i)] = \sum_{i=1}^{6} f(x_i)p(x_i) = (\$1)(1/6) + (\$2)(1/6) + \ldots + (\$6)(1/6) = \$3.50$$

6-25 A roulette wheel has 38 equal slots numbered 00, 0, 1, 2, 3, . . . , 36. If a player bets $1 on a number and that number comes up, he receives $36. If the player loses, he loses his dollar. Calculate the expected value of the game of roulette.

6-26 A man has 8 keys on a chain and doesn't know which one is needed to open his door. He picks a key at random and tries the lock. If it opens, fine; but if not, he gets so flustered he forgets which of the 8 keys he tried and must again select one of the 8 at random. What is the expected number of times this happens before he finds the correct key?

6-27 An election pollster makes the following predictions: Candidate A will definitely carry certain states whose total electoral vote is 150. There is no way candidate A can carry certain other states with an aggregate 175 electoral votes. Of the remaining states, the pollster feels there is a 0.2 chance of A winning a block of 100 votes, 0.7 chance of winning a block of 90 votes, and a 9-in-10 chance of winning a 25-vote block. According to the pollster, what is the expected number of electoral votes that candidate A will receive?

CENTRAL LIMIT THEOREM

6-28 Using the normal distribution model as an approximation to the binomial distribution, solve the following: Suppose that whenever we try to help someone, the probability of actually helping the person is 0.5. If we try to help 64 people in our lifetime, what is the probability that we actually help 30 or more of them?

6-29 In Exercise 6-19, what is the bound on the probability if the normal approximation is rejected and only the mean and variance of the random variable are accepted?

6-30 Compare the results of Exercises 6-19 and 6-29, and discuss.

7

DECISION-MAKING MODELS

7-1 INTRODUCTION

We said earlier that problem solving is a matter of appropriate selection. We select a desired goal and we select a process which leads to it. Whenever selection is involved, we must decide between alternatives. When decisions and corresponding behavior are guided strictly by emotion, the result is often erratic behavior, irrational and hysterical in character. On the other hand, decisions guided strictly by rationality are consistent and give due consideration to the possible consequences of alternative courses of action. However, strictly rational decisions may be sterile and void of the true nature of human behavior, which normally reflects all three modes of consciousness: sensation, affection, and logic.

The theory of decision that is emerging as an important discipline [134, 135] has attracted the interest of workers in many fields: economics, politics, medicine, law, engineering, to name some, and it incorporates human perception, emotion, and logic. Perception helps us translate the stimulus of the environment to an abstract model. Emotions guide our choice of values and associated objectives. Logic directs us to a rational process of selecting a course of action to realize our objectives. Decision theory provides a framework for behavior which admits the contribution of all three aspects of human consciousness.

It has been said that a wise man avoids getting into unpleasant situations from which the smart man manages to escape. The wise man is, therefore, a better decision maker because he normally has a better judgment of the consequences, has better foresight, and behaves in a way that increases the probability of outcomes which he desires (or decreases the probability of undesired outcomes).

The fundamental building blocks of decision theory include *utility theory* which permits us to construct a common basis for comparing our values, and *probability theory* to assess our state of knowledge.

Decisions can be made by an individual or a group. In this chapter we discuss primarily decision making of an individual. A brief discussion of the more difficult problem of group decision is also included.

Decision theory provides a basis for rational actions and consistent behavior. Although the actions may not solve our problems at times, we still believe that in general the framework of decision theory will guide us to improve our abilities to bridge the gap between where we are and where we wish to be, or between what we have and what we want, i.e., solve our problems.

7-2 DECISION MODELS

Model 1

Consider a farmer who must decide on what to do next year. After considering many alternative courses of action, he derives the following model which includes only three possible courses of action and three states of nature (weather conditions), as shown in Fig. 7-1. The model is abstract in a number

Courses of Action (Strategies)	States of Nature for Crops		
	Perfect	Fair	Bad
Plant Crop A	10	1	−2
Plant Crop B	8	4	0
Hothouse Crop	3	3	3

Figure 7-1

of ways. The number of courses of action included is certainly smaller than the number available in general. This could be due to constraints, such as availability of crops, the financial resources of the farmer, or other factors. The *states of nature* (such as rainfall, levels of temperature, winds, etc.) which are relevant to the success of his crops could be described to various degrees of detail. The simplification to three states *Perfect*, *Fair*, and *Bad* constitutes a high level of abstraction which includes many variable conditions within some bounds. For example, the state Perfect may imply that yearly rainfall is between certain bounds, and that the number of days of above 100°F temperature does not exceed a certain number, etc.

Almost needless to say, the listing of strategies and states of nature in the model both require decisions. But we lump these decisions, for simplicity, under the category of the modeling stage in our particular problem here.

The array of numbers which appears in Fig. 7-1 is called a *payoff matrix* and represents the degree of *satisfaction* or *utility* that the farmer believes he will derive. For example, he derives 4 units of utility when he plants crop B and the state of nature is Fair, and he derives no utility when the state of nature is Bad. You may now wonder how the farmer generates these numbers which represent his satisfaction or utility.

Normally, you would think of money (say, profit in dollars) as the measure of success in such an enterprise. This could be done if you wish, in which case profit is the measure of utility. Most people do not behave this way. They are concerned with many other factors even in what appear to be strictly matters of money. For example, the risk involved in realizing a profit, or the risk of going broke, are relevant; the amount of investment required compared to your total assets is important. In addition, many decisions are not at all matters of money, as the next two models will illustrate.

What must the farmer do in Fig. 7-1? Which strategy is he to choose? This, of course, will depend on what he wants, what his *objective* is. A statement of objective is central to the decision-making process. We shall discuss objectives and guide our farmer to a rational decision later in the chapter.

Model 2

You are a member of a petit jury. The hour of decision has arrived and you must cast your vote of guilty (G) or innocent (I) for the person on trial. These are the two strategies open to you. There are two corresponding states of nature (or true states of the world): innocent or guilty. The four entries in Fig. 7-2 designate the four possible outcomes. The first letter indicates

		True State Innocent I	True State Guilty G
You claim	Innocent I	II	IG
	Guilty G	GI	GG

Figure 7-2

what you say and the second what the true state is. As a rational man, you study the consequence of your decisions before acting. Thus, the outcome IG in which you claim the accused is innocent when in reality he is guilty may result in freedom for a guilty person. On the other hand, the outcome GI may result in punishment for an innocent man. II results in an innocent person freed, and GG results in the punishment of a guilty person. How do you assign a utility which will represent your satisfaction with each of the four outcomes? We suppose you might find it easier to rank the four outcomes.

For example, you might be most satisfied with II and least with GI. But how would you proceed to assess by numbers the strength of your feelings of satisfaction associated with each outcome? Your answer will depend on your values.

Consider a case in which you assign a prior probability $P(I) = 0.99$ to the probability that the person is innocent. After relevant evidence E is admitted, your state of knowledge may change. The relevance of the information submitted is decided by the judge; he also decides when to stop admitting evidence, and in so doing he participates in the modeling process (simplifying the case). You must now make a decision on how small $P(I/E)$ must be before you will claim, beyond reasonable doubt (and yet some doubt, however small), that the person is guilty.

Let evidence E be the event that a piece of red fabric was found at the site of a crime where a struggle took place, and that the accused has a red jacket of the same fabric with a piece of the same size torn from it. Now, we write

$$P(I/E) = P(I)\frac{P(E/I)}{P(E)}$$

in which $P(E) = P(E/I)\,P(I) + P(E/\bar{I})\,P(\bar{I})$

Suppose we have a reasonable way of calculating the relevance ratio $P(E/I)/P(E)$. How small must $P(I/E)$ be before you cast a vote for guilty?

This is a very subjective and individual matter. What is your attitude? How strong are your feelings of dissatisfaction when you are partner to the punishment of an innocent person? How strong are your feelings when a guilty person, who may endanger others, is freed?

Model 3

A medical doctor treats a five-year-old child who is admitted to the hospital complaining of back pain. X-rays reveal two collapsed vertebrae. Tests to establish possible causes of this phenomenon are all negative. Subsequent X-rays indicate that a third vertebra has collapsed and a fourth is beginning to deteriorate. This may cause paralysis because the spinal chord is likely to be severed. What action is the doctor to take? (This particular case is studied in detail in [60].) Consider the following alternatives:

1. Try more diagnostic tests.
2. Wait n days and watch for new symptoms.
3. Treat for one of the possible diseases.

The consequences of each alternative must be studied before they can be compared on a rational basis. A decision tree can be constructed, listing the various possible outcomes and associated conditional probabilities, as we did in Sec. 4-6, Examples 8 through 10.

For simplicity, let us develop the model of Fig. 7-3 for the case of a doctor treating a patient for a particular disease. Here we lump all possible treatments under *Treat* as a single category or a single alternative, and list only three states of nature.

		States of Nature		
		Cure	Paralysis	Death
Strategies	Wait			
	Treat			

Figure 7-3

What are the utilities for the six possible outcomes in our model? Could you rank them? Of course, there are uncertainties associated with the outcomes. Suppose past experience indicates that of the patients treated for the disease, 90% die, 4% are paralyzed (waist down), and 6% are cured. Of the patients who wait, 85% die within 2 years and 15% are cured. How does one establish the relative utilities of *cure*, *paralysis*, and *death*?

Main Elements of the Decision-Making Model

A decision-making model contains these five main elements:

Alternative actions that the decision maker controls because he can select whichever action he wishes.

States of nature that constitute the environment of the decision model. The decision maker does not control the states of nature.

Outcomes that are the results of a combination of an action and a state of nature.

Utilities that are measures of satisfaction or value which the decision maker associates with each outcome.

An Objective which is a statement of what the decision maker wants.

The prevailing objective in decision theory is *maximization of expected utility.* The objective constitutes the *decision rule* or *criterion* which guides the decision maker in his choice for a course of action.

The above elements can be identified in the three models discussed earlier. There is one more element of the decision model which is important and that is the assessment of the decision maker's *state of knowledge* regarding the states of nature. This can be achieved by assigning probabilities to the states of nature. Decision-making models are classified according to these assignments:

(a) *Decision under Certainty*. Each action results in one known outcome which will occur with certainty.

(b) *Decision under Risk.* Each state of nature has a known objective probability.
(c) *Decision under Uncertainty.* Each action can result in two or more outcomes, but the probabilities of the states of nature are unknown.
(d) *Decision under Conflict.* The states of nature are replaced by courses of action open to an opponent who is trying to maximize his objective function. Decision making under conflict is the subject of *game theory.*

These decision-making models are discussed in this chapter.

7-3 DECISION MAKING UNDER CERTAINTY

Decision making under certainty occurs when we know with certainty which state of nature will occur. In other words, there is a single column in the *payoff matrix* or the matrix of outcomes in the decision model. This seems a rather simple decision problem because all that is required is to select the action that leads to the maximum utility. However, there are many problems in which the number of possible courses of action open to the decision maker is so enormous that an exhaustive list of all of them is not feasible and the modeling process of reducing their number requires heuristic tools of problem solving.

For example, consider a factory with 20 different machines and 20 contracts for parts which these machines can produce [56]. Since the machines are different, they may require different costs for each contract. Hence, to compare costs and select the optimum assignment of contracts to machines, we should consider all possible assignments. The contracts can be assigned in $20 \times 19 \times 18 \times \ldots \times 2 \times 1$ ways, or a number which is larger than 2×10^{18}. Each of these assignments constitutes an alternative action or one row in the *payoff matrix.* Since the outcome is assumed certain, we have a single column. To impress you with the magnitude of the task required simply to compute the costs for all the alternatives, let us use some calculations. Suppose a computer can calculate the costs for each of $2{,}000{,}000 = 2 \times 10^6$ alternatives in 1 second (this is a high number even for the fastest computers available, because each alternative requires 20 multiplications and 20 additions). Since a year contains about 32×10^6 seconds (31,558,000 is closer), the computer can calculate the costs of about 64×10^{12} alternative actions in one year. To complete the calculations for all 2×10^{18} alternatives, $(200/64)10^4$ or about 30,000 years of computer time are required.

All this was merely to impress you with the fact that even decision making under certainty may become a problem that is not tractable. It also serves to illustrate that decision theory is a link in problem solving. Here the modeling aspect of problem formulation requires actions which cannot be placed on a strictly prescriptive and systematic basis; at best these actions are plausible and are guided by heuristics.

7-4 DECISION MAKING UNDER RISK

When the states of nature can be assigned objective probabilities of occurrence, we speak of decision making under risk. The probabilities may be the result of many years of experience during which a great deal of data was accumulated and probabilities established. This could be the case in Model 1 of Sec. 7-2, Fig. 7-1. Suppose the farmer uses the following probabilities for the states of nature:

$$P(\text{Perfect}) = 0.15,\ P(\text{Fair}) = 0.6,\ P(\text{Bad}) = 0.25$$

Since the objective is to maximize the expected value of utility, the farmer computes this expected value for each strategy and compares the results:

$$\begin{aligned}
&\text{Plant Crop A}: 0.15(10) + 0.6(1) + 0.25(-2) = 1.6\\
&\text{Plant Crop B}: 0.15(8) + 0.6(4) + 0.25(0) = 3.6\\
&\text{Hothouse}\quad\ \ : 0.15(3) + 0.6(3) + 0.25(3) = 3.0
\end{aligned}$$

The farmer's decision consistent with his objective should, therefore, be to plant crop B.

Of course, as the probability assignments to the states of nature change, the decision may also change. For example, with $P(\text{Perfect}) > 0.75$, the decision will be to plant crop A. On the other hand, with $P(\text{Bad}) > 0.75$, the decision will be to invest in hothouse crops. Verify these conclusions.

7-5 DECISION MAKING UNDER UNCERTAINTY

The most complicated case of decision making is that in which no objective probabilities are available for the occurrence of the states of nature. This could describe the situation of Model 2 in Sec. 7-2. It could also describe the situation when War and Peace are the relevant states of nature, and many other examples. How is the decision maker to proceed now?

We can distinguish a number of different attitudes to decision making under uncertainty as demonstrated by the following descriptions of decision makers. In order to make the discussion simpler to follow, let us consider Model 1 of the farmer's decision problem in Sec. 7-2 and maintain that the region where the crops are to be planted is unknown to the farmer and that no data are available to establish objective probabilities for the states of nature.

The Subjectivist

If the decision maker subscribes to the concept of a subjective probability as a measure of his state of knowledge, he can assign equal probabilities to all states of nature and thus indicate his highest state of ignorance or maximum entropy. He may assign unequal values when he has more knowledge. Using the criterion of maximum expected value of utility, he can then find a course of action consistent with this criterion. Thus, for the subjectivist the decision under uncertainty is reduced to one under risk.

A decision maker who does not subscribe to the concept of subjective probability can select a different rule or criterion for decision making under uncertainty, depending on his attitude as an experienced decision maker in the real world, i.e., the *pessimist*, the *optimist*, or the *regretist*.

The Pessimist

The pessimist reasons that if anything can go wrong, it is sure to happen to him. He may be the person who tells you that he is out of luck. When he buys stocks, they go down in value; when he sells, they go up; when it rains, his umbrella is home; and when he has it, it does not rain. A pessimist adopts the criterion of *maximin* or the maximum of the minima. Namely, he studies the consequences of his actions by considering what is the minimum utility associated with each action, and then he selects the action which yields the maximum of these minima.

For the decision model of Fig. 7-1 the pessimist proceeds with the analysis as shown in Fig. 7-4. On the right-hand side of the payoff matrix the

	States of Nature				
	Perfect	Fair	Bad	Row minima	
S_1: Plant crop A	10	1	−2	−2	
S_2: Plant crop B	8	4	0	0	
S_3: Hothouse	3	3	3	(3)	*maximin*

Figure 7-4 Maximin Criterion

minimum of each row is recorded. These numbers represent the worst results for each course of action. Since the pessimist takes the attitude that the worst always happens to him, he selects the best of the worst results, i.e., the maximum of the row minima which is referred to as the *maximin.* This is the value of 3 in our case (circled in Fig. 7-4) and, therefore, the decision is hothouse crops. Note that the pessimist is now guaranteed 3 units of utility.

If we were to remove the last course of action from our model (because all outcomes of 3 are replaced by −5 and are, therefore, dominated by the other two alternatives, for instance), then the pessimist would decide on crop B. In such a case he would be certain to realize no less than a zero utility. However, if nature is kind to him, he could realize as much as 8 units of utility.

The pessimist is, therefore, a risk averter. He is happy to settle for a guaranteed minimum satisfaction instead of taking a chance on realizing less, but also realizing more, than the guaranteed minimum.

The Optimist

The optimist is the risk lover. In each situation he has high hopes for the best outcome. His decision rule or criterion is to choose the best of the best. This is known as the *maximax criterion.* For each action he singles out the maximum and of all these, he selects the best, i.e., largest utility, as shown in Fig. 7-5. On the right-hand column the row maxima are listed and of

	States of Nature				
	Perfect	Fair	Bad	Row maxima	
S_1: Plant crop A	10	1	−2	(10)	*maximax*
S_2: Plant crop B	8	4	0	8	
S_3: Hothouse	3	3	3	3	

Figure 7-5 Maximax Criterion

these, the optimist selects the highest value, i.e., the *maximax.* Hence, he decides on strategy S_1: plant crop A.

* The "Inbetweenist"

The pessimist is too cautious, and the optimist is too audacious. Most people are in some category between these two extremes. Hurwicz [57] suggested a criterion which, by introducing an index of optimism, does not categorize the decision maker as either a pessimist or optimist. The optimist considers the best outcome in each strategy S_i, and the pessimist considers the worst. Since, normally, a person's attitude is somewhere between complete optimism and pessimism, a pessimism weighting factor α, which is a number between zero and 1, is assigned to the worst outcome, and $(1 - \alpha)$ to the best of each strategy. For any strategy S_i, we multiply the worst outcome by α, and the best outcome by $(1 - \alpha)$. The sum of the products is called the α index. For strategy S_i, then,

$$\alpha(\text{Worst of } S_i) + (1 - \alpha)(\text{Best of } S_i) = \alpha \text{ index of } S_i \qquad (7\text{-}1)$$

For example, for S_1 in Fig. 7-5 with $\alpha = 0.3$,

$$0.3(-2) + 0.7(10) = 6.4$$

and for S_2,

$$0.3(0) + 0.7(8) = 5.6$$

The α index in Eq. 7-1 is a linear function of α. We can, therefore, plot the α index as a function of α for each strategy S_i, as shown in Fig. 7-6. For each strategy we simply connect the best outcome and the worst. The best is a measure of the α index for $\alpha = 0$ (optimist), and the worst is a measure of the α index for $\alpha = 1$ (pessimist).

The closer α is to 1, the more of a pessimist you are. The closer α is to 0, the more of an optmist you are.

The "inbetweenist" criterion is to maximize the α index. Hence, from Fig. 7-6, for α less then 1/2, he uses S_1; for α between 1/2 and 5/8, he uses S_2; and for α larger than 5/8, he uses S_3. Thus, as α increases from 0 to 1, the "inbetweenist" goes from complete optimist to complete pessimist, or the larger α, the more of a pessimist he is.

The Regretist

The regretist is the person who is often heard saying: "I could kill myself for what I did. I could have done so much better had I selected a different course of action." Such a person is always full of regret for the

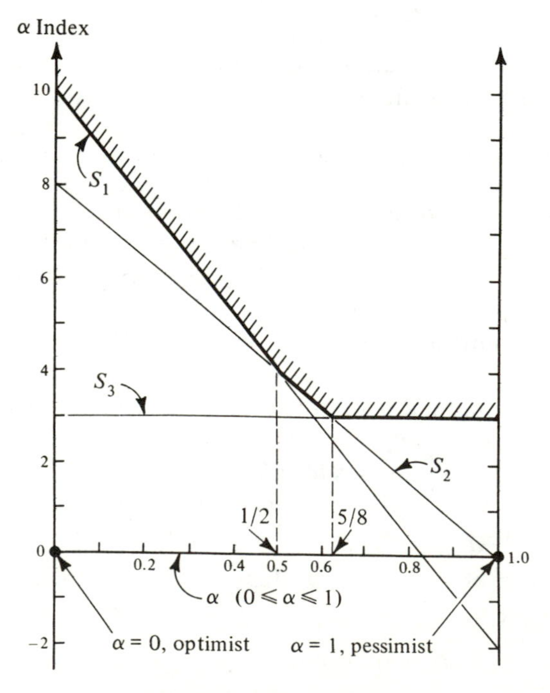

Figure 7-6 The α Index

difference between the outcome that he realizes and the maximum he could have realized for the particular state of nature which prevails. For example, in Fig. 7-4, if the state of nature turns out *Perfect* and he selects S_1, then he would have no regret. However, had he selected S_2 or S_3 he would regret the *loss of opportunity* of 2 or 7 units of utility, respectively, namely, the difference between the best possible for this state of nature and the outcome which he realizes.

The regretist's objective is to make his regret as small as possible. For this purpose, the original payoff matrix is rewritten with the outcomes representing his loss due to imperfect foresight. Such a matrix is called a *regret matrix*. The regret matrix for the model of Fig. 7-4 is shown in Fig. 7-7.

	States of Nature				
	Perfect	Fair	Bad	Row maxima	
S_1: Crop A	0	3	5	5	
S_2: Crop B	2	0	3	(3)	*minimax*
S_3: Hothouse	7	1	0	7	

Figure 7-7 Regret Matrix

To guarantee a lower bound for the maximum regret, we record the maximum regret for each row in a column on the right-hand side of the regret matrix, and select the strategy corresponding to the minimum of these maxima. We have then the criterion of *minimax* regret. The outcomes in the regret matrix are also called *opportunity costs*. In situations where it is possible to obtain more information on the states of nature, the opportunity costs provide the criterion for the upper bound on the value of the information. For example, to a decision maker in Fig. 7-4 who decided on S_2, information leading to certainty about the occurrence of the state *Bad* is worth no more than 3 units. This is the cost of opportunity for the outcome (S_2, Bad), as shown in Fig. 7-7.

* The Subjectivist Revisited

Now that we have discussed the attitudes of the nonsubjectivists in decision under uncertainty, let us return to the subjectivist One can view his decision to assign equal probabilities to all states of nature under conditions of complete ignorance (maximum entropy) as a *minimax* of regret in the assignment of the probabilities If we consider n states of nature, then, with a probability of $1/n$ for each, the maximum error or regret in assignment is $1 - (1/n)$, $(n \geq 2)$.

Suppose one of the states of nature is assigned a probability $(1/n) + \epsilon$. Then at least one of the remaining states must be assigned a decrease in probability of no less than $\epsilon/(n-1)$. The maximum decrease in assigned probability for one of the remaining states is ϵ. Thus, the maximum possible error in probability assignment becomes $1 - [1/n - \epsilon/(n-1)] = \epsilon/(n-1) + (1 - 1/n)$ when the probabilities of all $n - 1$ remaining states are reduced by an equal amount, and $\epsilon + (1 - 1/n)$ when the probability of only one of the remaining $n - 1$ states is reduced by ϵ. In either case, the maximum possible regret in probability assignment is increased from the case of equal assignment of probabilities when it is $(1 - 1/n)$.

7-6 UTILITY THEORY

The task of assigning numerical values to the possible outcomes in the decision model is within the domain of utility theory.

Famous mathematician and philosopher Daniel Bernoulli (circa 1730) suggested that utility $u(M)$ of a sum of money M could be measured by using the logarithm of M, i.e., $u(M) = \log(M)$. French naturalist Buffon suggested that if you have a sum M and it is increased by m, then the increase in your utility $u(m)$ should be computed from

$$u(m) = \frac{1}{M} - \frac{1}{M + m}$$

Cramer proposed a square root function for utility.

Von Neumann and Morgenstern [58] developed an axiomatic approach to utility theory. The axiomatic approach guarantees the existence of a cardinal* utility function if the decision maker subscribes to the axioms. The von Neumann-Morgenstern utility brings into consideration the decision maker's attitude toward risk by considering his preferences with respect to gambles which involve outcomes, in addition to his rational preferences for the outcomes. This will become more clear as we proceed to state the axioms. The subject is treated in detail in [59].

Axioms of Modern Utility Theory

Notation. Let the symbols A, B, C represent three outcomes or prizes resulting from a decision. These could be, for instance,

A: A tuition-free college education (4 years).
B: 3 summer trips for archeological diggings in Turkey, the Sinai desert and South America; all expenses paid.
C: A cash award of $18,000.

You may think, of course, of many other possible outcomes. They could all be cash awards of different sums or all noncash.

---

*A cardinal utility function assigns distinct numerical values to each outcome. This is distinguished from an ordinal utility which only ranks the outcomes.

We use the symbols $>$ and $\sim$ to indicate preference and indifference, respectively. Thus $A > B$ means that the decision maker prefers A to B. $B \sim C$ means that the decision maker is indifferent between the outcomes B and C.

AXIOM 1. Preferences can be stated and they are transitive; namely,

$$A > B \text{ and } B > C \text{ implies } A > C$$

and

$$A \sim B \text{ and } B \sim C \text{ implies } A \sim C$$

A rational decision maker faced with outcomes A, B, C will be able to state his preferences without violating the transitivity property.

The lady who cannot decide which she prefers, the blue or red dress, violates the axiom because she cannot state her preference which could be one of the following:

blue $>$ red
red $>$ blue
blue $\sim$ red

The lady who lists her preferences between red, blue, and green dresses as

blue $>$ green
green $>$ red
red $>$ blue

violates the transitivity property. No prejudice to women shoppers is implied here. We only wish to indicate that axiom 1 can be violated, but violaters cannot play our "game" for generating a utility for outcomes. It is only rational to insist that we know what our preferences are before we can speak of degree of satisfaction, or utility, derived from outcomes. For example, the irrationality which results from the violation of transitivity can best be illustrated by considering the second preference list of dress colors. Suppose the store owner gives the green dress to the lady. As she is about to leave, he calls her back and suggests that she can trade the green for the blue dress by paying him one dollar. Since she prefers the blue dress, she pays the dollar and is about to leave when the store owner offers to trade the blue for the red dress for one dollar. She agrees to that, too. But then he offers to exchange the red for the green dress for one dollar. Since she prefers the green to the red she accepts the offer. She has paid three dollars so far and she is back where we started with the green dress in her possession. The store owner can start a second cycle earning three dollars, and continue to "pump" the lady until, of course, she decides that something is wrong; namely, she behaves in an irrational manner because transitivity is violated.

Note, however, that preferences in general are an individual matter. An avid archaeology student who values the experience of the particular diggings in the three summers of outcome B mentioned earlier, may prefer B over A or C. A young man who does not intend to go to college, or one who is through with his college education, may prefer C. We consider A, B, C mutually exclusive.

Preliminary Concepts and Notations Applicable to Axioms 2 and 3. Consider a game in the form of a lottery, as shown in Fig. 7-8. A pointer spins in the center of a circle. When the pointer is set spinning, it is equally likely to be pointing in any direction when it comes to rest. A lottery ticket to play the game entitles you to prize A when the pointer stops in region A, and to prize B if it stops in region B. With each outcome there is an associated probability of occurrence. In Fig. 7-8 the probabilities are computed from the area of each region divided by the total area.

The lotteries of Fig. 7-8 could also be shown schematically by tree diagrams as in Fig. 7-9.

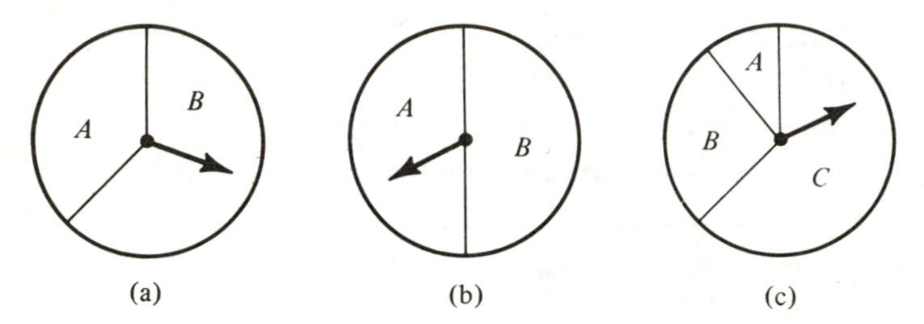

Figure 7-8 (a) Lottery L_1; (b) Lottery L_2; (c) Lottery L_3

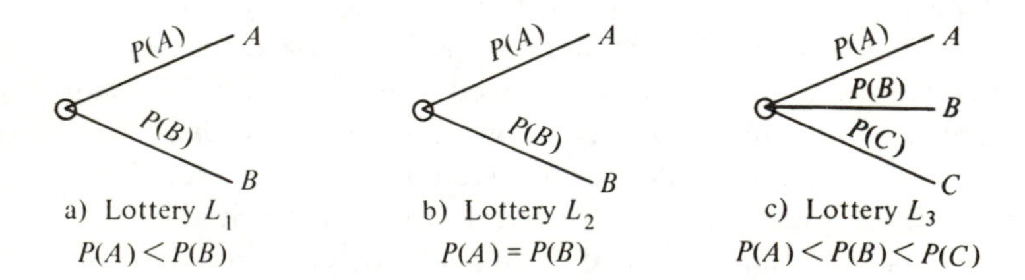

Figure 7-9 Schematic Representation of Lotteries: (a) Lottery L_1, $P(A) < P(B)$; (b) Lottery L_2, $P(A) = P(B)$; (c) Lottery L_3, $P(A) < P(B) < P(C)$

A further abstraction represents each lottery by symbols as follows:

$$L_1 = [P(A), A; P(B), B]$$
$$L_3 = [P(A), A; P(B), B; P(C), C]$$

Each outcome is separated by a comma from its probability of occurrence, and a semicolon separates the outcomes. Consider, for example, the following compound lottery. You are given a ticket to play lottery $L_1 = [P(A), A; P(L_2), L_2]$.

Here you receive a prize A with probability $P(A)$ and prize L_2 with probability $P(L_2)$. However, L_2 consists of a ticket marked L_2 which entitles you to play lottery $L_2 = [P(B), B; P(C), C]$. The complete compound lottery will finally result in one of the three prizes or outcomes A, B, C. This can be written as

$$[P(A), A; P(L_2), \{P(B/L_2), B; P(C/L_2), C\}]$$

The information between the braces { } is L_2. As can be verified from Fig. 7-10, the compound lottery can also be written in this form:

$$[P(A), A; P(B/L_2)P(L_2), B; P(C/L_2)P(L_2), C]$$

This represents symbolically also a single lottery in which outcomes A, B, C can occur with the probabilities indicated. Such a lottery L^* is shown in Fig. 7-11.

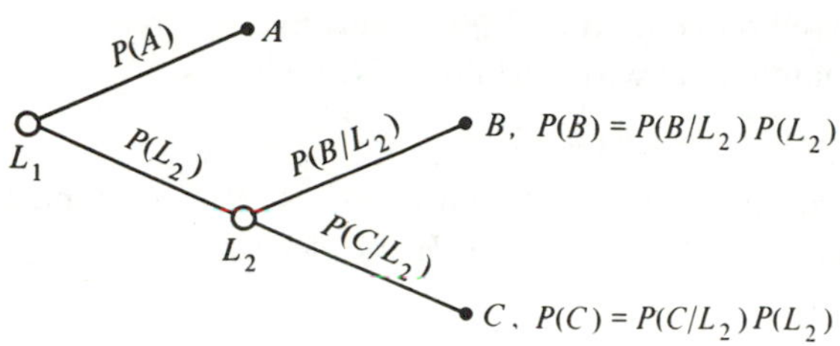

Figure 7-10 Compound Lottery Involving L_1 and L_2

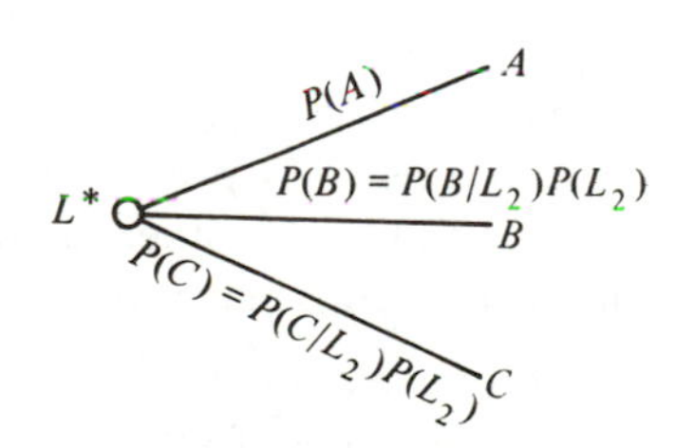

Figure 7-11 Representation of Compound Lottery of Figure 7-10 as a Simple Single Lottery: $L^* = [P(A), A; P(B), B; P(C), C]$

AXIOM 2. When a decision maker prefers an outcome A to an outcome B, i.e., when $A > B$, then

$$[P^*(A), A; (1 - P^*(A)), B] > [P(A), A; (1 - P(A)), B]$$

if, and only if, $P^*(A) > P(A)$. Namely, when $A > B$, then of two lotteries involving A and B, a decision maker prefers the one in which the probability of his preferred outcome is larger.

AXIOM 3. The decision maker is indifferent to the number of lotteries in a decision situation and is concerned only with the final outcomes and associated probabilities. Thus, the decision maker is indifferent to the compound lottery of Fig. 7-10 and the equivalent single lottery L^* of Fig. 7-11. In

Fig. 7-10 a pointer is spun twice, once in L_1 and once in L_2, and there may be fun in the excitement of suspense,* but not to our "rational" decision maker who would consider L^* of Fig. 7-11 equivalent although there a pointer is spun only once to yield one of three outcomes A, B, or C. In summary,

$$[P(A), A; P(L_2), L_2] \sim [P(A), A; P(B), B; P(C), C]$$

in which $L_2 = [P(B/L_2), B; P(C/L_2), C]$

$$P(B) = P(B/L_2)P(L_2)$$
$$P(C) = P(C/L_2)P(L_2)$$

AXIOM 4. If $A > B > C$, there exists a probability p for outcome A such that the decision maker is indifferent to a lottery $[p, A; (1 - p), C]$ involving his most preferred outcome A and least preferred outcome C, or receiving an intermediate outcome B with certainty. Namely,

$$B \sim [p, A; (1 - p), C]$$

As an example, consider the outcomes A, B, C to stand for $A =$ \$100 prize, $B =$ \$20 prize, $C =$ \$0 prize. If the decision maker's preferences are

$$A > B > C$$

then axiom 4 requires that the decision maker be able to select a probability p such that

$$B \sim [p, A; (1 - p), C]$$

This is reasonable, as can be deduced from the following argument. If we set $p = 0$ in the lottery above, then B is preferred because the lottery will yield C with certainty and $B > C$. On the other hand, for $p = 1$, the lottery is preferred because it will yield A with certainty and $A > B$. Thus,

For $p = 0$, decision maker prefers B
For $p = 1$, decision maker prefers lottery

At some point between $p = 0$ and $p = 1$, the decision maker is indifferent between the lottery and outcome B. The choice of p reflects the decision maker's attitude to risk.

Utility Function

Adhering to the constraints of the four axioms, it is possible to guarantee the existence of a *utility function* which is a rule for assigning a number to each outcome or to a corresponding equivalent lottery (axiom 4). The utility function is written in the form $u(A)$, or $u(B)$, etc., signifying the utility associated with the outcome in parentheses. The utility function has these properties:

*This fun can be included in the assessment of benefit (utility) associated with the outcomes.

Property 1. $u(A) > u(B)$

if, and only if, $A > B$

$u(A) = u(B)$

if, and only if, $A \sim B$

Property 2. The expected value of the utilities in a lottery L involving outcomes A and C is equal to the utility of the outcome B when $B \sim L$. Or, when

$$B \sim [p, A; (1 - p), C] = L$$

then $u(B) =$ expected value of $u(L)$

or $u(B) = p\, u(A) + (1 - p)\, u(C)$

The utility of lottery L is the expected value of the utility of the outcomes in the lottery.

The concept of a utility as a rational basis for comparing values or degrees of satisfaction applies to any type of outcome, as will be demonstrated by some examples.

Property 3. The choice of a reference value for the utility of the worst (or best) outcome, and the choice of a scale factor, which separates the utilities of the best and worst outcomes, are arbitrary. This is similar in concept to the reference value and scale in the measurements of temperature.

EXAMPLE 1. We return to Model 3 of Sec. 7-2 and consider the assignment of utilities to the three outcomes:

$A =$ Cure
$B =$ Paralysis (waist down)
$C =$ Death

Ginsberg and Offensend [60] devised a lottery named "A Game with a Witch Doctor" in order to assign utilities to the three outcomes. The witch doctor is all powerful. The decision-maker doctor playing the game is told that he can play only once as follows:

Make a choice between an outcome B guaranteed by the witch doctor or a lottery involving the other two outcomes. The lottery consists of an opaque bottle with a total of 100 pills. A game consists of drawing at random a pill from the bottle, with each pill being equally likely to be drawn. There are only two colors of pills, white and black. Drawing a white pill yields outcome A, and a black pill yields outcome C.

Now, consider a doctor decision-maker whose preferences are

$$A > B > C$$

From the properties of the utility function, we know that

$$u(A) > u(B) > u(C)$$

The question is: Where does $u(B)$ lie on the scale between $u(A)$, the best, and $u(C)$, the worst? For this purpose, $u(A)$ and $u(C)$ are assigned arbitrary values [consistent with transitivity, $u(A) > u(C)$] $u(A) = 100$, $u(C) = 0$.

Now, consider a lottery $[p, A; (1 - p), C]$ in which p is the fraction of white pills in the opaque bottle of 100 pills.

For $p = 0$ doctor prefers B
because lottery guarantees C(death) and $B > C$

For $p = 1$ doctor prefers lottery
because lottery guarantees A(cure) and $A > B$

For what value of p is the decision maker indifferent between B and the lottery?

Ginsberg and Offensend [60] report that one doctor selected $p = 0.4$. Namely, he was indifferent between paralysis (B) and a lottery of cure (A) and death (C) when the bottle had 40 white pills, see Fig. 7-12a,

$$u(B) = 0.4\, u(A) + (1 - 0.4)\, u(C) = 40$$

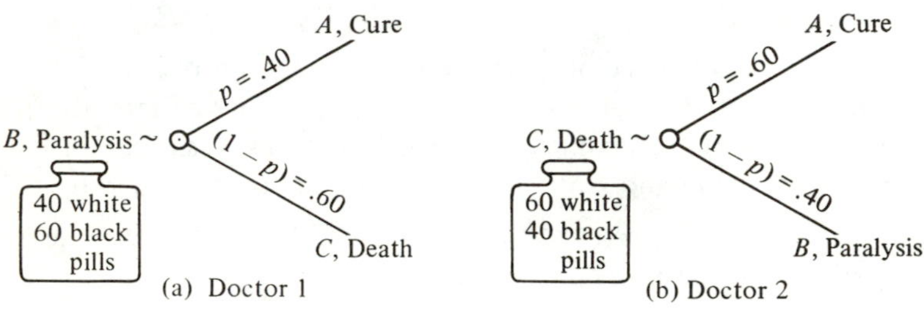

Figure 7-12

A second doctor arrived at $u(B) = -150$ when $u(A) = 100$ and $u(C) = 0$. Since the reference value for the utilities is arbitrary, the second doctor's utility assignment tells us that consistent with his utilities

$$u(A) > u(C) > u(B)$$

his preferences were

$$A > C > B$$

He was indifferent to death (C) or the lottery shown in Fig. 7-12b, in which the opaque bottle has 60 white and 40 black pills.* Drawing a white pill guarantees cure (A), and a black pill paralysis (B). If we assign $u(B) = 0$ and $u(A) = 100$, then

$$u(C) = 0.6\, u(A) + 0.4\, u(B)$$
$$= 60$$

If we subtract from all utilities 60, i.e., shift the reference basis so that $u(C) = 0$ for both doctors then $u(A) = 40$, $u(C) = 0$ and $u(B) = -60$. To make $u(A) = 100$ we multiply the utilities by 2.5 yielding $u(B) = -150$.

*Note that the black pill now represents paralysis.

In summary the utilities of doctor 2 were generated as follows (for comparison we also show the utilities of doctor 1):

	doctor 1	*doctor 2*	*−60*	*×2.5*
A, Cure	$u(A) = 100$	$u(A) = 100,$	40,	100
B, Paralysis	$u(B) = 40$	$u(B) = 0,$	−60,	−150
C, Death	$u(C) = 0$	$u(C) = 60,$	0,	0

Suggested Problem. To complete the example of Model 3 in Sec. 7-2, assign probabilities arbitrarily to the three states of nature on the basis of "wait" and "treat" separately. Compute the expected utility of "wait" and "treat" for each doctor in Example 1 and indicate his decision. Now, change the probability assignments so that the two doctors will (a) agree on the strategy of "treat"; (b) agree on "wait"; (c) disagree; (d) disagree in reverse to the disagreement in (c). One of these outcomes corresponds to your first probability assignment.

EXAMPLE 2. Consider the problem of self insurance [56] in which a businessman has a shipment of goods worth $10,000. There is a probability that the shipment may be destroyed in transit. The insurance company requires a $1500 premium to insure the shipment. How small must be the probability p of safe shipment for him to buy insurance? Suppose the total assets of the businessman are $15,000. Then, if he buys insurance, he is guaranteed 15,000 − 1500 = $13,500. Call this B. If he does not buy insurance, he is effectively playing a lottery

$$[p, \$15{,}000; (1-p), \$5000]$$

Namely, the shipment arrives safely with probability p, and he has $A =$ $15,000; the shipment is destroyed with probability $(1-p)$, and he has $C =$ $5000. Clearly,

$$A > B > C$$

To decide on a value for p, above which he does not insure and below which he does, he equates the utility of B, $u(B)$, to the utility of the lottery

$$u(B) = p\,u(A) + (1-p)\,u(C)$$

Now, suppose he uses a utility function which is the logarithm to the base 10 of the total assets. Then

$$\begin{aligned} \log 13{,}500 &= p \log 15{,}000 + (1-p) \log 5000 \\ 4.13 &= p(4.17) + (1-p)(3.70) \\ p &= \frac{0.43}{0.47} \\ &= 0.915 \end{aligned}$$

NOTE. When p is known from experience, we can use the above equation to establish what value is feasible for the premium B.

Problem. As the executive of an insurance company, would you insure the shipment for \$1500, considering that the company has total assets of \$100,000 and the probability of losing the shipment is 0.1?

EXAMPLE 3. Consider a decision maker whose utility function for money from zero to \$100 is curve (a) of Fig. 7-13. The satisfaction associated with each additional increment say of \$10, (above the first \$10) diminishes, as is reflected from the decrease in increments of utility in the figure. Thus, for example, from \$30 to \$40 the utility increases by 0.1, while from \$90 to \$100 the increase is 0.02. This is quite typical in the sense that the utility of an additional sum of money depends on how much you have.

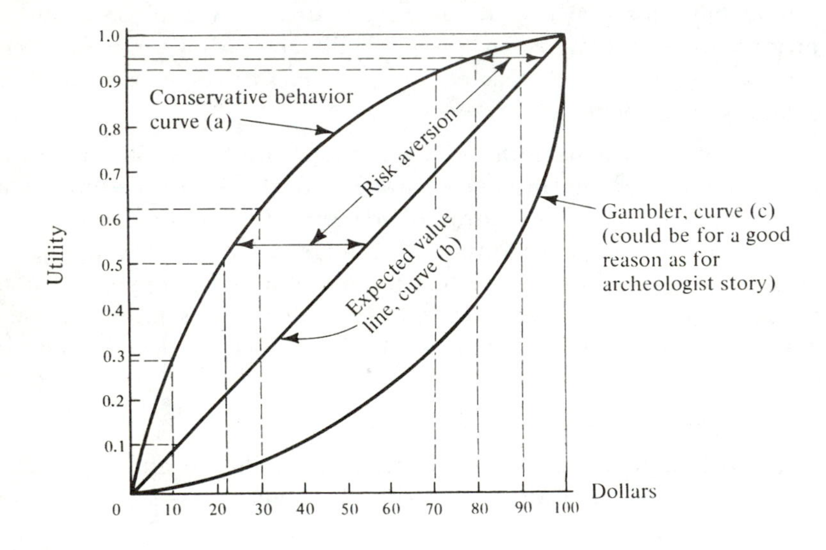

Figure 7-13 A Utility Function for Dollars from 0 to 100: Curve (a), Conservative; Curve (b), Expected Value; Curve (c), Gambler

The decision maker has \$30. He is offered the chance to pay the \$30 for a ticket to a lottery $L = [p, \$100; (1-p), \$0]$. How big must p be before he decides to buy the ticket and gamble?

Let $A =$ \$100 assets for best outcome

$C =$ \$0 assets for worst outcome

$B =$ \$30 guaranteed assets

$A > B > C$

The solution is obtained from the equation

$$u(B) = p\,u(A) + (1-p)\,u(C)$$

From Fig. 7-13, $u(100) = 1.0$, $u(0) = 0$; hence,

$$u(B) = p = 0.62$$

NOTE. By assigning to the best outcome A a utility of 1.0, and to the worst outcome C a utility of 0, the utility of an intermediate outcome B is equal in numerical value to the probability p when

$$B \sim [p, A; (1-p), C].$$

Namely,

$$u(B) = p\, u(A) + (1-p)\, u(C) = p$$

because $u(A) = 1$ and $u(C) = 0$.

EXAMPLE 4. In Example 3, suppose the lottery is $L = [1/2, \$70; 1/2, \$0]$. Would you pay \$30 to play the lottery?

Using curve (a) of Fig. 7-13,

$$u(30) = 0.62$$
$$u(L) = \frac{1}{2}u(70) + \frac{1}{2}u(0)$$
$$= 0.46$$

Since $u(30) > u(L)$, you should not play the lottery.

How much should you pay to play a lottery [1/2, \$70; 1/2, \$0], when you have \$30 in assets? Suppose you should pay a sum S ($S <$ \$30). Then the states of your assets are:

best outcome: $A = (30 - S) + 70$
worst outcome: $C = (30 - S) + 0$
guaranteed state: $B = 30$

$$u(30) = \frac{1}{2}u(100 - S) + \frac{1}{2}u(30 - S)$$

or

$$1.24 = u(100 - S) + u(30 - S)$$

From curve (a) of Fig. 7-13, by trial-and-error selection of values for S, we obtain $S = 20$ because

$$u(80) + u(10) = 0.95 + 0.29 = 1.24$$

or

$$u(30) = \frac{1}{2}u(80) + \frac{1}{2}u(10)$$

EXAMPLE 5. Consider a young archeologist who, in the course of his studies, discovers an ancient manuscript that describes a statue and discloses its location on a remote island. The archeologist travels to the island to search for the statue. After a year of effort in vain, frustrated and with total assets of \$30, he decides to leave. As he is ready to depart, one of the natives approaches him with the authentic statue which he sought. When the archeologist offers to buy the statue, the native insists on a price of \$31 for it. Our distressed archeologist runs to the island "casino" and pays his \$30 for the only game on the island, a lottery $L = [p, \$31; (1 - p), \$0]$.

What are the best (A), worst (C), and guaranteed (B) outcomes in this case? Suppose $p = 1/4$. Is our archeologist rational? Does the utility curve (a) of Fig. 7-13 apply in the present case? If you were the archeologist, what utilities would you assign the three outcomes?

This example should serve to illustrate the subjective nature of utility.

The three utility curves of Fig. 7-13 are distinctly different in terms of the rate of change of utility as a function of wealth (dollars):

In curve (a), the rate of increase in utility gets smaller with wealth.

In curve (b), the rate of increase in utility is constant regardless of wealth.

In curve (c), the rate of increase in utility gets larger with wealth.

Which of the three utility curves applies to our young archeologist?

Why is it reasonable, in general, to label the curves above and below the expected value line in Fig. 7-13 as *conservative behavior* and *gambler*, respectively? When is the label *gambler* not appropriate? Give examples.

EXAMPLE 6. This example is an adaptation of the famous Truel story [61]. In the days when the duel was a common and honorable way of settling matters there lived three gentlemen A, B, and C (we give only the initials of their first names to honor anonymity). All three were desperately in love with a beautiful princess; none of them could win her hand. They, therefore, decided to commit suicide. When this sad development reached the King, he sought the advice of his chancellor and together they contrived an honorable solution in the spirit of the times. The three gentlemen were invited to participate in a *truel* in accordance with the following rules. Each man was to be given a pistol and two bullets and position himself at the vertex of an equilateral triangle. A was to fire one bullet first, next B, and then C. They were to proceed in this strict order, until the bullets were used up or one man survived. The sole survivor was to win the hand of the princess. If more than one man survived, neither one was to win the princess. Each man could choose as his target either of the other two contenders.

The three gentlemen accepted the offer for the truel, and even considered the rules fair in view of their abilities as marksmen. The probabilities of each hitting a target were established on the basis of years of observation to be:

A: 2/3, hit; 1/3, miss
B: 3/4, hit; 1/4, miss
C: 1, hit; never miss

Let us consider the deliberations of gentleman A on the eve of the truel. The outcomes of the truel from his point of view could be:

Outcome 1: He wins the princess.

Outcome 2: He survives, but does not win the princess.

Outcome 3: He does not survive.

A was indifferent to outcomes 2 or 3. He considered life not worthwhile should outcome 2 be the result. He, therefore, assigned outcomes 2 and 3 a utility of zero, and outcome 1 a utility of one. His expected utility $E(u)$ was

$$\begin{aligned} E(u) &= P(\text{Outcome 2 or 3})u(\text{Outcome 2 or 3}) \\ &\quad + P(\text{Outcome 1})\,u(\text{Outcome 1}) \\ &= P(\text{Outcome 2 or 3}) \cdot 0 + P(\text{Outcome 1}) \cdot 1 \\ &= P(\text{Outcome 1}) \end{aligned}$$

Thus, *A*'s objective of maximizing his expected utility was identical with the objective of maximizing the probability of his being the sole survivor, i.e., maximizing P(Outcome 1).

Put yourself in *A*'s position. What would you do with the first bullet? What is P(Outcome 1) for *A*?

Please do not look at the solution below before you attempt to answer these questions. You may find it helpful to review Exercise 2-15 first.

Solution (with a little psychology). In the spirit of Chapter 5 on modeling, *A* reasoned that it was too complicated to consider the entire tree of all possible outcomes, and some aggregation to reduce complexity was in order. After all, there were six bullets in the truel. Each bullet could lead to five (yes, five!) outcomes: aim at one opponent and hit or miss; aim at the other opponent and hit or miss; or shoot in the air, deliberately aiming at neither opponent to make sure that neither is likely to be hit. If *C* were not a perfect marksman and hits were not deadly, then $5^6 = 15{,}625$ possible outcomes would appear at the end branches of a probability tree, and even with each of *C*'s bullets leading to three outcomes (*C* does not miss when he aims), the total number of outcomes would still be large.

A decided to apply a little psychology and "humanize the problem" in a way which may appear cruel to us, but was quite reasonable in the days of the great duels. He reasoned as follows:

1. "If I fire at *B*, two things may happen:
   (a) I hit and kill *B*, then *C* is sure to kill me.
   (b) I miss *B*, but then I may get *B* mad and he may act irrationally by firing at me instead of firing at the more formidable opponent *C*."
2. "If I fire at *C*, two things may happen:
   (a) I hit and kill *C*, and then I become the sole target for *B*.
   (b) I miss, and then *C* may get mad and fire at me in case he survives *B*'s turn. Here *C* will act irrationally because *B*

is more formidable than I am, but when you fire at people they do not always act rationally."

3. "If I shoot in the air, I get neither opponent mad, and *B* will fire at *C*. After all, at this stage *B* has no reason to act in an irrational manner."

The above deliberations caused *A* to decide on shooting his first bullet in the air. In the process he reduced the problem considerably to the point that the tree diagram is as shown in Fig. 7-14. The subscripts *m* and *h* stand for miss and hit, respectively, and the letter at the arrowhead identifies the target. For example, $B_h \rightarrow C$ signifies that *B* is aiming at *C* and the outcome is a hit (a hit implies a kill in our example here). $A_m \rightarrow B$ signifies that *A* is aiming at *B* and the outcome is a miss. The numbers on the links of the tree indicate the probabilities of the events.

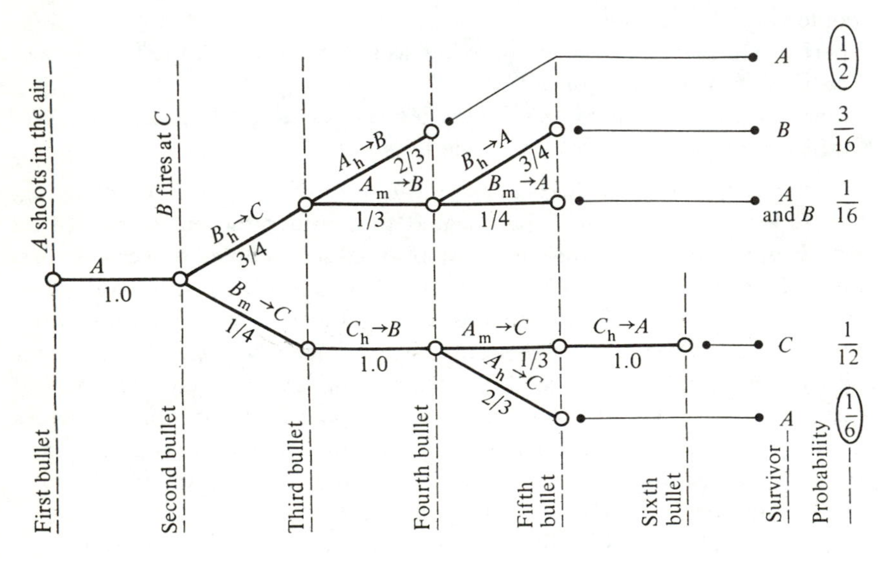

Figure 7-14 *A*'s Reduced Model ("Using a Little Psychology") in the Form of a Tree Diagram in the Truel

The probability of A being the sole survivor, and thus winning the princess, is $1/2 + 1/6 = 2/3$.

NOTE. The truel example can serve as a simple model for competition between three manufacturers, sales organizations, other business enterprises, or governments. In such applications we may increase the number of "bullets" (i.e., damaging actions) per competitor.

How would the tree diagram of Fig. 7-14 change if each of the three gentlemen had an unlimited supply of bullets and the truel continued in strict order of firing until a sole survivor remained?

7-7 UTILITY ASSIGNMENTS AND THE DECISION MODELS

In a decision model involving many outcomes, the decision maker can proceed to assign utilities by first ranking the outcomes. The most preferred outcome, A, is assigned a utility $u(A) = 1.0$, and the least preferred, C, is assigned $u(C) = 0$. Any intermediate outcome B_i is compared to a lottery $[p, A; (1 - p), C]$. The value of p for which the decision maker is indifferent to B_i or the lottery is equal to the utility $u(B_i)$ of outcome B_i.

All this sounds, of course, much simpler in theory than in practice. People are not accustomed to thinking in terms of probabilities. In addition, experience of many investigations indicates that people do not behave as if their objective is to maximize the expected value of the outcomes, but rather to maximize the expected value of some function of the outcomes, i.e., a utility function. For example, when people are asked to bid on a game with a lottery [1/2, \$0; 1/2, \$100], the average bid is about \$20 [62], and not \$50 as would appear to be the case on the basis of expected value of the outcome,

$$\frac{1}{2}(0) + \frac{1}{2}(100) = 50$$

A conservative decision maker (curve (a) of Fig 7-13) with zero assets would be indifferent between a cash prize of \$22, $u(22) = 0.5$, and the lottery $L = [1/2, \$0; 1/2, \$100]$, $u(L) = 0.5$.

In general, it is difficult for people to distinguish between various values of very small probabilities, such as 0.0003 or 0.00006. The same is true for values close to 1, 0.9973, 0.979, etc. It is, therefore, simpler at times to establish utilities by fixing the probabilities in the lottery L as 1/2, because people do have a feeling for a 50–50 chance, and ask them to select a guaranteed outcome which is equivalent to the lottery. That is, find B for which $B \sim [1/2, A; 1/2, C]$, when $A > B > C$.

Prescriptive (or Normative) and Descriptive Utilities

Utilities can be established by selecting a mathematical function which is derived axiomatically, and which tells people essentially what rule they should follow if they wish to behave in a rational, consistent manner. Such a utility is called a normative or a prescriptive utility.

The actual behavior of people in the market place may be quite different, however. Establishing the utilities in the market place as they are, and not as they ought to be, leads to a descriptive utility. However, in many instances when people become aware of their inconsistencies, they may change their utilities accordingly and thus come closer to not violating the axioms of rational behavior. The framework of decision and utility theory can uncover such inconsistencies, and make decision makers aware of how they behave in the decision-making phase of problem solving. Such awareness will bring about more consistency, more rationality, and, we hope, more success in meeting objectives in a manner which is compatible with the decision maker's attitude toward risk and his value system.

Some Results of Descriptive Utilities. Swalm [63] studied the utilities of money for executives. He devised hypothetical situations and asked the executives to make choices between lotteries involving risk and guaranteed outcomes. On the basis of these questions, he plotted utility functions.

For example, consider an executive who is indifferent to $B =$ \$1,000,000 with certainty or a lottery $L = [1/2, \$0; 1/2, \$5 \times 10^6]$. The lottery might represent a contract in which there is a 50–50 chance of outcome 0 or $\$5 \times 10^6$. We write, then,

$$u(\$10^6) = 1/2\, u(\$0) + 1/2\, u(\$5 \times 10^6)$$

We now arbitrarily assign $u(\$0) = 0$ and $u(5 \times 10^6) = 10$ utiles (units of utility). On that basis, $u(\$10^6) = 5$ utiles. This information provides 3 points on the executive's utility curve for money.

Next, the executive is asked to choose between a contract with 50–50 chance of making $\$10^6$ or nothing, or a cash prize B for selling the contract. Suppose he is indifferent between $L = [1/2, \$0; 1/2, \$10^6]$ and $B = \$300{,}000$. Then,

$$\begin{aligned} u(\$300{,}000) &= \frac{1}{2}u(\$0) + \frac{1}{2}u(\$10^6) \\ &= 0 + \left(\frac{1}{2}\right)(5) \\ &= 2.5 \end{aligned}$$

Now, a situation is considered where a loss is involved. The executive is asked to bid on a contract. If successful, the contract will net \$300,000; if unsuccessful, it will cost \$200,000. Of course, he could elect not to bid, i.e., $B = 0$. For what probability p of success would he be indifferent to $L = [p, \$300{,}000; (1 - p), -\$200{,}000]$ or $B = 0$, i.e.,

$$u(0) = p \cdot u(300{,}000) + (1 - p) \cdot u(\$-200{,}000)$$

Suppose the excutive is indifferent for $p = 3/4$. Then we can compute $u(-\$200{,}000) = -7.5$ utiles. The utility curve obtained by fitting a curve to the points we have generated so far is shown in Fig. 7-15.

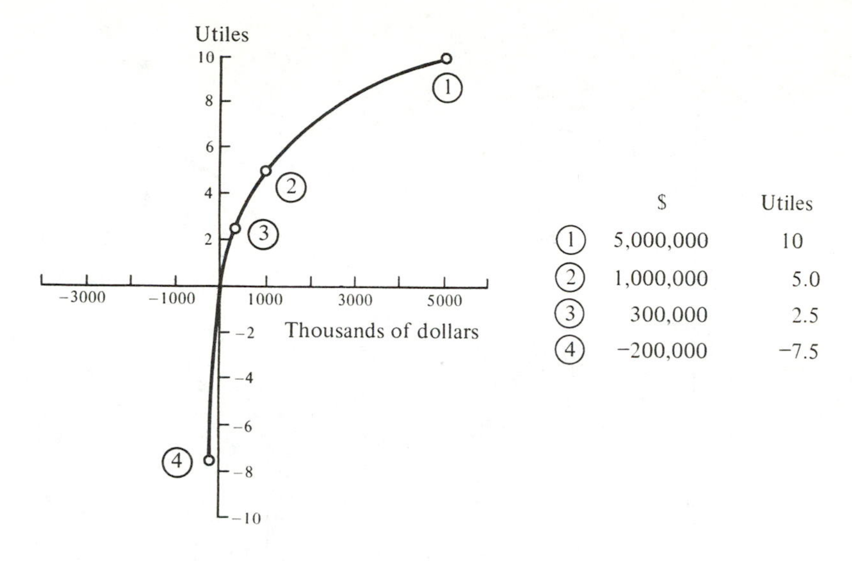

Figure 7-15 Descriptive Utility Curve

Some of the results reported by Swalm [63] are shown in Fig. 7-16. The executives whose utility functions are described by these curves range all the way from what appears to be an extremely cautious executive to a gambler. Swalm has found that attitudes toward risk vary widely for executives within the same company. In general, he speculates on the basis of his studies that USA executives in the business world are not the big risk takers they are thought to be. In many cases, they consistently avoid risks which in the long run, viewed from the overall company welfare, would be considered worth taking.

7-8 DECISION MAKING UNDER CONFLICT—GAME THEORY

There are decision situations in which an opponent is involved in place of states of nature. The opponent could be a business competitor, an enemy in a conflict, or even the states of nature for that matter, if the decision maker wishes to construct a model in which he stipulates that "nature is out to get him" and it plays the role of a rational competitor or enemy. *Game theory* is the study of such situations, in which more than one decision maker is involved, with each trying to maximize his utility or minimize his disutility. The name "game theory" is borrowed from similar situations which arise in common parlor games, such as chess, bridge, tic-tac-toe, or matching pennies. The model of game theory can be applied in many areas remote from parlor games, such as military strategy, politics, economics, and many more.

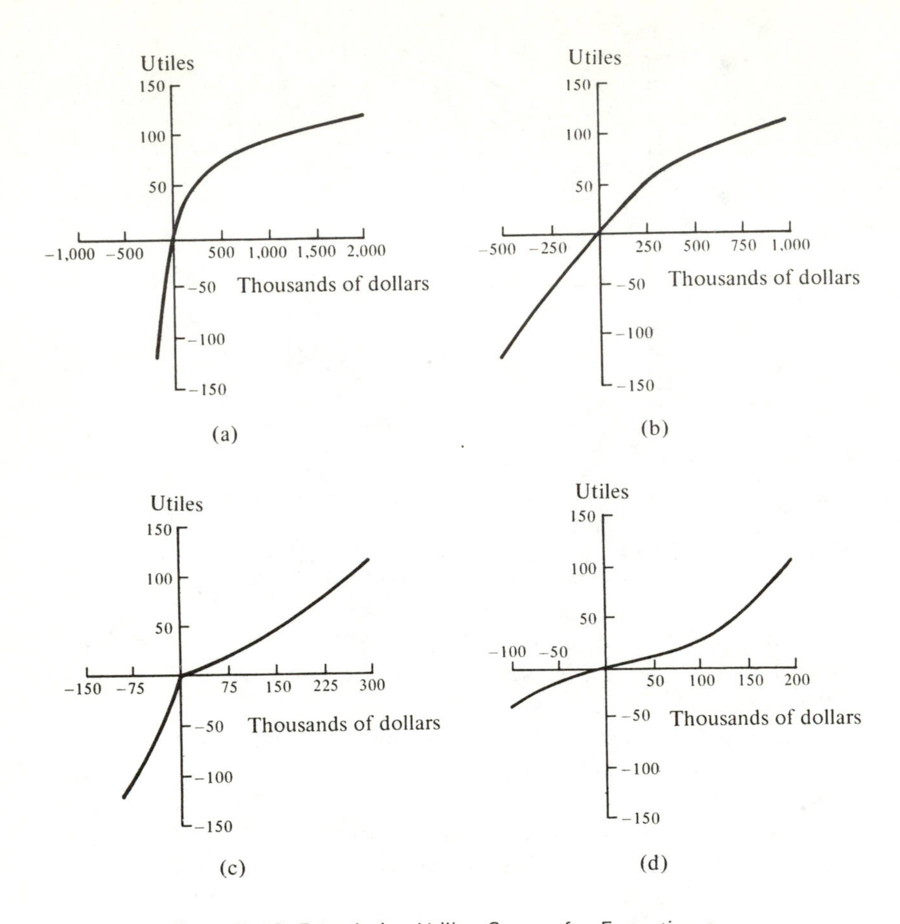

Figure 7-16 Descriptive Utility Curves for Executives: (a) Extremely Conservative; (b) Conservative; (c) Inclined Toward Risk; (d) Gambler. Adapted with permission from the author and *Harvard Business Review,* "Utility Theory—Insight into Risk Taking," Ralph O. Swalm, November-December 1966.

Games can be classified by the number of players as two-person games, three-person games, or n-person games for n people. In games with three or more players, coalitions can be formed where the members of the coalition act for their common interest against an opponent or opponents.

Another classification distinguishes *finite games,* with a finite number of strategies for the players, from *infinite games* in which an infinite number of strategies are present.

A third classification is related to the outcomes. *Zero-sum games* are games in which the total outcome for the players is zero. For example, in a two-person game, the gain of one player is always identical to the loss of the other. *Nonzero-sum games* are games in which this is no longer true; namely, the gain of one player, in a two-person game, for example, is not equal to the loss of the other. This can often be the case when the players associate different utilities with the same outcomes.

The Game Theory Model

To fix ideas and introduce the terminology of game theory, consider a two-person model of game theory in Fig. 7-17. The model represents three

		Player 2				
		OP_1	OP_2	OP_3	OP_4	Row minima
	S_1	12	3	9	8	3
Player 1	S_2	5	4	6	5	(4) ← *maximin*
	S_3	3	0	6	7	0
Column maxima		12	(4) *minimax*	9	8	

Figure 7-17 A Model of a Zero-Sum Two-Person Game

strategies S_i ($i = 1, 2, 3$) open to our decision maker, labeled player 1, and four strategies, OP_j ($j = 1, 2, 3, 4$), open to the opponent, labeled player 2. The twelve numbers in the matrix, called the *payoff matrix*, indicate the amount paid by player 2 to player 1. The amounts may represent any rewards or the utility of the rewards corresponding to the outcomes (S_i, OP_j) ($i = 1, 2, 3$; $j = 1, 2, 3, 4$).

Game theory assumes that the decision maker and the opponent are rational, and that they subscribe to the *maximin* criterion as the decision rule for selecting their strategies. Note that no claim is made that this is the best criterion; rather, this is the assumed criterion in game theory. Although this criterion was viewed earlier as the cautious pessimist attitude in decision under uncertainty, it is important to point out that the situation is somewhat different in the game-theory model. In decision under uncertainty, we have no reason to believe that nature is out to get us, and that a state "will be selected" on the basis of how much it could hurt the decision maker. On the other hand, the opponent in the game-theory model is viewed as a malevolent opponent, and this justifies caution. Therefore, the cautious maximin criterion has much appeal.

Consistent with the maximin criterion, player 1 records the row minima and selects the maximum of these, as shown in the column on the right of the payoff matrix in Fig. 7-17. He, therefore, selects strategy S_2. The opponent wishes to minimize the amount he must pay player 1 and wishes also to achieve this by the cautious maximin approach in which he guarantees, by his choice of a strategy, the worst that can happen to him. Since he is at the paying end of the game, he records the column maxima in a row below the payoff matrix in Fig. 7-17, and of these he selects the minimum or the minimax. On that basis, player 2 selects strategy OP_2.

Although player 2 finds the minimum of the column maxima (minimax), it is essentially a maximin criterion. The reason he computes the minimax is because he is at the paying end, the way the payoff matrix is represented, but he still attempts to get *the best of the worst.*

Note that both players use the pessimist's cautious maximin approach in the sense that they cannot be worse off. Should one of them abandon this approach and select a different strategy than that dictated by the maximin criterion, then the other player will be even better off. For example, should player 2 select a strategy different from OP_2, then player 1 may obtain 5 or 6 instead of 4. Similarly, if player 1 abandons the maximin criterion and player 2 sticks to his, player 2 may pay 3 or 0 instead of 4.

The value 4 in Fig. 7-17 is called the *value of the game*, or the *solution of the game*. The intersection point of strategies S_2 and OP_2, which leads to the outcome 4 (solution of the game), is called a *saddle point** because at the saddle point the outcome is the maximum of its column and the minimum of its row. A game with a saddle point is also referred to as a game with an *equilibrium solution.*

The game of Fig. 7-17 is not *a fair game* because player 2 is losing and player 1 is winning 4 units of utility each time the game is played. This is not to be confused with the concept of *zero-sum game* which it still is because what player 1 is winning and what player 2 is losing adds up to zero. The game of Fig. 7-17 can be changed into a *fair game* by subtracting 4 units from each entry in the *payoff matrix*. The value of the game would then be zero. A *fair game* is a game whose value is zero.

Games with No Saddle Point—Mixed Strategies

Consider the game of Fig. 7-18. Here there is no saddle point because the maximum of the row minima does not coincide with the minimum of the column maxima. Or there is no outcome entry in the payoff matrix which is both its row minimum and its column maximum. By the maximin criterion, the players choose strategies S_2 and OP_1. However, player 2 can reason that since player 1 plays by the criterion, he would be better off playing OP_2, in which case he will pay only 4 instead of 5. Similarly, player 1, knowing what player 2 might think, will play S_1 to obtain 6 instead of 4. Therefore, there seems to be no equilibrium.

---

*Because player 1 reaches it from below (maximin) and player 2 from above (minimax), like walking on a hyperbolic paraboloid to the saddle point. (See Chapter 1.)

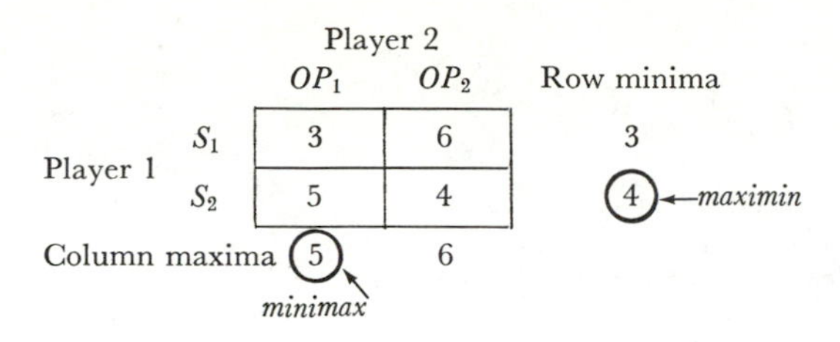

Figure 7-18 Game with No Saddle Point

A famous von Neumann theorem of game theory proves that for every two-person zero-sum game, equilibrium can be reached if mixed strategies are allowed. Consider a game with n strategies S_i for player 1. A *pure strategy* is a scheme in which *only one* of these strategies is played with probability of 1 and all others are not played. A *mixed strategy* consists of a scheme in which a probability different from 1 is associated with the use of each available pure strategy. For example, a mixed strategy for player 1 could be a scheme in which two coins are tossed, and if two heads appear he plays S_1; otherwise, he plays S_2. The mixed strategy in this case is denoted in the form $(1/4, S_1; 3/4, S_2)$. Similarly, player 2 could use a mixed strategy $(1/8, OP_1; 7/8, OP_2)$ by playing OP_1 when 3 heads appear in the toss of three coins, and playing OP_2 otherwise.

Generating the Mixed Strategy. In Fig. 7-18, player 1 will try to increase the maximin of 4, and player 2 will try to decrease the minimax of 5. It is reasonable that equilibrium, if reached at all, will be somewhere between these two values. A mixed strategy will reach the equilibrium state. Such a strategy is generated on the basis of the following reasoning:

PLAYER 1. Since I have no idea what player 2 will do, but I know that he is out to get me, let me be cautious and assure myself that, no matter which strategy he plays, I shall realize the same *expected value of utility*. This, then, is my criterion for a game with no saddle point. Hence, suppose I play S_1 with probability p and S_2 with probability $1 - p$. Then, if player 2 plays OP_1, my expected value of utility, $E(u)$, will be

$$E(u) = p3 + (1 - p)5$$

and if he plays OP_2,

$$E(u) = p6 + (1 - p)4$$

Since I want these two results to be equal,

$$p3 + (1 - p)5 = p6 + (1 - p)4$$

$$p = \frac{1}{4}$$

Hence, my *mixed strategy* is

$$\left(\frac{1}{4}, S_1; \frac{3}{4}, S_2\right)$$

and my expected value of utility is

$$E(u) = \left(\frac{1}{4}\right)3 + \left(\frac{3}{4}\right)5 = 4.5$$

PLAYER 2. Since player 1 is my opponent, I had better be cautious and assure myself the same expected value of loss in utility, irrespective of which strategy he plays. This is my criterion. So, suppose I play OP_1 with probability q, and OP_2 with probability $1 - q$. Then, if player 1 plays S_1, I have expected loss $E(u)$:

$$E(u) = q3 + (1 - q)6$$

and if he plays S_2,

$$E(u) = q5 + (1 - q)4$$

Since I want these results to be the same,

$$q3 + (1 - q)6 = q5 + (1 - q)4$$

$$q = \frac{1}{2}$$

Hence, my *mixed strategy* is

$$\left(\frac{1}{2}, OP_1; \frac{1}{2}, OP_2\right)$$

and $$E(u) = \left(\frac{1}{2}\right)3 + \left(\frac{1}{2}\right)6 = 4.5$$

Thus, the value of the game at equilibrium is 4.5.

NOTE. The fact that the value of the game came out to be 4.5, i.e., exactly midway between the maximin of 4 and minimax of 5, is a coincidence. For example, in the game in Fig. 7-19,

	OP_1	OP_2	Row minima
S_1	0	6	0
S_2	5	4	(4) ← *maximin*
	(5) *minimax*	6	

Figure 7-19

Player 1 uses a mixed strategy $(1/7, S_1; 6/7, S_2)$.

Player 2 uses a mixed strategy $(2/7, OP_1; 5/7, OP_2)$.

The value of the game is 30/7.

Problem. Verify the strategies and the value of the game in the above note.

Another Interpretation of a Mixed Strategy

There are situations in which the strategies are not mutually exclusive. In such situations it may be possible to play a mixture of two or more strategies at the same time. For example, instead of playing the mixed strategy $(1/4, S_1; 3/4, S_2)$ by tossing two coins and playing S_1 when two heads appear and S_2 otherwise, we may in certain situations play both strategies each time in a mix of 1 : 3. An example will illustrate such a situation.

Consider a student who makes a living selling sunglasses and umbrellas at a football stadium [64]. From experience he knows that he can sell 500 umbrellas when it rains. When it shines, he can sell 1000 sunglasses and 100 umbrellas. An umbrella costs $0.50 and sells for $1; glasses cost $0.20 and sell for $0.50 each. He can invest $250 in the business for next Sunday. Everything that isn't sold is a total loss. The decision model for this situation is as given in Fig. 7-20.

		State of Nature Rain	Shine
He buys for	Rain, S_1	250	−150
	Shine, S_2	−150	350

Figure 7-20

In this payoff matrix the positive numbers indicate gain (profit) and the negative numbers are loss (the dollars were not transformed into utilities in the present case). Thus, when he buys for *Shine*, he pays $50 for 100 umbrellas and $200 for 1000 sunglasses, or a total expenditure of $250. If it rains, he sells only the 100 umbrellas for $100; the sunglasses are a loss. The net loss is $150. If it shines, he sells the umbrellas for $100 and the sunglasses for $500 and his net profit is 600 − 250 = $350. When he buys for *Rain* and it rains, his net profit is $250, but if it shines, the net loss is $150.

Let us suppose that our student views nature as an opponent. He, therefore, models the situation in terms of a game theory model. The game has no saddle point (verify). Hence, he computes a mixed strategy, $[p, S_1; (1-p), S_2]$, from the equation

$$250p + (1-p)(-150) = (-150)p + (1-p)(350)$$
$$p = \frac{5}{9}$$

He could, therefore, use strategies S_1 and S_2 with probabilities 5/9 and 4/9, respectively. Instead, he could invest 5/9 of his $250 in S_1 (5/9 of 500 umbrellas) and 4/9 of $250 in S_2 (4/9 of 1000 sunglasses + 4/9 of 100 umbrellas), i.e., purchase (5/9) 500 + (4/9) 100 umbrellas and (4/9) 1000 sunglasses. The expected profit, rain or shine, would be

$$(250)\left(\frac{5}{9}\right) + \left(\frac{4}{9}\right)(-150) = \$72.22$$

Problem 1. What should the student do if he knows that the probability of rain is 1/4? Consider the case in which all his assets amount to $250. Replace the outcomes in the payoff matrix by utilities of the outcomes to the student. Sketch roughly a utility curve for this purpose.

Problem 2. Repeat Problem 1, considering that the probability of rain is 1/2, but in addition it is known from experience that when it shines the probability of selling 100 umbrellas is 3/4 and the probability of selling 1000 sunglasses is 19/20; the probability of selling 500 umbrellas when it rains is 9/10.

Dominance

There are situations in which one strategy is equal or superior to another on an entry-by-entry basis in the payoff matrix (with at least one entry superior). In such cases, we say that the superior strategy is *dominant* and we eliminate the *dominated* strategy from the matrix. For example, when we compare S_1 and S_3 in the following payoff matrix, we note that $3 > 2$ and $6 > 5$ so that S_3 is *dominant* over S_1:

	OP_1	OP_2
S_1	2	5
S_2	4	3
S_3	3	6

Hence, we eliminate S_1 from the matrix, and reduce the game to

	OP_1	OP_2
S_2	4	3
S_3	3	6

Similarly, in the matrix

	OP_1	OP_2	OP_3	OP_4	OP_5	OP_6	OP_7
S_1	−6	−1	1	4	7	4	3
S_2	7	−2	6	3	−2	−5	7

OP_2 dominates OP_3, OP_4, OP_5, OP_7, because player 2 is at the paying end and, therefore, he reduces the game to Fig. 7-21.

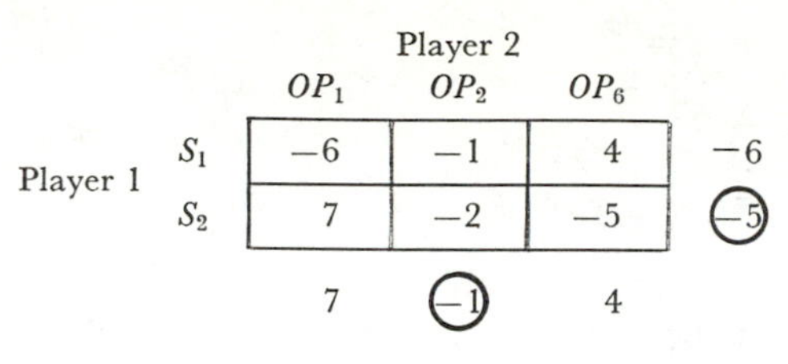

		Player 2 OP_1	OP_2	OP_6	
Player 1	S_1	−6	−1	4	−6
	S_2	7	−2	−5	(−5)
		7	(−1)	4	

Figure 7-21

* Graphical Solution of 2 × *n* Games

The optimal mixed strategies in $m \times n$ games ($m \geq 2$, $n \geq 2$) can be computed by the simplex method of linear programming [64]. When one of the players has only two pure strategies, a graphical solution is also possible. For example, consider the game of Fig. 7-21. The game has no saddle point. Player 1 will play S_1 with probability p, and S_2 with probability $1 - p$. To compute p graphically, we proceed as follows. In Fig. 7-22, the vertical axis indicates the payoff to player 1, and the horizontal axis indicates the value of p from 0 to 1. At $p = 0$ he uses only S_2 (pure strategy), and at $p = 1$ he uses only S_1 (pure strategy). Between these two extremes he uses a mixed strategy. His payoff depends on the value of p and on what player 2 does. For example, if player 2 uses OP_1, then the payoff to player 1 is given by:

$$\text{for } OP_1: \quad \text{Payoff} = -6p + 7(1 - p) \tag{7-2}$$

This is the equation of a straight line shown in Fig. 7-22. Similarly, if player 2 uses OP_2 or OP_6, we have, respectively,

$$\text{for } OP_2: \quad \text{Payoff} = -1p + (-2)(1 - p) \tag{7-3}$$

and

$$\text{for } OP_6: \quad \text{Payoff} = 4p + (-5)(1 - p) \tag{7-4}$$

These equations are also shown in Fig. 7-22. The minimum payoffs are represented by the heavy line in Fig. 7-22. Player 1 wishes to maximize the minimum payoff, to obtain a *maximin* solution; therefore, he selects the highest point on the heavy line. Since the solution (point marked *solution* in Fig. 7-22) lies at the intersection of the lines for OP_1 and OP_2, it can also be obtained algebraically by equating Eqs. 7-2 and 7-3, and solving for p:

$$-6p + (1 - p)7 = -1p + (1 - p)(-2)$$

$$p = \frac{9}{14}$$

Hence, player 1 uses a mixed strategy (9/14, S_1; 5/14, S_2), and his payoff is

$$\left(\frac{9}{14}\right)(-6) + \left(\frac{5}{14}\right)(7) = -\frac{19}{14}$$

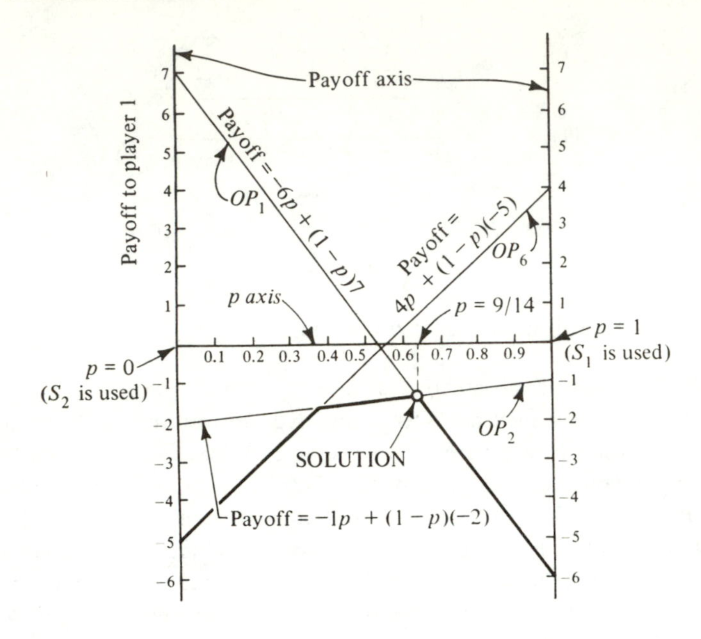

Figure 7-22 Graphical Solution to 2 × *n* Game

when player 2 plays OP_1 or OP_2. However, should player 2 use OP_6, then player 1 obtains a higher payoff:

$$\left(\frac{9}{14}\right)(4) + \left(\frac{5}{14}\right)(-5) = \frac{11}{14} > -\frac{19}{14}$$

Player 2, therefore, considers the reduced game (after eliminating OP_6):

	OP_1	OP_2
S_1	−6	−1
S_2	7	−2

and obtains

$$-6q + (1 - q)(-1) = 7q + (1 - q)(-2)$$

$$q = \frac{1}{14}$$

which yields the strategy ($1/14$, OP_1; $13/14$, OP_2) with a payoff of

$$\left(\frac{1}{14}\right)(-6) + \left(\frac{13}{14}\right)(-1) = -19/14$$

This payoff is identical to the payoff for player 1 because equilibrium has been reached, i.e., a solution has been obtained.

Problems 3 and 4. Solve the games in Figs. 7-23 and 7-24. Indicate the strategy of each player and the payoff.

		Player 2 OP_1	OP_2
Player 1	S_1	4	−3
	S_2	2	1
	S_3	−1	3
	S_4	0	−1
	S_5	−3	0

Figure 7-23

		Player 2 OP_1	OP_2	OP_3
Player 1	S_1	6	3	−2
	S_2	−4	4	5

Figure 7-24

Problem 5. Solve the rain-shine game of Fig 7-20 by using a graphical method of solution.

Problem 6. What happens to strategies which are dominated by others in a $2 \times n$ or $m \times 2$ game when a graphical solution is employed?

Two-Person Nonzero-Sum Games

Zero-sum games do not apply to situations in which one player's gain is not equal to the other's loss. Such situations arise quite often. For example, when you intentionally lose to an opponent in a parlor game, or in a sport, both parties may gain in utility in the outcome. Such games in which the gain of one player *is not equal* to the loss of the other are called nonzero-sum games. Two examples will illustrate the nature of nonzero-sum games.

EXAMPLE 1. The classical illustration of the nonzero-sum game is known as the *prisoner's dilemma*. In this game, the "players" are two prisoners who each have a choice to confess or not to confess a crime. The prisoners are not in communication with each other, but each has full knowledge of the consequences of his decisions. The payoff matrix is shown in Fig. 7-25. Each

		Prisoner 2 Not Confess	Confess
Prisoner 1	Not Confess	(3, 3)	(10, 1)
	Confess	(1, 10)	(8, 8)

Figure 7-25 Prisoner's Dilemma Payoff Matrix

outcome is represented by two numbers, indicating the number of years in prison for prisoner 1 and 2, respectively. For example, if prisoner 1 confesses and 2 does not, 1 gets a one-year sentence and 2 gets a ten-year sentence. The outcomes could, of course, represent utilities to the prisoners.

Prisoner 1 reasons as follows: If prisoner 2 does not confess, then he is better off confessing. But he is also better off confessing when 2 confesses. On this basis, he should decide to confess. By similar reasoning, prisoner 2 decides to confess because, irrespective of what 1 does, he is better off confessing. Therefore, they will settle for the outcome (8, 8). However, if they considered the situation not from a selfish, but rather from an altruistic, point of view and placed trust in each other's good motives, they would both not confess and settle for (3, 3). Note that in this case, if each considered doing the best for the other, they would still end up with outcome (3, 3).

To resolve the dilemma, it becomes necessary to consider Immanuel Kant's categorical imperative which stipulates the following criterion for action: "Act in such a way that if others did the same, you would benefit." This kind of group, rather than individual, rationality would lead to a choice of *not confess* in the above game. However, to consider such collective rationality would imply a strong commitment to mutual trust as a high value. This is not simple nor easy to attain.

EXAMPLE 2. This example is similar to a nonzero-sum game discussed by Kassouf [65]. The model is a simple abstract game between two world powers, say, ROW and COLUMN. The strategies of ROW(R_1, R_2) and of COLUMN (C_1, C_2) are:

R_1 or C_1: Disarm and divert saved resources to rehabilitation of city slums.

R_2 or C_2: Continue to arm at a cost of 5 utiles to each power.

The utilities to each power are shown between parentheses for each outcome, with the first number applying to power ROW and the second to COLUMN, as shown in Fig. 7-26. The game is a nonzero-sum game.

For example, (6, −11) implies that when ROW continues to arm and COLUMN disarms, power ROW takes over COLUMN. COLUMN loses

		COLUMN	
		C_1 (Disarm)	C_2 (Arm)
ROW	R_1 (Disarm)	(5, 5)	(−11, 6)
	R_2 (Arm)	(6, −11)	(−5, −5)

Figure 7-26 World Power Nonzero-Sum Game

11 utiles (goods, services, etc.) to ROW. However, since ROW continued to arm, the expenditure of 5 utiles is subtracted from 11 for a net gain of 6.

Reasoning for ROW and COLUMN as we did for the prisoners in the prisoner dilemma, they continue to arm with the resulting payoff of (−5, −5). Again, this is the result of mistrust. Of course, if only one party is altruistic, it will suffer the consequences of the other party's selfishness. Communication between the parties can help lead to rationality in behavior which will result in (5, 5). However, questions of credibility, mutual trust, and utility assessment in different cultures must be resolved before effective communication could resolve the world-power dilemma. The United Nations could make a novel effort to create an arena for communication that might lead to rational behavior [66].

7-9 GROUP DECISION MAKING

When a group involving two or more members is faced with a decision problem, we can still apply the tools that we discussed for a single decision maker, provided they agree on:

1. The model, i.e., how many alternatives should be considered.
2. The assessment of probabilities.
3. Utilities of outcomes.

Seldom does a group agree on all of these items. As a basis for joint action, a majority rule could be adopted to establish group preferences. However, a simple example will illustrate that the transitivity property may be violated, leading to inconsistent results, when the majority rule is applied to three persons, P_1, P_2, P_3, faced with the problem of choosing one of three strategies, S_1, S_2, S_3. Suppose the three persons rank the strategies as shown in Fig. 7-27. Using the majority rule, we proceed as follows:

	PERSON		
	P_1	P_2	P_3
First Choice	S_1	S_2	S_3
Second Choice	S_2	S_3	S_1
Third Choice	S_3	S_1	S_2

Figure 7-27

Decide on group preference between S_1 and S_2. Since both P_1 and P_3 prefer S_1 to S_2, the group preference is $S_1 > S_2$. Considering next S_1 and S_3, P_2 and P_3 prefer S_3 to S_1. Hence, majority rule preference is $S_3 > S_1$. Then considering S_2 and S_3, we conclude that $S_2 > S_3$. This violates transitivity because

$$S_2 > S_3,\ S_3 > S_1, \text{ and } S_1 > S_2$$

The fact that transitivity is violated leads to the possible choice of any of the three strategies, S_1, S_2, S_3, depending on the order in which we make choices between pairs of strategies by the majority rule. Thus:

1. Between S_1 and S_2, the group eliminates S_2. Between S_1 and S_3, the group prefers S_3. Hence, S_3 is the group choice.
2. Between S_1 and S_3, the group eliminates S_1. Between S_2 and S_3, the group prefers S_2. Hence, S_2 is the group choice.
3. Between S_2 and S_3, the group eliminates S_3. Between S_1 and S_2, the group prefers S_1. Hence, S_1 is the group choice.

In summary: Beginning with S_1 and S_2, the choice is S_3.
Beginning with S_1 and S_3, the choice is S_2.
Beginning with S_2 and S_3, the choice is S_1.

Therefore, the group choice by majority rule is arbitrary. Such choice is irrational because it leads to an inconsistency in the decision process. The above example is known as the *Arrow Paradox* [67].

The problem of establishing societal or group utilities as contrasted with individual utilities is a problem of great importance in our democratic society which operates on majority rule. There is growing interest in this area and studies are being conducted to establish societal utilities by rational procedures. For example, the problem of an individual and societal utility function applied to welfare programs is examined in [68].

7-10 SUMMARY

Decision theory provides a basis for rational actions and consistent behavior when a choice must be made between alternative courses of action.

A decision model includes these main elements:

Alternative actions: Actions which the decision maker controls.

States of nature: Actions not controllable by the decision maker.

Outcomes: Results from combinations of actions and states of nature.

Utilities: Measures of satisfaction to the decision maker for each outcome.

Objective: Decision rule (say, maximizing utility) as a criterion for choice of action.

Probability assessments: Decision maker's assessment of probabilities of states of nature. Objective or subjective probabilities.

Decision-making models are classified according to the assignment of probabilities to the states of nature as:

Decisions under certainty
Decisions under risk
Decisions under uncertainty
Decisions under conflict—game theory

The prevailing criterion in decision making is the maximization of the expected utility. Other criteria, such as *maximin*, *maximax*, and *minimax regret* are possible for decision under uncertainty.

For decisions under conflict—game theory, the maximin criterion is assumed.

Games are classified by the number of players and by the number of strategies, or as zero-sum and nonzero-sum games.

Utility theory guarantees that a utility function for outcomes exists if the following axioms hold true:

AXIOM 1. Transitivity is not violated:

$A > B$ and $B > C$ implies $A > C$
$A \sim B$ and $B \sim C$ implies $A \sim C$

AXIOM 2. When $A > B$,

$[p^*, A; (1 - p^*), B] > [p, A; (1 - p), B]$
if, and only if, $p^* > p$

AXIOM 3. The decision maker is indifferent to the number of lotteries if the final outcomes and associated probabilities are the same:

$$\underbrace{[P(A), A; P(L_2), L_2]}_{\text{Two lotteries}} \sim \underbrace{[P(A), A; P(B), B; P(C), C]}_{\text{One lottery}}$$

$L_2 = [P(B/L_2), B; P(C/L_2), C]$
$P(B) = P(B/L_2)\, P(L_2)$
$P(C) = P(C/L_2)\, P(L_2)$

AXIOM 4. If $A > B > C$, then there exists a lottery involving A and C with a probability p for which

$B \sim [p, A; (1 - p), C]$

Properties of Utility Function:

Property 1. $u(A) > u(B)$ if, and only if, $A > B$
$u(A) = u(B)$ if, and only if, $A \sim B$

Property 2. When $B \sim [p, A; (1 - p), C]$,
$u(B) = p\, u(A) + (1 - p)\, u(C)$

Property 3. The choice of reference value and scale for units of utility is arbitrary.

Both normative (prescriptive) and descriptive utilities can be generated.

Group decision making may often lead to inconsistencies because the transitivity property is violated. Group decision making is more complex than individual decision making.

EXERCISES

7-1 A farmer has to decide what crop to plant in a given area. Suppose that he has four strategies: to plant asparagus, peas, tomatoes, or do nothing; and the states of nature can be summarized in four possibilities: perfect weather, good weather, variable weather, and bad weather. The utilities of the outcomes under these conditions are summarized below in a payoff matrix:

	STATES OF NATURE			
STRATEGIES	N_1: Perfect weather	N_2: Good weather	N_3: Variable weather	N_4: Bad weather
S_1: Plant asparagus	12	5	4	0
S_2: Plant peas	10	6	6	3
S_3: Plant tomatoes	9	8	7	3
S_4: Plant nothing	4	4	4	4

What strategy should be selected if the farmer uses:
(a) the maximin criterion?
(b) the maximax criterion?
(c) the regret criterion?

7-2 It is Monday night and you have to decide whether to study hard for your Tuesday morning class, just skim through the notes, or watch TV. You have no idea whether there will be a quiz the next day, and if there is, whether it will be easy or difficult.
What would your strategy be (study, skim, or watch TV), given the following payoff matrix showing the utility of each outcome, if you were:
(a) an optimist?
(b) a pessimist?
(c) a regretist?

	No quiz	Easy quiz	Difficult quiz
Study hard	7	5	8
Skim notes	6	8	7
Watch TV	10	4	0

7-3 Three different utility functions are graphed in Fig. 7-13. Depending on Mr. Jones' attitude toward risk, his own personal utility function may be described by one of these three curves.

Mr. Jones is offered a *free* ticket to a lottery in which he stands to win \$70 with a probability of 1/10, and lose \$5 with a probability of 9/10.

(a) Answer the following for *each* of the three utility curves shown in Fig. 7-13: Should Mr. Jones accept the above lottery if his current assets are \$30? (That is, how does Mr. Jones' decision to play the lottery change with different attitudes toward risk?)

(b) Again, answer the following for each of the three utility curves: For what probability of winning \$70 should Mr. Jones be indifferent to the lottery if his current assets are \$30? \$10?

7-4 Work the problem at the end of Example 2 in Sec. 7-6.

7-5 Explain briefly, using the concepts of utility theory, why life insurance policies increase the utility of both the seller (the insurance company) and the buyer. You may assume that an average individual can buy life insurance worth \$100,000 for \$150 per year. Make assumptions on total assets of both the buyer and the seller of insurance policies, and roughly sketch the utility curves for both parties.

7-6 Show how the transitivity property may be violated when 3 people rank 3 outcomes.

7-7 Suppose you are in Las Vegas and have only enough money left to buy a ticket to play one of the 2 games in Figs. 7-28 and 7-29. Which game would you choose (assuming you subscribe to the axioms of utility)?

GAME I: Spin the pointer on the wheel in Fig. 7-28 and win the prize in the region in which the pointer comes to rest. It is equally likely to come to rest pointing in any direction.

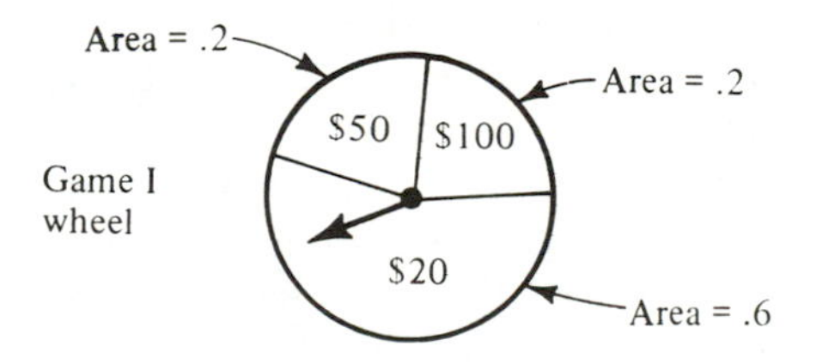

Figure 7-28

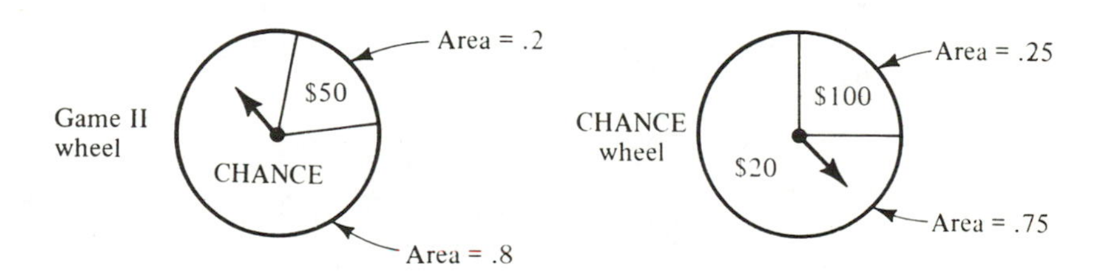

Figure 7-29

The prizes associated with the 3 regions, and the fraction of total area encompassed by each region, are indicated.

GAME II: Same rules, except the wheel has only 2 prizes (as shown in Fig. 7-29), one of which is marked "CHANCE." This is a ticket to spin another wheel with prizes as indicated.

7-8 Two doctors are trying to decide whether they should treat a certain patient or wait. Their respective utilities of the 3 outcomes, cure, death, and paralysis, are as follows:

	u(CURE)	u(PARALYSIS)	u(DEATH)
Doctor 1	100	40	0
Doctor 2	100	0	−200

The probabilities of the 3 outcomes, based on the action taken, are as follows:

	CURE	PARALYSIS	DEATH
Treat	$p = 0.6$	$p = 0.1$	$p = 0.3$
Wait	$p = 0.4$	$p = 0.5$	$p = 0.1$

Given the above information (show all your work):

(a) Will Dr. 1 decide to wait or treat?

(b) Will Dr. 2 decide to wait or treat?

7-9 As we have seen in Exercise 7-8, two doctors may arrive at different decisions due to the different utilities that they assign to outcomes.

(a) Discuss ways in which the two doctors of Exercise 7-8 may reach agreement on whether or not to treat the patient.

(b) Discuss situations in which the two doctors of Exercise 7-8 would agree on an action (whether to treat or wait).

7-10 Suppose Dr. 1 of Exercise 7-8 assigns two different utility values to each state of nature, one value for "treat" and one for "wait."

(a) Discuss why (or why not) this is a reasonable approach.

(b) Select utility values associated with each "action-state-of-nature" combination and solve the problem of finding which course of action the doctor should take. Use the probabilities given in Exercise 7-8.

7-11 For the following game, determine if there are dominated strategies and cross them out. Then use the graphical method with the resulting payoff matrix to determine the following:

		Player *B* 1	2	3
Player *A*	1	1	3	11
	2	8	5	2
	3	10	6	3

(a) The optimal strategy for Player A.
(b) The value of the game.
(c) Is the value of the game determined in (b) the same for Player B?
(d) Can A obtain a higher payoff than the value determined in (b)? Explain.

7-12 Work Problem 1, page 344 Sec. 7-8.

7-13 Assume the doctor assigns the following utilities to the outcomes in Table 6-4 for Example 1 in Sec. 6-9.

	Pills Are Good	Pills Are Not Good
S_1: Give pills	100	0
S_2: Do not give pills	30	100

(a) Suppose the diameter of the pill is found to be 10.2 mm. If the doctor makes his decision on the basis of the outcome with largest expected utility, what will he do?
(b) Outline a trial-and-error procedure, using the above utilities, for determining a decision diameter such that if a pill were smaller than this diameter, the doctor would use it, and if a pill were larger than this diameter, he would reject it.
(*Hint:* We are trying to find a decision diameter to make equal the expected values of utility from accepting and rejecting.)

7-14 For Model 3 in Sec. 7-2, how would you go about assigning utility values to each state of nature? Justify the procedure you use in assigning these values. What strategy would you adopt, once these values have been assigned?

7-15 Discuss the five elements of the decision-making model as they apply to the project you selected in Exercise 1-1.

7-16 Even though the "odds" are against the player in casino games, millions of people play these games each year. Discuss this phenomenon in terms of utility (to the player and to the casino) and nonzero-sum games.

7-17 Consider a game in which each of the two players, A and B, writes on a piece of paper H or T, and then they compare their choices. If both have H, player A receives one dollar from player B. If both have T, A receives three dollars from B. If their choices do not match, A pays B two dollars.
(a) Complete the following payoff matrix:

		Strategy of Player B: H	T
Strategy of Player A	H		
	T		

(b) Find the mixed strategy for Player A.
(c) Find the value of the game.

8

OPTIMIZATION MODELS—"SELECTING THE BEST POSSIBLE"

8-1 INTRODUCTION

When we attempt to solve a problem we often discover that there are several possible solutions. In such cases, we are faced with the added task of selecting the *best* of all *possible solutions*. A class of models for selecting the best, also known as the *optimum*, is referred to here as *Optimization Models*. It is necessary to be more explicit with regard to the meaning of two words before optimization models can be developed. These are the word *possible* in the context of *possible solutions*, and the word *best* in the context of *the best of all possible solutions*.

Let us illuminate the significance of these words through an example. Figure 8-1 shows a network of links between nodes. The number on each link designates the cost of travel along it between the two adjacent nodes. A man wishes to get from node A to node B. We *constrain* him to move between the nodes heading only north or east. There are $5!/2!3! = 10$ different *possible routes* which do not violate the constraints on the directions of travel. Hence, a *possible solution is one which achieves the desired goal state from an initial state, without violating the constraints*, namely, without doing that which is not to be done in the problem considered. Possible solutions in this context are also referred to as *feasible* or *acceptable* solutions.

Now that we have identified the possible solutions, we must select the best of these in terms of a *reference criterion*. In the present example, we stipulate the criterion of *least cost*. Hence, *the word best has meaning with reference to a criterion which identifies what we want explicitly*. So in Fig. 8-1 our man wishes to get from A to B by using the route of least cost. The best solution with respect to a reference criterion is referred to as the *optimum*.

To recapitulate, the *possible solutions* (feasible or acceptable) are iden-

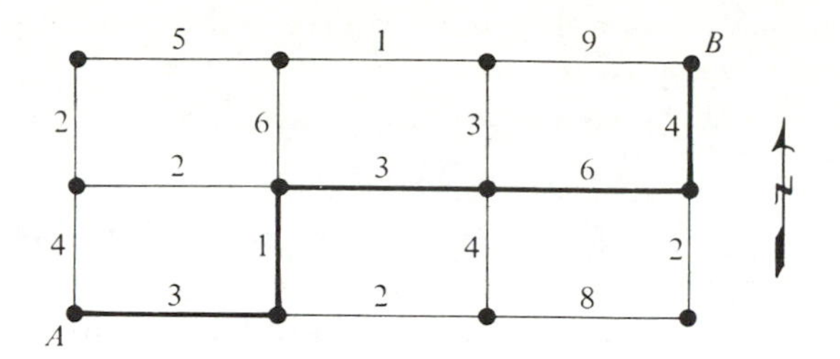

Figure 8-1 Least Cost Path From *A* to *B*

tified in terms of what we can or cannot do in the solution to a problem, namely, the constraints. The *optimum*, or the best of the possible solutions, is selected by reference to what we want to achieve explicitly, namely, a clearly stated objective with an identifiable criterion for a value judgment on what is best.

To select the optimum route in Fig. 8-1, one approach is to compute the cost of each of the ten possible routes and find the one of least cost. The optimum is marked by heavy lines in the figure and represents a cost of 17 units. In more complex problems, such a direct search for the optimum may not be possible with the constraints of reasonable time and effort for the search. Consider, for example, a grid of 20×20 instead of the 2×3 in Fig. 8-1. In such a grid there are $40!/20!20!$ possible routes from A to B. This is a number larger than 13×10^{10}. To obtain the cost of each route will require the addition of 40 numbers (one for each of the 40 links), and a computer that can perform 2×10^6 additions per second will require about one month of continuous computations. For a grid of 100×100, the computations are not possible in a lifetime. In view of these enormous numbers of possible solutions, methods are needed to reduce the extent of the search and associated computations and still identify the best of all possible solutions, i.e., the optimum, in reasonable time and effort. Such methods are called programming methods (the word programming here is not related to computer programming). We distinguish among *linear programming*, *nonlinear programming*, *and dynamic programming* models.* Programming methods facilitate a reduced effort in the *search for the best*, i.e., solution of optimization problems. This chapter discusses these methods of solution and some aspects of constructing the corresponding optimization models.

8-2 LINEAR FUNCTIONS

As a preliminary discussion to the treatment of linear programming models, let us consider, first, linear functions in general, and more specifically

*For instance, problems of the type shown in the example of Fig. 8-1 are suitable to dynamic programming.

linear functions of two variables, so we can resort to a geometric description of the functions in the plane of the paper.

A *linear function* of n variables $x_1, x_2, \ldots, x_n$ has the form

$$F(x_1, x_2, \ldots, x_n) = a_0 + a_1x_1 + a_2x_2 + \ldots + a_nx_n$$

The coefficients $a_0, a_1, a_2, \ldots, a_n$ are constants, and the notation $F(x_1, x_2, \ldots, x_n)$ on the left-hand side signifies compactly that we have a function (symbol F) of variables $x_1, x_2, \ldots, x_n$. The word *function* implies that for each set of values for the n variables x_i $(i = 1, 2, \ldots, n)$, we have a unique value —that is, one, and only one, value—for the function F. The linearity aspect is signified by the fact that each variable x_i in a linear equation appears to a power of one, that there are no products of variables, and the total value of the function is the sum of such linear (first power) terms plus a reference constant a_0. Linearity implies, therefore, that the value of the function changes in proportion to the change in the variables. For example, consider a change Δx_k in variable x_k only. Then the corresponding incremental change ΔF in the value of the function is $a_k\Delta x_k$. If the change in x_k is $3\Delta x_k$, then $\Delta F = 3a_k\Delta x_k$. Namely, for a threefold increase in the change of x_k, there is a threefold increase in the incremental change of F. The linearity property tells us also that an incremental change of an amount Δx_k produces the same incremental change ΔF, *regardless* of what the values of variable x_i are when the change Δx_k takes place. This is referred to as the *superposition property*, and it uniquely characterizes linear functions as distinct from nonlinear functions.

For example, in these three functions,

$$\begin{aligned} F_1(x_1, x_2) &= x_1x_2 \\ F_2(x_1, x_2) &= 2x_1^2 + x_2^2 \\ F_3(x_1, x_2) &= 5 + 4x_1 + 7x_2 \end{aligned}$$

F_3 is linear, F_1 and F_2 are nonlinear. This means that superposition holds for F_3 only.

Problem. Show that superposition does not hold for the nonlinear functions F_1 and F_2.

Linear functions of two variables can be described by models of straight lines, and linear functions of three variables can be described by models of planes. Beyond three variables, geometric models are not possible. While this limits our ability to visualize or model the higher-order linear function through geometry, it does not interfere with our ability to solve problems in which such functions occur.

Linear Function Models

Model 1. A dental assistant in a dentist's office devotes, on the average, 1 1/2 hours for each pedodontic service (infant dental care), and 1 hour for each general dental service to adults. The assistant works only 6 hours per day. If we designate by x_1 and x_2 the daily number of pedodontic and general services, respectively, and by s_1 the number of assistant hours per day which are unused, we can write

$$1.5x_1 + x_2 + s_1 = 6$$

In other words, if the assistant has four pedodontic cases in any given day, $x_1 = 4$, then the six hours are used up with zero time for general cases, $x_2 = 0$, and no unused time, $s_1 = 0$. Similarly, for six general cases, $x_2 = 6$, $x_1 = 0$, and $s_1 = 0$. When no cases occur, $x_1 = x_2 = 0$ and $s_1 = 6$; and if $x_1 = 2$ and $x_2 = 1$, then $s_1 = 2$.

Model 2. A man has 12 units of a *resource* that could be hours of work, money, objects of all kinds, and so on. He can exchange 3 units of his resource for each object 1, and 2 units for each object 2. Designating by x_1 and x_2 the number of units for objects 1 and 2, respectively, and letting s_1 stand for the number of units of unused resource in the exchange, we can write this linear function:

$$3x_1 + 2x_2 + s_1 = 12$$

Geometric Interpretation of Linear Functions

Let us now study the geometric interpretation of linear functions. If we set $s_1 = 0$ in the equation for Model 2 above, we have the following linear equation in variables x_1 and x_2:

$$3x_1 + 2x_2 = 12 \qquad (8\text{-}1)$$

This equation can be modeled by a straight line, as shown in Fig. 8-2. In the figure, the x_1 axis is identified as the line $x_2 = 0$ because x_2 is zero everywhere along axis x_1, regardless of what value x_1 has. Similarly, the axis x_2 is identified as $x_1 = 0$. The line for Eq. 8-1 can be drawn by identifying two points on it in the x_1, x_2 coordinate system. To do so, set $x_1 = 0$ and compute $x_2 = 12/2 = 6$. This yields the point (0, 6) in which the two numbers in parentheses represent the values of x_1 and x_2 (x_1, x_2). To get a second point, set $x_2 = 0$ and compute $x_1 = 12/3 = 4$ to yield (4,0). These two points are shown in the figure, and the straight line of Eq. 8-1 goes through them. Of course, we could identify two different points on the line by selecting an arbitrary value for x_1, say, $x_1 = 2$, yielding the point (2,3), and then making x_2 any arbitrary value different from 3 so as not to get the same point on the line. For example, for $x_2 = 1.5$, $x_1 = 3$ and we have the point (3,1.5).

Suppose now that we change the right-hand side of Eq. 8-1, replacing 12 by 24:

$$3x_1 + 2x_2 = 24$$

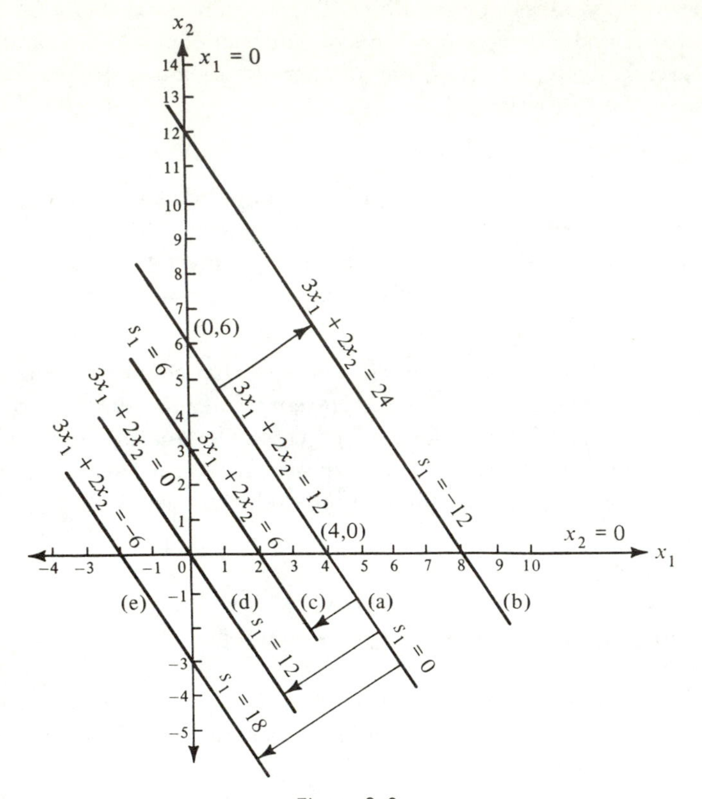

Figure 8-2

Then the line of Eq. 8-1 moves in a direction perpendicular to itself, away from the origin of the coordinate system as shown in Fig. 8-2. This can be verified by establishing the points (8,0) and (0,12) on the line. Thus, the slope is not altered when the free constant in the linear equation is changed; the line is merely shifted. The slope is established by the coefficients of the variables x_1 and x_2. If we now replace the free coefficient 12 in Eq. 8-1 by a smaller number, 6, the straight line moves closer to the origin; and when we replace 12 by 0, it goes through the origin. The lines we have so far in Fig. 8-2 are:

$$\left.\begin{array}{ll} 3x_1 + 2x_2 = 12 & \text{(a)} \\ 3x_1 + 2x_2 = 24 & \text{(b)} \\ 3x_1 + 2x_2 = 6 & \text{(c)} \\ 3x_1 + 2x_2 = 0 & \text{(d)} \\ 3x_1 + 2x_2 = -6 & \text{(e)} \end{array}\right\} \quad (8\text{-}2)$$

Equation 8-2(a) is the original Eq. 8-1, and the other equations are derived from it by assigning the appropriate variable s_1 in this form of Eq. 8-1:

$$3x_1 + 2x_2 + s_1 = 12 \quad (8\text{-}3)$$

To generate Eqs. 8-2(a-e), we assign s_1 the values 0, -12, 6, 12, and 18, respectively. Thus, using Eq. 8-3, we can identify the straight line for Eq. 8-1 by writing $s_1 = 0$, as shown in Fig. 8-2.

To study the significance of Eq. 8-3, we return to Model 2 and consider the 12 units of resource to represent the 12 hours per day available in a TV shop. Suppose the shop requires 3 hours for the assembly of TV set type 1 (object 1), and 2 hours for TV set type 2 (object 2). x_1 and x_2 represent the number of TV sets of types 1 and 2, respectively, assembled per day, and s_1 stands for the number of unused available hours of the shop per day. For example, if we have 3 unused hours, $s_1 = 3$, then

$$3x_1 + 2x_2 = 9$$

If $s_1 = 0$, then our resource of shop hours is completely used up and we have

$$3x_1 + 2x_2 = 12$$

Indeed, if we accept the premise or constraint that only 12 hours of the shop are available, then we can write

$$3x_1 + 2x_2 \leq 12 \qquad (8\text{-}4)$$

The inequality ($<$) will apply whenever the shop is not used for the full 12 hours, namely, $s_1 > 0$, and the equal sign ($=$) will apply when $s_1 = 0$.

But suppose we question the constraint of 12 hours and wish to study the possibility of expanding the capacity of the shop to 3 shifts, making 24 hours available daily. Then in Eq. 8-3 we assign $s_1 = -12$, yielding

$$3x_1 + 2x_2 = 24$$

Problem 1. How does the constraint Eq. 8-4 change if TV type 2 takes 90 minutes to assemble? (Show on a figure and compare.)

Problem 2. How does the constraint Eq. 8-4 change if type 1 and type 2 take 135 minutes and 90 minutes, respectively? (Show on a figure and compare.)

8-3 LINEAR PROGRAMMING—EXPOSURE

EXAMPLE 1. The owner of the TV shop in Sec. 8-2 makes a profit of \$40 for each set type 1, and \$100 for each set type 2. His *objective* is to maximize the profit (reference criterion). How many sets of type 1 (x_1) and type 2(x_2) is he to assemble per day?

Stated more compactly, we can use the models of linear equations to write:

Problem Statement. Find the values of x_1 and x_2 which maximize the *objective function*

$$F = 40x_1 + 100x_2 \qquad (8\text{-}5)$$

subject to the *constraints*

$$\left.\begin{aligned} 3x_1 + 2x_2 &\leq 12 \\ x_1 &\geq 0 \\ x_2 &\geq 0 \end{aligned}\right\} \tag{8-6}$$

Feasible Solutions. The constraints represent the bounds on feasible or acceptable values for x_1, x_2. Namely, no more than 12 hours of shop time are available per day, and the number of assembled sets of type 1 or type 2 must be equal to or larger than zero.

Figure 8-3 shows the *feasible region* for acceptable values of x_1 and x_2. The boundaries are marked by $x_1 = 0$, $x_2 = 0$, and $s_1 = 0$; s_1 is referred to as a *slack variable* which is constrained to values of zero or larger, $s_1 \geq 0$. Thus, the first inequality of Eq. 8-6 can be transformed to the equation

$$3x_1 + 2x_2 + s_1 = 12$$

The slack variable represents the amount of unused resource. For $s_1 = 0$, we obtain the boundary line shown in Fig. 8-3. The interiór of the feasible region is identified by the $x_1 > 0$, $x_2 > 0$, and $s_1 > 0$. Thus, in Fig. 8-3, $s_1 = 0$ plays the same role as $x_1 = 0$ and $x_2 = 0$, namely, it identifies a boundary line of the feasible region.

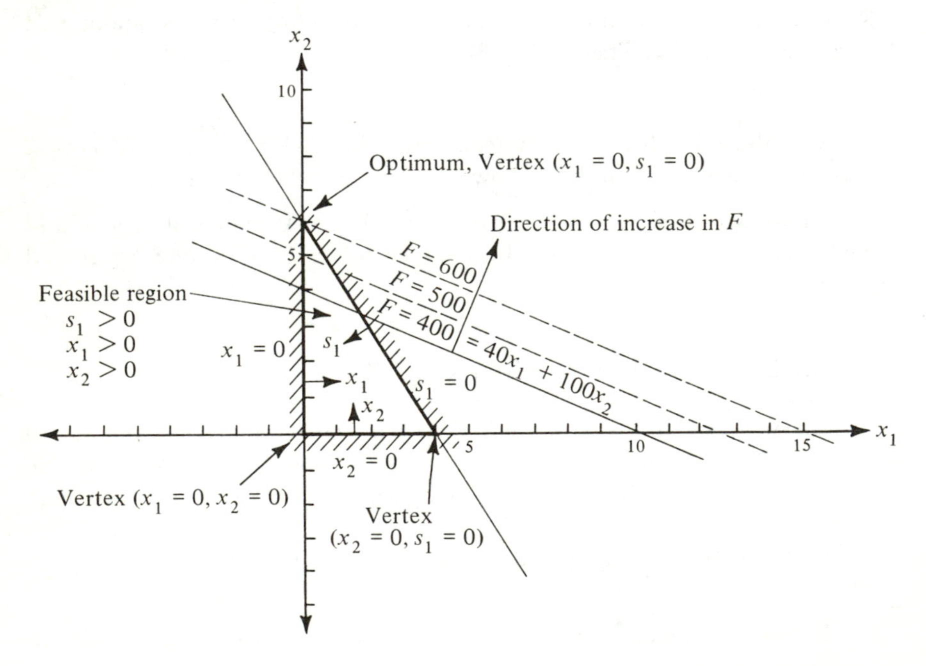

Figure 8-3

Solution. The solution consists of a choice of values x_1 and x_2 within the *feasible* region which will maximize F in Eq. 8-5. Arbitrarily setting $F = 400$ in this equation, we can plot it as shown in Fig. 8-3. Any pair of values along this line in the feasible region constitute a solution for which the profit is $F = \$400$. However, from our discussion of the last section, we know that F gets larger if we translate the line $40x_1 + 100x_2$ perpendicular to itself in the direction shown by the arrow in the figure. We, therefore, move the line until it is about to have no points in the feasible region. This occurs for $F = \$600$, and the point in the feasible region and on the line of the objective function is (0, 6), representing the *optimum* solution.

Problem. Explain the connection between unused resources and a slack variable. Give two examples.

EXAMPLE 2. Suppose the profits in Example 1 are \$70 for TV type 1 and \$40 for TV type 2. The objective function is then

$$F = 70x_1 + 40x_2$$

For the same constraints as in Example 1, the optimum solution is (4,0) which yields the maximum profit of \$280. Verify.

EXAMPLE 3. If the profits in Example 1 are \$60 and \$40 for sets types 1 and 2, respectively, then the objective function is

$$F = 60x_1 + 40x_2$$

and the optimum solution consists of any point on the boundary $s_1 = 0$. The maximum profit is $F = \$240$. Verify.

Solution of Example 1 by the Simplex Algorithm

Let us now restate Example 1, but solve it by using the following reasoning. The optimum solution to a linear programming problem lies on a vertex of the feasible region. This is apparent in Fig. 8-3, but can be shown to apply in general, regardless of the number of variables. In some instances, as in Example 3 above, the solution is not unique and lies on a boundary that contains two vertices and all points on the boundary between them. Using this conclusion, we can inspect each vertex by substituting its coordinates into the objective function, and find that vertex which maximizes the objective function. A more efficient approach, called the *simplex algorithm*, proceeds from any vertex to the one of its two neighbors which will lead to a larger increase in the objective funtion.* This process is continued at each vertex until a vertex is reached from which a move to any neighboring vertex will result in no increase in the objective function. Such a vertex represents the solution.

*This is true here in an idealized version of the simplex method. In general, the largest *rate* of increase is used instead of largest increase in the objective function.

To apply the *simplex algorithm* to Example 1, we write

$$\text{Maximize} \quad F = 40x_1 + 100x_2$$
$$\text{or} \quad \text{Max } F = 40x_1 + 100x_2 \tag{8-7}$$

subject to the constraints

$$\begin{aligned} 3x_1 + 2x_2 &\leq 12 \\ x_1 &\geq 0 \\ x_2 &\geq 0 \end{aligned} \tag{8-8}$$

Step 1. Define a slack variable $s_1 \geq 0$ so that the first inequality is reduced to an equation:

$$3x_1 + 2x_2 + s_1 = 12 \tag{8-9}$$

with $x_1 \geq 0, x_2 \geq 0, s_1 \geq 0$

Step 2. Identify one vertex of the feasible region. Such vertices occur at the intersections of two boundary lines of the feasible region, namely, the points ($x_1 = 0, x_2 = 0$), ($x_1 = 0, s_1 = 0$), and ($x_2 = 0, s_1 = 0$); see also Fig. 8-3. The variables which identify such a vertex by virtue of their zero values, are called *nonbasic variables*. The remaining variables (s_1 in our case) are called *basic variables*. For example, let us begin at the vertex

$$x_1 = 0, x_2 = 0$$

At this point the remaining variable (slack variable s_1) and the objective function F are:

$$s_1 = 12 \qquad \text{(Eq. 8-9)}$$
$$F = 0 \qquad \text{(Eq. 8-7)}$$

The nonbasic variables are x_1 and x_2 and the basic variable is s_1.

Step 3. Express the basic variables (s_1 in our case) and the objective function F in terms of the nonbasic variables (i.e., x_1 and x_2 in the present example):

$$s_1 = 12 - 3x_1 - 2x_2 \qquad \text{(Eq. 8-9)}$$
$$F = 40x_1 + 100x_2 \qquad \text{(Eq. 8-7)}$$

Step 4. Move to a vertex adjacent to the vertex selected in Step 2. Considering each nonbasic variable separately, we establish:

(a) by how much it can increase within the feasible region.
(b) the increase of the objective function corresponding to the largest increase in the variable in (a).

On the basis of our findings in (a) and (b), we select the next vertex which is the one corresponding to the largest increase in F.

In the present example, from the equations for s_1 and F in terms of x_1 and x_2 in Step 3, we reason as follows:

Nonbasic Variable x_1:

(a) Since s_1 and x_2 are constrained to nonnegative values (Eq. 8-9), then, keeping $x_2 = 0$, the largest increase in x_1 is 4 units, at which point s_1 reaches its smallest possible value of zero, $s_1 = 0$. Verify.

(b) For a 4-unit increase in x_1, F increases by $40 \times 4 = 160$.

Nonbasic Variable x_2:

(a) Keeping $x_1 = 0$, the maximum increase for x_2 is 6 which again drives s_1 to zero, $s_1 = 0$. Verify.

(b) For a 6-unit increase in x_2, F increases by $100 \times 6 = 600$.

On the basis of the above values, it is more productive to move along the boundary $x_1 = 0$ to the vertex $x_1 = 0$ and $s_1 = 0$, because then the objective function increases by 600. Moving along the boundary $x_2 = 0$ to the vertex $x_2 = 0$, $s_1 = 0$ yields an increase of only 160 units for F.

The new *basic solution* (new vertex) is, therefore,

$$x_1 = 0,\ s_1 = 0, \text{ with } x_2 = 6 \text{ and } F = 600$$

The *nonbasic variables* are now x_1 and s_1, and the *basic variable* (nonzero) is x_2.

Step 5. Transform the objective function and each new basic variable* to linear functions of the nonbasic variables. From the equations in Step 3,

$$x_2 = 6 - \frac{3}{2}x_1 - \frac{1}{2}s_1$$

$$\begin{aligned} F &= 40x_1 + 100(6 - 1.5x_1 - 0.5s_1) \\ &= 600 - 110x_1 - 50s_1 \end{aligned}$$

If the function F in Step 5 has variables with positive coefficients, we return to Step 3 and repeat the process through Step 5 until, finally, all variables in the function F in Step 5 have negative coefficients. We then have the solution because any move away from this last vertex will be accompained by a decrease in the objective function. In the present example, Step 5 has F with negative coefficients for all variables; hence, the solution is at $x_1 = 0$, $s_1 = 0$; any move from there implies either an increase in x_1 or s_1, and in either case F gets smaller. Thus, the optimum is at $x_1 = 0$, $s_1 = 0$ for which $F = 600$.

NOTE. We selected a rather simple example; however, the basic five steps above apply to problems with n variables in which case a geometric solution is not possible. To gain further insight into the simplex algorithm, we shall use it to solve another example in the next section.

In the previous example we have three variables x_1, x_2, s_1 and one equation in terms of these variables, Eq. 8-9. Generally, in linear programming problems the number of variables n is larger than the numbers of equations m, $n > m$. A *basic solution*

*In the present case we have a single new basic variable. In general all new basic variables are transformed.

in such a situation is defined as a solution obtained by solving for m variables in terms of the remaining $n - m$ variables and then assigning the value zero to each of these remaining variables. The solved m variables are the *basic variables*, and the remaining $n - m$ variables set to zero are the *nonbasic variables*.

Question the Premises

The inequality

$$3x_1 + 2x_2 \leq 12$$

establishes a constraint in the form of the boundary $s_1 = 0$ in Fig. 8-3. In the spirit of the *will to question* or the *will to doubt*, let us ask: *What if. . .*? What if the constraint is changed, i.e., make s_1 negative so that the feasible region is increased? Suppose $s_1 = -3$. Then,

$$3x_1 + 2x_2 = 15$$

We can study how sensitive F is to such a change by inspecting Fig. 8-3. However, it is more useful to get this information from Step 5 of the simplex algorithm which applies in general, even when geometry can no longer be used as a model for the problem. In Step 5 we have

$$F = 600 - 110x_1 - 50s_1$$

For every unit decrease of s_1 (from zero), F increases by \$50, $(-50)(-1) = 50$. In other words, for each additional hour we keep the TV shop in operation over the present 12-hour limit, profit increases by \$50. This describes the sensitivity of F to a change of one unit in the constraint on the number of available shop hours. Now we need to compare the cost of keeping the shop open for another hour with the associated profit in order to decide on a course of action.

Problem. What will be the change in F if $3x_1 + 2x_2 \leq 12$ in Example 1 is changed to $2x_1 + 2x_2 \leq 12$?

8-4 LINEAR PROGRAMMING—AN APPLICATION TO DENTAL PRACTICE [69]

A dentist employs three assistants in his practice which uses two operatories. His services fall into two categories: pedodontic and general dentistry. Experience indicates that his resources are used as follows for each of the two services.

A pedodontic service requires:

0.75 hr of an operatory

1.50 hr of an assistant's time
0.25 hr of dentist's time

A general dentistry service requires:

0.75 hr of an operatory
1.00 hr of an assistant's time
0.50 hr of dentist's time

Net profit from each service is:

$8 for a pedodontic service
$6 for a general dental service

Constraints. The dentist and the three assistants each work an 8-hour day.

Problem Objective. The dentist wishes to use his time, the assistants' time, and the equipment in the most productive way to maximize profit.

Linear Programming Model. Let us use the following notation for compactness and ease of identification:

p = number of *pedodontic* services per day
g = number of *general* dentistry services per day
s_d = number of *unused dentist's* hours per day
s_a = number of *unused assistants'* hours per day
s_o = number of *unused operatory* hours per day

The statement of the problem takes the following form, using a linear programming model:

Maximize the objective function:

$$F = 8p + 6g \tag{8-10}$$

Subject to these constraints:

$$\begin{array}{ll} \text{8 dentist's hours} & 0.25p + 0.50g \leq 8 \\ \text{24 assistants' hours} & 1.50p + g \leq 24 \\ \text{16 operatory hours} & 0.75p + 0.75g \leq 16 \\ p \geq 0, \quad g \geq 0 & \end{array} \tag{8-11}$$

Using the slack variables s_d, s_a, s_o, we can write the constraints in the form

$$\begin{array}{lll} 0.25p + 0.50g + s_d = 8 & & \text{(a)} \\ 1.50p + g + s_a = 24 & & \text{(b)} \\ 0.75p + 0.75g + s_o = 16 & & \text{(c)} \\ p \geq 0, g \geq 0, s_d \geq 0, s_a \geq 0, s_o \geq 0 & & \end{array} \tag{8-12}$$

Solution by Graphical Method*

Figure 8-4 identifies the feasible region in terms of the constraints. The line $s_o = 0$, which represents the upper bound on operatory time, lies outside the feasible region and is therefore not an active bounding constraint. Thus, the solution will include some unused operatory time, because it occurs where $s_o > 0$. The objective function is plotted for $F = 24$. To increase F, the line is translated up as far as possible without completely leaving the feasible region. This occurs when F goes through the vertex ($s_a = 0$, $s_d = 0$) for which $p = 8$ and $g = 12$.

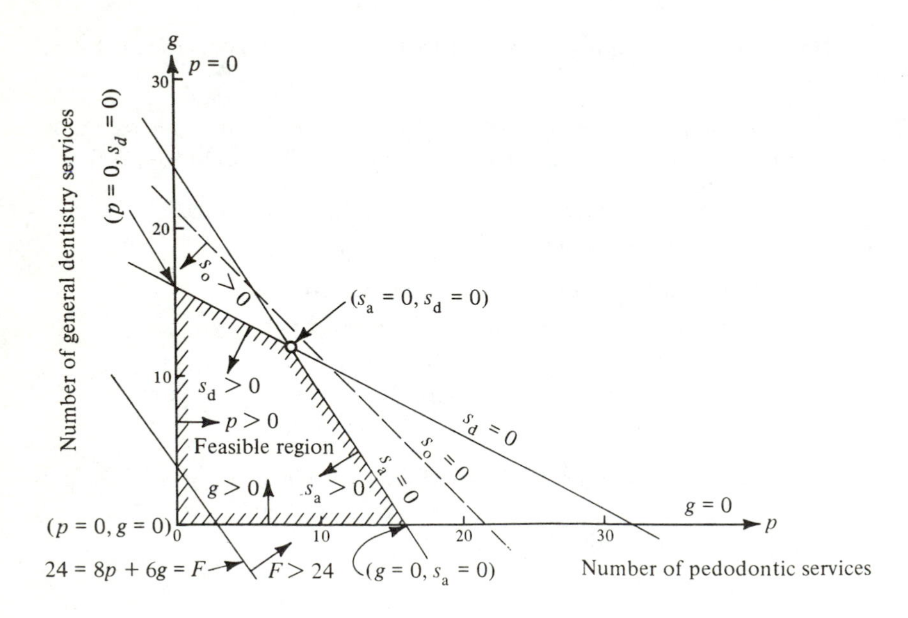

Figure 8-4

The corresponding value for F is:

$$F = (8 \times 8) + (6 \times 12) = \$136$$

Problem. Change operatory time to 8/9 of an hour per each type of service, and solve the problem.

---

*Some linear programming problems can be transformed to the form of a game insofar as the computations are concerned. See *Graphical Solution of Games* in Sec. 7-8 to note the possible analogy between game theory and linear programming. For an example of a linear programming problem brought to the form of a game (a diet problem), see [64], pp. 210–213.

Solution by the Simplex Algorithm*

Step 1. Define slack variables s_d, s_a, s_o, yielding Eq. 8-12.

Step 2. Select the vertex ($p = 0$, $g = 0$) with p and g as the nonbasic variables. At this vertex we have (see Eqs. 8-10 and 8-12):

$$p = 0, g = 0, s_d = 8, s_a = 24, s_o = 16, F = 0$$

Step 3. Using Eqs. 8-12 and 8-10, express the basic variables and F in terms of the nonbasic variables. From Eqs. 8-12 and 8-10:

$$\left.\begin{aligned} s_d &= 8 - 0.25p - 0.50g && \text{(a)} \\ s_a &= 24 - 1.50p - 1.00g && \text{(b)} \\ s_o &= 16 - 0.75p - 0.75g && \text{(c)} \\ F &= 8p + 6g && \text{(d)} \end{aligned}\right\} \qquad (8\text{-}13)$$

$$p \geq 0, g \geq 0, s_d \geq 0, s_a \geq 0, s_o \geq 0$$

Step 4. Move to the vertex adjacent to vertex ($p = 0$, $g = 0$) for which we realize the largest increase for F. This move is accomplished by studying the changes in F corresponding to the maximum possible changes in the nonbasic variables p and g, i.e., moving away from vertex ($p = 0$, $g = 0$).

Nonbasic Variable p:

(a) From Eq. 8-13, if we set $g = 0$, p can change by 32 in (a), by 16 in (b), and by $16 \times 4/3$ in (c). Therefore, we can change p by 16 units at most without violating the nonnegativity of s_d, s_a, s_o. For $p = 16$: $s_d = 4$, $s_a = 0$, $s_o = 4$.

(b) For a 16-unit change in p, F increases by $8 \times 16 = 128$.

Nonbasic Variable g:

(a) From Eq. 8-13, if we set $p = 0$, g can change by 16 in (a), by 24 in (b), and by $16 \times 4/3$ in (c). Therefore, a change in g by at most 16 units is possible without violating the constraints on s_d, s_a, and s_o. For $g = 16$: $s_d = 0$, $s_a = 8$, $s_o = 4$.

(b) For a 16-unit change in g, F increases by $6 \times 16 = 96$.

Conclusion. Move along the boundary $g = 0$, i.e., change p because the corresponding increase in F is larger than if g is changed. Hence, we move from vertex ($p = 0$, $g = 0$) to ($g = 0$, $s_a = 0$). The nonbasic variables are now $g = 0$, $s_a = 0$ with the remaining variables: $s_d = 4$, $s_o = 4$, and $F = 128$.

---

*The solution procedure here represents an *idealized version* of the simplex algorithm. Here we look for the variable which leads to the largest increase in F. In the *practical version,* the variable with largest coefficient is selected (p here because $8 > 6$), or the variable which leads to the largest *rate* of increase in F (i.e., largest increase in F per unit change in variable). This reduces considerably the computations when many variables are involved.

Step 5. Transform the objective function and the basic variables to linear functions of the nonbasic variables g and s_a. This can be done by an elimination process, working with Eqs. 8-12 and 8-10.

From Eq. 8-12(b): $p = 16 - \frac{2}{3}g - \frac{2}{3}s_a$

Substituting for p in Eq. 8-12(a) and solving for s_d,

$$s_d = 8 - \frac{1}{4}\left(16 - \frac{2}{3}g - \frac{2}{3}s_a\right) - \frac{1}{2}g = 4 - \frac{1}{3}g + \frac{1}{6}s_a$$

Substituting for p in Eq. 8-12(c) and solving for s_o,

$$s_o = 16 - 0.75\left(16 - \frac{2}{3}g - \frac{2}{3}s_a\right) - 0.75\,g = 4 - \frac{1}{4}g + \frac{1}{2}s_a$$

Substituting for p in Eq. 8-10,

$$F = 8\left(16 - \frac{2}{3}g - \frac{2}{3}s_a\right) + 6g = 128 + \frac{2}{3}g - \frac{16}{3}s_a$$

Rewriting these equations compactly, we have:

$$\left.\begin{aligned} p &= 16 - \frac{2}{3}g - \frac{2}{3}s_a && \text{(a)} \\ s_d &= 4 - \frac{1}{3}g + \frac{1}{6}s_a && \text{(b)} \\ s_o &= 4 - \frac{1}{4}g + \frac{1}{2}s_a && \text{(c)} \\ F &= 128 + \frac{2}{3}g - \frac{16}{3}s_a && \text{(d)} \end{aligned}\right\} \qquad (8\text{-}14)$$

We have now the constraints and the objective function in terms of the nonbasic variables g and s_a. We now repeat Step 4, using Eqs. 8-14 in search for the next vertex from $(g = 0, s_a = 0)$.

Nonbasic Variable g:

(a) From Eq. 8-14, if we set $s_a = 0$, g can change by 24 in (a), by 12 in (b) and by 16 in (c). Hence, 12 is the maximum change. For $g = 12$: $p = 8$, $s_d = 0$, $s_o = 1$.

(b) For a 12-unit change in g with $s_a = 0$,

$$F = 128 + \frac{2}{3}(12) = 136.$$

Nonbasic Variable s_a:

(a) From Eq. 8-14, if we set $g = 0$, s_a can change by 24 in (a), by any positive value in (b) and (c). Thus, 24 is the maximum change for s_a. For $s_a = 24$: $p = 0$, $s_d = 8$, $s_o = 16$.

(b) For a 24-unit change in s_a with $g = 0$,

$$F = 128 - \frac{16}{3}(24) = 0$$

Hence, we change nonbasic variable g. This could be concluded by inspection from Eq. 8-14(d) because s_a has a negative coefficient. Since s_a and g must be positive, it is most advantageous to keep $s_a = 0$, and increase g in order to increase F.

The new vertex is, therefore, that which leads to $F = 136$, i.e., $s_a = 0$, $s_d = 0$.

Next, we repeat Step 5 beginning with a transformation of Eqs. 8-14 to express the basic variables and F in terms of the new nonbasic variables s_a and s_d. From Eq. 8-14(b), we express g in terms of s_a and s_d, and then, substituting in equations (a), (c), and (d), we obtain these transformed equations:

$$\left.\begin{aligned} g &= 12 + \frac{1}{2}s_a - 3s_d && \text{(a)} \\ p &= 8 - s_a + 2s_d && \text{(b)} \\ s_o &= 1 + \frac{3}{8}s_a + \frac{3}{4}s_d && \text{(c)} \\ F &= 136 - 5s_a - 2s_d && \text{(d)} \end{aligned}\right\} \qquad (8\text{-}15)$$

The objective function in Eq. 8-15(d) has negative coefficients for both nonbasic variables s_a and s_d. Therefore, since these variables must assume positive values in the model of the problem, any move from the current vertex ($s_a = 0$, $s_d = 0$) will cause a decrease in F. Hence, this vertex is the optimum. At this vertex,

$$p = 8,\ g = 12,\ s_o = 1,\ \text{and}\ F = 136$$

Equation 8-15(d) also tells us how sensitive F is to a change in the constraints. For $s_a = -1$, namely, for each additional assistants' hour, the profit F increases by \$5; and for $s_d = -1$ (each additional dentist's hour), F increases by \$2. It may be productive for our problem-solving-minded dentist to consider the possibility of lifting some constraint, such as increasing assistant help, and considering the trade off between the cost of such a move and the corresponding increase in profit.

Problem. How many hours of assistants' time can the dentist add before the operatory hours become a boundary to the feasible region?

8-5 LINEAR PROGRAMMING—GENERALIZATION OF METHOD

The examples of Secs. 8-3 and 8-4 will now serve as a basis for generalizing the method of linear programming.

Linear programming was developed during World War II. Dantzig [136] is credited with the prime contribution in the development of the method in a form that led to its widespread use in government, industry, agriculture, and the military. Applications cover transportation, water resource management, diet planning, personnel allocation, building design, and many more. The basic building blocks of a linear-programming model are:

1. A *linear* input-output relationship must exist between the available resources (input) and the products or activation (output). In the example of Sec. 8-4, the resources, input, were the time of dentist, assistants, and operatories. The products, output, were the pedodontic and general dental services.
2. A *linear* objective function must exist, which constitutes a measure of performance that is to be maximized (or minimized). In Sec. 8-4 this was the profit F as a linear function of output of dental services.

The linear programming model represents a problem in which limited resources must be allocated to alternative products or activities in a way that will maximize or minimize* the performance criterion. In general, the model takes the form:

Maximize (or minimize) the objective function:

$$F = \sum_{j=1}^{n} c_j x_j \tag{8-16}$$

Subject to the constraints:

$$\begin{aligned}
a_{11}x_1 + a_{12}x_2 + \dots + a_{1n}x_n &\leq b_1 \\
a_{21}x_1 + a_{22}x_2 + \dots + a_{2n}x_n &\leq b_2 \\
\vdots \qquad\qquad\qquad & \quad \vdots \\
a_{i1}x_1 + a_{i2}x_2 + \dots + a_{in}x_n &\leq b_i \\
\vdots \qquad\qquad\qquad & \quad \vdots \\
a_{m1}x_1 + a_{m2}x_2 + \dots + a_{mn}x_n &\leq b_m \\
\text{with } x_j \geq 0 \quad \text{for all } j &
\end{aligned} \tag{8-17}$$

The coefficients b_i represent the limitation of available resources.

The solution consists of finding the optimal mix of the variables x_j, which are at the control of the decision maker, so as to maximize or minimize F by using the available resources. Therefore, variables x_j are referred to as *decision* or *control variables.*

The solution lies at an extreme point, or an intersection, of the boundaries which identify the feasible region. This is so because the linear-pro-

*The performance criterion may be such that the objective is to minimize it, say, cost instead of profit.

gramming solution applies to a feasible region identified as a convex set. Figure 8-5 shows examples of convex and nonconvex sets, respectively. Geometrically, a set of points in a plane is *convex* if all points on a straight line segment joining any two points of the set belong to the set.

The solution to a problem, such as shown graphically in Fig. 8-6, can be obtained algebraically without resorting to geometry by finding the

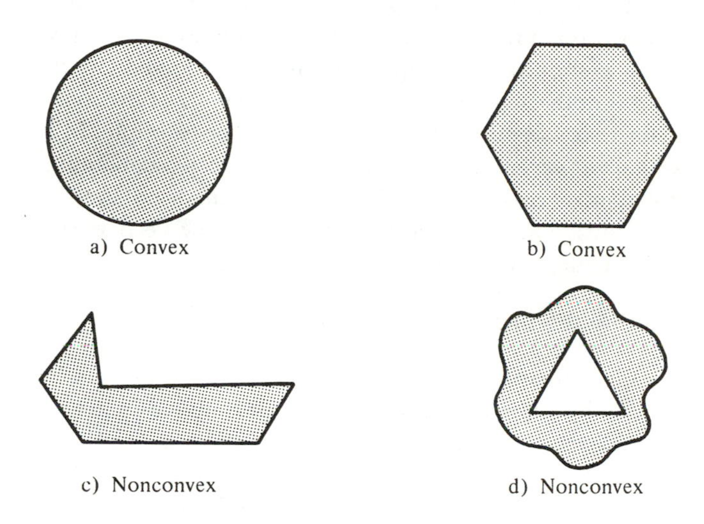

Figure 8-5 Convex and Nonconvex Sets: (a) Convex; (b) Convex; (c) Nonconvex; (d) Nonconvex

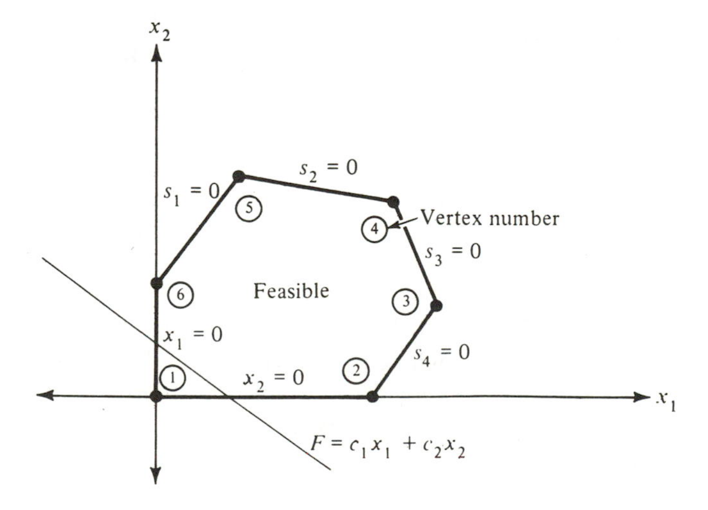

Figure 8-6

values of x_1 and x_2 at the vertices, inspecting the corresponding value of the objective function F at each vertex, and selecting that which yields the maximum F. The figure shows four slack variables s_1, s_2, s_3, and s_4 in addition to the variables x_1 and x_2. The objective function is linear in x_1 and x_2. While the above process to search for the optimum is possible, it requires that *each* vertex be identified in terms of x_1 and x_2 and the corresponding value of F be determined. The identification of each vertex requires the solution of two equations in two unknowns (intersection of two lines). For a problem with n variables x_j, n simultaneous equations must be solved to identify each extreme point (another name for a vertex). The power of the *simplex algorithm* lies in the fact that it reduces the search effort. For example, in Fig. 8-6 the search may start at vertex ① then proceed to ② or ⑥, depending on which results in a larger increase in F. This decision is made in the simplex algorithm. For a solution at ④, the algorithm will follow vertices ①, ②, ③, ④ or ①, ⑥, ⑤, ④ depending on the decision at ①.

* Formalizing the Simplex Algorithm

The simplex algorithm employed in Secs. 8-3 and 8-4 can be presented more compactly by using a tableau that is manipulated in accordance with the following instructions which are essentially the same as those developed in detail in Sec. 8-4. To be explicit, we refer again to the example of Sec. 8-4; Eqs. 8-10 and 8-12 are rewritten here in the form of Fig. 8-7.

	p	g	s_d	s_a	s_o		
Row 1	0.25	0.50	1	0	0	8	← Eq. 8-12(a)
Row 2	(1.50)	1.00	0	1	0	24	← Eq. 8-12(b)
Row 3	0.75	0.75	0	0	1	16	← Eq. 8-12(c)
Row 4	−8	−6	0	0	0	0	← Eq. 8-10

Position for value of F

Figure 8-7 Tableau 1

Rows 1, 2, 3 represent Eq. 8-12. Each coefficient in the table multiplies the corresponding variable in the column above it. The last column consists of the free coefficients. The last row represents the objective function written as $-8p - 6g + F = 0$. The value for F is in the last column of row 4. Initially, this position is assigned the value of zero. The initial tableau consists then of one column for each control variable (p and g here) and each slack variable, and one column for the free constants representing the limits on resources. Each row represents one equation of constraint, and the last row is the objective function.

The numbers in the last row (except the position for the value of F) are called *indicators*. The *simplex algorithm* proceeds with the tableau of Fig.

8-7 as follows:

Step 1. Select any column with a negative indicator. Suppose we choose the first column. Determine the ratios 8/0.25, 24/1.50, 16/0.75, i.e., elements of the last column divided by corresponding elements in the respective rows in the first column. The results are 32, 16, and 64/3, which are identical with the numbers obtained at the beginning of Step 4 in Sec. 8-4. Of these, select the smallest, i.e., 16, which corresponds to the coefficient 1.50 in the first column. We circle this number in the Tableau of Fig. 8-7 and call it the *pivot.*

Step 2. Factor out the value 1.50 from each number in the second row, yielding a new second row.

New row 2: $1 \quad \frac{2}{3} \quad 0 \quad \frac{2}{3} \quad 0 \quad 16$

This row is now used to generate a new form for all other rows in the tableau in such a way that they will each have a zero value in the first column. To do this we proceed as follows. To reduce to zero the value 0.25 in the first row, we multiply new row 2 by 0.25 and subtract it term by term from row 1:

Original row 1:	0.25	0.50	1	0	0	8
0.25(New row 2):	0.25	$\frac{1}{6}$	0	$\frac{1}{6}$	0	4
Subtract						
New row 1:	0	$\frac{1}{3}$	1	$-\frac{1}{6}$	0	4

Similarly, for new row 3:

Original row 3:	0.75	0.75	0	0	1	16
0.75(New row 2):	0.75	$\frac{1}{2}$	0	$\frac{1}{2}$	0	12
Subtract						
New row 3:	0	$\frac{1}{4}$	0	$-\frac{1}{2}$	1	4

For new row 4:

Original row 4:	−8	−6	0	0	0	0
−8(New row 2):	−8	$-\frac{16}{3}$	0	$-\frac{16}{3}$	0	−128
Subtract						
New row 4:	0	$-\frac{2}{3}$	0	$\frac{16}{3}$	0	128

The new tableau is shown in Fig. 8-8. The current value of $F = 128$ is shown in the lower right corner. The new nonbasic variables are now g and s_a.

	p	g	s_d	s_a	s_o	
New Row 1	0	$\frac{1}{3}$ (circled)	1	$-\frac{1}{6}$	0	4
2	1	$\frac{2}{3}$	0	$\frac{2}{3}$	0	16
3	0	$\frac{1}{4}$	0	$-\frac{1}{2}$	1	4
4	0	$-\frac{2}{3}$	0	$\frac{16}{3}$	0	128

Figure 8-8 Tableau 2

Step 3. Repeat steps 1 and 2, beginning with Tableau 2 of Fig. 8-8. Namely, select a column with a negative indicator, i.e., column 2. Compute the ratios 4/(1/3), 16/(2/3), 4/(1/4) and find the smallest, 4/(1/3); 1/3 becomes the pivot (circled in figure). Now factor out 1/3 from the first row and use this new row 1 to generate a new form for all other rows so that each will have a zero element in the column containing the pivot element:

New row 1: 0 1 3 $-\frac{1}{2}$ 0 12

Calculations for new row 2:

Row 2 from Tableau 2:	1	$\frac{2}{3}$	0	$\frac{2}{3}$	0	16
2/3(New row 1):	0	$\frac{2}{3}$	2	$-\frac{1}{3}$	0	8
Subtract						
	1	0	−2	1	0	8

Calculations for new row 3:

Row 3 from Tableau 2:	0	$\frac{1}{4}$	0	$-\frac{1}{2}$	1	4
1/4(New row 1):	0	$\frac{1}{4}$	$\frac{3}{4}$	$-\frac{1}{8}$	0	3
Subtract						
	0	0	$-\frac{3}{4}$	$-\frac{3}{8}$	1	1

Calculations for new row 4:

Row 4 from Tableau 2:	0	$-\frac{2}{3}$	0	$\frac{16}{3}$	0	128
−2/3(New row 1):	0	$-\frac{2}{3}$	−2	$\frac{1}{3}$	0	−8
Subtract						
	0	0	2	5	0	136

The results are shown in Tableau 3 of Fig. 8-9. Since all indicators are positive, the solution has been reached with the maximum value of $F = 136$ shown in the lower right corner. The corresponding values for the variables

p	g	s_d	s_a	s_o	
0	1	3	$-\frac{1}{2}$	0	12
1	0	-2	1	0	8
0	0	$-\frac{3}{4}$	$-\frac{3}{8}$	1	1
0	0	2	5	0	136

Figure 8-9 Tableau 3

p, g, and s_o, which have zero coefficients in the last row, appear in the last column on the right. For p, look under the column for p, find the nonzero element 1, and proceed to the rightmost element in the row which contains it, namely, 8. Similarly, for g we have 12, and for s_o, 1. The values 2 and 5 in the last row below s_d and s_a, respectively, represent the sensitivity of F to a change in the contraints on dentist's time and assistants' time. See last paragraph of Sec. 8-4.

The procedure outlined above can be followed for large problems with many variables and a large number of constraints. Computer programs for the simplex algorithm have been developed with a great deal of efficiency and economy in conducting the calculations.

* NOTE. The reader familiar with linear algebra may find it of interest to note that the above transformations from tableau 1 to tableau 2 and then to 3 are essentially equivalent to an elimination process to solve for p and g in terms of s_a and s_d with s_a and s_d finally set equal to zero. p and g are the basic variables and s_a, s_d the nonbasic variables in the final solution. (Variable $s_o = 1$, because $s_o = 0$ is not an active constraint, see Fig. 8-4).

This solution for p and g in terms of s_a and s_d can be identified in the tableaus as follows. If rows 1 and 2 in tableau 1 of Fig. 8-7 are interchanged and then columns 3 and 4 are interchanged, the tableau takes the form shown in Fig. 8-10.

Tableau 1

p	g	s_a	s_d	s_o	
(1.50)	1	1	0	0	24
0.25	0.50	0	1	0	8
0.75	0.75	0	0	1	16
-8	-6	0	0	0	0

Figure 8-10

The tableaus of Figs. 8-8 and 8-9 (using the same interchange of rows and columns) then take the forms shown in Figs. 8-11 and 8-12.

Tableau 2

p	g	s_a	s_d	s_o	
1	$\frac{2}{3}$	$\frac{2}{3}$	0	0	16
0	$\frac{1}{3}$ (circled)	$-\frac{1}{6}$	1	0	4
0	$\frac{1}{4}$	$-\frac{1}{2}$	0	1	4
0	$-\frac{2}{3}$	$\frac{16}{3}$	0	0	128

Figure 8-11

Tableau 3

p	g	s_a	s_d	s_o	
1	0	1	−2	0	8
0	1	$-\frac{1}{2}$	3	0	12
0	0	$-\frac{3}{8}$	$-\frac{3}{4}$	1	1
0	0	5	2	0	136

Figure 8-12

The 2×2 identity matrix in the upper left corner of the last tableau was originally occupied by the 2×2 matrix

$$\begin{bmatrix} 1.50 & 1 \\ 0.25 & 0.50 \end{bmatrix}$$

in the first tableau. The computations yielded the inverse of this matrix, which is separated by the dashed lines in the last tableau, namely,

$$\begin{bmatrix} 1.50 & 1 \\ 0.25 & 0.5 \end{bmatrix} \begin{bmatrix} 1 & -2 \\ -0.5 & 3 \end{bmatrix} = \begin{bmatrix} 1 & 0 \\ 0 & 1 \end{bmatrix}$$

Thus, through pivoting (elimination), p and g were solved in terms of s_a and s_d so that F was transformed to a linear function of s_a and s_d from a function of p and g.

Concluding Remarks—Question the Premises

It is important to emphasize again that a problem represented by linear programming is a *model*. As such it suffers from the errors of omission and commission discussed in Chapter 5. The model may, in some cases, represent a strong motivation to be efficient, or minimize costs or use of resources, in a portion of a total operation or in a subsystem. In such cases, great caution must be exercised. The wrong problem may be solved because of errors of omission, i.e., the rest of the system relevant to the problem at hand. Churchman [70] relates the story of the efficiency-minded manager who decided to minimize the cost of transporting material from the factories of a company to its warehouses. Linear programming was applied, resulting in a small saving of several thousands of dollars which hardly covered the cost of the analysis. Further study, in depth, which doubted the basic premise that minimizing transportation costs was the parameter most relevant and sensitive to the efficient operation of the company, led to saving of millions of dollars. When the total system was viewed and a model constructed, it appeared to consist of materials going into factories, from the factories to warehouses, from there to retail outlets, and finally to the customer. It became apparent that the existing constraints (company policies) on the quantities stored in the warehouses were hurting the total system. When these constraints, which were employed in the earlier linear programming model of the subsystem (factory-to-warehouses transportation), were lifted, it became most productive to ship certain items to some warehouses which before had never received them, and to stop the shipping of certain items to others. Minimization of cost to the company's operation required that the expenditure for transportation between factories and warehouses be increased rather than decreased.

In a sense, this illustrates that an optimization on a local basis from a subsystem point of view may prove wrong from the total system point of view.

* The Dual

To every linear programming problem there corresponds a *dual problem*. The original problem is referred to as the *primal problem*. If the primal problem is a maximization problem with the equations of constraint using $\leq$ in inequalities, then the *dual problem* is a minimization problem with equations of constraint using $\geq$ in inequalities, and vice versa. It is, therefore, arbitrary as to which is to be labeled the primal and which the dual.

In compact form, using matrix notation, we can write

Primal Problem: (decision variables x_i)

$$\max F = \{c\}^{\mathrm{T}}\{x\} \qquad \text{subject to } [A]\{x\} \leq \{b\},\ \{x\} \geq \{0\}$$

Dual Problem: (decision variables y_i)

$$\min G = \{y\}^{\mathrm{T}}\{b\} \qquad \text{subject to } \{y\}^{\mathrm{T}}[A] \geq \{c\}^{T},\ \{y\} \geq \{0\}$$

For example, Eqs. 8-10 and 8-11 take the following form of a *primal problem*

(p and g are replaced by symbols x_1 and x_2, respectively):

$$\max F = \{8, 6\}\begin{Bmatrix} x_1 \\ x_2 \end{Bmatrix}$$

$$\text{subject to} \begin{bmatrix} 0.25 & 0.50 \\ 1.50 & 1.00 \\ 0.75 & 0.75 \end{bmatrix}\begin{Bmatrix} x_1 \\ x_2 \end{Bmatrix} \leq \begin{Bmatrix} 8 \\ 24 \\ 16 \end{Bmatrix}$$

$$\begin{Bmatrix} x_1 \\ x_2 \end{Bmatrix} \geq \begin{Bmatrix} 0 \\ 0 \end{Bmatrix}$$

The *dual problem* is:

$$\min G = \{y_1 \quad y_2 \quad y_3\}\begin{Bmatrix} 8 \\ 24 \\ 16 \end{Bmatrix}$$

$$\text{subject to } \{y_1 \quad y_2 \quad y_3\}\begin{bmatrix} 0.25 & 0.50 \\ 1.50 & 1.00 \\ 0.75 & 0.75 \end{bmatrix} \geq \{8, 6\}$$

$$\begin{Bmatrix} y_1 \\ y_2 \\ y_3 \end{Bmatrix} \geq \begin{Bmatrix} 0 \\ 0 \\ 0 \end{Bmatrix}$$

When the simplex algorithm is employed in the solution of the primal problem, the coefficients of the slack variables in the final form of the objective function are the optimal values of the dual variables with the signs reversed. The variables y_i above are the slack variables s_d, s_a, and s_o of the primal problem in See 8-4. Both the primal and dual problems lead to the same solution. Therefore, *the optimum values of the dual variables represent a measure of sensitivity of the objective function F to a unit change in the constants b_i of the equations of constraints in the primal problem.*

Example of a Primal Problem and Its Dual:

Consider the primal problem:

$$\max F = 3x_1 + 5x_2$$
$$\text{subject to } 6x_1 + 8x_2 \leq 48$$
$$5x_1 + 10x_2 \leq 50$$
$$x_1 \geq 0, x_2 \geq 0$$

In matrix notation this *primal problem* is

$$\max F = \{3, 5\}\begin{Bmatrix} x_1 \\ x_2 \end{Bmatrix}$$

$$\text{subject to} \begin{bmatrix} 6 & 8 \\ 5 & 10 \end{bmatrix}\begin{Bmatrix} x_1 \\ x_2 \end{Bmatrix} \leq \begin{Bmatrix} 48 \\ 50 \end{Bmatrix}, \quad \begin{Bmatrix} x_1 \\ x_2 \end{Bmatrix} \geq \begin{Bmatrix} 0 \\ 0 \end{Bmatrix}$$

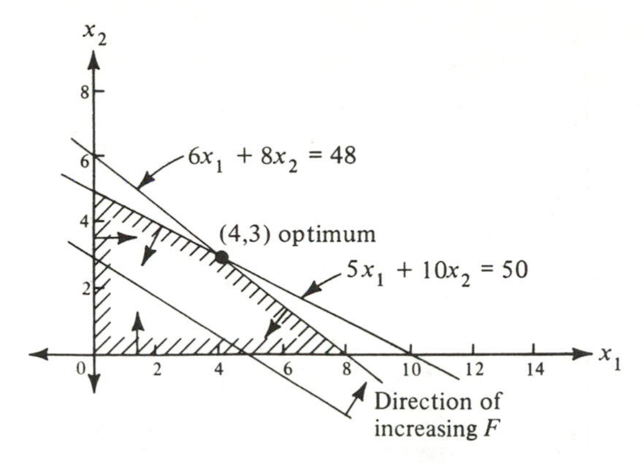

Figure 8-13 Graphical Solution to Primal Problem

The graphical solution to this primal problem is shown in Fig. 8-13. Thus the primal solution is $x_1 = 4$, $x_2 = 3$; therefore,

$$\max F = 3(x_1 = 4) + 5(x_2 = 3) = 27$$

Now consider the dual to the primal problem of Fig. 8-13.

Dual problem:

$$\min G = \{y_1 \ y_2\}\begin{Bmatrix} 48 \\ 50 \end{Bmatrix} = 48y_1 + 50y_2$$

$$\text{subject to } \{y_1 \quad y_2\}\begin{bmatrix} 6 & 8 \\ 5 & 10 \end{bmatrix} \geq \{3, 5\}, \quad \begin{Bmatrix} y_1 \\ y_2 \end{Bmatrix} \geq \begin{Bmatrix} 0 \\ 0 \end{Bmatrix}$$

$$\text{or} \quad 6y_1 + 5y_2 \geq 3$$
$$8y_1 + 10y_2 \geq 5$$
$$y_1 \geq 0, \ y_2 \geq 0$$

The graphical solution to this dual problem is shown in Fig. 8-14. The values $y_1 = 0.25$ and $y_2 = 0.30$ at the optimum represent the sensitivity of F to unit changes in $b_1 = 48$ and $b_2 = 50$, as shown by the following two examples.

EXAMPLE 1. Suppose b_1 is changed by 12 units from 48 to 60. This will change the feasible region so that the optimum is at (10,0),

$$\max F = 3(x_1 = 10) + 5(x_2 = 0) = 30$$

and we have a 3-unit increase in F. But $y_1 = 0.25$ represents the increase in F due to a unit increase in b_1, and since b_1 was increased by 12 units, the total corresponding increase in F should be $0.25 \times 12 = 3$, which agrees with the graphical solution.

EXAMPLE 2. Suppose b_2 is changed from 50 to 60. The optimum is then at (0, 6) with

$$\max F = 3(x_1 = 0) + 5(x_2 = 6) = 30$$

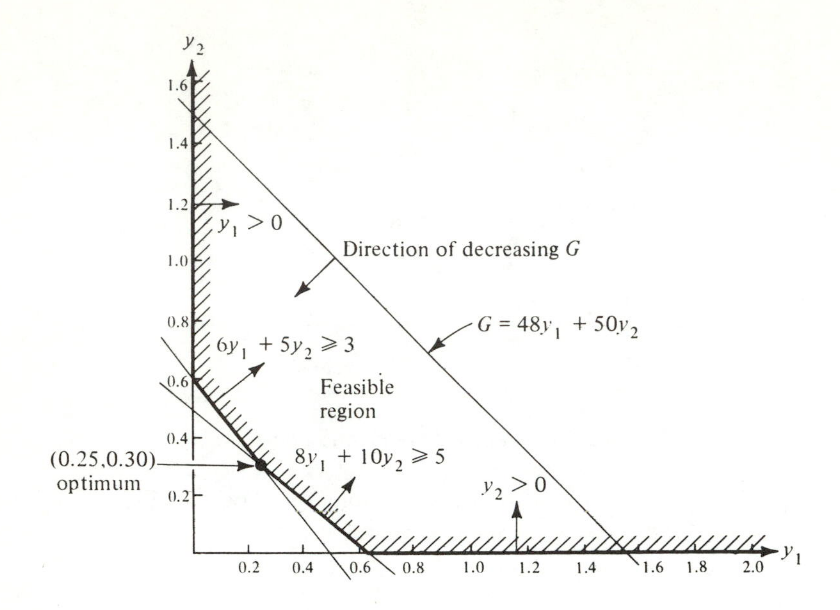

Figure 8-14 Graphical Solution to Dual Problem

Again, a 3-unit increase (merely a coincidence). But $y_2 = 0.3$ represents 0.3 unit of increase in F for each unit increase in b_2. For b_2 increased by 10, $0.3 \times 10 = 3$, which agrees with the graphical solution.

∗ **Problem 1.** Solve the problem of Fig. 8-13 by using the simplex algorithm.

∗ **Problem 2.** Solve the problem of Fig. 8-14 by using the simplex algorithm.

∗ **Problem 3.** Compare the solutions above in problems 1 and 2, and discuss the significance of the slack variables in each solution.

8-6 NONLINEAR PROGRAMMING

Nonlinear programming models are similar in form to linear programming models, except that at least one of the equations in the model, such as an equation of constraint or the objective function, is a nonlinear function of the control or decision variables x_j. Three examples illustrate nonlinear programming models.

EXAMPLE 1 (Nonlinear objective function). Minimize the objective function

$$f = (x_1 - 20)^2 + (x_2 - 12)^2$$

subject to the constraints

$$3x_1 + 2x_2 \leq 48,$$
$$x_1 + 2x_2 \leq 32,$$

or

$$3x_1 + 2x_2 + s_1 = 48$$
$$x_1 + 2x_2 + s_2 = 32$$
$$x_1 \geq 0, x_2 \geq 0, s_1 \geq 0, s_2 \geq 0$$

The feasible region and the objective function F are shown in Fig. 8-15. The constraints are linear, but the objective function is nonlinear and has the form of a circle. For $x_1 = 20$ and $x_2 = 12$, the circle degenerates to a point where $F = 0$. As the radius of the circle centered at (20, 12) gets larger, F gets larger because it is equal to the square of the radius. The smallest value for F which does not violate the constraints occurs at the point where the circle, representing the function F, makes first contact with the feasible region. In Fig. 8-15 this occurs at the point where $s_1 = 0$ is tangent to the circle. We leave it as an exercise to the reader to compute the optimum value of F and the corresponding values of x_1 and x_2. Also, how sensitive is the optimum F to an increase in one unit of s_1 from $s_1 = 0$? Why doesn't the optimum F change by the same amount when s_1 changes from $s_1 = 1$ to $s_1 = 2$, as it does when $s_1 = 0$ changes to $s_1 = 1$?

EXAMPLE 2 (Nonlinear constraint). Maximize the objective function

$$F = 4x_1 + 3x_2$$

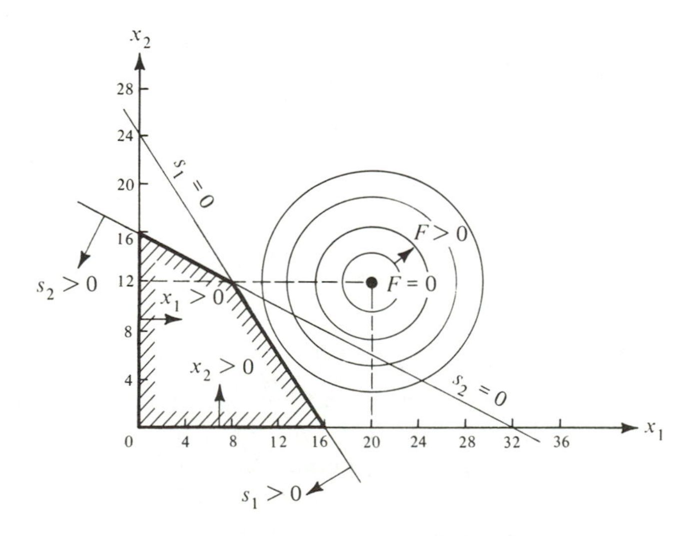

Figure 8-15 Nonlinear Objective Function and Linear Constraints

subject to the constraints

$$(x_1 - 20)^2 + (x_2 - 10)^2 \leq 25$$

or

$$(x_1 - 20)^2 + (x_2 - 10)^2 + s_1 = 25$$

$$x_1 \geq 0,\ x_2 \geq 0,\ s_1 \geq 0$$

The geometric model for this example is shown in Fig. 8-16. Find the optimum solution, i.e., x_1 and x_2 which maximize F.

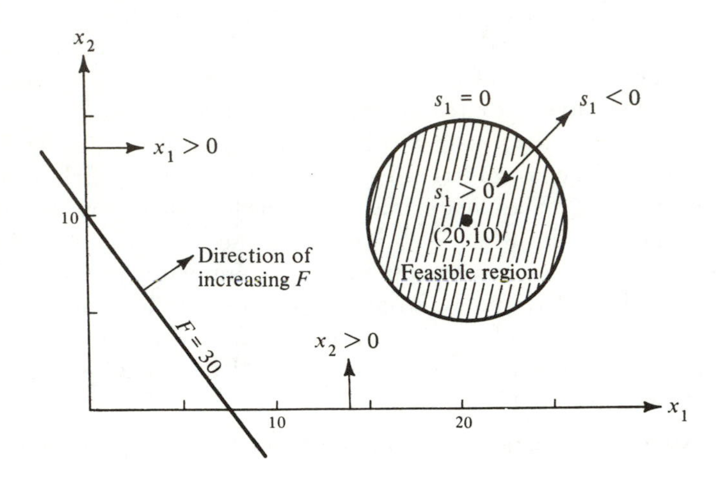

Figure 8-16 Linear Objective Function and Nonlinear Constraint

EXAMPLE 3 (Nonlinear objective function and one nonlinear equation of constraint). Maximize the objective function

$$F = x_1^2 + x_2^2$$

subject to the constraints

$$-(x_1 - 3)^2 + x_2 \leq 7$$

$$x_1 \leq 5$$

or

$$-(x_1 - 3)^2 + x_2 + s_1 = 7$$

$$x_1 + s_2 = 5$$

$$x_1 \geq 0,\ x_2 \geq 0,\ s_1 \geq 0,\ s_2 \geq 0$$

* NOTE. The choice of nonlinear functions in the form of circles in Examples 1 and 2 is merely a matter of convenience in showing graphical models. In general, the nonlinear function may assume a variety of forms. For instance, in Fig. 8-17 we have a linear objective function and nonlinear constraints. The figure represents a graphical solution to the following design problem [71]: synthesize a truss, i.e., establish the cross-sectional areas

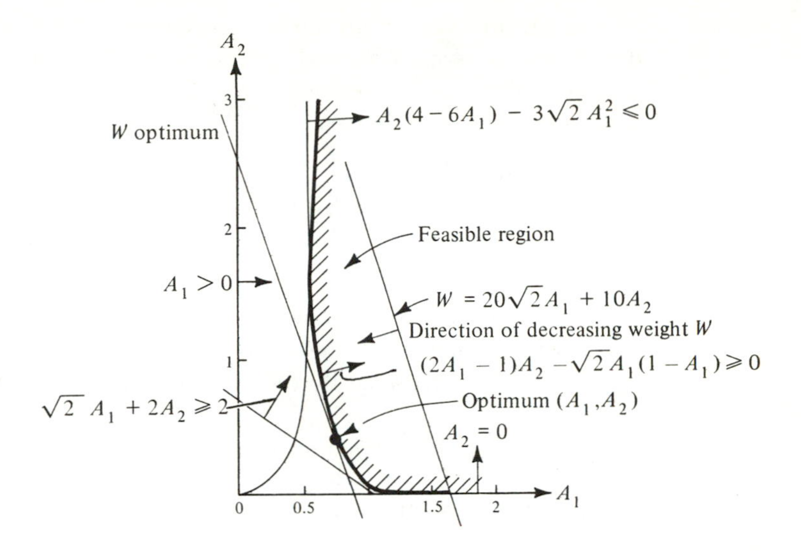

Figure 8-17 Graphical Solution to Truss Design (Area Assignment; Truss Shown in Fig. 8-18)

A_1, A_2, A_3 for the three members under the load conditions shown in Fig. 8-18. Because of symmetry of the loading conditions, design variables A_1 and A_3 are identical.

The objective function is linear and it represents the weight W of the truss,

$$W = 2\gamma l_1 A_1 + \gamma l_2 A_2$$

in which γ is the material density and l_1, l_2 are the lengths of each of the inclined members and the vertical member, respectively. For $\gamma = 1$, $l_1 = \sqrt{2}\,10$ in., $l_2 = 10$ in. we have

$$W = 20\sqrt{2}\,A_1 + 10A_2 \tag{8-18}$$

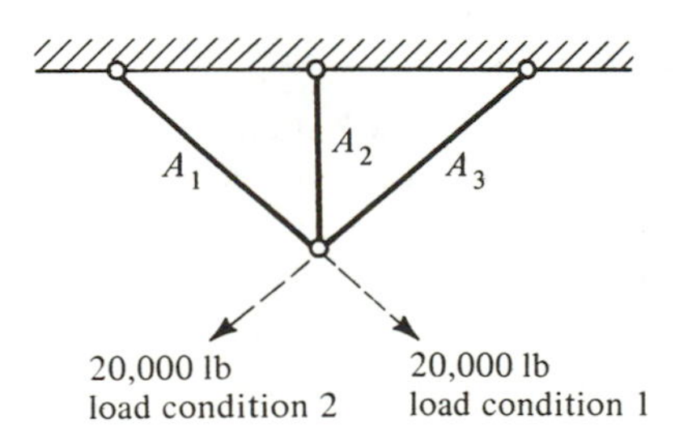

Figure 8-18 Truss Subject to Two Load Conditions

The constraints are not all linear and represent the limits on induced stress in terms of allowable stress so as to prevent instability, or failure of the truss members (no details are given here on how these equations were derived):

$$\begin{aligned} &\sqrt{2}\,A_1 + 2A_2 \geq \sqrt{2} && \text{(a)} \\ &A_2(4 - 6A_1) - 3\sqrt{2}\,A_1^2 \leq 0 && \text{(b)} \\ &(2A_1 - 1)A_2 - \sqrt{2}\,A_1(1 - A_1) \geq 0 && \text{(c)} \\ &A_1 \geq 0 \qquad A_2 \geq 0 \end{aligned} \tag{8-19}$$

The optimum solution is shown in Fig. 8-17 as the point where the objective function is tangent to the boundary of the feasible region, which is identified by the shaded portion of the figure.

Solution of Nonlinear Programming Problems

Nonlinear programming problems can be solved by a number of techniques. One technique approximates the nonlinear functions as a combination of linear function segments (piecewise linear functions) and proceeds by an algorithm similar to that of linear programming. Some problems require, however, much more sophisticated techniques. While there is no single efficient solution method for the general nonlinear programming model, many efficient algorithms have been developed for certain classes of models such as, for example, quadratic programming which applies to a quadratic objective function and linear equations of constraint. For two or three variables, a graphical solution analogous to that used in linear programming may be employed, as illustrated in the examples of this section.

8-7 DYNAMIC PROGRAMMING

The method of dynamic programming is based on the mathematical concept of recursion. It essentially converts a problem in which an optimum policy involving n components is desired, to a sequential decision process of n separate stages dealt with in a recursive fashion. The method is based on Bellman's [72] *principle of optimality* which states:

> An optimal policy has the property that, whatever the initial state and initial decision are, the remaining decisions must constitute an optimal policy with respect to the state resulting from the first decision.

What the principle is telling us is that if we know how to proceed with the best course of action from any intermediate state to a final desired goal state, we should follow this course of action regardless of how we get to the intermediate state from an initial state. This is illustrated conceptually in

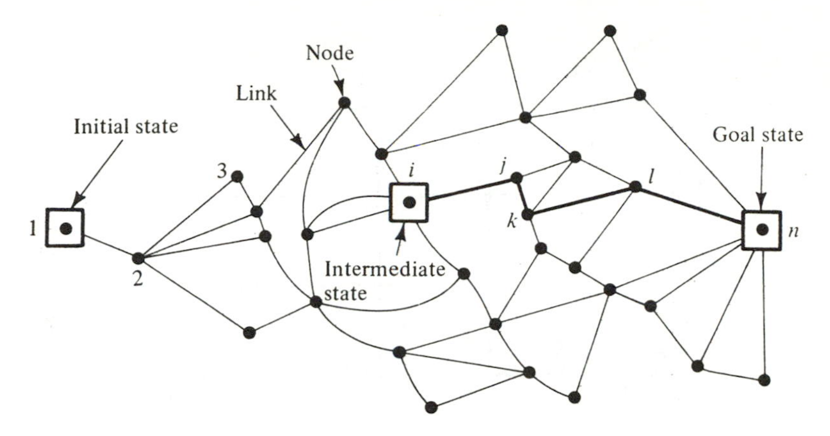

Figure 8-19

Fig. 8-19. If we know the best* path, between nodes, from an intermediate state, say node i, to the desired state, node n, then, regardless of how we get to node i from node 1, we should proceed along this best path from i to n, say, path $ijkln$ in the figure. The principle of optimality points out the recursive nature of the method. We can now say that if we know the best path from node 1 to node n in Fig. 8-19, then, regardless of how we arrive at node 1 from some earlier states preceding it, we should proceed along this best path. The conceptual diagram of Fig. 8-19 should not be misconstrued as one which applies to problems of travel *only*, although it certainly applies to such problems, as will be illustrated in Example 1 which follows. Each node should be viewed as a *state* where a decision must be made. The *decision* is characterized conceptually by selecting a link from the node to a neighboring adjacent node. The sequence of links between nodes, i.e., the path from the initial state to the final desired goal state, constitutes a *decision policy*. The object of dynamic programming is to find the optimal policy with respect to a reference criterion by using a sequential decision-making process which is computationally efficient in that it reduces greatly the search for the best.

We shall illustrate the method by first using an example in which the outcomes of our decisions are known with certainty. Then we shall follow it by examples in Secs. 8-8 and 8-9 in which the outcomes of the decisions are probabilistic.

EXAMPLE 1. We return now to an example similar to the one discussed in Sec. 8-1. A man wishes to reach node B from node A in Fig. 8-20, heading only north and east, along a path of maximum utility. The utility associated

*Best with respect to a reference criterion such as cost, time, distance, etc.

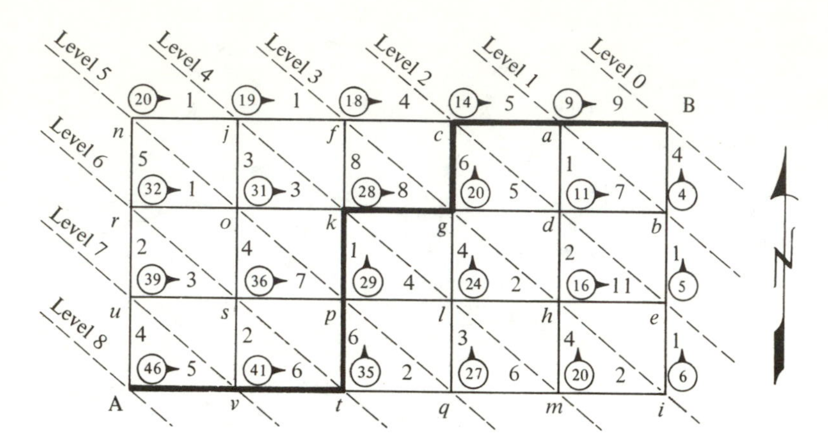

Figure 8-20

with each link is marked on the links between nodes. The designation of levels along the dashed inclined lines is used to identify all the nodes at a particular stage in the decision process. At level 8, a decision leads to one of the nodes at level 7, and a decision at one of the nodes of level 7 leads to level 6, and so on from level i to level $i - 1$ until the destination is reached at level 0. Thus, eight decisions must be made between A and B. Each level number identifies the number of decisions which must still be made before the destination is reached. The model appears similar to that of Fig. 5-16 and also Fig. 8-1, of course. There are a total of $(5 + 3)!/3!5! = 56$ different paths between A and B, each consisting of eight links. Which is the best? We can, of course, identify all paths, compute the associated utilities, and select the best. This procedure will prove rather inefficient as the number of north and east blocks increases, and will become an impossible task as the grid is of the order of, say, 100×100 as indicated in Sec. 8-1. Dynamic programming provides us with a method which not only makes manageable the impossible task of dealing with a large grid, but it also reinforces a basic concept of problem solving: do not always start at the beginning; starting at the end and working backwards may prove rather productive.

Dynamic Programming Solution. Let us go to the end and position ourselves at level 1, only one level away from the goal B. At node a we can travel only east to B with a utility of 9. This is indicated by the circled number at node a. The arrow on the circle indicates the direction of travel. At b we have 4 in the circle. Now we move back to level 2. At node c we can proceed only to node a with 5 units of utility; but since we have 9 units from a to B, the total from c is 14. At d we can move to a or b. But going to a yields 1 unit and from a we have 9, so the total is 10. On the other hand, going to b, the total is

11. We, therefore, select the best of the two regardless of how we get to *d*. The best, which is 11, is entered in the circle, and the arrow points to *b*. At *e* we have 5. Now we move to level 3 and proceed the same way. For example, at node *g* we have 6 units going to node *c*, but from *c* on we have 14, as recorded in the circle at *c*; the total is 20. Going to *d*, the total is 16; therefore, we go to *c*. Proceeding this way from level to level, we inspect each node and make a *binary decision*, namely, a choice between two options: going north or east to the next level. Each option requires adding two numbers, the utility of the link to the adjacent node and the circled number at the node ahead. To emphasize this point again, suppose we are at level 6 node *s*. Going to *o* we have 4 units and from *o*, regardless of how we get there, the best path has 31 units; the total is 35. Going to *p*, the total is 36, so we go to *p*, and the best or maximum utility path from *s* to *B* is 36 units regardless of how we get to *s*. This best path from *s* follows the nodes *spkgcaB*, as indicated by the arrows on the circles at each node, starting from node *s*. Finally, by working backwards we reach level 8, node *A*, and find that the optimum path from *A* to *B* yields a utility of 46 units and goes through the sequence of nodes *AvtpkgcaB* marked by the heavy line in the figure.

EXAMPLE 2. Let us suppose that the numbers on the links of Fig. 8-20 represent units of distance, and we wish to minimize the distance of travel from *A* to *B*. The solution is shown in Fig. 8-21 in which the circled numbers represent the shortest distances to *B*. The optimum path, marked by the heavy line in the figure, is *AvtqmiebB* with length of 27 units of distance.

Note that the shortest distance from *A* to the adjacent level is 4, along link *Au*, and yet the shortest path from *A* to *B* goes through *v* and not *u*. This demonstrates that beginning the decisions at the initial state, instead of working backwards, would have led to a nonoptimal path. It also

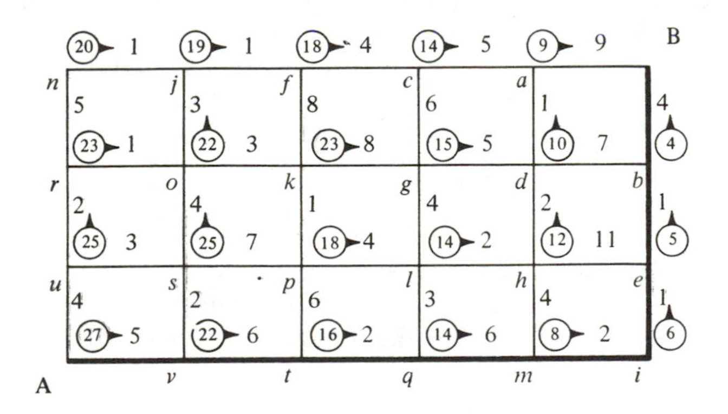

Figure 8-21

demonstrates that dynamic programming leads to an optimum for the entire problem of n levels or stages from initial to goal state, and in the process local optima between intermediate stages may be sacrificed.

Problem 1. What is the optimum path of maximum utility in Fig. 8-20 subject to the constraint that the path must go through node l?

Problem 2. What is the optimum path of shortest distance in Fig. 8-21 subject to the constraint that it must go through node g?

Problem 3. Verify the optimum path in Fig. 8-1.

Recursive Relationship of Dynamic Programming

The method of dynamic programming is often presented in the form of a compact recursive relationship which contains the essence of the sequential decision process traced backwards from the destination to the initial state. Let us use Example 1 of this section to write such a recursive relationship.

Suppose our man is at level n, state s (n represents the number of levels or stages to go, and a state is a node), and selects node x_n in going to level $n-1$ (x_n can be one of two nodes in our problem). For example, for $n = 4$ and $s = l$, $x_4 = g$ or h. Let us denote by $F_n(s, x_n)$ the total utility for the best path the man can follow from level n to the goal state, when he is at state s and selects x_n for the next state (node). Let $x_n{}^*$ be the choice of x_n which leads to an optimum $F_n{}^*(s) = F_n{}^*(s, x_n{}^*)$ from state s to the goal state. For example, at level $n = 4$ and $s = l$, $x_4 = g$ yields $F_4(l, g) = 24$, and $x_4 = h$ yields $F_4(l, h) = 18$; therefore $x_4{}^* = g$, with $F_4{}^*(l) = F_4{}^*(l, x_4{}^*) = 24$.

The objective in dynamic programming problems is to find $F_n{}^*$ (initial state) and the corresponding decisions at each state (i.e., policy). The recursive relationship of the dynamic programming method proceeds by finding $F_1{}^*(s), F_2{}^*(s), \ldots, F_n{}^*$ (initial state). In Example 1 above we have at level 1

$$s = a, b$$
$$F_1{}^*(a, x_1{}^*) = F_1{}^*(a) = 9, \qquad x_1{}^* = \text{node B}$$
$$F_1{}^*(b, x_1{}^*) = F_1{}^*(b) = 4, \qquad x_1{}^* = \text{node B}$$

Let $u(s, x_n)$ designate the utility along the link between state s and the immediate next node x_n at level $n-1$. For example, $u(d, x_2) = 1$ or 7 for $x_2 = a$ or b respectively. Using this notation we can write for the states $s = c, d, e$ at level 2

$$F_2(s, x_2) = u(s, x_2) + F_1{}^*(x_2)$$

in which $F_1{}^*(x_2)$ is the optimum $F_1{}^*(s)$ at node x_2. The optimum $F_2{}^*(s)$ is

$$F_2{}^*(s) = \max_{x_2} \{u(x, x_2) + F_1{}^*(x_2)\}$$

In general, at level n, we have

$$F_n(s, x_n) = u(s, x_n) + F^*_{n-1}(x_n)$$
$$F_n^*(s) = \max_{x_n} \{u(s, x_n) + F^*_{n-1}(x_n)\} \quad (8\text{-}20)$$
$$n = 1, 2, \ldots, N \qquad N = \text{total number of levels}$$

in which $F^*_{n-1}(x_n)$ is the optimum from state x_n to the goal state. Equation 8-20 is a recursive relationship in which we begin with $n = 1$, then set $n = 2$ and so on, until we reach $n = N$, i.e., the total number of levels. The symbol x_n under *max* in the equation designates that we find the node x_n for which the expression in brackets is a maximum. For a minimization problem x_n is written under *min*.

Consider for example the application of Eq. 8-20 to level 6 of Fig. 8-20 with the state being node r. Then we have

$$F_6(\text{node } r, x_6) = u(\text{node } r, x_6) + F_5^*(x_6)$$

Here x_6 can be node n or node o. For $x_6 = \text{node } n$, $u(\text{node } r, \text{node } n) = 5$, $F_5^*(\text{node } n) = 20$ and thus

$$F_6(\text{node } r, x_6 = \text{node } n) = 5 + 20 = 25$$

For $x_6 = \text{node } o$, we have

$$F_6(\text{node } r, x_6 = \text{node } o) = 1 + 31 = 32$$

From Eq. 8-20 we have then

$$F_6^*(\text{node } r) = \max_{x_6} \{u(\text{node } r, x_6) + F_5^*(x_6)\} = 32$$

with $x_6^* = \text{node } o$.

For the problem of Example 2 above (Fig. 8-21), Eq. 8-20 takes the form:

$$F_n^*(s) = \min_{x_n} \{d(s, x_n) + F^*_{n-1}(x_n)\} \quad (8\text{-}21)$$
$$n = 1, 2, \ldots, N \qquad N = \text{total number of levels}$$

in which $d(s, x_n)$ designates the distance along the link between state s and the immediate next node x_n at level $n - 1$.

*8-8 SEQUENTIAL DECISIONS WITH RANDOM OUTCOMES

Consider a man who wishes to sell his home. On the basis of much data which he has acquired, he expects to get ten independent offers, arriving in sequence one at a time, over a period of a year. Since he wishes to sell the house within a year, but does not want to accept just any offer, he makes further studies that lead to the probability distribution of potential offers as shown in Fig. 8-22. The probability is zero for offers x in the range

$$x > \$120{,}000$$
$$x < \$\ 20{,}000$$

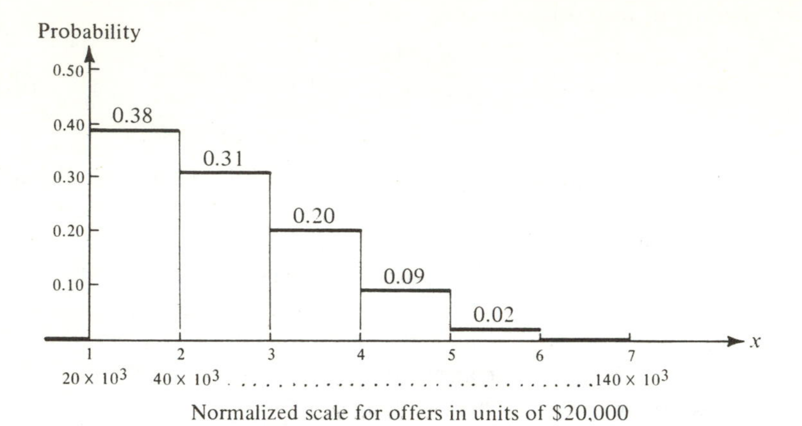

Figure 8-22

For offers in the range $20,000 $\leq x \leq$ $120,000 the probabilities are:

$$\begin{array}{lll} 0.38 & \text{for} & \$20{,}000 \leq x \leq \$40{,}000 \\ 0.31 & \text{for} & \$40{,}000 < x \leq \$60{,}000 \\ 0.20 & \text{for} & \$60{,}000 < x \leq \$80{,}000 \\ 0.09 & \text{for} & \$80{,}000 < x \leq \$100{,}000 \\ 0.02 & \text{for} & \$100{,}000 < x \leq \$120{,}000 \end{array}$$

Figure 8-22 shows a normalized scale for the offers so as to make the distribution a probability density function, namely, of unit total area. Each unit of the scale represents $20,000.

Let us suppose that the man has committed himself to accept the last of the ten offers whatever it is, if he has rejected all preceding offers. We emphasize again that it is assumed one offer at a time is being considered, and that the offers are independent and have the probability distribution of Fig. 8-22. What is the optimal policy, i.e., rule for decisions, if the objective is to maximize the expected value of the accepted offer? What the man needs is a policy in the form of a sequence of ten numbers, one for each offer other than the last one which he is committed to accept. Each number will specify the boundary between accepting the current offer or waiting further. For example, if the first offer is $120,000, then he should accept it; and if it is $20,000, it is plausible that he should wait. But for what amount is he to become indifferent between accepting and waiting? It is reasonable that he should compare future expected value with the value of the current offer being considered. If desired, the actual dollar value may be transformed to corresponding utilities, and then the expected utility of waiting should be compared with the utility of the current offer with the objective of maximizing expected utility. In the present example, we shall consider the expected value of the offer accepted, not its utility.

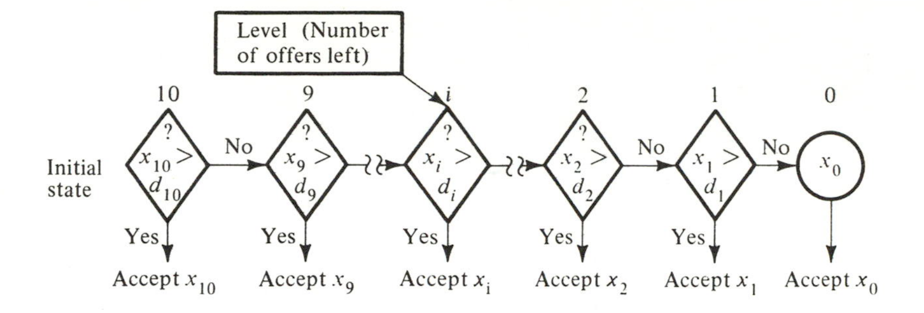

Figure 8-23 Schematic Description of Sequential Decision Process

The problem is shown schematically in Fig. 8-23. At each level the level number represents the number of offers left in the future, considering a total of ten offers. x_i is the value of the offer made at level i when there are i offers left out of ten. The decision rule d_i at level i is a value in dollars (or units of utility when utility of money is used) such that if the offer x_i at state i is equal to or larger than d_i, x_i is accepted; if smaller, x_i is rejected in favor of waiting for future offers. Namely, at each level i,

for $\quad x_i \geq d_i \quad$ accept offer x_i

for $\quad x_i < d_i \quad$ do not accept offer x_i, wait

But, of course, after the ninth offer has been rejected, our man accepts whatever the tenth offer is, i.e., whatever x_0 is.

Dynamic Programming Solution

Starting at level zero, the expected value of the last offer x_0 is given by (see Fig. 8-22):

$$\begin{aligned} E(x_0) &= (0.38 \times 30 \times 10^3) + (0.31 \times 50 \times 10^3) \\ &\quad + (0.20 \times 70 \times 10^3) + (0.09 \times 90 \times 10^3) \\ &\quad + (0.02 \times 110 \times 10^3) \\ &= \$51.2 \times 10^3 \\ F_0(x_0) &= x_0 \end{aligned}$$

in which $F_0(x_0)$ represents the value of the objective function at level zero. Therefore, *at Level* 1, $d_1 = \$51.2 \times 10^3$. Namely, when the offer x_1 is larger than or equal to d_1, x_1 should be accepted; and when $x_1 < d_1$, wait for x_0 which has a larger expected value of d_1. Thus, we can write

$$F_1(x_1) = \max\,[x_1;\, E\{F_0(x_0)\} = d_1] \tag{8-22}$$

where $F_1(x_1)$ is the function to be maximized at level 1 and it is the maximum, i.e., the larger of the two values in the brackets, the offer x_1 or the expected value $F_0(x_0)$ in the future, as shown in Fig. 8-24.

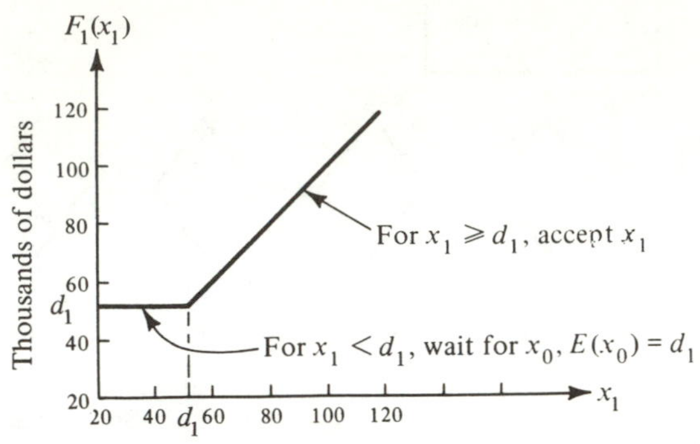

Figure 8-24 Plot of $F_1(x_1)$. One Offer Left

At Level 2, two offers are left. The future prospects from this level appear as follows. At level 1, any offer larger than d_1 will be accepted and a lower offer will be rejected because then d_1 will be expected (see Fig. 8-24). Hence, the expected value of these future options is the expected value of $F_1(x_1)$ in Eq. 8-22 or in Fig. 8-24 in which

$$F_1(x_1) = x_1 \quad \text{for } x_1 \geq d_1$$
$$F_1(x_1) = d_1 \quad \text{for } x_1 < d_1$$

For $x_1 \geq d_1$:

Probability of x_1 in range above d_1	*Contribution to expected value*
$P(51.2 \times 10^3 \leq x_1 \leq 60 \times 10^3) = 0.31 \times \frac{8.8}{20} = 0.136$	$0.136 \frac{51.2 + 60}{2} 10^3 = 7.6 \times 10^3$
$P(60 \times 10^3 < x_1 \leq 80 \times 10^3) = 0.20$	$0.20 \times 70 \times 10^3 = 14.0 \times 10^3$
$P(80 \times 10^3 < x_1 \leq 100 \times 10^3) = 0.09$	$0.09 \times 90 \times 10^3 = 8.1 \times 10^3$
$P(100 \times 10^3 < x_1 \leq 120 \times 10^3) = 0.02$	$0.02 \times 110 \times 10^3 = 2.2 \times 10^3$
	$\$31.9 \times 10^3$

For $x_1 < d_1$:

$$P(x_1 < d_1) = 0.38 + 0.31 \times \frac{11.2}{20} = 0.554$$

But when $x_1 < d_1$, x_1 is rejected and the value expected in future offers is d_1. However, since the probability is 0.554 for x_1 to be rejected, it contributes $0.554 \times d_1$ to the expected value at level 2. The total expected value at level 2, which is the expected value of $F_1(x_1)$, is, therefore,

$$E\{F_1(x_1)\} = 31.9 \times 10^3 + 0.554 \times 51.2 \times 10^3 = \$60.2 \times 10^3$$

Hence, $d_2 = \$60.2 \times 10^3$

and $$F_2(x_2) = \max\,[x_2;\, E\{F_1(x_1)\} = d_2] \tag{8-23}$$

Namely, accept x_2 when $x_2 \geq d_2$

reject x_2 when $x_2 < d_2$

A plot of $F_2(x_2)$ is shown in Fig. 8-25.

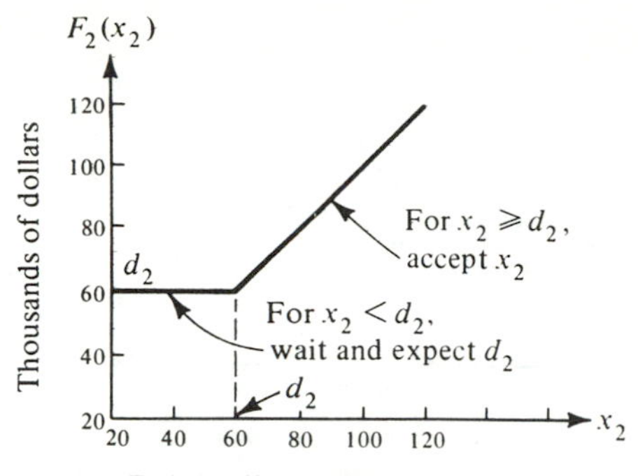

Figure 8-25 Plot of $F_2(x_2)$. Two Offers Left

At Level 3 we compute the expected value of $F_2(x_2)$ as the expected value of the future. From Eq. 8-23 and Fig. 8-25,

$$F_2(x_2) = x_2 \quad \text{for} \quad x_2 \geq d_2$$

and

$$F_2(x_2) = d_2 \quad \text{for} \quad x_2 < d_2$$

For $x_2 \geq d_2$:

Probability of x_2 in range above d_2	*Contribution to expected value*
$P(60.2 \times 10^3 \leq x_2 \leq 80 \times 10^3) = 0.20\dfrac{19.8}{20.0} = 0.198$	$0.198 \times 70.1 \times 10^3 = 13.9 \times 10^3$
$P(80 \times 10^3 < x_2 \leq 100 \times 10^3) = 0.09$	$0.09 \times 90 \times 10^3 = 8.1 \times 10^3$
$P(100 \times 10^3 < x_2 \leq 120 \times 10^3) = 0.02$	$0.02 \times 110 \times 10^3 = 2.2 \times 10^3$
	$\$24.2 \times 10^3$

For $x_2 < d_2$:

$$P(x_2 < d_2) = 0.38 + 0.31 + 0.20 \times \frac{0.2}{20} = 0.692$$

$$0.692 \times d_2 = 0.692 \times 60.2 \times 10^3 = \$41.6 \times 10^3$$

The total expected value at level 3, which is the expected value of $F_2(x_2)$, is

$$E\{F_2(x_2)\} = (24.2 + 41.6) \times 10^3 = \$65.8 \times 10^3$$

Hence, $d_3 = \$65.8 \times 10^3$

and

$$F_3(x_3) = \max\,[x_3;\ E\{F_2(x_2)\} = d_3] \tag{8-24}$$

The plot for $F_3(x_3)$ is similar to those of Figs. 8-24 and 8-25, except the position of the decision value d_3 is at $\$65.8 \times 10^3$ as shown in Fig. 8-26. Therefore, we leave it to the reader to plot the values of F_i at the remaining levels.

At Level 4, from Eq. 8-24,

$$F_3(x_3) = x_3 \quad \text{for} \quad x_3 \geq d_3$$

$$F_3(x_3) = d_3 \quad \text{for} \quad x_3 < d_3$$

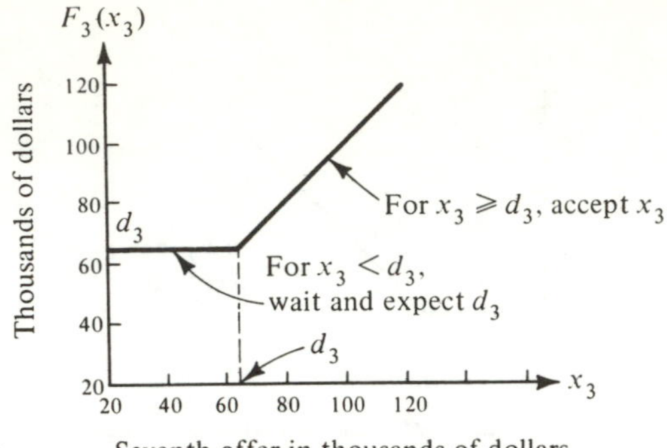

Figure 8-26 Plot of $F_3(x_3)$, Three Offers Left

For $x_3 \geq d_3$:

$$P(65.8 \times 10^3 \leq x_3 \leq 80 \times 10^3) = 0.20 \times \frac{14.2}{20.0} = 0.142$$

$$0.142 \times \frac{65.8 + 80}{2} \times 10^3 = 10.4 \times 10^3$$

$$.09 \times 90 \times 10^3 = 8.1 \times 10^3$$

$$.02 \times 110 \times 10^3 = \underline{2.2 \times 10^3}$$

$$\$20.7 \times 10^3$$

For $x_3 < d_3$:

$$P(x_3 < d_3) = 0.38 + 0.31 + 0.20\frac{5.8}{20} = 0.748$$

$$0.747 \times d_3 = 0.748 \times 65.8 + 10^3 = \$49.1 \times 10^3$$

$$E\{F_3(x_3)\} = (20.7 + 49.1) \times 10^3 = \$69.8 \times 10^3$$

$$d_4 = \$69.8 \times 10^3$$

and $$F_4(x_4) = \max\,[x_4; E\{F_3(x_3)\} = d_4] \tag{8-25}$$

Continuing this way with the recursive relation,

$$F_i(x_i) = \max\,[x_i; E\{F_{i-1}(x_{i-1})\} = d_i] \tag{8-26}$$

we obtain the following solution for the optimal policy:

Solution. When an offer x_i is received with i offers to go,

accept the offer if $x_i \geq d_i$
reject the offer if $x_i < d_i$

$d_1 = \$51,200$
$d_2 = \$60,200$
$d_3 = \$65,800$

$$d_4 = \$69{,}800$$
$$d_5 = \$72{,}900$$
$$d_6 = \$75{,}400$$
$$d_7 = \$77{,}500$$
$$d_8 = \$79{,}300$$
$$d_9 = \$80{,}900$$
$$d_{10} = \$82{,}200$$

Problem. Verify the above results from d_4 to d_{10} inclusive.

*8-9 SEQUENTIAL DECISIONS WITH NORMAL DISTRIBUTION OUTCOMES

Consider the following game from Sasieni, Yaspan, and Friedman [73]. Ten numbers are drawn in sequence from a standardized normal distribution. We view each number drawn as an offer in Sec. 8-8. If you accept a number, you are paid an amount equal to that number. You can accept only one number, and if you reject the first nine, you must accept the tenth number whatever its value. At each level i of the ten levels, the decision of whether to accept the value x_i drawn, or reject it and wait for future draws, depends on the value of x_i and the expected value of following draws. The objective is to determine a policy which will maximize the expected value of the game.

Solution. Let $F_i(x_i)$ denote the value of the game at level i when there are i more numbers to be drawn out of the total of ten draws, and x_i is the value of the draw at level i. Then at level zero, last draw, $F_0(x_0) = x_0$ and $E(x_0) = 0$, or $E\{F_0(x_0)\} = d_1 = 0$.

At level $i = 1$,

$$F_1(x_1) = \max\,[x_1;\, E\{F_0(x_0)\} = d_1] \qquad \text{(8-27)}$$

Therefore, $F_1(x_1)$ can take on these values:

$$F_1(x_1) = d_1, \quad \text{if} \quad x_1 < d_1$$
$$F_1(x_1) = x_1, \quad \text{if} \quad x_1 \geq d_1$$
$$d_1 = 0$$

$F_1(x_1)$ is shown graphically in Fig. 8-27.

At level $i = 2$, two draws are left and

$$F_2(x_2) = \max\,[x_2;\, E\{F_1(x_1)\} = d_2] \qquad \text{(8-28)}$$

$$E\{F_1(x_1)\} = d_1 \times \int_{-\infty}^{d_1} p(x_1)\, dx_1 + \int_{d_1}^{\infty} x_1\, p(x_1)\, dx_1$$

$$E\{F_1(x_1)\} = 0 \int_{-\infty}^{0} p(x_1)\, dx_1 + \int_{0}^{\infty} x_1\, p(x_1)\, dx_1$$

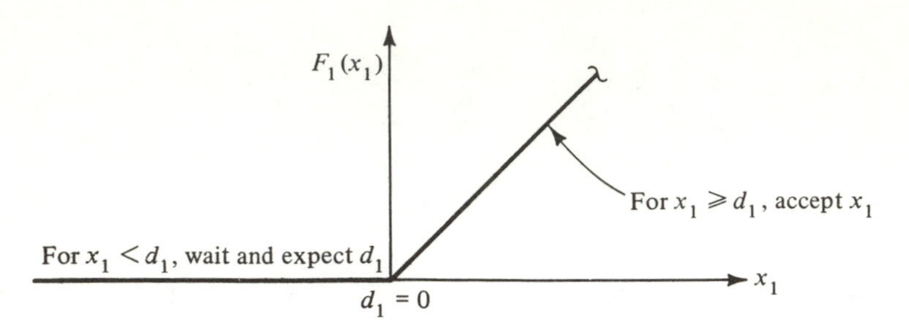

Figure 8-27

For the standardized normal distribution,

$$p(x) = \frac{1}{\sqrt{2\pi}} e^{-x^2/2}$$

and the second integral above can be evaluated by noting that $xe^{-x^2/2}$ is the derivative of $-e^{-x^2/2}$, or $-e^{-x^2/2}$ is the integral of $xe^{-x^2/2}$,

$$\begin{aligned} \int_\alpha^\infty xp(x)\,dx &= \int_\alpha^\infty x\frac{1}{\sqrt{2\pi}}e^{-x^2/2}\,dx \\ &= -\frac{1}{\sqrt{2\pi}}e^{-x^2/2}\Big|_\alpha^\infty \\ &= \frac{1}{\sqrt{2\pi}}e^{-\alpha^2/2} \\ &= p(\alpha) \end{aligned}$$

Namely, the integral $\int_\alpha^\infty x\,p(x)\,dx$ is equal to the ordinate $p(\alpha)$ of the standardized normal distribution. Therefore,* $\int_0^\infty x\,p(x)\,dx = p(0) = 0.3989$, and $E\{F_1(x_1)\} = d_2 = 0.3989$, say, 0.399. Substituting this result in Eq. 8-28, we have

$$F_2(x_2) = \max[x_2;\ 0.399]$$

$F_2(x_2)$ can take on the following values:

$$\begin{aligned} F_2(x_2) &= d_2, \quad \text{if} \quad x_2 < d_2 \\ F_2(x_2) &= x_2, \quad \text{if} \quad x_2 \geq d_2 \\ d_2 &= 0.399 \end{aligned}$$

A plot of $F_2(x_2)$ vs. x_2 is shown in Fig. 8-28.

At level $i = 3$, three draws are left and

$$F_3(x_3) = \max\,[x_3;\ E\{F_2(x_2)\} = d_3] \tag{8-29}$$

*See Table 6-2.

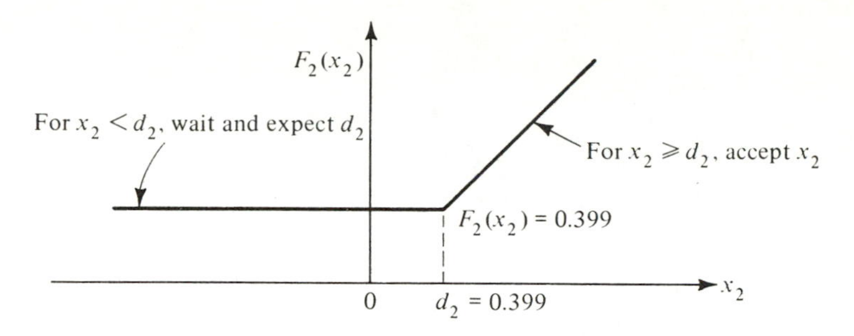

Figure 8-28

$$\begin{aligned} E\{F_2(x_2)\} &= d_2 \times \int_{-\infty}^{d_2} p(x_2)\, dx_2 + \int_{d_2}^{\infty} x_2\, p(x_2)\, dx_2 \\ &= 0.399 \times 0.655 + p(0.399) \\ &= 0.261 + 0.369 \\ &= 0.630 \\ E\{F_2(x_2)\} &= d_3 = 0.630 \end{aligned}$$

Thus,

$$F_3(x_3) = \max\,[x_3;\, 0.630]$$

$F_3(x_3)$ can take on the values

$$\begin{aligned} F_3(x_3) &= d_3, \quad \text{if} \quad x_3 < d_3 \\ F_3(x_3) &= x_3, \quad \text{if} \quad x_3 \geq d_3 \\ d_3 &= 0.630 \end{aligned}$$

as shown in Fig. 8-29.

Proceeding this way backwards from state to state with the general recursive relation,

$$F_i(x_i) = \max\,[x_i;\, E\{F_{i-1}(x_{i-1})\} = d_i] \tag{8-30}$$

we obtain for the optimum policy of playing the game the following decision

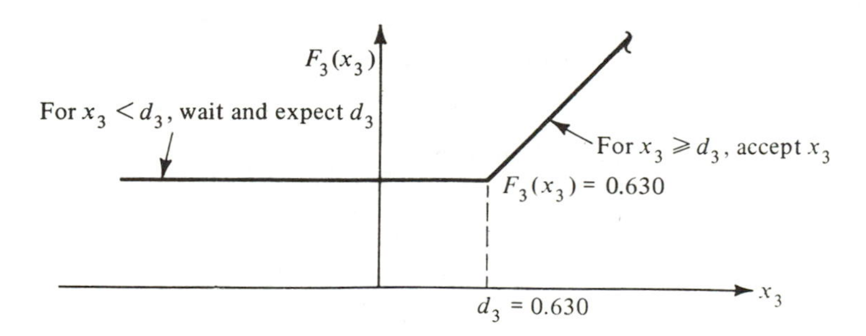

Figure 8-29

values at each state:

$$
\begin{aligned}
d_1 &= 0 \\
d_2 &= 0.399 \\
d_3 &= 0.630 \\
d_4 &= 0.790 \\
d_5 &= 0.912 \\
d_6 &= 1.01 \\
d_7 &= 1.09 \\
d_8 &= 1.16 \\
d_9 &= 1.22 \\
d_{10} &= 1.28
\end{aligned}
$$

NOTE. Dynamic programming can be used for problems in which the feasible region is not convex and the objective function and the equations of constraint are nonlinear and discontinuous. There is, however, a restriction on the form of the objective function. The objective function is not expressed in the form of an explicit function of the decision variables x_i such as in linear and nonlinear programming, but instead each term in the objective function is itself a function of the decision variable.

8-10 SUMMARY

When there exist several possible solutions to a problem, programming methods can be used to find the optimum with respect to a desired reference criterion. The programming methods reduce the search for the best solution to a manageable task in many problems that cannot otherwise be solved.

Three programming methods are discussed: linear, nonlinear, and dynamic. In linear programming, the functions of the decision variables x_i are linear, namely,

$$F(x_1, x_2, \ldots, x_n) = a_0 + a_1x_1 + a_2x_2 + \ldots + a_nx_n$$

Linear programming models have the general form:
maximize (or minimize) the objective function

$$F = \sum_{j=1}^{n} c_jx_j$$

subject to linear equations of constraint

$$
\begin{aligned}
a_{11}x_1 + a_{12}x_2 + \ldots + a_{1n}x_n &\leq b_1 \\
a_{21}x_1 + a_{22}x_2 + \ldots + a_{2n}x_n &\leq b_2 \\
\vdots \qquad\qquad &\qquad \vdots \\
a_{m1}x_1 + a_{m2}x_2 + \ldots + a_{mn}x_n &\leq b_m \\
x_j \geq 0 \quad \text{for all} \quad j &= 1, 2, \ldots, n
\end{aligned}
$$

Using matrix notation, the above can be written compactly as

$$\max F = \{c\}^{T}\{x\}$$

subject to $[A]\{x\} \leq \{b\}, \{x\} \geq \{0\}$.

Each linear programming problem has a *dual problem* (the original problem is called the *primal problem*). The model of the *dual* has the following form in terms of the coefficients a_{ij}, b_i, and c_j of the *primal*:

$$\min G = \textstyle\sum_{i=1}^{m} y_i b_i$$

subject to the constraints

$$\begin{array}{l} a_{11}y_1 + a_{21}y_2 + a_{31}y_3 + \ldots + a_{m1}y_m \geq c_1 \\ a_{12}y_1 + a_{22}y_2 + a_{32}y_3 + \ldots + a_{m2}y_m \geq c_2 \\ \quad\vdots \\ a_{1n}y_1 + a_{2n}y_2 + a_{3n}y_3 + \ldots + a_{mn}y_m \geq c_m \\ \qquad y_i \geq 0 \quad \text{for all} \quad i = 1, 2, \ldots, m \end{array}$$

Or, using matrix notations, the dual becomes

$$\min G = \{y\}^{T}\{b\}$$

subject to $\{y\}^{T}[A] \geq \{c\}^{T}, \{y\} \geq \{0\}$.

Variables y_i are the slack variables in the primal problem.

Linear programming problems can be solved graphically for two or three variables. For more variables, the simplex algorithm is used.

The simplex algorithm begins with a tableau which includes the coefficients a_{ij}, b_i, and c_j in the linear programming model in Fig. 8-30 for n variables x_j and m equations of constraint. To the right of coefficients a_{ij} an identity matrix of order $m \times m$ is inserted, and the elements of the last row following the last coefficient $-c_n$ of the objective function are all set to zero. The elements of the last row are called *indicators*.

	Decision variables				*Slack variable*				
	x_1	x_2	$\cdots$	x_n	s_1	s_2	$\cdots$	s_m	
Coefficient of constraint equations	a_{11}	a_{12}	$\cdots$	a_{1n}	1	0	$\cdots$	0	b_1
	a_{21}	a_{22}	$\cdots$	a_{2n}	0	1	$\cdots$	0	b_2
	$\vdots$	$\vdots$		$\vdots$	$\vdots$	$\vdots$		$\vdots$	$\vdots$
	a_{m1}	a_{m2}	$\cdots$	a_{mn}	0	0	$\cdots$	1	b_m
Coefficients of objective function (with sign reversed)	$-C_1$	$-C_2$	$\cdots$	$-C_n$	0	0	$\cdots$	0	0

Figure 8-30

The *simplex algorithm* consists of the following steps:

Step 1. Select any column k with a negative indicator. Find the smallest ratio b_i/a_{ik} for $i = 1, 2, \ldots, m$. The element a_{rk}, which appears in the denominator of the smallest ratio, is identified and labeled as the *pivot element.*

Step 2. Factor out the pivot element a_{rk} from row r (namely, divide all elements of row r by a_{rk}) and enter the resulting row as row r in the second tableau. Multiply this row r by a_{1k} and subtract from row 1. The result is recorded as row 1 in the second tableau. Repeat this process for all rows (except row r). Namely, for row l, multiply row r by a_{lk}, subtract from row l, and record the result as row l in the second tableau. The process is also applied to the last row of coefficients c_j. Thus, the second tableau contains a column k with 1 in position rk and zeros elsewhere.

Step 3. Repeat steps 1 and 2, generating a new tableau each time, until all *indicators* (elements of last row) are positive. The value in the lower right corner of the last tableau with all positive indicators is the optimal value of the objective function. The corresponding values of variables x_j are in the last column where coefficients b_i are entered in the first tableau. The values of variables x_j are identified as follows. If, in the column under x_j in the last tableau, all elements are zero except for a 1 in row s, then the value of x_j for the optimum solution appears in row s of the last column.

The nonzero elements of the last row, except for that in the last column, represent the sensitivity of the objective function to unit changes in the constraints. A value in the column under s_i $(i = 1, 2, \ldots, m)$ represents the increase in F due to a unit increase in b_i of the ith equation of constraint.

Nonlinear programming models are similar in form to linear programming, except that at least one of the equations in the model, a constraint or the objective function, is a nonlinear function of the decision variables x_j.

The solution of nonlinear programming models requires more complex procedures than linear programming. However, for two variables a graphical solution, similar in form to that of linear programming, is possible. Also, piecewise linearization of the nonlinear functions can transform a nonlinear problem to one that can be solved by an algorithm similar to that used in the linear problem.

Dynamic Programming is useful in problems of sequential decisions and is based on the *principle of optimality* which states:

> An optimal policy has the property that, whatever the initial state and initial decision are, the remaining decisions must constitute an optimal policy with respect to the state resulting from the first decision.

The solution procedure in dynamic programming works backwards, using a recursive relationship. Examples of such relationships are given by Eqs.

8-20 and 8-26:

$$F_n^*(s) = \max_{x_n} \{u(s, x_n) + F_{n-1}^*(x_n)\}$$
$$F_i(x_i) = \max [x_i; E\{F_{i-1}(x_{i-1})\}]$$

Dynamic programming is more of a fundamental element in the philosophy of problem solving than an algorithm for the solution of a class of problems. Therefore, formulation of a problem as a dynamic programming model is a truly creative task.

EXERCISES

8-1 Are the following equations linear or nonlinear?

(a) $2x_1^2 + 3x_2 = 16$
(b) $3xy = 7$
(c) $3x + 3y = 7$
(d) $\dfrac{x+y}{3} = 5$
(e) $\dfrac{3}{x+y} = 5$
(f) $4x_1 - \left(\dfrac{1}{2}\right)x_2 + 5x_3 = 0$

8-2 Graph the regions identified by the following inequalities, and indicate whether or not each is convex.

(a) $x + 2y \leq 4$, $x \geq 0$, $y \geq 0$
(b) $x + y \leq 6$, $5x + y \leq 10$, $x \geq 0$, $y \geq 0$
(c) $x - y^2 \leq 0$, $0 \leq x \leq 9$, $0 \leq y \leq 3$
(d) $2x + y \geq 5$, $x + y \leq 2$, $x \geq 0$, $y \geq 0$
(e) $x^2 + y^2 \leq 1$, $x \geq 0$, $y \geq 0$
(f) $x^2 + y^2 \geq 1$, $x \geq 0$, $y \geq 0$
(g) $x - 2y \leq -2$, $2x + y \geq 6$, $3x + 2y \leq 18$, $x \geq 0$, $y \geq 0$
(h) $x + y \leq 2$, $x + y \geq 4$

8-3 Describe in your own words why the optimum of a linear programming problem lies on a vertex of the convex feasible region. (*Hint:* Use a graphical argument.)

8-4 For the following constraint equations, graph the feasible region.

(i) $-x_1 + x_2 \leq 1$
(ii) $6x_1 + 4x_2 \geq 24$

(iii) $2x_1 + 3x_2 \leq 24$
(iv) $x_1 \geq 0$
(v) $x_2 \geq 0$

Taking the above feasible region into consideration, optimize the following objective functions:

(a) Minimize $x_1 + x_2$
(b) Maximize $2x_1 + 3x_2$
(c) Maximize $x_1 - 2x_2$

8-5 Rework Exercise 8-4 [parts (a), (b), (c)], eliminating constraint equation (iii).

8-6 Suppose that a TV manufacturer turns out only two types of TVs: a black-and-white model, selling at a profit of $30 each, and a color set, selling at a profit of $50 each. The factory has an assembly line for each model, but their capacity is limited. It is possible to produce at most 8 black-and-white TVs per day on the first line and 5 color TVs per day on the second. Due to limited labor supply, there are only 12 employees, so the available labor amounts to 12 man-days per day. To assemble a black-and-white set requires one man-day, and a color set takes two man-days.

The manufacturer wishes to know how many TVs of each type he should produce to maximize his profit. Let B represent the number of black-and-white TVs and C the number of color TVs.

(a) Write the objective function.
(b) Write the constraint equations.
(c) For each constraint equation, what would the slack variable stand for?
(d) Obtain the optimal point graphically. What is the profit?

8-7 Suppose the head of the TV manufacturing company in Exercise 8-6 is offered a contract that calls for a minimum of 15 total TVs per day.

(a) Rework parts (a), (b), and (d) of Exercise 8-6 with this added constraint.
(b) Assuming that the manufacturer wishes to accept the contract, what actions might he take in order to insure that he can fulfill the demands of the contract?

8-8 A company makes two types of leather belts. Belt A is a high-quality belt, and belt B is a lower quality. The respective profits are 40¢ and 30¢ per belt. Each belt of type A requires twice as much time as a belt of type B; and, if all belts were type B, the company could make 1000 per day. The supply of leather is sufficient for only 800 belts per day total (both types A and B). Belt A requires a fancy buckle, of which only 400 per day are available. There are 700 buckles per day available for belt B.

If x_1 is the number of type A belts and x_2 the number of type B belts,

(a) Write the objective function to maximize profit P.
(b) Write the constraint equations.
(c) Sketch the feasible region and obtain the optimal solution graphically (i.e., the number of belts of each type which maximize profit).

8-9 An airline company is considering the purchase of three types of jet passenger airplanes. The cost of each 747 is $15 million, $10 million for each DC 10, and $6 million for each 707. The company has at its disposal $250 million.

The net expected yearly profits are \$800,000 per 747, \$500,000 per DC 10, and \$350,000 per 707. It is predicted that there will be enough trained pilots to man 25 new airplanes. The maintenance facilities can handle 45 707s. In terms of their use of the maintenance facilities, each DC 10 is equivalent to 1 1/3 707s, and each 747 is equivalent to 1 2/3 707s.

Given the above information, the company wishes to know how many planes of each type to purchase in order to maximize profit. Write the objective function and constraint equations for this problem.

8-10 A company which supplies laboratory animals to researchers has mice of a certain strain at three of its laboratories: 200 mice in Atlanta, 150 mice in Los Angeles, and 100 mice in Houston. A biologist in Chicago needs 250 of these mice and a biochemist in New York wants the remaining 200.

Find the shipment schedule that would minimize the shipping cost to the animal supply company. Individual shipping costs in cents per mouse are shown below.

	Atlanta	Los Angeles	Houston
Chicago	38¢	10¢	18¢
New York	34¢	22¢	25¢

***8-11** Using the simplex algorithm as developed in the chapter and demonstrated step by step for the dentist example (Sec. 8-4), solve Exercise 8-6 and compare the solution with that which you have found graphically.

8-12 Referring to the dentist example in this chapter,
(a) Work the problem on page 368.
(b) Work the problem on page 371.

8-13 The graph in Fig. 8-31 shows the solution to the following linear programming problem:
Maximize $10x + y$

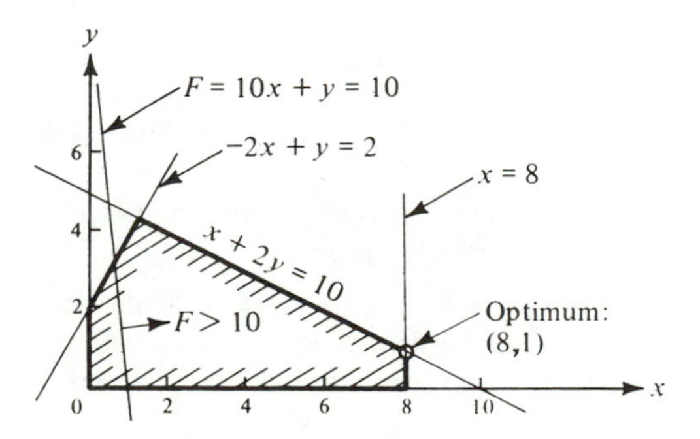

Figure 8-31

subject to:

(i) $x + 2y \leq 10$

(ii) $x \leq 8$

(iii) $-2x + y \leq 2$

(iv) $x \geq 0, y \geq 0$

For each of the following changes, describe whether or not the optimum will change and how.

(a) Change objective function to $F = -3x + y$.

(b) Change objective function to $F = x$.

(c) Change constraint (ii) to $10x + y \leq 80$.

(d) Remove constraint (i).

(e) Change constraint (ii) to $x \leq 12$.

8-14 A classic linear programming problem is the diet problem. The goal is to ascertain what quantities of certain foods should be eaten to meet the nutritional requirements at a minimal cost. The following table lists the vitamin content (A, C, and D) in mg of each of three foods and the unit cost in dollars, and the daily vitamin requirements in mg. For this problem assume the daily requirements can be satisfied by only these three foods.

Vitamin	Eggs (dozen)	Beef (pound)	Milk (gallon)	Minimum Daily Requirement
A	a_{11}	a_{12}	a_{13}	b_1
C	a_{21}	a_{22}	a_{23}	b_2
D	a_{31}	a_{32}	a_{33}	b_3
Cost	c_1	c_2	c_3	

Let E represent the dozens of eggs;
B represent the pounds of beef;
M represent the gallons of milk.

(a) Write the objective function.

(b) Write the constraint equations.

8-15 A large bakery has m plants located throughout the state. Daily bread production at the ith plant is limited to S_i loaves ($i = 1, 2, \ldots, m$). Each day the bakery must supply to its n distributing warehouses D_j ($j = 1, 2, \ldots, n$) loaves to meet demand requirements. The cost of transportation per loaf from plant i to warehouse j is C_{ij}. Let x_{ij} be the number of loaves shipped from plant i to warehouse j.

Write the objective function for minimum cost and the constraint equations for this transportation problem.

8-16 Given a wire of length 60 inches, design a rectangle bounded by wire using no more than the available length and having maximum area. Formulate the problem as a nonlinear programming problem and use a graphical solution.

8-17 Generate the radius r and height h of a cylinder with open ends which will have no less than 400π cubic inches of volume, using a minimum amount of material; i.e., make the surface area ($2\pi rh$) a minimum. Formulate the problem and show graphically the feasible region as well as the objective function for an arbitrary surface area of 200π. Discuss how you would proceed with a solution and indicate the difficulties that may be encountered. (*Hint:* Volume of a cylinder $= \pi r^2 h$.)

8-18 Determine the shortest path (and its length) if you were travelling from San Francisco to Washington along the indicated arcs of Fig. 8-32. Assume that you can travel only from west to east, that is, in the forward direction only.

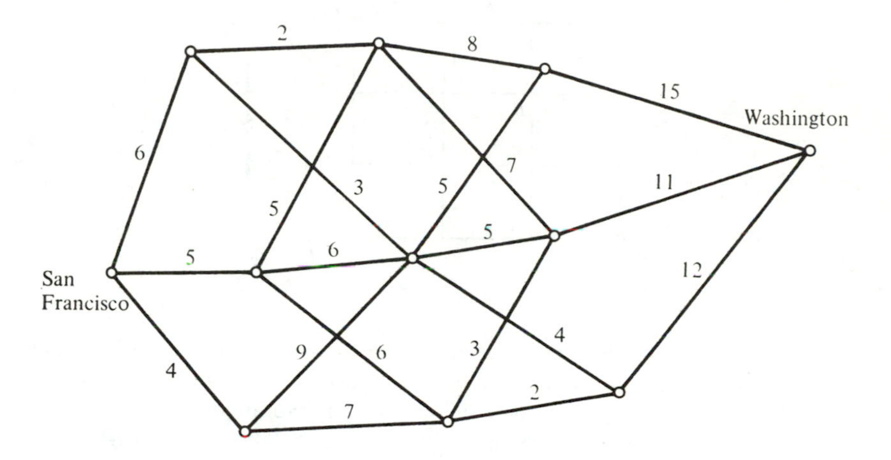

Figure 8-32

8-19 Given the diagram in Fig. 8-33. You are travelling from A to B and at each node you may travel only north or east.

(a) Determine the shortest path and its length.

(b) Determine the number of distinct paths between A and B.

(c) Explain the advantage of "working backwards" as opposed to an unordered search for the shortest path.

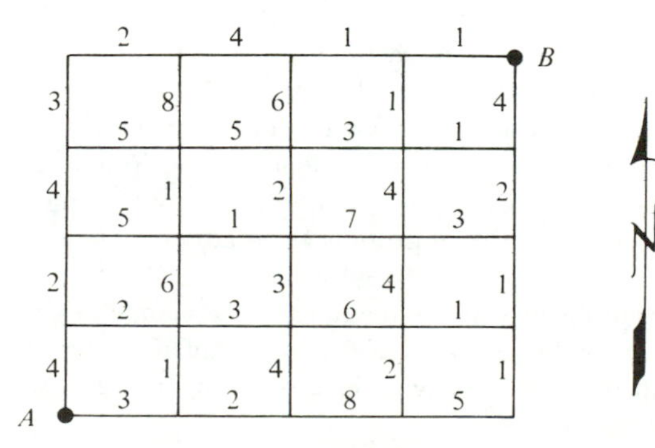

Figure 8-33

8-20 (a) Determine the shortest path from A to C via B in Fig. 8-34. At each node, you may travel only north or east.

(b) Will the distance determined in (a) be the same as the shortest distance from A to C? Explain your answer.

(c) How many distinct paths exist from A to B? from B to C? Is the product of these the same as the number of distinct paths from A to C? What does this product represent?

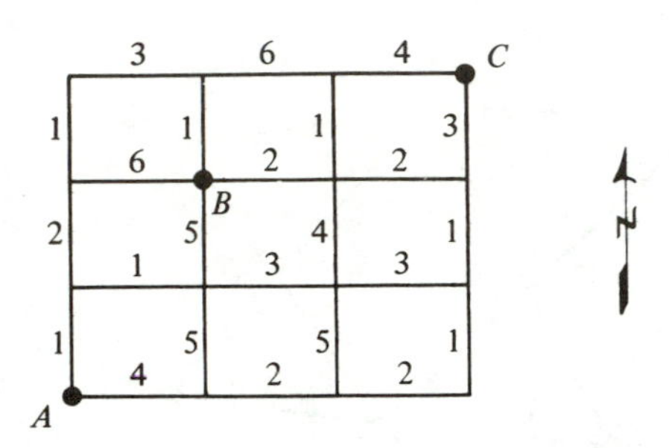

Figure 8-34

8-21 Referring to Sec. 8-8, the man described in the section is trying to make sequential decisions of whether or not to accept offers for his house. The offers ("outcomes") are random, following a probability distribution which is shown in Fig. 8-22 for offers from \$20,000 to \$120,000. The man is aware of this distribution.

Suppose we normalize these monetary values to a scale from 0 to 100. We do this by subtracting 20,000 from all monetary values and dividing by 1000. For example:

$$\$120{,}000 \longrightarrow \left(\frac{120{,}000 - 20{,}000}{1000}\right) = 100$$

Discuss on a qualitative basis how the man's decisions to accept or reject offers would differ (if at all) if the monetary values are scaled down in this way and the corresponding utility values in Fig. 7-13 are used in their place. Do this for curves (a), (b), and (c) of Fig. 7-13.

8-22 Referring to the project you chose in Exercise 1-1:

(a) What is the objective?

(b) Can this be expressed as an objective function? (Identify what your decision variables, i.e., x_j's, represent in your problem.)

(c) Identify the constraints on the solution of your problem.

(d) Can you express the constraints as equations in terms of the same x_j's as in (b)?

(e) Is your problem a linear, nonlinear, or dynamic programming problem? If it cannot be formulated as any one of these types, can you see any advantages and/or dangers in attempting to approximately model the problem as one of these?

9

DYNAMIC SYSTEM MODELS

9-1 INTRODUCTION—AN EXPOSURE

This section is to be considered as an exposure to concepts which are treated in more detail later in the chapter. The repeated treatment will reinforce the concepts which are the foundation of dynamic system models.

Systems

A *system* is a collection of elements aggregated by virtue of the links of form, process, or function which tie them together and cause them to interact. All that is excluded from the system constitutes the *environment*. The criterion for the choice of system size in terms of what elements to include or exclude is based on the same considerations as those which guide the modeling process in general as discussed in Chapter 5. A system model is constructed to help *understand* relationships between elements, forms, processes, and functions, and to enhance our ability to *predict* system response to inputs from the environment. This understanding may enable us to *control* system behavior by adjusting inputs to achieve a desired output whenever such control is possible.

System models have been constructed for the world population, for the economy of a country, for the learning process, for education in general, for a corporation, for an airplane, an automobile, transportation in general, for the heart, the brain, and the list may be extended to great length. Systems can be characterized in terms of an *input-output model*, as shown in Fig. 5-12. A system is usually abstracted and idealized in the modeling process so that its *state* can be described at any time t by a finite number of quantities $x_1(t), x_2(t), \ldots, x_n(t)$. These quantities are called the *state variables* of the system. The temperature of an oven may represent a state variable of the oven as a system with a single state variable. The position and velocity of a moving body may represent the state variables of a vibrating mass suspended from a spring. In a model of a population plagued by an epidemic, the system state variables may be the number of people susceptible, the number infected, and the number immune (see Sec. 9-8).

Dynamic Systems

A *dynamic system* is defined as a system whose state depends on the input history. The change in state is represented by the change in the state variables. The *state variables* are, therefore, in the nature of *levels* which result from the accumulation of change as a function of time. For example, in Fig. 9-1, the amount of water $x(t)$ in the tank can be considered the level of the system. If a quantity of α liters is added to the tank per unit time, and a quantity of β liters is removed per unit time, then after Δt units of time the amount of water in the tank, or the *level* of the system, will change by an amount $(\alpha - \beta)\Delta t$. Starting with a history of level $x(t_i)$ at time t_i, then after Δt units of time the level $x(t_i + \Delta t)$ becomes:

$$x(t_i + \Delta t) = x(t_i) + (\alpha - \beta)\Delta t \tag{9-1}$$

Note that $x(t)$ does not represent the height or level of water in the tank directly, although it could easily be converted to such a measure of the state. The word *level*, used to describe the state of a system, has a much broader meaning than its limited use to describe height.

The state or level of a system can be traced dynamically if we know the *rate* at which the state variable changes. In Eq. 9-1, the *rate* of change of $x(t)$ is $(\alpha - \beta)$; hence, from level $x(t_i)$ at time t_i we can derive level $x(t_i + \Delta t)$ at a later time.

A *static system*, in contrast to a dynamic system, is one whose state depends only on the current input, and not on the history of inputs. In

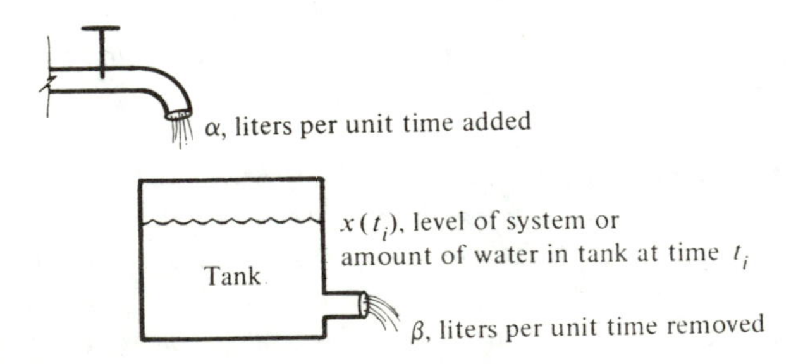

Figure 9-1 Amount of Water in Tank as a State Variable

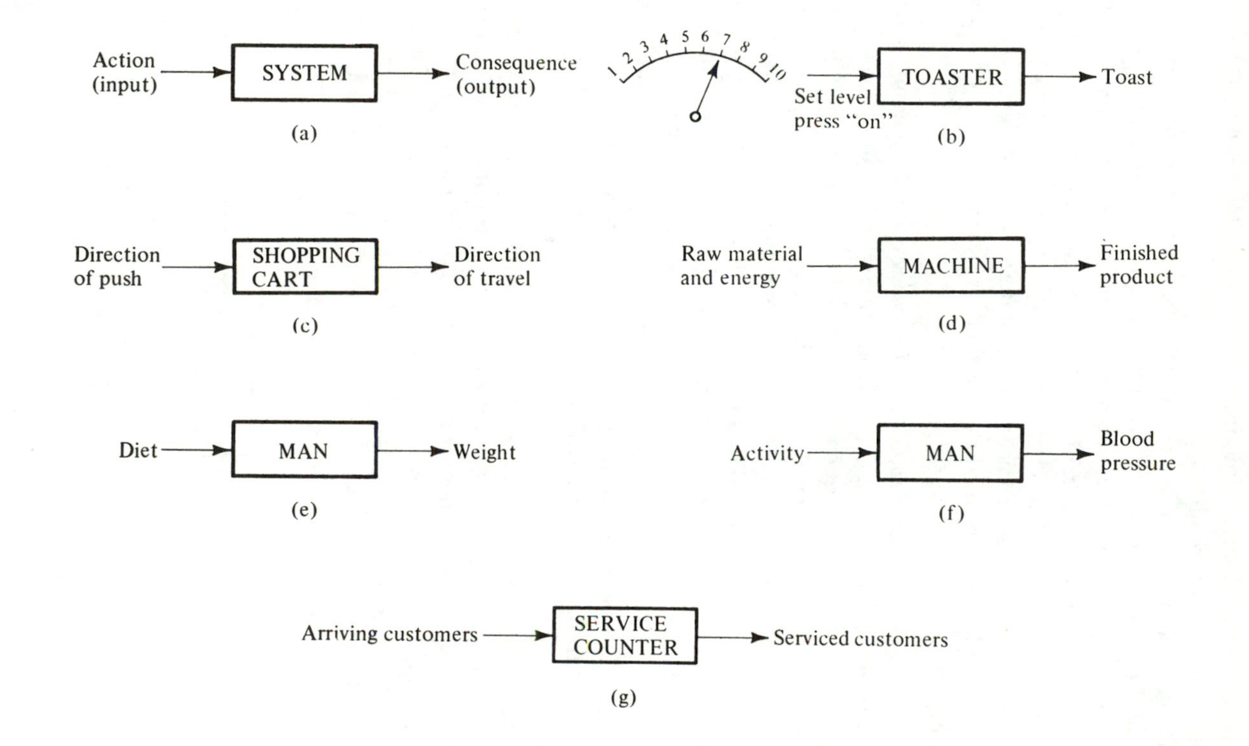

Figure 9-2 Examples of Dynamic Systems

reality, it is difficult to conceive of a static system, because nothing ever happens instantaneously. However, we often model a dynamic system as a static system over a limited time interval in order to simplify the analysis and still obtain valid results.

Dynamic systems characterize much of the natural and the man-made world, and man himself. The ubiquity of dynamic systems is manifested in the human body (in which body temperature, blood pressure, and other indicators are state variables of importance), and in human patterns of behavior. The activity of problem solving is a dynamic system process in which we endeavor to go from an initial state to a desired goal state. The state, whether initial, intermediate, or final, is described by state variables which change as we progress to a solution.

Flow of Energy and Information in a Dynamic System. A dynamic system is shown schematically in a most general form in Fig. 9-2(a) in which an action, input, is the *cause* of a consequence, output. The examples in the rest of the figure show the generality of the model in part (a).

Let us consider the toaster of part (b) in which a dial indicator can be set at 10 different levels of darkness of the toast, from light to dark in varying degrees. Suppose we set the dial at level 7 which should result in medium dark toast. If the toast is too dark, we set the dial lower the next time until we develop, with experience, a relationship between the dial setting (input) and the darkness of the toast (output). The input causes *energy* to flow in the form of heat from the electric wiring in the toaster to the bread. The observation of the consequence, namely, the darkness of the toast, constitutes *information* which we use to change the input in the next step. So while *energy* flows from input to output, *information* flows from output to input. The input triggers a flow of energy, *causing* an output; and the output triggers a flow of information, causing a new input. Schematically, these flows are shown in Fig. 9-3. Thus, when we have a desired goal in a dynamic system, an understanding of the input-output relationship will enable us to select an input most compatible with the desired output.

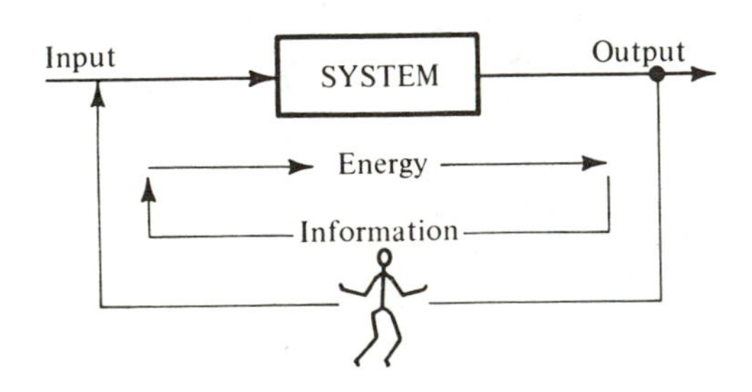

Figure 9-3 Flow of Energy and Information Between System Input and Output

Problem 1. Describe the kind of information that changes the input in each part of Fig. 9-2.

Control

There are two fundamentally different ways of controlling a system to achieve a desired goal output:

1. Select an input and wait for an output with no interference during the waiting period.
2. Select an input, observe the trend or direction of the output, and modify the input accordingly to get as close as possible to the desired goal output.

For example, when a helmsman steers a boat head-on in the direction of a lighthouse, he may do it in one of the two ways described in the above schemes:

1. Set the boat on the desired course and retire to his cabin.
2. Set the boat on the desired course, observe the actual direction taken by the boat, and adjust the direction accordingly.

In both cases the steersman exercises control over the direction of the boat. However, in Case 1 the input triggers a flow of energy, causing an output, but there is no flow of information from output to a new input to close the "loop". Hence, such control of consequences or output from a system is called an *open loop control.* Case 2 is a *closed loop control* also known as *feedback control*, because the output provides information that is fed back to the input.

For the toaster example considered earlier, we may say that each time we set the dial and wait for toast without further interference, we have an open loop control. However, considering repeated use of the toaster, trying to get a desired level of darkness by adjusting the dial on the basis of the results from each open loop operation, we have feedback and, hence, a closed loop control for the total operation. The closed loop control is exercised, however, only at discrete time intervals.

Ubiquity of Control. Feedback control is a key to survival; it has endowed all living organisms with the system characteristics most productive for survival. The main feature in control with feedback is the flexibility which can be exercised by the controller to vary the input, leading to a change in output. The species with a high degree of specialization were less likely to adapt to disturbances from the environment and, therefore, less likely to survive.

Control can be exercised at various levels. Man, for example, with the supreme goal of survival, makes a choice of lower-level goals which may enhance survival. In making such choices, he exercises control over his destiny. The lower-level goals are achieved by lower-level feedback control systems. Thus, control can be exercised and amplified in stages. Control can be voluntary, such as guiding a shopping cart or steering a boat, or it can be involuntary, such as the control of temperature or blood pressure in the human body.

The emergence of an era of automation was marked in essence by the ability to monitor a process continuously by feedback control without resorting to a human operator. Automation has advanced greatly with the advent of the digital computer that can process information so accurately and rapidly that the output has little time to deviate from a desired value before the information channel in the feedback loop effects a corrective change to cancel or diminish the deviation.

The feedback control mechanism is evident in government. For instance, a government may introduce economic policy regulations (control) to achieve desired levels of output such as employment, inflation, gross national product, etc. When the actual output differs from that desired, policies and regulations are changed. The ecological system is balanced by feedback, anywhere from the balance of the number of rabbits and predators, to germs and population susceptible to them in an epidemic.

Dynamic systems with feedback in the natural and man-made world seem to be striving to achieve goals which in turn serve as inputs for the purpose of achieving higher goals yet. This hierarchy of goals and controls to achieve them can be extended to progressively higher levels until we reach a level where science, philosophy, and theology converge and meet at a junction to pose the most profound questions: To what extent are we masters of our destiny? Is there a *Grand Steersman* who exercises continuous closed loop control by making observations and changing the input to achieve a *Grand Goal*? Or is the control open loop, in which case a flow of energy was started at genesis to achieve a grand goal and no interference has followed? Is there a supreme goal? If so, was it fixed at genesis, or is it subject to change? If it is subject to change, what frame of reference guides such change?

The Bible offers models of the Grand Steersman, but we must stop at this point and permit each to reflect on his view of the *Grand Model*. Such reflection can be most exciting and most helpful in putting in perspective our day-to-day activities and giving the proper weight to deviations from our goals (and resulting stress) of a lower level, which are subordinate to survival and to the grand goal.

Cybernetics. Norbert Wiener (1894–1964), one of the great mathematicians of this century, recognized the ubiquity of control and its dependence on the controlling "helmsman" and made outstanding contributions to the development of a theory of control. He named this field of study *cybernetics*, which comes from the Greek word for helmsman, *Kybernet*. His book on the subject of cybernetics [74], published in 1948, is still a classic. In view of the above discussion, it is not surprising that Wiener also published a book entitled *God & Golem, Inc.*, (1964) in which Golem refers to a legendary creation of a Robot by Rabbi Judah Low of Prague in the 16th century. The legend claims that the rabbi blew the breath of life into a clay structure and made it the instrument for achieving desired goals, such as protecting the lives of the persecuted Jewish people in Prague.

Problem 2. Describe Fig. 3-11 as a dynamic system with feedback.

Problem 3. Suppose a man tells his wife that she can purchase a new automobile for no more than $4000. She proceeds to a show room to make a choice. Can you relate this process to government regulation and individual behavior in terms of levels of control?

9-2 BUILDING BLOCKS IN DYNAMIC SYSTEM MODELS

In this and the following section we describe a number of dynamic systems, using block diagrams that identify the elements of the dynamic systems and the flow of energy and information. The examples will serve to emphasize the generality of dynamic system models.

Block Diagrams

Block diagrams are used to describe the components of a dynamic system and the flow of actions and consequences, or causes and effects, which link the components to form a total system. For example, the half-adder of Fig. 3-9, repeated here as Fig. 9-4 for easy reference, is an example of a block diagram. The language of block diagrams employs an alphabet of four symbols (see Fig. 9-5):

1. An arrow
2. A branch point
3. A block
4. A comparator or summation junction

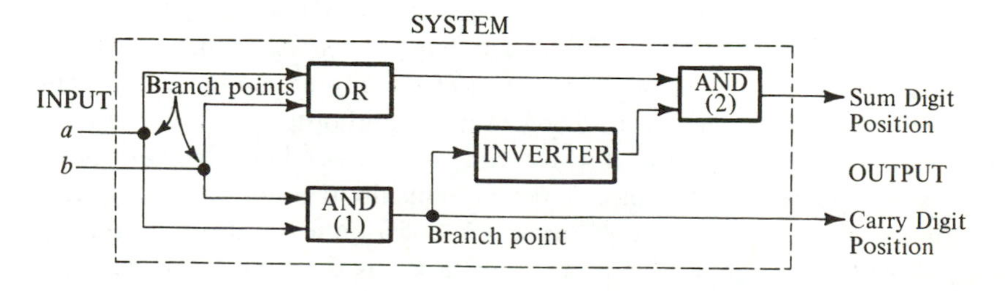

Figure 9-4 A Half-Adder as an Example of a Block Diagram

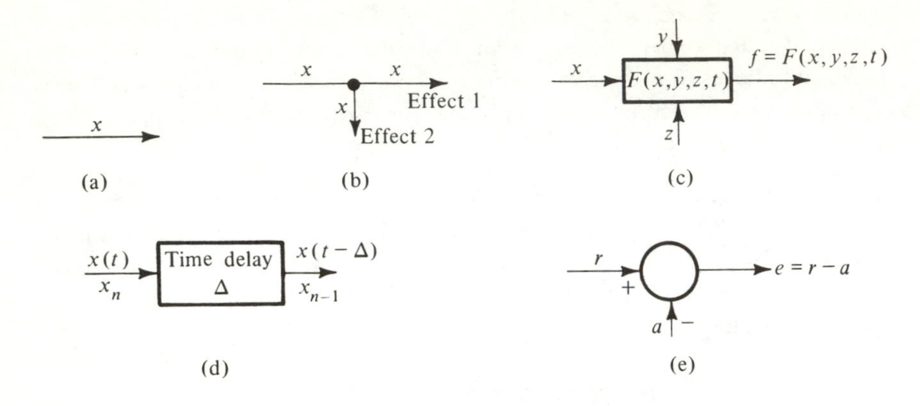

Figure 9-5 Symbols in Block Diagrams of Dynamic Systems: (a) Signal; (b) Branch Point; (c) Block with Output-Input Relationship F; (d) Block with Time Delay (Unit Delay for Discrete-Time Systems); (e) Comparator

An arrow is used to designate a signal from cause to effect. The signal represents a variable acting in the direction of the arrow. The arrows in the half-adder of Figure 9-4 are signals.

A *branch point* is used to describe a situation in which a signal causes more than one effect. In Fig. 9-4, the output signal from the box *AND*(*1*) has a branch point, causing an effect in the *INVERTER* box and in the *Carry Digit* position.

A *block* in the form of a rectangular box represents a functional relationship between the incoming (inputs, causes) and outgoing (outputs, effects) signals. The input signals are the *independent variables*, and the output signals are the *dependent variables*. Each of the boxes in Fig. 9-4 is a block. The block can be represented mathematically by expressing the output signal in terms of the input signals. The relation may be algebraic, or it may be a differential equation relationship. A simple delay is also a possible relationship, as indicated in Fig. 9-5(d). For certain types of blocks (linear relationships) it is customary to speak of a *transfer function*, which relates the Laplace transform of the output to that of the input. Similarly, the *system function* relates the Fourier transforms of input and output. When the system is not linear or when the use of transforms is not desired, we describe the box by its *input-output* relationship. This relationship transforms the inputs to outputs in a specified manner. Thus, in Fig. 9-5(c), the input-output relationship depends on signals x, y, and z, and also on the time t. In some cases the time dependence does not appear. The block may represent a constant multiplier of an input (amplifier), an integrator to accumulate changes in level to yield new levels, or other types of functions. Of value in some systems representation (discrete-time systems) is the unit delay Δ, where Δ is the interval between successive observations of the signals in the system. For example, in a compound interest problem, the principal at the end of year n, P_n, is obtained as follows. The principal at the end of the preceding year, P_{n-1}, is delayed $\Delta = 1$ year, and then multiplied by the annual interest rate, i. The accrued interest $i \cdot P_{n-1}$ is added to P_{n-1} to generate the principal at the end of the current year, n. In Fig. 9-5(d) the time delay box has a signal $x(t)$ or x_n entering the box, and a signal $x(t - \Delta)$ or x_{n-1} leaving the box. Namely, at time t, the signal leaving the box is that signal which entered Δ units of time earlier.

A *comparator* or *summation junction* in the form of a circle designates that the outgoing signal is the sum of the incoming signals. The incoming signals may be added or subtracted and this is indicated by the sign (+ or −) next to the signals entering the comparator. For example, if signal r enters a comparator with a signal a as shown in Fig. 9-5(e), then the outgoing signal e is

$$e = r - a$$

This explains the reason for the name comparator. If r is a desired reference temperature set as the goal in a home-heating system, and a is the actual temperature measured in the home, then a is *compared* with the reference r in the comparator. If they agree, $e = 0$, namely, the deviation or *error* e is zero; otherwise, e differs from zero and causes a signal to flow and effect a change that will bring the prevailing temperature closer to the desired temperature. The comparator also appears sometimes with a summation sign or two diagonally crossed lines in the circle, $\textcircled{\Sigma}$ or $\otimes$

Models of Dynamic Systems Using Block Diagrams (Examples from Early History)

We shall now show some examples of dynamic system models and illustrate the use of block diagrams.

EXAMPLE 1. Heron (also known as Hero), who lived in Alexandria in the first century A.D., was one of the early contributors to feedback control. In one of his creations, a float valve was employed to maintain a constant level of wine in an "Inexhaustible Goblet" [75], as shown in Fig. 9-6.

The level of wine in the goblet is maintained by the following mechanism. A tube connects the goblet to a sensing vessel so that the level of wine in both is the same. The float arm is adjusted in length so the float is at the desired level of the wine. Then when wine is taken out of the goblet, the float goes down an amount equal to the decrease in the level of the wine in the goblet. This causes the lever to lift the valve arm and open the valve so wine will pour from the reservoir of wine into the goblet. As the desired level is approached the valve opening decreases until it is finally shut.

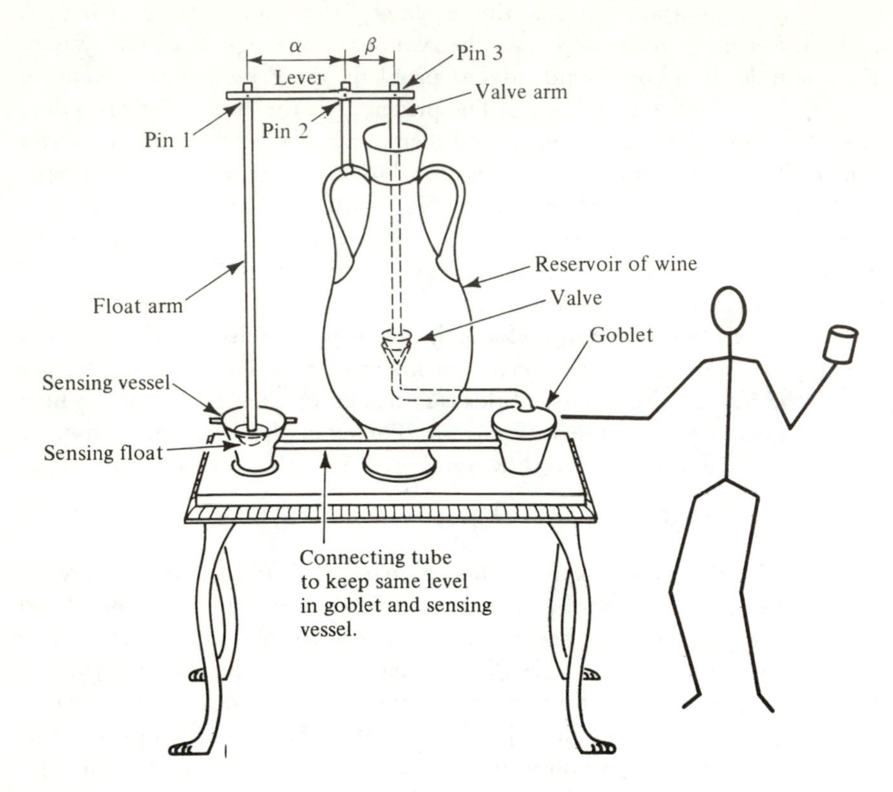

Figure 9-6 Heron's Inexhaustible Goblet

A block diagram for the "Inexhaustible Goblet" is shown in Fig. 9-7. All four elements of Fig. 9-5 appear in the diagrams. The input is a reference r which represents the level of the float, or the length of the float arm, based on the desired constant level of wine in the goblet. The output h is the actual level of wine in the goblet. The branch at the output indicates that this actual level h is also transmitted to the sensing vessel through the connecting tube, and thus becomes the input to the float. The block labeled *Sensing float* represents the float with a transfer function of 1 (recorded inside the block) because the float assumes exactly the level h with no amplification or change of any form. Once the actual level h is sensed by the float, it is subtracted from the reference level by the downward motion. If $r = h$, then the error e at the comparator is zero and there is no input to the block *Lever* and, therefore, the output v, valve opening, is also zero.

The transfer function for the block *Lever* is designated by the symbol k which represents the amplification of input e, namely,

$$v = ke$$

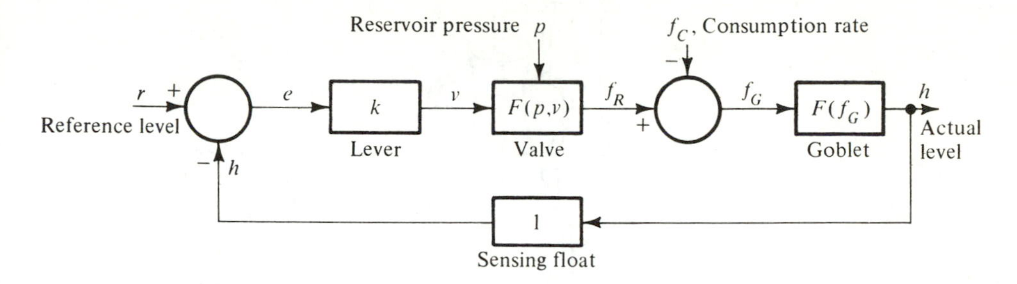

Figure 9-7

r Desired Reference Level (When Valve is Closed)
h Actual Level
v Valve Opening
p Pressure at Valve (Depends on Level of Wine in Reservoir)
f_R Flow Rate into Goblet (From Reservoir)
f_C Flow Rate Out of Goblet (From Consumption)
f_G Net Flow into Goblet

k is the ratio of the two arms in the lever mechanism of Fig. 9-6, i.e., $k = \beta/\alpha$. When wine is consumed from the goblet, the float goes down an amount e and, therefore, lifts the valve arm by an amount ke and opens the valve by this amount.

The block *Valve* has two inputs, the amount of opening v and the pressure p of the wine in the reservoir which depends on the level of wine in it. The flow rate f_R from the valve depends on these two inputs; hence, the block indicates a transfer function $F(p, v)$, namely,

$$f_R = F(p, v)$$

The actual form of the function is not shown here, but it can be derived.

The net flow rate of wine into the goblet, f_G, depends on the flow rate, f_R, from the valve in the reservoir and on the consumption rate, f_C, from the goblet. Thus, at the second comparator of the diagram, we have

$$f_G = f_R - f_C$$

Finally, the block *Goblet* has a net input, f_G, and a state variable, h, which are functionally related. The shape of the goblet and the deviation of h from the desired reference level r, will contribute to the actual form of this transfer function, $F(f_G)$.

Figure 9-8 shows Heron's alternate scheme for a wine dispenser. The desired reference quantity r of wine is controlled by the position of a regulating weight on the left of the lower lever. Moving the regulating weight further to the left is analogous to shortening the float arm in Fig. 9-6.

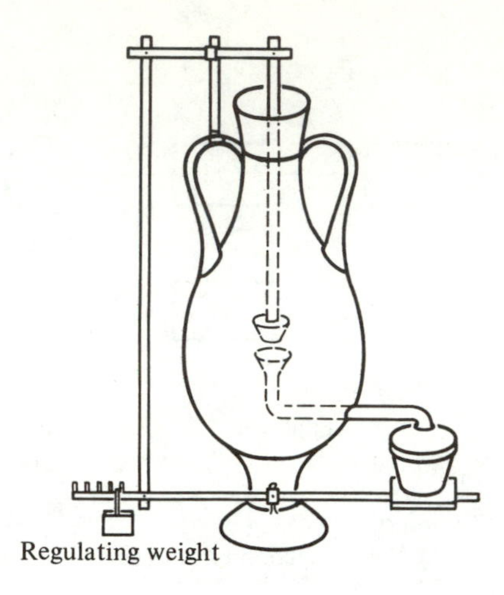

Figure 9-8 Heron's Wine Dispenser Controlled by a Weight

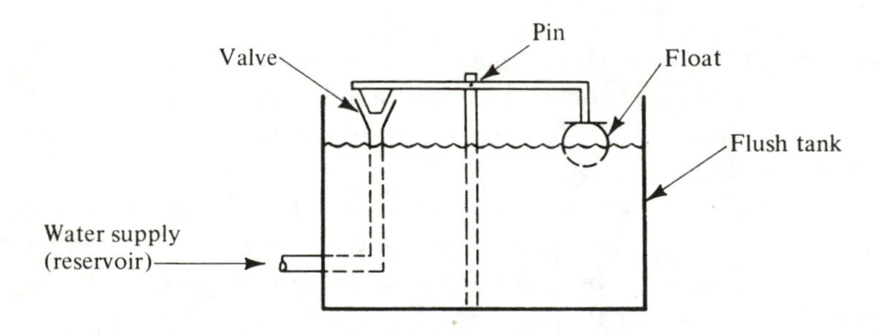

Figure 9-9 Flush Tank Level Control Mechanism

Problem. Show that the block diagram of Fig. 9-7 applies to the dispenser of Fig. 9-8.

The block diagram of Fig. 9-7 applies to flush tanks for toilets used today, as shown in Fig. 9-9. No such mechanisms were used in toilets in Heron's days. First-century man was more busy keeping his wine goblets full. . . .

EXAMPLE 2. Before proceeding to discuss the first known feedback control device that dates back to the days of King Ptolemy II of Egypt in the 3rd century B.C., it will help us to consider an example of an open loop control system known as Heron's floating syphon [75].

The rate of flow f of water from a container through an orifice, as shown in Fig. 9-10, is given by

$$f = A\sqrt{2gh}$$

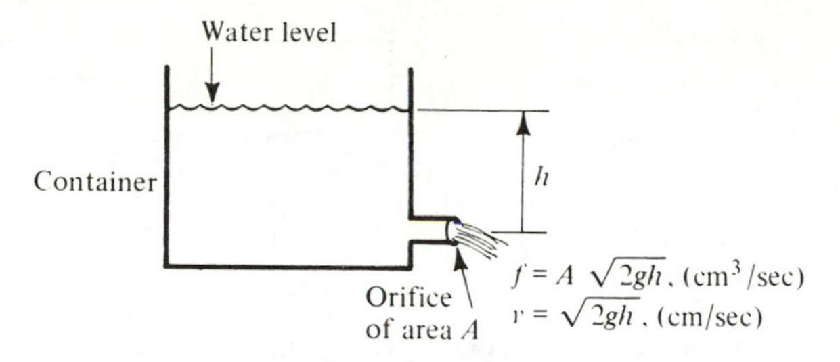

Figure 9-10 Flow through an Opening from a Tank

in which h is the level of water above the orifice, A is the cross-sectional area of orifice opening, and $g = 981$ cm/sec^2 is the acceleration of gravity.

The rate equation can be derived by equating the loss in potential energy, mgh, of water of mass m, as it drops h units of height from the top of the container to the orifice, with the gained kinetic energy at the exit from the orifice $1/2\, mv^2$, in which v is the velocity of exit in units of cm/sec:

$$mgh = \frac{1}{2}mv^2$$

$$v = \sqrt{2gh}$$

The rate of flow f is obtained by multiplying velocity v by the cross-sectional area A to yield units of volume per second, say, cm^3/sec. As the flow continues through the opening, the level of water in the container recedes and the rate of flow f gets smaller.

To maintain a constant flow rate f through an orifice despite decreasing water level in a container, Heron contrived the floating syphon scheme shown in Fig. 9-11. The distance h is maintained despite the receding level of water in the container.

The block diagram for the floating syphon is shown in Fig. 9-12. The control is open loop. The head h is maintained between tube opening and water level by the float by virtue of the original input of head h. The objective is to maintain a constant flow rate f (output); however, nowhere in the system is the flow rate f sensed. True, this variable f is determined by h,

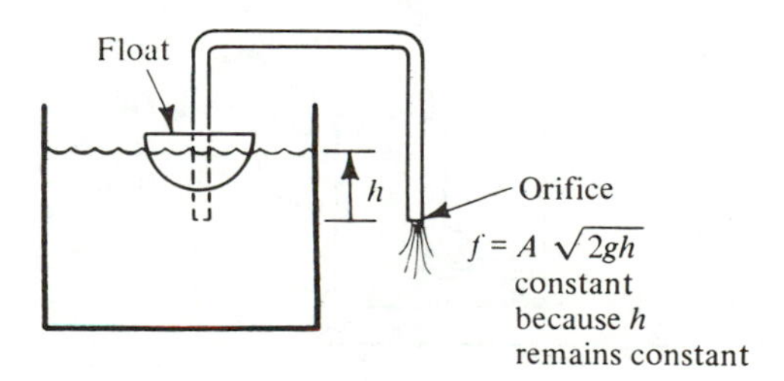

Figure 9-11 Heron's Floating Syphon

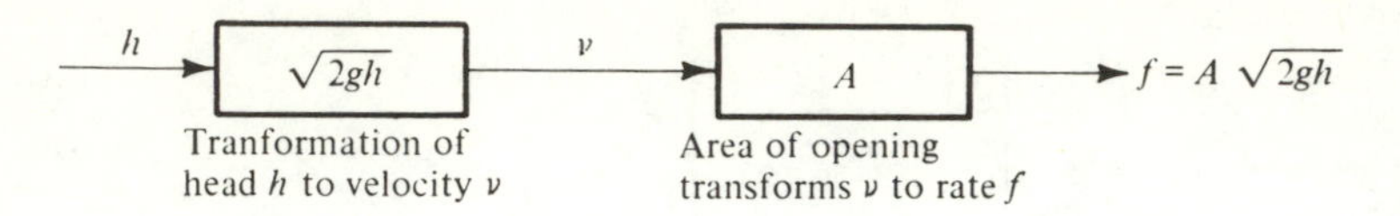

Figure 9-12 Block Diagram of Heron's Floating Syphon

but there is no feedback. The flow f will remain constant, not because of feedback control but only if no disturbance is present. For example, should an obstruction block the flow from the tube, or should the float be rocked from its stable position, there is no automatic feedback that will sense the *change in rate of flow* and cause a correction.

Using the present example as background information, we can now proceed to the water-clock of the 3rd century B.C.

EXAMPLE 3. One of the earliest automatic feedback control systems is the water-clock of Ctesibius (or Ktesibios) of Alexendria, a contemporary of Euclid and Archimedes who lived in the 3rd century B.C. The Ctesibius water-clock is considered by some [75] to be the first known feedback device. The clock, shown in Fig. 9-13, consists of a column C mounted on a tank float TF and connected to the base of drum D. As water flows at a constant

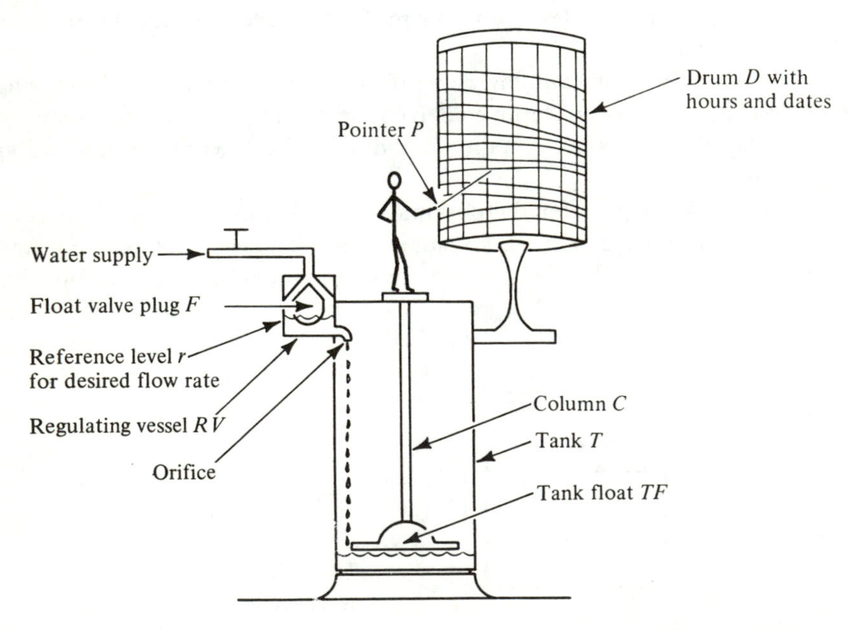

Figure 9-13 Ctesibius (or Ktesibios) Water-Clock System

rate into the tank *T*, the column *C* rises at a constant rate and causes the drum *D* to rotate. The pointer *P* on the column points at the date and hour marked on the drum. The control or regulating mechanism resides in the regulating vessel *RV* that connects the tank *T* to the water supply (reservoir of water). The regulating vessel has a cone-shaped float valve plug *F*, so designed that when the level of water in the vessel *RV* is at the reference level *r*, the desired constant flow of water trickles into the Tank *T*. If the level of water in the regulating vessel rises, causing the flow rate into the tank *T* to increase, then the float *F* rises, reducing the size of the valve opening and decreasing the amount of water flowing into the vessel. If the level is lower than the reference level *r*, the flow from the orifice into the tank *T* decreases, but then the float *F* drops and more water flows from the opening into *RV*. The float *F* will stay undisturbed only if the rate of flow into tank *T* is at the desired rate. This arrangement results in a control mechanism that seeks a constant level of water in the regulating vessel *RV* and, thus, a constant flow through the orifice into the tank *T*. The constant flow rate is maintained despite possible changes in water pressure *p* into vessel *RV* because of changes in water level in the reservoir upstream. The block diagram for the control system is shown in Fig. 9-14.

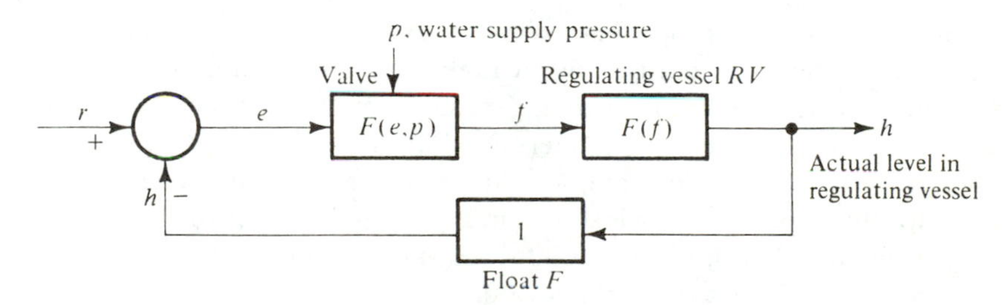

Figure 9-14 Block Diagram for Regulator in Regulating Vessel *RV* in Figure 9-13:
r Desired Reference Level in Regulating Vessel *RV*
h Actual Level in Regulating Vessel *RV*
e $r - h$, Valve Opening
p Water Supply Pressure
f Flow Rate of Water into Regulating Vessel *RV*

The float plug valve *F* serves the dual function of sensor and controller. As a float it senses the level *h* in the regulating vessel, and as a cone-shaped plug it controls this level by increasing or decreasing the size of valve opening and, hence, the flow from the water supply upstream into the tank *T*.

EXAMPLE 4. The drinking-straw regulator, Fig. 9-15, designed to enforce a uniform rate of drinking from a wine barrel, has its origin in China of the 12th century A.D. [75]. The bamboo tube is about two feet long and has in it a float with cone-shaped ends. When the straw is not in use, the

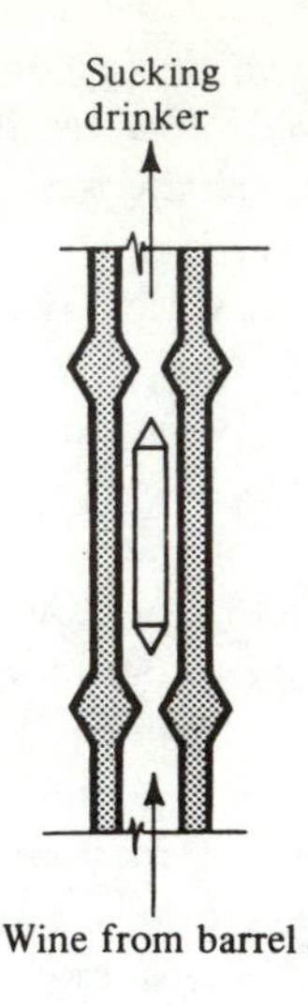

Figure 9-15 Vertical Section of Drinking-Straw Regulator

float sinks, closing the opening. When a drinker sucks from the top, the float rises. If he sucks too rapidly, it makes the top opening smaller; if he sucks too slowly, the lower opening gets smaller. The optimum rate, which is the desired reference rate, is determined by the weight of the float, and is obtained when the suction rate and flow rate are the same. The block diagram for this drinking straw is shown in Fig. 9-16. The control scheme is not fully automatic because the equilibrium state of flow is achieved by the drinker who optimizes the sucking rate.

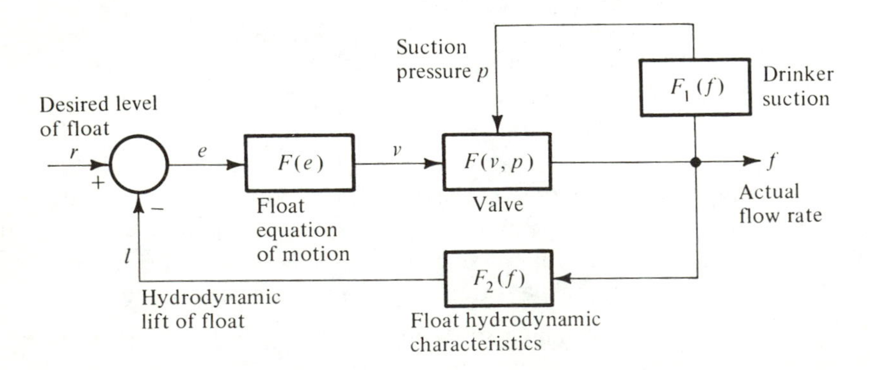

Figure 9-16 Block Diagram for Drinking Straw of Figure 9-15

9-3 MORE RECENT MODELS OF DYNAMIC SYSTEMS

EXAMPLE 1. The house-heating control system used in many dwellings can be described by the block diagram of Fig. 9-17 which is analogous to those described in the preceding examples of wine- and water-level control systems. We leave it as an exercise to the reader to identify the elements of the system from the block diagram.

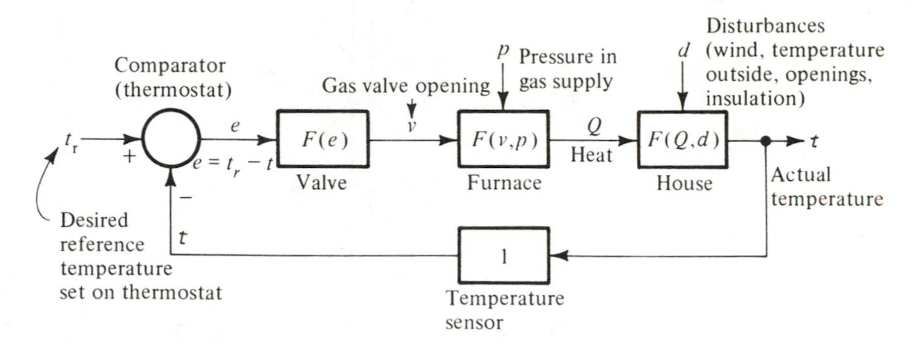

Figure 9-17 Block Diagram for House Heating Control System

EXAMPLE 2. In 1788, James Watt invented the flyball governor to control the speed of rotation of a shaft driven by a steam engine. The constant speed of rotation was required in steamships, generators, and other devices. The control mechanism is shown in Fig. 9-18. Gears transmit shaft rotation to rotation of the governor. As the speed of rotation of the shaft increases, the governor spin increases, and so do the centrifugal forces acting on the flyballs. The flyballs move outward, pulling down the sleeve S with lever L, and thus close the steam valve opening V. This causes less steam to flow and thus the speed of shaft rotation is reduced. When shaft speed gets too low, the process is reversed, i.e., the valve opens more than under the desired reference conditions, more steam flows in, and the speed is increased toward the desired level. A block diagram of the governor is shown in Fig. 9-19.

EXAMPLE 3. Consider a dynamic pricing model in terms of supply and demand. The law of supply and demand claims that a constant market price is achieved only when supply and demand are equal. This is based on the hypothesis that when prices rise, the market demand for goods decreases and the supply increases. The reverse happens when prices decline. The consumer (buyer) and manufacturer (seller) enter a feedback control loop which restores equilibrium to permit transactions. The model discussed here is not

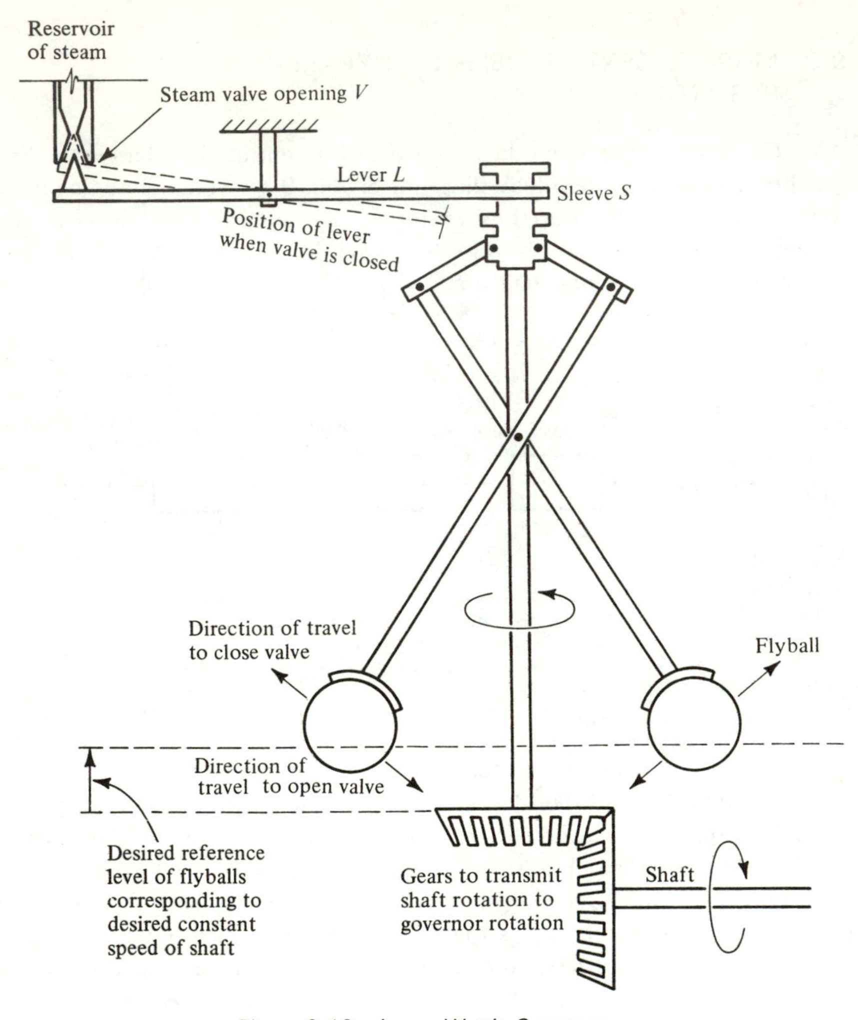

Figure 9-18 James Watt's Governor

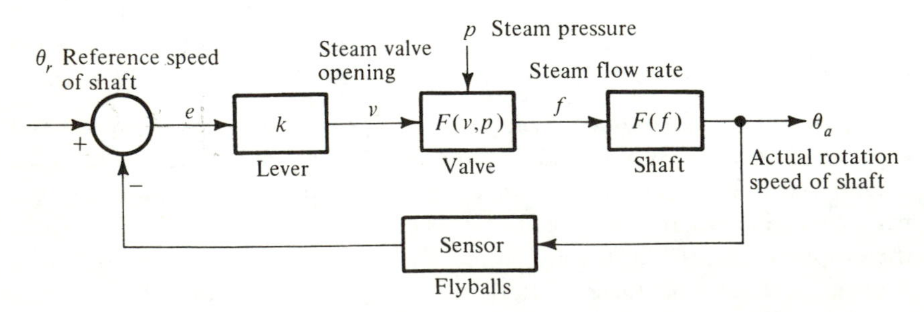

Figure 9-19 Block Diagram of Watt's Governor

a model for the actual exchange of goods for money, but rather a model for arriving at a satisfactory price so an exchange can take place.

The amount that the consumer buys depends on the price—the higher the price, the less purchased. The manufacturer increases the price with increased demand on the number of goods. This is shown in Fig. 9-20. In order for the price to converge to price P_e where a transaction takes place, the slope of the consumer price curve must be less steep than that of the manufacturer curve. Namely, the manufacturer should increase the price with increase in demand at a higher rate than the consumer readiness to pay a higher price increases with decrease in supply. In other words, in the equations for the consumer and manufacturer price curves (see Fig. 9-20),

$$P = \alpha_c + cQ \qquad \text{and} \qquad P = \alpha_m + mQ$$

the coefficients c and m, representing the slopes of the lines, must be such that $|c| < |m|$. P is the price, Q the quantity, and α_c, α_m are the intercepts of the lines with the P axis.

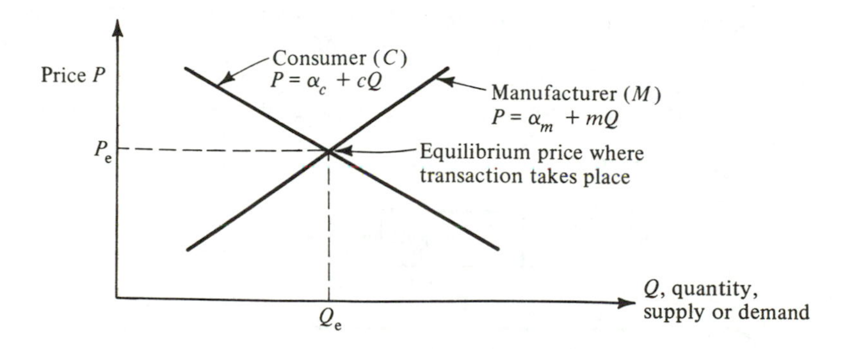

Figure 9-20

To understand why the above inequality must hold, we study Fig. 9-21. Part (a) shows the loop to equilibrium when $|c| < |m|$, and part (c) shows a runaway situation in the negotiation for a price when $|c| > |m|$. In part (b) there is an impasse in a closed loop where no change takes place and, hence, the situation may be viewed as a stable oscillation in price negotiation. In part (a), the consumer (C) first wishes a quantity Q_1 at a price P_1, point 1 in the figure. The manufacturer (M) offers C a larger quantity Q_2 at the price P_1 that C is willing to pay, point 2 in the figure. But for such a quantity, C is ready to pay only a price P_2 (point 3). At price P_2, M offers only a quantity Q_3 (point 4). For quantity Q_3, C is ready to pay P_3 (point 5). From point 5 we proceed with the same argument as we started at point 1 until we converge to the point where the lines for M and C intersect.

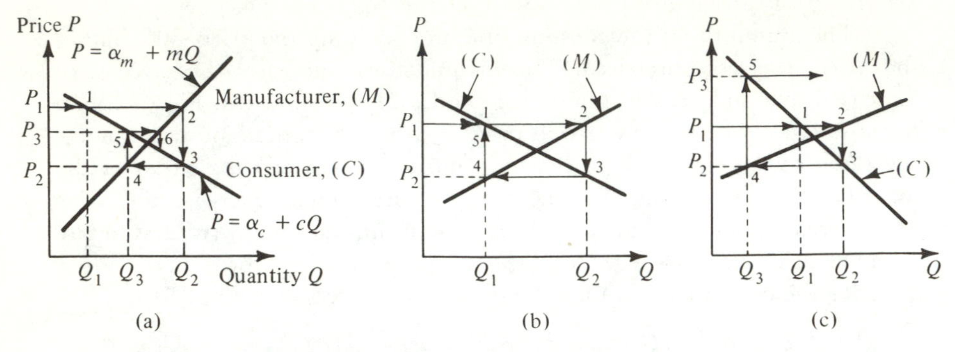

Fig. 9-21 (a) $|c| < |m|$; (b) $|c| = |m|$; (c) $|c| > |m|$

Starting at point 1 in Fig. 9-21(c) and following the reasoning used for part (a), we generate points 2, 3, 4, 5 and there is no convergence to a transaction price. Note that P_3 falls outside the range between P_1 and P_2, and so does Q_3 with respect to Q_1 and Q_2. This is not the case in part (a) where convergence is achieved. In part (b), the loop of points 1, 2, 3, 4 is continued without converging or diverging.

The block diagram for the pricing models of Fig. 9-21 is shown in Fig. 9-22.

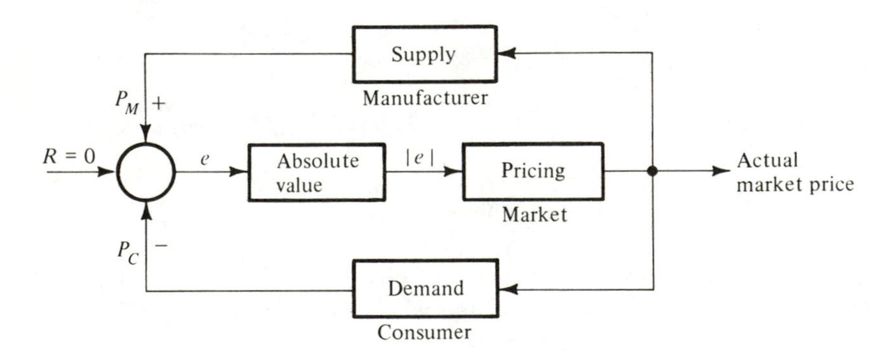

Figure 9-22 Block Diagram for Law of Supply-Demand in Pricing Model:
$e = P_M - P_C$
Manufacturer Price, P_M, Increases with Quantity
Consumer Price, P_C, Decreases with Quantity
R = Transaction Price Fluctuation = 0

EXAMPLE 4. Figure 9-23 shows a block diagram for a dynamic system model of a person trying to achieve a desired goal. Suppose the *goal* is to get the grade A in a course on patterns of problem solving. The *plan* calls for certain levels of grades in quizzes, homework, the midterm, the final; it may

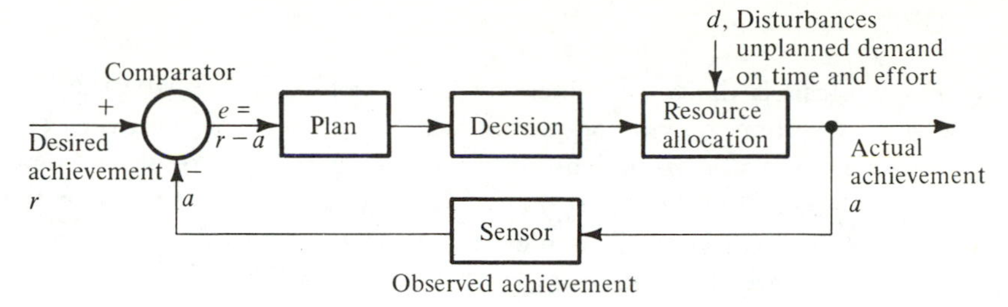

Figure 9-23 Dynamic System with Feedback to Achieve a Goal

include meeting the instructor periodically and impressing him, class participation, etc. The *decision* block will select the best strategy from the available actions, and the *resource allocation* of time and effort will be made accordingly. The *achievements* are monitored (sensed) by the student and compared with the desired goal which may be translated to an effective overall score of 90 or better (to secure an A). The *plan, decisions,* and *resource allocations* are adjusted to reduce the error e between the desired (reference) and the actual record of achievements.

General Block Diagram for a Dynamic System with Feedback. Figure 9-24 is a general block diagram model of a dynamic system with feedback. This model applies to many situations in the natural and man-made world, and is not limited to physical systems. Wherever a cause-and-effect relationship can be identified, conceptually, at least, a dynamic system model can be generated. In the next section we consider a model of a biological system.

> NOTE. Caution must be exercised in the measurement of output. As we mentioned in Sec. 5-8 in connection with the uncertainty principle, the act of measurement may change the quantity we wish to measure. We may also find that the measurement procedure is not effective because it does not measure the output of interest. Recall Sec. 1-8, "Will to Doubt," and the story on the measurement of student reading ability.

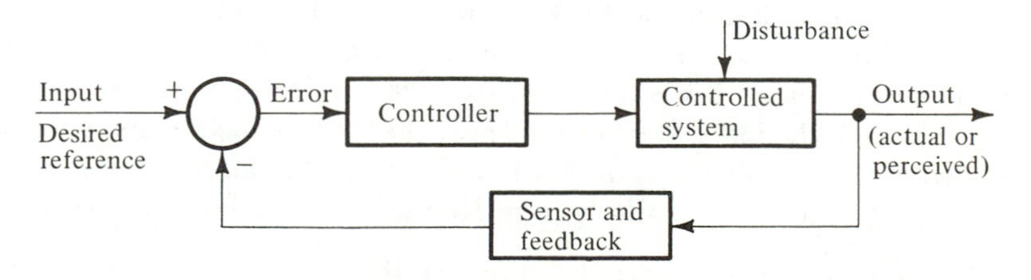

Figure 9-24 General Block Diagram for a Dynamic System with Feedback

9-4 HOMEOSTASIS—CONTROL IN LIVING ORGANISMS

Homeostasis is the word coined by Cannon [76] to describe the biological control system of living organisms which regulates the state variables of the internal environment such as body temperature, blood pressure, sugar level in the blood, body water content, concentration of chemicals in the tissues, etc. Biological control enables living organisms to function over a range of environmental disturbances and changes in the biological system itself.

As an example of homeostasis we describe the elements of the blood pressure control system in the human body [77, 78]. Blood pressure, just as temperature, in a human body remains within prescribed bounds throughout life. This is achieved by the homeostatic mechanism that is shown schematically in Fig. 9-25. The basic components are a *sensor* (also called a *receptor*) to detect the deviation of blood pressure from the desired level, and a feedback mechanism to convey the message to an *effector* or *controller* that attempts to restore the system state to the desired level. The feedback mechanisms may be the nerves or, for some purposes, the hormones circulating in the blood.

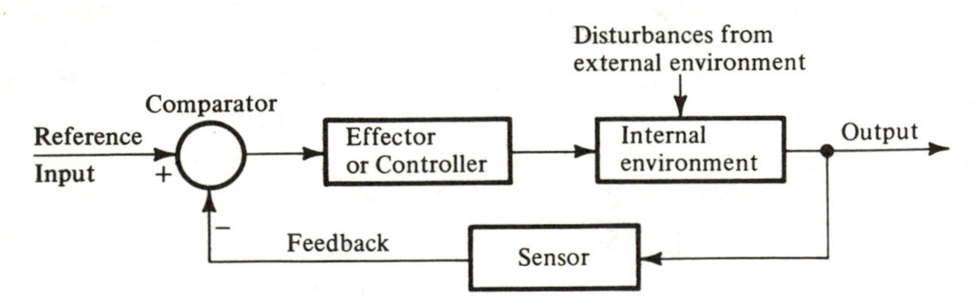

Figure 9-25 Fundamental Elements of Homeostasis

The Heart and Blood Circulation. The average blood pressure level in humans is about 100 mm Hg.* The heart pumps the blood through the system, and with each heart beat the pressure in the arteries varies from a low of 80 mm Hg to a high of 120 mm Hg. The human heart, shown schematically in Fig. 9-26, consists of two pumps. Each pump has two chambers, an atrium where blood enters and a ventricle from which blood is pumped out. The right pump drives the blood through the vessels in the lungs. As it moves through the lungs, the blood releases the carbon dioxide and is replenished with oxygen. The replenished blood returns to the left pump, which is the stronger of the two, and is circulated by it throughout the rest of the

*Atmospheric pressure at sea level is 760 mm Hg. (Hg is the symbol for the metal mercury.)

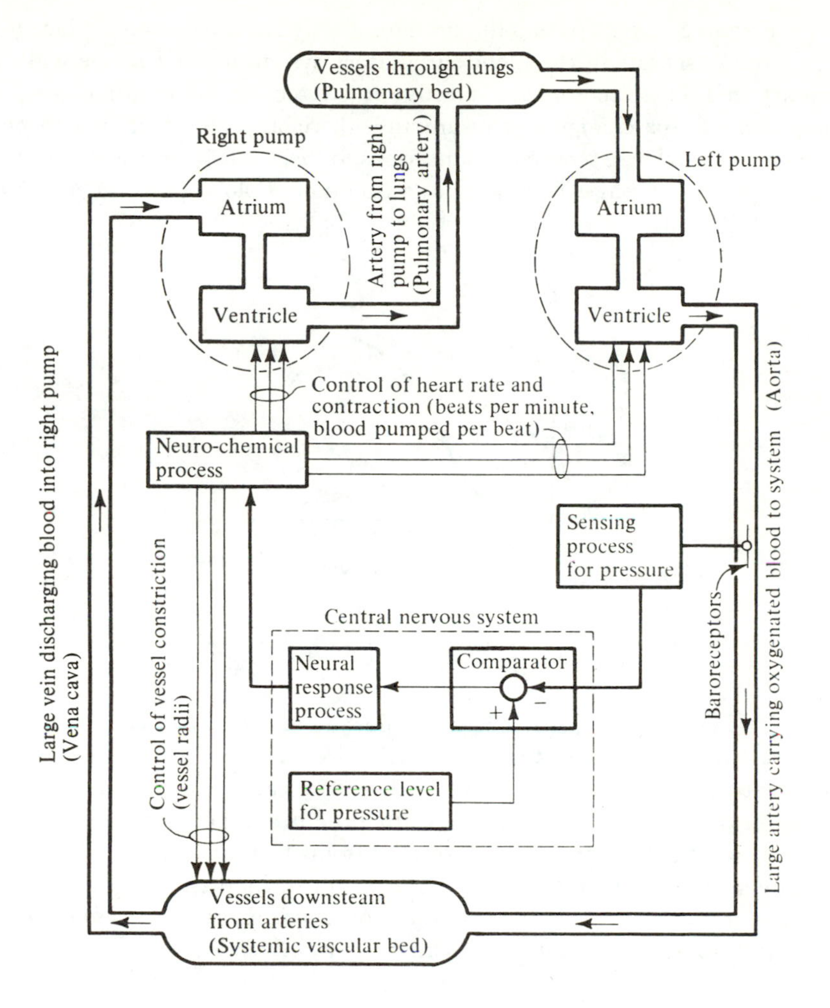

Figure 9-26 Blood Pressure Control System—Schematic Diagram [79]

body to transport oxygen until it is spent, and then returns to the right pump to be replenished again, thus completing the cycle. The pressure in the shorter circulation (called *pulmonary circulation*) driven by the right pump is much lower than that of the longer circulation (called *systemic circulation*) driven by the left pump. The physician normally checks the systemic pressure.

At rest, the left pump pumps 5 to 6 liters of blood per minute, and the heart beats about 70 times per minute.* Under the demand of heavy exercise, the rate of pumping may increase to 30 liters per minute.

*For an athlete it is lower, 60 to 50 beats per minute.

Pressure Sensors. To keep the system informed of the demands placed on it, there are sensors in the blood circulatory system located in the walls of the arteries. These sensors are sensitive to stretch (or deformation) so that they respond to a change in pressure and, therefore, are called *pressoreceptors* or *baroreceptors*. These sensors send messages to the central nervous system with the number of messages increasing as a function of blood pressure, as shown in Fig. 9-27.

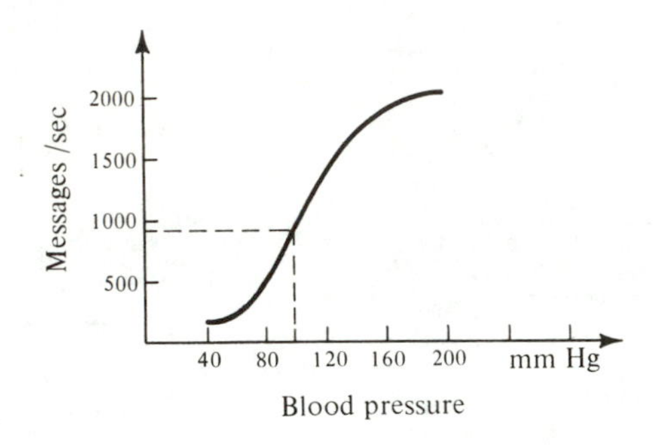

Figure 9-27 Sensor (Pressoreceptor) Response in Messages per Second as a Function of Blood Pressure

The heart is a unique organ in the human body in that, even when removed from the rest of the system, it will continue to beat at a constant inherent rate. The nerves leading to the heart cause a deviation from this constant rate. One set of nerves can cause it to increase the rate, and another to decrease it. The increased rate of flow from the heart can be achieved by pumping more than the normal 80 ml per beat, increasing the number of beats from the normal of 70 per minute, or a combination of these changes. The decrease in rate of flow is achieved by reversing these changes. The sensors increase the number of messages to the brain as a function of blood pressure (Fig. 9-27). When the pressure is high, they cause a decrease in the flow of blood and, thus, a decrease of pressure in the arteries. When the pressure is low, the smaller number of messages from the sensors to the brain causes an increase in flow from the heart and, thus, an increase in blood pressure in the arteries.

The blood pressure must be above a minimum level in order to cause adequate circulation through the body. The upper limit on pressure prevents the rupture of tissues in the blood vessels and, in general, safeguards against "overworking" the system.

Vessel Constriction and Blood Pressure. Pressure p of blood going through a blood vessel is related to flow rate f, blood viscosity v, vessel length l, and its radius r:

$$p = \text{constant} \times \frac{f \times v \times l}{r^4}$$

This relationship was first suggested by a French physiologist, Poiseuille, more than 100 years ago. The viscosity of the blood, v, and the length of vessels in the system remain constant. However, the radius r and flow rate f vary. For the same pressure p, the flow through a vessel of radius $r = 2$ mm is 16 times that of a vessel with $r = 1$ mm because of the fourth power for r in the equation. Since the smaller vessels downstream from the arteries have circular muscles that can cause a change in the vessel diameter, pressure upstream can be affected by them. There are nerves that regulate the radius of the vessels.

Blood Pressure Control. Thus, the arterial pressure is controlled by heart action and by the action of the vessels downstream from the arteries. When the blood pressure is high, sensors send messages to trigger a decrease in flow from the heart and an increase in the diameters of the vessels. When the blood pressure is low, the changes are reversed. Changes in heart rate (beats per minute), vigor of contraction (amount of blood pumped per beat), and vessel constriction are effected through neurochemical processes. The complete blood pressure control system is shown schematically in Fig. 9-26. A more abstract model of the same system is shown in Fig. 9-28.

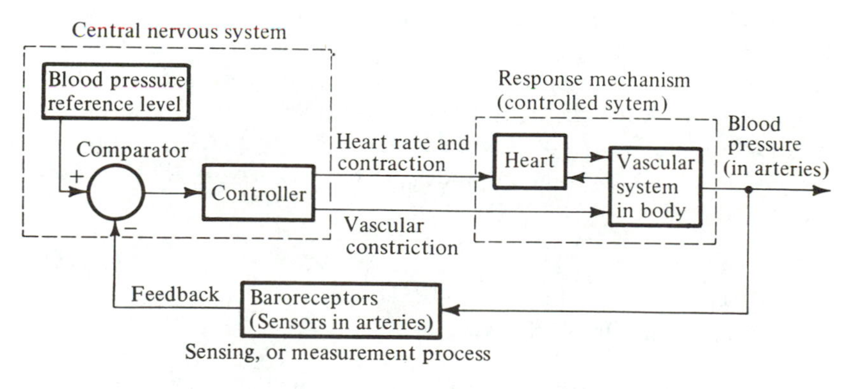

Figure 9-28 Abstraction of Blood Pressure Control System of Figure 9-26 [79]

Experiments with animals have shown that if the regulating mechanism is destroyed by cutting the nerves transmitting the signals from the sensors to the brain and bleeding causes a loss of 10% of blood in volume, the blood pressure becomes too low to maintain life. On the other hand, with the regulatory homeostatic mechanism intact, the same animals can survive a loss of 40% of blood in volume. In case of excessive bleeding, or hemorrhage, the system responds by constricting the blood vessels in the extremities (hence, the pale appearance of a person who loses a great deal of blood), and by increasing the heart pumping rate (observed by feeling a faster pulse for a bleeding person). Hence, an average person who normally has over 5 liters of blood can survive the loss of up to 2 liters of blood.

The blood pressure control system is, therefore, important in the event of bleeding (hemorrhage). It also regulates the more common event of change in position as we get out of bed. Gravity exerts a force on the blood downward and the sensors of blood pressure must fire signals to the brain to insure a flow appropriate to supply the blood vessels in the brain, or else consciousness could be lost each time we get out of bed. The system responds to demand. When the muscles are busy in vigorous action in a portion of the body, blood flow will increase there, while vessels elsewhere will be constricted to help meet the demand for higher pressure and increased flow where needed. During relaxation or sleep the activity is low, demand for oxygen supply is reduced, and, therefore, the pressure in the system is lower.

Breakdown in the System. When the signals keeping the heart beating at its normal inherent rate are disrupted due to a breakdown in the information transmission system, a person may not be able to function. The beats may go down to 30 to 40 beats per minute, not sufficient to transport oxygen and sugar to the body. A breakdown of this kind has been dealt with by implanting a battery-operated pacemaker in the body connected so as to transmit the signals to the heart for a normal beat rate. However, the nervous system that affects the change of resistance to blood flow in the vascular bed, and varies the rate of heart beat and contraction (flow per beat), may not be operative. This limits the activities of the person in need of a pacemaker to those compatible with the demands which can be placed on a heart with a normal beat and contraction, and rules out intensive activities of long or moderate duration.

9-5 CONTROLLABILITY AND OPEN VS. CLOSED LOOP CONTROL

The examples of the preceding sections will provide us now with a foundation for generalizing some key concepts which characterize the behavior of dynamic systems.

In the first section we discussed the toaster and the steering of a boat, and made a distinction between two types of control in dynamic systems: *open loop* and *closed loop* or *feedback control.* In open loop control, the input causes an output, but there is no flow of information from output to input. In closed loop control, there is a flow of information from output to input. The fundamental difference between these two modes of control resides in the fact that the feedback control system is *goal seeking.* A reference input is selected to achieve a desired goal output. When the system deviates from the desired goal, and the feedback is automatic, the very act of deviating from the goal becomes the operator or signal for corrective action. The coupling between the move away from the desired goal and the operating control is complete, and any move toward "freedom" from the constraint to achieve a prescribed goal triggers a corrective input. This is true for the mechanical control systems and the homeostasis systems discussed earlier. However, since so far in our discussions we repeatedly pointed out the ubiquity of control systems and their application even to human affairs, it is important to emphasize that control in this arena is not as automated or effective, at least not yet, as in biological systems and man-made systems. Prohibition was a form of control on liquor consumption which proved ineffective. This is not to say that liquor consumption cannot be controlled at all; it merely indicated that the control may be effective on a more limited basis. In the case of prohibition, the understanding of the system was lacking; the hypothesis was that an input in the form of a law and control through punishment would achieve the desired goal. The hypothesis was proven *not true* and, consequently, rejected. The controllability of the process failed.

Controllability. Although we discussed controllability above with reference to a social system, we hasten to add that controllability is a concept which attends all control systems. For each dynamic system with control we must establish the region in which control can be achieved and thus be aware of the uncontrollable region. For example, in the control of temperature inside a closed environment, we may have a control system that will keep the temperature at a reasonably constant level of 70°F so long as the temperature outside the environment is in the range of 30° to 72°. But the system is not controllable outside this range, and above 72° the temperature inside is about equal to that outside, as shown in Fig. 9-29. The limited controllability may be due to a heat source of limited capacity and inadequate insulation of the inside environment.

When Open Loop and When Feedback. Once we know that a system is controllable within a region of interest, we then concern ourselves with other characteristics of its behavior. Among those, we may decide on whether to use open or closed loop control to achieve the desired goal behavior. When open loop control is used, much more caution must be exercised in the choice of input because the system is out of control once the input sets the system in motion to achieve a desired goal. Hence, change in the system itself or disturbance from the environment outside the system cannot be dealt with. For example, once we set the dial on the toaster, disturbances resulting from changes in the current flowing through the heating wires (change in input), or a change in the system resulting from a break of one or more wires, will lead to a deviation from the desired output.

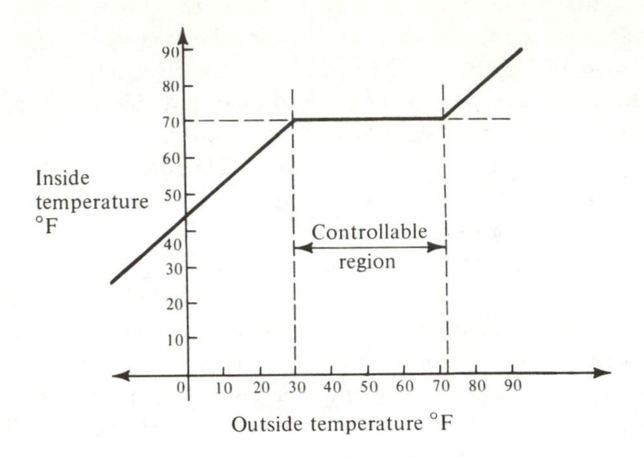

Figure 9-29 Region of Controllable Temperature in an Enclosed Environment

In a closed loop system the input is changed as the output is monitored. For example, in a "closed loop" toaster, the heat generated near the toast can be sensed and the dial adjusted accordingly, thus correcting for disturbances in the power supply or changes in the system.

Cooking from a recipe is an open loop control operation, while the preparation of a soup with a desired level of seasoning which is monitored by tasting and adding incremental amounts of salt, pepper, and other ingredients is a closed feedback loop operation.

The dynamic response system that transforms input to output in an open loop control must be less vulnerable than a closed loop system to disturbances from the environment and variations in characteristics from within, in order to achieve comparable control of output. A closed loop system can manage to achieve desired outputs despite disturbances from the environment and changes in the controlled system. This means that when feedback is used, the basic transformation system can be more vulnerable and, therefore, less costly than that of the open loop system. However, the closed loop system requires the addition of a feedback mechanism that will include a sensor, comparator, and actuator or effector of change in input. The economic comparison of the two systems will depend on the problem at hand and is in the domain of the control theory specialist.

The main advantage of feedback is the partial commitment of initial input to effect a final consequence, and the attending flexibility in adaptation to a hostile disturbing and unyielding environment with the aid of a vulnerable system subject to change with time. This is an essential advantage in the fight for survival as mentioned earlier. However, few advantages are pure and unaccompanied by undesired "side effects," and feedback is no exception, as will be shown in the following sections.

9-6 AMPLITUDE AND PHASE IN THE RESPONSE OF DYNAMIC SYSTEMS

Input-Output Relationship. One of the fundamental characteristics of a dynamic system is described in terms of its transformation of an input into an output. This transformation characteristic is given in the form of a function called a *transfer function* mentioned earlier. For example, an operator of a crane may push a control lever upward by an amount of I inches (input), causing the crane to lift a heavy weight a distance of $8I$ inches (output). The input I is thus transformed to an output $O = 8I$. The system transfer function, S, is the number 8 in this case. In general, we can write

$$O = SI \tag{9-2}$$

The output O and input I may vary with time, as usually is the case in a dynamic system, and the output depends on the history of inputs.

* *More Complex Transfer Function.** Let us consider the mass, spring, and dash-pot† model of a dynamic system shown in Fig. 9-30. Equation 9-2 takes the following form‡

$$O(t) = S(\Omega)\, I(t)$$

in which

$$S(\Omega) = \frac{1}{1 - (\Omega/\omega)^2 + i2\zeta(\Omega/\omega)} \tag{9-3}$$

$I(t)$ is a sine or cosine function related to the input force $F(t)$, $F(t) = k \cdot I(t)$; k is the stiffness of the spring in units of force per unit elongation. Ω is the frequency of the input, ω is the natural frequency of the system. ζ is related to the dashpot characteristics or the damping of the system as follows: $\zeta = c/(2\sqrt{km})$ in which c = damping coefficient, k = spring stiffness, m = mass, and $i = \sqrt{-1}$. The damping force F_d in the dashpot is equal to the product of damping coefficient c and the velocity of mass m, or $F_d = c(du/dt)$.

---

*The reader who finds this subsection difficult should proceed first to the *Simple Experiment* on page 439, then read this part.

†The dash-pot is a device that represents a model for energy absorption such as hydraulic brakes or other forms of energy dissipation through friction, in which the braking force is a linear function of the velocity of the piston in the dash-pot. The spring and mass are energy storage elements. The spring can store potential energy and the mass, kinetic energy.

‡This form holds true for an oscillating input force which is modeled by a sine or cosine function when only the *steady state response* is considered. The *steady state* response is the response after the initial *transient response* of the system decays. The transient response reflects the early adjustment of the system to the input disturbance. See page 440.

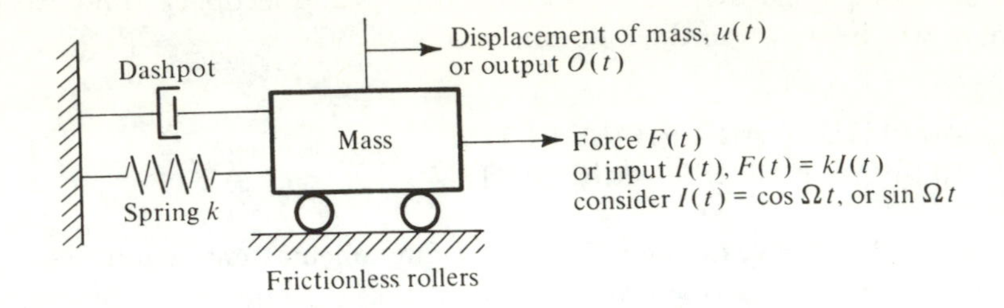

Figure 9-30 Mass, Spring, Dashpot Model of a Dynamic System

$S(\Omega)$ is a complex number and, therefore, it contains *two independent pieces of information:* a *magnitude* and a *phase.* Let us explain what this means. Suppose we have the complex number $z = a + ib$; a is the real part and b is the imaginary part. Figure 9-31 locates these components of the complex

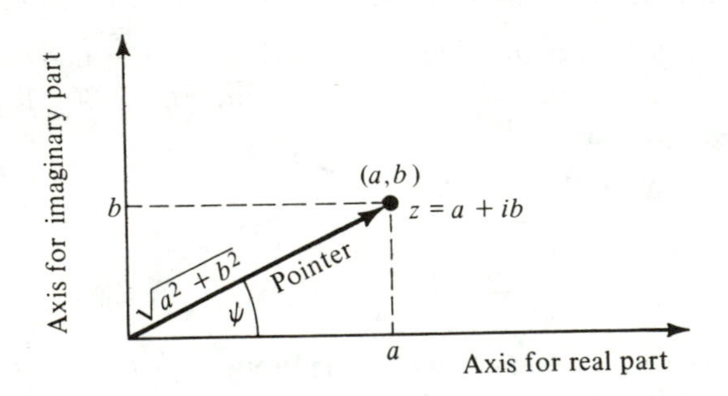

Figure 9-31 The Components of a Complex Number

number z on the real and imaginary axes. The magnitude of complex number z is the length of the pointer from the origin to point (a, b). The phase is the angle ψ which the pointer makes with the axis for imaginary parts. Hence,

$$\text{Magnitude of } z = |z| = \sqrt{a^2 + b^2}$$

$$\text{Phase of } z = \psi = \tan^{-1}\frac{b}{a}$$

$$\text{or } \tan\psi = \frac{b}{a}$$

The magnitude and phase are independent, namely, we can have different phase angles for the same magnitude, or different magnitudes for the same phase.

For the transfer function $S(\Omega)$ of Eq. 9-3, the magnitude and phase are

$$\left.\begin{aligned} |S(\Omega)| &= \frac{1}{\sqrt{[1-(\Omega/\omega)]^2+(2\zeta\Omega/\omega)^2}} \\ \psi &= \tan^{-1} - \frac{2\zeta\Omega/\omega}{1-(\Omega/\omega)^2} \end{aligned}\right\} \tag{9-4}$$

The magnitude $|S(\Omega)|$ is a *magnification* or amplification in response or output. For example, in the system of Fig. 9-30 when $F(t) = F_{max} \sin \Omega t$, the maximum response $O(t)$ in each cycle of oscillation will be $|S(\Omega)|$ times the static response of the system. The static response is the displacement when the maximum load* F_{max} is applied gradually with no oscillation, or with zero frequency Ω. The magnification or amplification in response changes as a function of frequency ratio Ω/ω and is shown in Fig. 9-32 for various values of critical damping coefficient ζ.

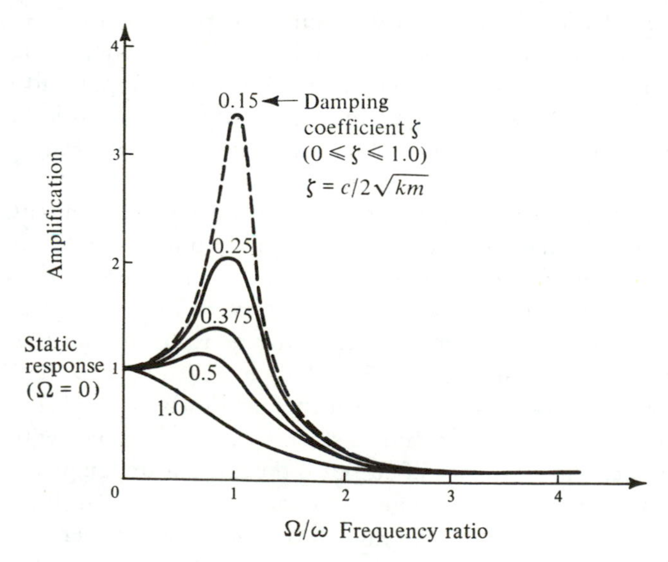

Figure 9-32 Amplitude as a Function of Frequency Ratio (Input Frequency to Natural Frequency) for Different Values of Damping

A Simple Experiment. To make the above more concrete, let us try a simple experiment. Take a number of rubber bands and tie them together to form a string of about 8 to 10 inches. Tie an object at one end. Hold the object and let the string of rubber bands hang free. Measure the length of

*The maximum value of $F(t)$ occurs when $\sin \Omega t = 1$; therefore, $\max F(t) = F_{max} \cdot 1 = F_{max}$.

the string in this unstretched position. Now get hold of the free end of the string and let the object slowly out of your hand until it rests at an equilibrium position, stretching the string in the process. The elongation is the *static response*. Now pull the object further down, say, by 1 or 2 inches (while holding the free end so it does not move), and then suddenly let it go. The object will start vibrating up and down with respect to the equilibrium position. Count the number of cycles of oscillation per second. This number is the *natural frequency* ω of this dynamic system. Bring the object to rest at the equilibrium position, and then suddenly start moving your hand holding the string up and down at about the same frequency as the natural frequency ω. Note the *initial delay* in the response of the object and then the *steady state* response at the frequency of the moving hand. Note also the *transient response* in the initial instances as the system adjusts to the input. Stop the oscillation and then start moving the hand holding the string very rapidly at a very high frequency Ω with $\Omega \gg \omega$. Note that the object hardly moves. This agrees with Fig. 9-32 which indicates a very small magnification for $\Omega \gg \omega$. The magnification is expressed by $|S(\Omega)|$, and the time *delay* in system response, namely, the time interval between the peak values of input and output, is related to the phase ψ. Each dynamic system has characteristic time delays to accommodate disturbances, and each system has an inherent magnification nature; it is more receptive to some inputs and less to others. Is it any different with people? Well, dynamic systems, even the inanimate, have a "personality" which is characterized by a transfer function in terms of amplification (receptivity) and phase (delay, or how much in tune it is with input).

Usually, in dynamic systems the output oscillation lags behind the input. The system lags behind in beginning its response and then, once it starts responding, it lags behind the input in reaching a peak value. For instance, when we boil water and turn the gas slowly up and down with a certain frequency, then the amount of steam generated changes with the same frequency (i.e., the frequency of output is the same as the frequency Ω of the input), but the maximum amount of steam comes later than the maximum level of gas because the system requires time for heating; hence, the delay. The lagging behind of the output increases with the increase in frequency of the input. The more rapidly we vary the input, the more delayed is the response.

* Transient and Steady State Response [80]

The system of Fig. 9-30 has, in general, the following response $u(t)$ to a forcing function $F(t)$:

$$u(t) = A_1 e^{-\zeta\omega t} e^{-i\omega\sqrt{1-\zeta^2}t} + A_2 e^{-\zeta\omega t} e^{i\omega\sqrt{1-\zeta^2}t} + |S(\Omega)| \cos(\Omega t + \psi_2) \tag{9-5}$$

where

$$A_1 = \frac{\omega}{2\sqrt{1-\zeta^2}}\left[\frac{\omega\sqrt{1-\zeta^2}(\Omega^2-\omega^2)-i\zeta\omega(\Omega^2+\omega^2)}{(\Omega^2-\omega^2+2\zeta\omega^2)^2+4\zeta^2\omega^4(1-\zeta^2)}\right]$$

$$A_2 = \frac{\omega}{2\sqrt{1-\zeta^2}}\left[\frac{\omega\sqrt{1-\zeta^2}(\Omega^2-\omega^2)+i\zeta\omega(\Omega^2+\omega^2)}{(\Omega^2-\omega^2+2\zeta\omega^2)^2+4\zeta^2\omega^4(1-\zeta^2)}\right]$$

$$S(\Omega) = \frac{1}{1-(\Omega/\omega)^2+i2\zeta(\Omega/\omega)}, \quad \psi_2 = \tan^{-1}\frac{2\zeta\omega\Omega}{\omega^2-\Omega^2}$$

Equation 9-5 contains three terms. The first two represent the *transient* part of the response. Although these terms contain imaginary components which could not represent real physical quantities, the two terms can be combined mathematically to obtain a single transient term which is a decaying oscillation. In fact, the complete response can be written as

$$u(t) = \frac{\omega^2}{\sqrt{(\omega^2-\Omega^2)^2+4\zeta^2\omega^2\Omega^2}} \times \left[\frac{1}{\sqrt{1-\zeta^2}}\cos(\sqrt{1-\zeta^2}\,\omega t+\psi_1)+\cos(\Omega t+\psi_2)\right]$$

where

$$\psi_1 = \tan^{-1}\frac{\zeta(\Omega^2+\omega^2)}{\sqrt{1-\zeta^2}(\Omega^2-\omega^2)}$$

and the term out in front is just $|S(\Omega)|$.

For damping below a *critical value* for which $\zeta = 1.0$, called *critical damping*, and at which no oscillation results, the transient response is seen to be an oscillation with frequency $\omega\sqrt{1-\zeta^2}$ and an amplitude which decreases exponentially with time, by the factor $e^{-\zeta\omega t}$. $1/\zeta\omega$ is called the *time constant*, and is the time τ in which the envelope of the transient response decreases to 0.37 of its initial value. Initially, $t = 0$ and $e^{-\zeta\omega t} = e^0 = 1$. When $e^{-\zeta\omega t} = e^{-1} = 0.37$, $t = \tau = 1/\zeta\omega$, which is the time constant. The transient response disappears with time, being less than 1% of its initial value after 5 time constants have elapsed.

The third term of Eq. 9-5 is the *steady state* response and it persists with a frequency Ω of the exciting force $F(t)$ as long as $F(t)$ persists. Equation 9-3 is identical to Eq. 9-5 when the transient part disappears, with $u(t) = O(t)$ and $I(t) = \cos \Omega t$. Figure 9-33 shows the transient and steady state response of a dynamic system such as that of Fig. 9-30.

A steady state response may be characterized by a constant steady value rather than an oscillation at a constant frequency and amplitude. For instance, let us consider a heating system and suppose our input is a setting of a new temperature level of 70°F at time t_0 from a setting of 66°F, as shown in Fig. 9-34(a). The ideal response would be an output temperature identical to the input in part (a). However, it takes time (delay) to achieve the desired level of temperature, and this can be achieved in two ways. In part (b), the

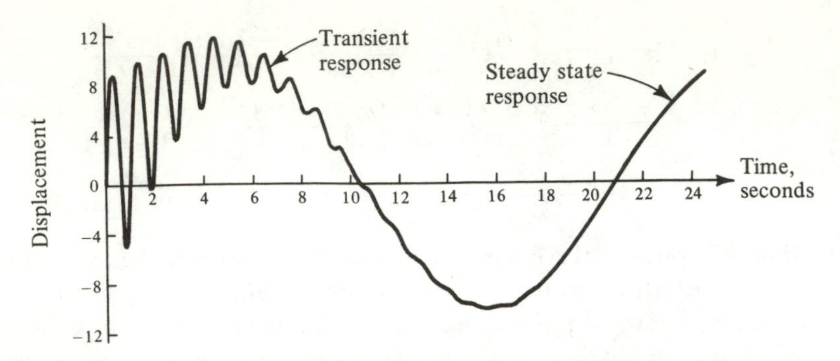

Figure 9-33 Transient and Steady State Response

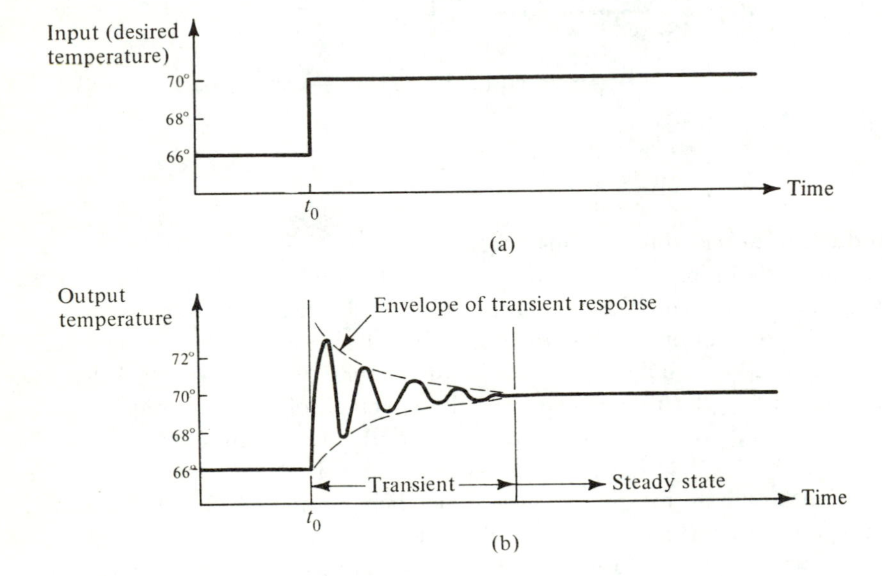

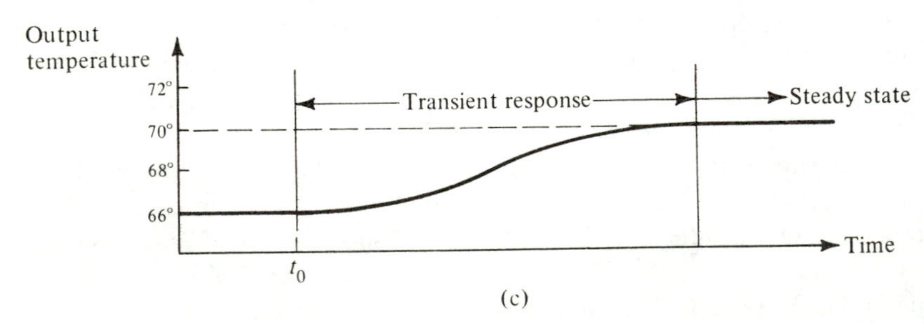

Figure 9-34 Input and Possible Corresponding Outputs (a) Input; (b) Output of an Underdamped System; (c) Output of an Overdamped System

system is *underdamped* (i.e., less than critical) and it oscillates about the desired level, exceeding it (overshooting) first, then overcorrecting the other way until it reaches a steady state. In Fig. 9-34(c), the system is *overdamped* (more than critical damping), which means that the system does not oscillate because of a high braking capacity, and slowly reaches the desired level without overshooting it. This system generally takes longer to approach the steady state.

Importance of Phase or Time Delay. Oscillatory behavior of the kind described earlier is of great importance in feedback control, not only in man-made systems but also in human behavior patterns, economic booms and depressions, and other areas. To control a system for the purpose of achieving a desired output, the delay between input and output is of prime importance. Consider, for example, the university supply of engineers. When demand for engineers slackens, the information is fed back to new high school graduates who consider an engineering career, and many of them change their plans. Four years later the result is that, when the demand for engineers rises, there are not enough graduating engineers. This information is fed back and leads to an increase in enrollment that may lead again to an oversupply of engineers a few years later. These oscillations are also typical of the economic arena in general in which the levels of economic activity and employment depend on rate of investment. The rate of investment depends on the expected profit which, in turn, is a function of the trend in economic activity, actual and perceived. Thus, the loop is closed and oscillation results, because of the time delays and the attending difference between predicted and actual outcomes. In general, investment and profit are separated by a time delay; the profits (outputs) of the present are the delayed response to investments (inputs) of the past.

When the delay in output is half a period T of the oscillation, continuous input may lead to the outputs shown in Fig. 9-35 (a) and (b). In part (a), the input is smaller than the output, which precedes it by $(1/2)T$, and opposite to the output which occurs with it (i.e., input and output are out of phase); therefore, it will cause the oscillation to decay. In Fig. 9-35(b), the input is larger than the output, which precedes it by $(1/2)T$, but still opposite to the output occurring with it; therefore, the output is amplified. In part (c), the output is in phase with the input or is one period T delayed from it; this causes a reinforcing of the output.

Negative and Positive Feedback. Figures 9-35(a) and (b) can be viewed in the category of *negative feedback*, because the input feedback is opposite to the output deviations from the reference level. Figure 9-35(c) represents *positive feedback*, because it reinforces the output deviation from the reference level. A transition between (a) and (b) is a sustained *stable oscillation* of a constant amplitude when the input (feedback) generates a correction which

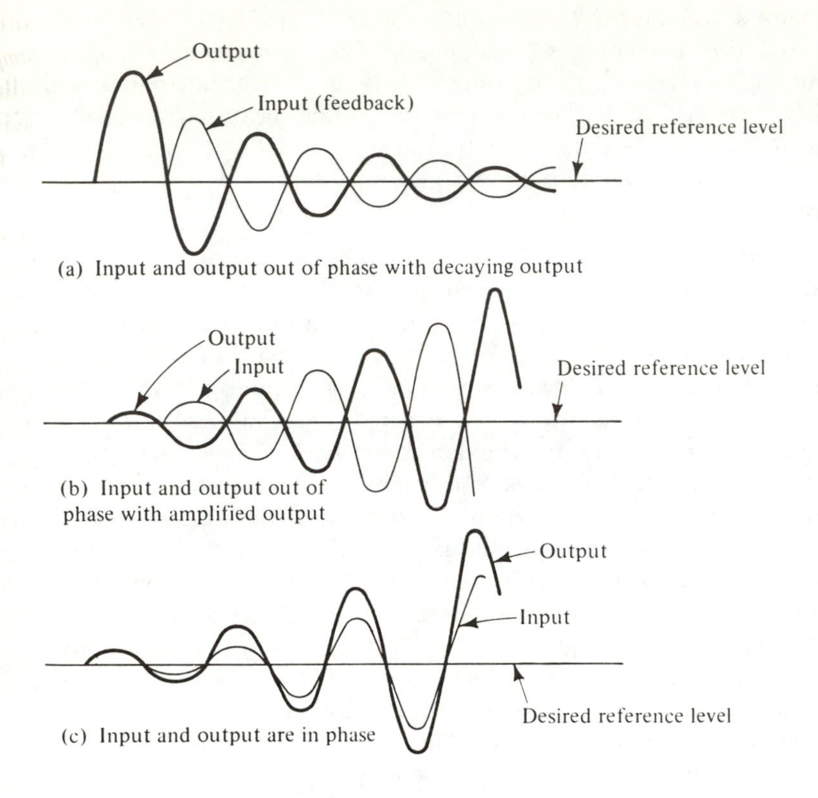

Figure 9-35 Results of Phase Relationships between Input and Output: (a) Input and Output out of Phase, with Decaying Output; (b) Input and Output out of Phase, with Amplified Output; (c) Input and Output are in Phase.

is of the same amplitude as the detected output deviation, but opposite in phase (half a period T delay). Section 9-7 treats these concepts in more detail.

9-7 CHARACTERISTICS OF FEEDBACK SYSTEMS

Transfer Function of a Feedback System

Let us consider now the general model of a feedback control system of Fig. 9-36. For each block in the block diagram we can write a relationship of the form given by Eq. 9-2, namely,

$$\text{Block Output} = (\text{Block Transfer Function}) \times (\text{Block Input}) \tag{9-6}$$

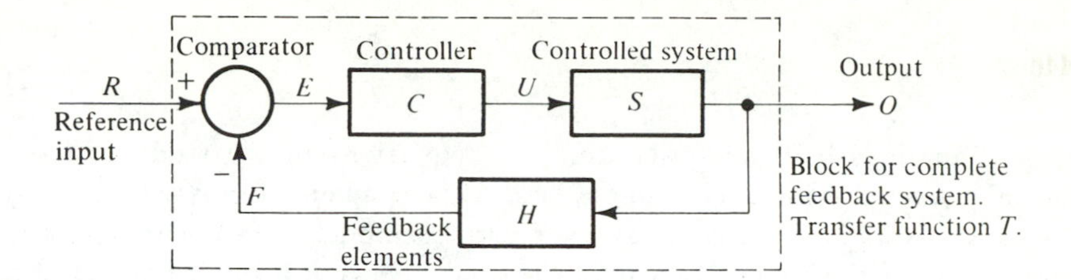

Figure 9-36 Block Diagram for General Feedback Control System

The transfer functions for the three blocks in Fig. 9-36 are identified by the letters in the blocks.

The input and output to each block are identified by the symbols on the ingoing and outgoing arrows, respectively. We can, therefore, write Eq. 9-6 for each block:

$$\left.\begin{array}{lll} \text{Controller:} & U = CE & \text{(a)} \\ \text{Controlled system:} & O = SU & \text{(b)} \\ \text{Feedback elements:} & F = HO & \text{(c)} \end{array}\right\} \qquad (9\text{-}7)$$

The feedback elements with transfer function H usually include the properties of the sensors and the transducers* associated with the measurement of the output. The comparator yields the following relationship between error E, reference input R, and feedback F:

$$E = R - F \qquad (9\text{-}8)$$

To compute the transfer function T for the complete closed loop feedback system, we view the content of the dashed line block of Fig. 9-36 as a single block with input R, output O, and unknown transfer function T:

$$O = TR \qquad (9\text{-}9)$$

Start with Eq. 9-7(b), which has output O on the left-hand side, and substitute for U from Eq. 9-7(a). Then substituting for E from Eq. 9-8 and finally for F from Eq. 9-7(c), we have

$$O = SC(R - HO)$$

Solving for O in terms of R yields

$$O = \frac{CS}{1 + CSH} R \qquad (9\text{-}10)$$

Comparing Eqs. 9-9 and 9-10, we obtain the following expression for the

*A transducer transforms energy from one form to another, such as sound to electric current, etc.

transfer function T:

$$T = \frac{CS}{1 + CSH} \tag{9-11}$$

Equations 9-10 and 9-11 are basic equations in the study of *linear control system* behavior. A dynamic system is linear when effect is proportional to cause. Nonlinear dynamic systems can be modeled as linear dynamic systems when the range of variation is small, because a small portion of a curve can be modeled as a straight line. The same consideration is employed in Chapter 8 to deal with nonlinear programming problems by using a piecewise linearization approach. Here our discussion is limited to linear control systems.

Using the transfer function T of Eq. 9-11, we can now examine the quantitative difference between open loop and closed loop control systems with regard to variations in the controlled system characteristics, disturbances from the environment, and stability of performance.

Sensitivity to Variations in the Controlled System S [81, 82]

Figure 9-37 shows two block diagrams of control configurations for controlled system S. Part (a) is an open loop control and (b) is a closed loop feedback control. The transfer function S for the controlled system is taken

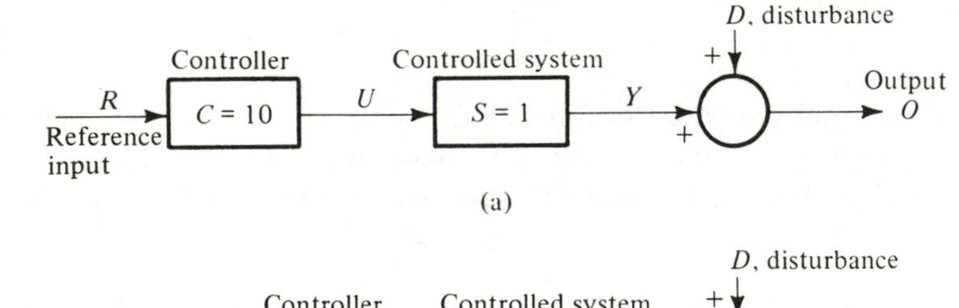

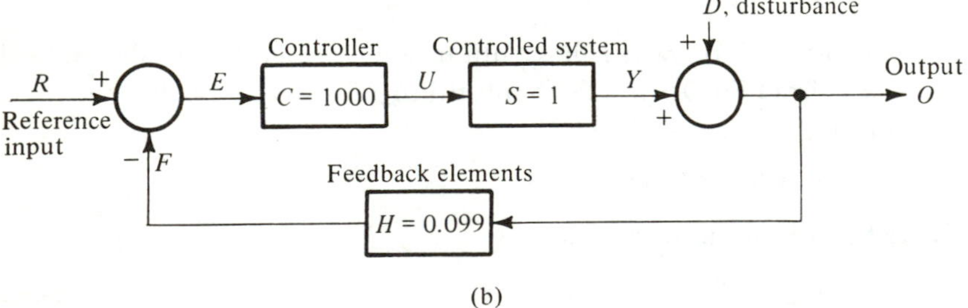

Fig. 9-37 (a) Open Loop Configuration with Disturbance D; (b) Closed Loop Configuration with Disturbance D

as unity for simplicity. The disturbance D, due to disturbances from the environment, is added to the output Y from the controlled system to give a total output $O = D + Y$. The sensitivity of each control configuration to disturbance D will be treated later; at present we consider D equal to zero and, therefore, $O = Y$.

Suppose we wish to design a controller C to generate an output O which will be ten times larger than the reference input R, namely, a transfer function $T = 10$, or

$$O = 10R \tag{9-12}$$

Considering each of the control configurations of Fig. 9-37, for the open loop (with $D = 0$),

$$O = SCR \tag{9-13}$$

and for $S = 1.00$, the transfer function C for the controller must be 10, $C = 10$, to comply with Eq. 9-12.

For the closed loop control, using Eq. 9-11, we have

$$T = \frac{CS}{1 + CSH} = 10 \tag{9-14}$$

The feedback control configuration provides us with an additional degree of freedom in design since we can choose both the feedback block transfer function H, and the controller transfer function C; $S = 1.00$ was established earlier. Let us choose $C = 1000$ and $H = 0.099$. These values satisfy Eq. 9-14 so that output O will be ten times the reference input R.

Now suppose the controlled system S deteriorates and, instead of $S = 1.00$, it becomes $S = 1.10$, i.e., a 10% variation in the controlled system transfer function. The effect on the controlled variable, output O, differs for each control configuration as shown by the following.

Open loop. Using Eq. 9-13 with $S = 1.10$ and $C = 10$,

$$\begin{aligned} O &= (1.10)(10)R \\ &= 11.00R \end{aligned} \tag{9-15}$$

Therefore, a 10% change in the system due to deterioration (or error in assessment of performance) leads to the same 10% change, or error, in desired output O.

Closed loop. Using Eq. 9-10 with $S = 1.10$, $C = 1000$, and $H = 0.099$,

$$\begin{aligned} O &= \frac{(1000)(1.10)}{1 + (1000)(1.10)(0.099)}R \\ &= \frac{1100}{109.9}R \\ &= 10.009R \end{aligned} \tag{9-16}$$

Thus, a 10% variation in the controlled system S results in only a 0.09% deviation of the actual from the desired output.*

A deviation of 10% in output, as in the open loop configuration, would require a much more drastic change or deterioration in the controlled system. Specifically, even for a smaller deviation in output of 1%, $O = 10.10R$ in place of the desired $O = 10.00R$, we have

$$T = \frac{CS}{1 + CSH} = 10.10$$

or

$$\frac{1000S}{1 + (1000)(0.099)S} = 10.10$$

and

$$S = 101$$

Namely, S would have to change from $S = 1$ to $S = 101$ before a 1% error is introduced in the desired output when feedback is used.

The example used here illustrates dramatically the advantage of feedback. A closed loop control can be designed so that the output is extremely insensitive to variation in the controlled system. In the example, we considered only algebraic quantities for the transfer functions and no frequency dependence was present; however, the same type of relationship applies in the presence of dynamic elements with frequency dependence.

Sensitivity to Disturbance from the Environment—Noise [81, 82]

A second important advantage of closed over open loop control systems is their insensitivity to large disturbances from the environment (also referred to as noise). Let us introduce $D \neq 0$ in Fig. 9-37 and consider $S = 1$ and $C = 10$ as in the preceding example.

Open loop. The output O is computed as follows:

$$\left.\begin{aligned} U &= CR \qquad &\text{(a)} \\ Y &= SU \qquad &\text{(b)} \\ O &= Y + D \qquad &\text{(c)} \end{aligned}\right\} \tag{9-17}$$

Substituting from (a) into (b), and then from (b) into (c),

$$O = SCR + D \tag{9-18}$$

Comparing this equation with Eq. 9-13, it is apparent that the disturbance D adds linearly to the output produced by the combination of controller C and controlled system S. In fact, with the additive disturbance entering the dynamic system, as shown in Fig. 9-37 (a), there is no way of designing con-

*The percent deviation is computed as follows:

$$\%\ \text{deviation} = \frac{10.009R - 10R}{10R}100 = 0.09\%$$

troller C in the open loop configuration to reduce the error produced in the output.

Closed loop. In the closed loop configuration of Fig. 9-37 (b), the output O in the presence of disturbance D is given by (reader verify):

$$O = \frac{CS}{1 + CSH}R + \frac{1}{1 + CSH}D \tag{9-19}$$

Using $S = 1$, $C = 1000$, and $H = 0.099$ as in the preceding example, we obtain

$$O = 10R + \frac{1}{100}D \tag{9-20}$$

Thus, the feedback reduces the effect of the disturbance D on the desired output by a factor of 100.

For instance, for a disturbance $D = R$ we have in the open loop, with $S = 1$ and $C = 10$ in the preceding example,

$$O = 11R$$

instead of the desired output of $10R$. In the closed loop with $S = 1$, $C = 1000$, and $H = 0.099$,

$$O = 10.01R$$

Thus, a 10% error is introduced in the open loop and only a 0.1% error in the closed loop.

Stability

A system is defined as *stable* if a disturbance from an equilibrium state which it occupies creates in its wake a tendency to return to equilibrium after the disturbance stops. For example, when a pendulum is perturbed from its vertical equilibrium position, the perturbation generates a tendency to return to the original position of equilibrium. The motion towards equilibrium is normally an oscillation about the equilibrium state with gradually decreasing amplitudes as energy is dissipated in each cycle because of friction at the support point and friction in the motion of the pendulum through the surrounding air. As long as the motion is damped out, the system is stable. The same holds true for a marble located at the lowest point of a circular container, as shown in Fig. 9-38(a). When the marble is disturbed by being pushed up the container wall, it will oscillate about the equilibrium state until it comes to rest. The marble in Fig. 9-38(b) is in an unstable state of equilibrium; a small perturbation will not generate in its wake a tendency to return to this state. Stability must, then, first be discussed with respect to a state of equilibrium and, in addition, the disturbance must be of a magnitude which is in the range of normal performance. Note that a strong enough disturbance imparted on the marble in Fig. 9-38(a) may cause it to fly out

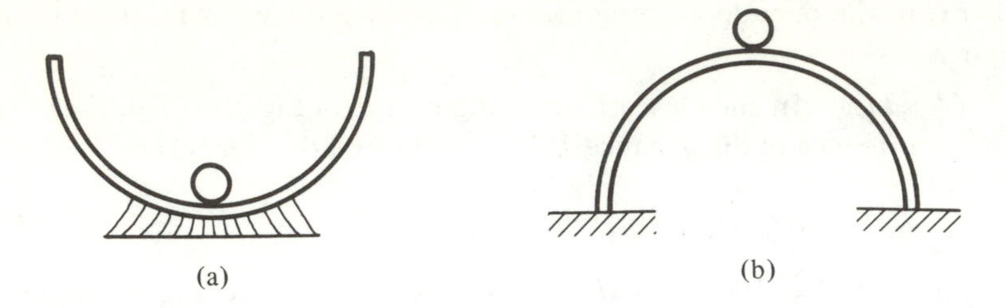

Figure 9-38 Positions of (a) Stable Equilibrium; (b) Unstable Equilibrium

of the container. If such disturbances are very common, then indeed the state of equilibrium in Fig. 9-38(a) cannot be considered stable.

The problem of stability is of great importance in feedback control systems, because the dynamic characteristics of the controller C and feedback elements H govern the stability of the controlled system S. Thus, a desired output may constitute a *stable*, *unstable*, or *neutrally stable* state of equilibrium, depending on the magnitude and phase (or time delay) of the feedback in relation to the sensed output. A neutrally stable state of equilibrium is a sustained oscillation of constant amplitude with respect to a reference state.

Consider the output of a system to be subject to disturbances in the form of an oscillation from a desired reference state. Suppose that to counteract this oscillation, feedback causes an input to the system that is 180° out of phase, or delayed half a period, with respect to the output deviations. If, in addition, the feedback input has amplitudes smaller than the sensed output deviations, then the deviations will decay and the desired reference state is *stable*. This kind of feedback is a *negative feedback*. If the delay is half a period, but the amplitudes caused by the feedback are of the same magnitude as the output deviations from the desired reference, the oscillation will persist and we have a *neutrally stable state*. For a feedback with amplitudes larger than the observed disturbance and half a period delay, the oscillations will increase and the output will get progressively farther from the desired state; hence, the state is *unstable*.* As long as the feedback opposes the deviations in output from the desired goal output, i.e., is 180° out of phase with respect to it, or is delayed a half a period with respect to it, the feedback is a *negative feedback*.

*Special devices called phase-advancers can be introduced in the feedback system to produce a time lead in anticipation of a time delay. For example, alternating current in an electric capacitor circuit leads the voltage applied to it, i.e., maximum output current occurs before the maximum in the input voltage. Evolution has incorporated anticipatory response mechanisms in human and animal nervous systems to avoid oscillation and overshooting in output. When a sensory organ is stimulated, the pulses are produced at an increasing rate if the stimulation is increasing in magnitude. The signal or output maximum occurs before the input or stimulus maximum.

When the feedback is in phase, i.e., delayed a complete period and is in the same direction as the detected output deviation, we have *positive feedback*, and the oscillation amplitudes continue to build up.* Some of these points were discussed briefly in the last section and are repeated here to reinforce the concepts.

The models of Fig. 9-21 can be viewed as stable in part (a), neutrally stable in (b), and unstable in (c). Personal modes of behavior may also be classified according to the response to a disturbance. A stable person can recover fast from a disturbance to return to the equilibrium state from which he is shaken and proceed with the normal routines of daily life. An unstable person fails to return to equilibrium for long periods under rather slight disturbances. Feedback control can be wisely used to control the performance of an inherently unstable system. For example, using visual feedback of position, we can balance a long stick upright in our open hand. A rocket lifted off a launch pad is inherently unstable, but it can be guided by using feedback control to force it to assume a desired course.

Instability is one of the disadvantages that can attend the introduction of feedback control into a dynamic system that is otherwise stable. We, therefore, must first know the system. Hence, it is not surprising that we are still trying to understand the nature of the economic system. Is it inherently stable and, if left alone, will it find its inherent stable equilibrium? Or should government feedback control be exercised by monetary policies, taxes, interest rates, money supply, spending, etc? An understanding of the system will permit us to minimize the ill effects of wild oscillations of booms and depressions and the attendant human suffering. This need for understanding the system to enhance predictions of behavior is a fundamental purpose of the modeling process discussed in Chapter 5. Better understanding of dynamic system behavior will improve our ability to control systems where control is needed, and will spare our energies where control may be a detriment to achieving the desired goals.

Brief Recapitulation—Advantages and Disadvantages of Feedback Control

Feedback control systems are much less sensitive to controlled system variations and to disturbances from the environment than open loop control systems. However, a price must be paid for these advantages. A comparison of the controllers C used in the control configurations of Fig. 9-37 serves to illustrate that the feedback system requires a much higher amplification or

*Although positive feedback seems to cause a move away from the desired reference input goal, in some applications or models the resulting exponential growth in output may indeed be the *long run goal*. English [137] claims that this is the objective of the investments in an economy.

gain controller ($C = 1000$) than the open loop ($C = 10$). In addition, the closed loop system requires sensors, transducers, and a comparator.

Stability is another area for comparing the open and closed loop control configurations. The controller and controlled system in an open loop system may display stable behavior characteristics; however, the introduction of feedback may lead to conditions of instability. Thus, in feedback systems, stability cannot be taken for granted even when the controller and controlled system separately display stable behavior.

Optimal Control

The tools and concepts of optimization discussed in Chapter 8 are also employed in control systems. Dynamic programming and linear and nonlinear optimization models are used to achieve an optimum control output from the controller into the controlled system. The optimum has, of course, meaning only with respect to a reference criterion. In control theory, such a reference criterion of performance is referred to as an *index of performance*. One index of performance (IP) used is the *mean square error*. Namely, the objective is to minimize the average sum of the squares of the deviation e (referred to as error earlier) from the desired output. For a desired output $x_d(t_i)$ at time t_i, and actual output $x_a(t_i)$ at time t_i persisting over a time interval Δt_i, we have

$$\text{Increment of Deviation Measure} = [x_d(t_i) - x_a(t_i)]^2 \Delta t_i \tag{9-21}$$

Over a period of time T, the sum of the squares of the deviations becomes

$$\text{Total Deviation Measure} = \sum_i e^2(t_i) \Delta t_i \tag{9-22}$$

in which

$$e(t_i) = x_d(t_i) - x_a(t_i)$$

and

$$\sum_i \Delta t_i = T$$

The average or mean square error of e is obtained by dividing the total sum of squares of deviations in Eq. 9-22 by time T:

$$\text{Mean Square Error} = \frac{1}{T} \sum_t e^2(t_i) \Delta t_i \tag{9-23}$$

In the limit, for $\Delta t_i \longrightarrow 0$, Eq. 9-23 takes the form of an integral

$$\text{Mean Square Error} = \frac{1}{T} \int e^2(t)\, dt \tag{9-24}$$

The objective is to minimize this mean square error over the time T in which the system is in use. The mean square error is conceptually similar to the variance discussed in Chapter 6.

Another index of performance is minimizing the time for the deviation e to be reduced to zero. Other indices of performance not discussed here have been considered and used in the design of control systems [83].

Adaptive Control

From the earlier discussions it is apparent that dynamic systems can deteriorate with time (or may be inaccurately modeled), changing the basic input-output characteristics. Systems are also subject to disturbances from the environment which are random and can best be characterized by the methods of Chapter 6. Since both the changes in the dynamic system and the disturbances cannot be completely predicted, it is not conceivable that a controller with a fixed transfer function C which is optimum for one set of conditions, say, controlled system S and disturbance D, will also be optimal for an entire spectrum of S and D values as the system changes and the noise D varies. Therefore, a second feedback loop, called an *adaptive controller*, can be introduced to adjust the parameters of the controller (i.e., change its transfer function) on the basis of changes in the controlled system S and the detected disturbances D, so it adapts to the new conditions.

Figure 9-39 shows a feedback control system with an adaptive controller. The adaptive controller performs three main functions [83]:

Identification of the controlled system characteristics, or measurement of an index of performance for the controlled system on the basis of the input and output to it, including the effect of disturbance from the environment.

Decision for action based on a comparison of measurements in the identification stage with the desired optimum performance.

Modification of controller parameters adapting to change in the controlled system and disturbances from the environment.

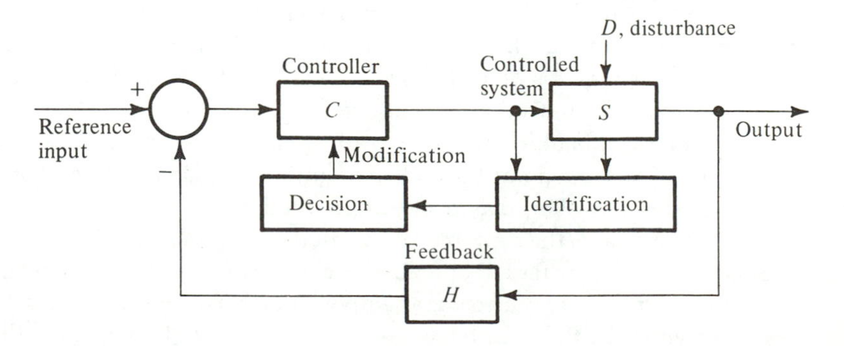

Figure 9-39 Feedback Control System with an Adaptive Controller

Adaptive control has advanced greatly with the development of the digital computer. To adjust the controller before the need for adjustment is dissipated requires rapid action in the above three functions, which may require extensive measurements and computations. The computer has the required speed in many applications.

Adaptive control is being used to devise learning machines in which a punishment-and-reward scheme is employed. A reward is given in the form of a positive contribution to an index of performance whenever the performance is the desired performance (say the correct weather prediction), and punishment in the form of a negative addition to the index of performance is used when the performance differs from that desired. The objective of the adaptive learning system is to establish a properly balanced scheme of reïnforcement so that the performance will be statistically better than a comparable machine with no adaptive learning built in. Adaptive learning has been applied to weather prediction, to playing chess, and to a machine designed to balancc an inherently unstable rod upright on a moving cart, to name some uses. The extension of adaptive learning to model the human learning process may prove the most novel and revolutionary advance in history. The capacity of the human nervous system is far from having been exhausted; some researchers doubt that we use as much as 10% of it. Thus we are far from having reached the constraining boundaries which may eventually limit our capacity for learning. Progress in the understanding of the human learning process could produce an era in which man, using the untapped capacity of his brain, will contrive novel ways to model himself, society, the environment, and the man-made world in a form that will make the decisions on *what ought to be* our values easier to generate, and will make *what is* more compatible with the image of *what it was perceived to be.*

What Ought to be

Adaptive control and feedback control in general can be applied at various levels. For example, the student who sets a reference goal of scoring the grade A may change his plans in Fig. 9-23 as he adapts to disturbances from random demands on his time and effort, and to changes in his learning system such as new difficulties in concentration, emotional stress, and other factors. However, at some point his ability to adjust his plans and decisions, i.e., changes in the controller, may prove unproductive, and goals of a higher level, of which survival is the supreme goal, will dictate a need to change the reference goal, namely, try for a B or a C. The change of the desired reference input which prescribes a desired goal is an important form of adaptation which is a key to survival. In this chapter we treat the concepts of control for the purpose of achieving goals. What the goals ought to be in human affairs in the first place, and how they ought to change dynamically, depends on human values. This is the subject of the next chapter which may be best characterized as the chapter on *what ought to be.*

Problem. Show that Eq. 9-10 for positive feedback has the form

$$O = \frac{CS}{1 - CSH} R$$

9-8 SIMULATION OF DYNAMIC SYSTEMS

Preview

Effective control to achieve desired goals depends strongly on understanding the dynamic system that is to be controlled. In addition, if feedback is used in the control process, knowledge of time delays and possible instabilities in the system response is important. Simulation provides a modeling tool to gain such knowledge and understanding. In dealing with industrial systems, it is possible to study the influence of hiring policies, sales policies, and pricing policies as controls to achieve a desired level of business volume, a desired level of return on investment, a desired level of inventory, etc. Forrester discusses such simulation models of dynamic systems in his celebrated book, *Industrial Dynamics* [45]. A simple and powerful digital computer simulation language [84] called DYNAMO has also been developed. The simulation permits answers to questions of the type: *What if. . . .* Thus we can simulate, sort of act out, many years of dynamic phenomena by using the digital computer. This is conceptually similar to generating experiences at an unusually rapid pace, with the exception that the simulated experience is not the real thing but rather a model of it. We mentioned this repeatedly in Chapter 5, but we must continue to do so to avoid the common mistake of taking the model so seriously that it becomes the real thing. The danger is that, rather than using the model as a guide for understanding and prediction, we become enslaved to our own creation and adjust our behavior so as to make the model be the real thing.

One more note. In applying control to dynamic systems, we must question the goals and change them when appropriate. Let us briefly consider a dynamic model in which a change in goal has been suggested. Forrester in another book, *Urban Dynamics* [85], applies simulation to study the well being of a city. His model includes the city population aggregated in three groups: professional workers and management, labor or skilled workers, and unskilled workers or underemployed. It also includes industry, housing, and real estate taxes on housing and industry. The flow of people into and out of the city is taken as a function of the economic opportunities and the housing situation. The steady state solution in Forrester's simulation model shows deterioration with time to a 45% rate of unemployment among the unskilled, a shortage of skilled workers, and high real estate taxes. The high taxes are necessary to support the unemployed. To upgrade the well being of the city, Forrester studies the effect of alternative controls introduced through policies adopted by the city government. Two such alternative policies, among other public policies considered by Forrester, are slum clearance and job training to increase mobility from unskilled to skilled labor ranks. The simulation model is applied to each of the alternative policies.

Slum Clearance. For slum clearance, 5% of the slum housing is considered removed (demolished) each year, forcing the unskilled to seek housing elsewhere. This makes room for new industries, creates new job opportunities, encourages the arrival of new skilled workers, but discourages the unskilled because of lack of low-cost housing. This, in turn, causes a relative decrease in the unemployed population and thus less support funds are required, which means lower taxes and more encouragement to the development of more industry. Note that lower taxes are not as important to the unskilled who pay very little to begin with, if at all.

Job Training. In the alternate policy of job training, Forrester assumes that each year 5% of the unskilled population acquire skills and become skilled workers. This policy increases the skilled workers' population, and initially decreases the unskilled population of workers. However, as the word about the training program policy gets around, unskilled workers from other cities are attracted, thus leading to only a small change in the ratio of skilled to unskilled, and resulting in only a small relief in tax rates.

On the basis of the simulation model, Forrester concludes that control through the slum clearance policy is more desirable for the city's revival than the job training program policy. The key question is, of course: What are the goals in the first place? Forrester points out that although the job training program should be rejected in favor of slum clearance for the goal of urban revitalization, as a service to society it might be considered a success. Forrester's model considers *the city* as an entity with the objective of legislating policies to reduce the city real estate taxes and increase the proportion of skilled workers occupying the *pigeon holes* (houses) in the city blocks. This objective may not concern itself with the fate of the unskilled workers who are discouraged and leave the city to become a burden elsewhere while enduring additional suffering during the migration stage. The goal ought to be concerned with *people* not *pigeon holes* occupied by people. However, this is not to be misinterpreted as criticism of Forrester's endeavors. To the contrary, Forrester has made a contribution of great value in his models and studies. Such studies provide the insight for asking the right questions which ultimately will permit a closer match between model and reality. The conclusions of Forrester's work led to questions regarding the goals, and thus to improved models such as the one studied by Kadanoff [87].

Extended Model. Kadanoff extended Forrester's model of urban dynamics to larger regions and considered a nationwide training program. This was motivated by Forrester's claim that as unskilled workers in a city are trained they are replaced by an immigration of unskilled workers from outside the city, thus reducing the benefits of the job training policy. A nationwide job training program would avoid this kind of a migration trend. With the extended model Kadanoff concludes that, on a nationwide basis, the goal of a more attractive nation for its inhabitants can be better achieved by a policy of job training rather than slum clearance. Again the model of the subsystem (city) may lead to solutions which are not desirable (let alone optimal) from the total system (nation) point of view. This was also discussed in Chapter 8, Sec. 8-5, in connection with optimization models.

World Population Model. Forrester's work in *World Dynamics* [86] provided the foundation for the model to predict the future of the world population described in *Limits to Growth* [37] and mentioned in Chapter 5, Sec. 5-3. Here, too, the objectives must be studied more carefully. Also the level of aggregation is perhaps too high. Aggregation of the entire world population may hide the fundamental differences between cultures and their goals. A social group dominated by scarcity has a different value associated with population growth and family size than one dominated by affluence. The affluent view children as a liability (in addition to an asset), requiring investment in emotions, material, time, and effort, and thus for optimal benefit and optimal use of resources the number of children per family is not large. The family in the deprived social group is preoccupied with survival, and children may be viewed primarily as an asset to help the cause of survival by working in the fields and helping support the elders of the family. Again, this is not to be misconstrued as criticism of the Limits to Growth Model. The model has made an outstanding contribution in raising issues and generating a great deal of thought and work which are destined to put in better perspective the search for what man's goals ought to be, and to improve his understanding of how to achieve them.

Example of a Dynamic System Simulation

To get a more concrete feeling for what a simulation of a dynamic system involves, let us consider a problem which is likely to be of concern to all of us: the dynamic simulation of an epidemic resulting from an infectious disease.

Influenza, which keeps recurring periodically with new strains of virus, caused the death of about 20 million people in 1918 as it swept through the world, and has been a cause of discomfort for many since and before. The virus of the influenza invades the tissues lining the respiratory passages and then enters the saliva. The disease spreads by the evaporation of saliva when a susceptible individual comes in contact with one who is infected. A person who recovers from the influenza is normally *immune* for two years or more, but of course immunity is ultimately lost. To study the dynamics of an epidemic we, therefore, consider the population in three aggregates or blocks:

SUSceptible Population,	SUSP
INFected Population,	INFP
IMMune Population,	IMMP

Level Equations. Each aggregate or block has a state or level which is influenced by an input and output. Using the notation employed by Forrester,* we designate the levels of SUSP, INFP, and IMMP at time J as SUSP. J, INFP. J and IMMP. J, respectively. Following a time difference "delta t," DT, the time J + DT is designated as K, then K + DT is designated as L, and so on. The levels or states of the population aggregates at time K can be computed as the sum of their respective levels at time J plus the incremental changes over the time interval DT. In equation form, we have:

$$\left.\begin{aligned} \text{SUSP.K} &= \text{SUSP.J} + (\text{DT})(\text{LR.JK} - \text{IR.JK}) && \text{(a)} \\ \text{INFP.K} &= \text{INFP.J} + (\text{DT})(\text{IR.JK} - \text{RR.JK}) && \text{(b)} \\ \text{IMMP.K} &= \text{IMMP.J} + (\text{DT})(\text{RR.JK} - \text{LR.JK}) && \text{(c)} \end{aligned}\right\} \quad (9\text{-}25)$$

In these equations:

IR = Infection rate, number of people infected per day
RR = Recovery rate, number of people recovering per day
LR = Loss of immunity rate, number of recovered people losing immunity per day

The symbols JK following each of the rate symbols in Eq. 9-25 designate that these rates hold for the time interval between J and K. The rates are considered constant during each time interval but may, of course, vary from one time interval to another.

The computations for the three population aggregates can be conducted so as to simulate the system behavior for a period of a few weeks, using small time increments DT. The simulation is started by assigning initial values to each population aggregate and to the rates IR, RR, and LR.

Rate Equations. The key to the use of Eq. 9-25 is the establishing of the rates of change in levels. For IR we consider two factors, the probability p of a new infection when an infected and susceptible person come in contact, and the number of such possible contacts, say, N per day. The probability p multiplied by the possible number of contacts constitutes an estimate to the expected value of the number of new people infected each day. The

*Forrester's notation for the simulation language DYNAMO is used throughout this example to impress the reader with its simplicity and usefulness.

number of possible paired contacts N at time K is equal to the product SUSP.K*INFP.K (*here is a symbol for multiplication). The probability p can be viewed as the fraction F of people infected per contact or the ratio of the number of new infections to the number of contacts. Thus, infection rate for a time interval KL between time K and L is given by

$$\text{IR.KL} = \text{F*SUSP.K*INFP.K} \tag{9-26}$$

The recovery rate RR is computed as the ratio of the infected population to the average number of days to recovery, or period of disease (PD). The recovery rate in the time interval KL between time K and L is based on the infected population at time K, INFP.K

$$\text{RR.KL} = \text{INFP.K/PD} \tag{9-27}$$

Namely, if we have, say, INFP.K = 1600 people, and PD = 8 days, then one-eighth of the people (200) recover daily. The rate, of course, changes as INFP changes; therefore, small time intervals DT such as 0.1 day, may be considered.

The rate of loss of immunity LR is the ratio of the immune population to the period of immunity (PIMM):

$$\text{LR.KL} = \text{IMMP.K/PIMM} \tag{9-28}$$

The closed loop block diagram for the three aggregates of the population is shown in Fig. 9-40 which is another representation of Eq. 9-25.

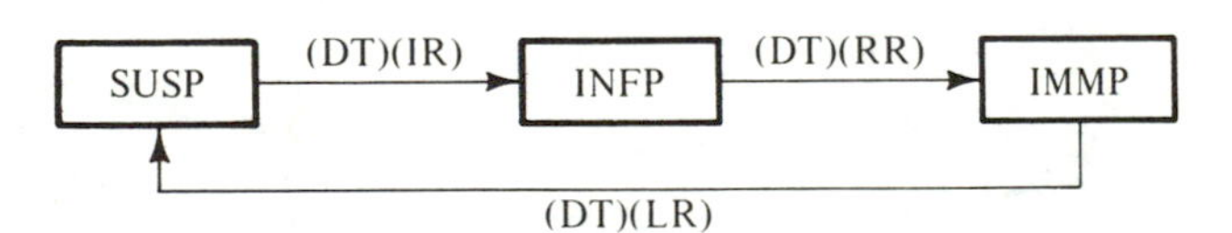

Figure 9-40 Block Diagram for Epidemic (Described in Eq. 9-25)

DYNAMO Simulation Program. The DYNAMO [84] simulation language program for the model of Fig. 9-40, supported by Eqs. 9-25 to 9-28, is given in Table 9-1. The simulation must begin with at least one infected person, possibly introduced from outside because, with INFP = 0, IR will be zero (see Eq. 9-26) and the disease cannot spread. Each statement in Table 9-1 is designated by a letter symbol on the left, signifying the following:

- L = Level or state equation, Eq. 9-25
- R = Rate equation, Eqs. 9-26, 9-27, 9-28
- N = Initial condition, initial values for SUSP, INFP, IMMP taken here as 1000, 1, and 0, respectively
- C = Constant, values for F, PD, PIMM taken here as 0.001, 8, and 1000, respectively.

Table 9-1 DYNAMO Program for Epidemic Model

```
L     SUSP.K = SUSP.J + (DT)(LR.JK - IR.JK)
N     SUSP = 1000  (PERSONS)
L     INFP.K = INFP.J + (DT)(IR.JK - RR.JK)
N     INFP = 1  (PERSONS)
L     IMMP.K = IMMP.J + (DT)(RR.JK - LR.JK)
N     IMMP = 0  (PERSONS)
R     IR.KL = F*SUSP.K*INFP.K
C     F = 0.001
R     RR.KL = INFP.K/PD
C     PD = 8  (DAYS)
R     LR.KL = IMMP.K/PIMM
C     PIMM = 1000  (DAYS)
PLOT  SUSP = S/INFP = F/IMMP = M
SPEC  DT = .1/LENGTH = 40/PLTPER = 1/PRTPER = 0
```

The last two lines in the table specify the output from the simulation calculations. The line before last

PLOT SUSP = S/INFP = F/IMMP = M

indicates a command to plot the desired variables with the identification symbols indicated following the equal sign. Thus, SUSP will be plotted as a function of time and identified by the symbol S on the curve, INFP will be identified by symbol F, and IMMP by symbol M (see Fig. 9-41). The last line in Table 9-1,

SPEC DT = .1/LENGTH = 40/PLTPER = 1/PRTPER = 0

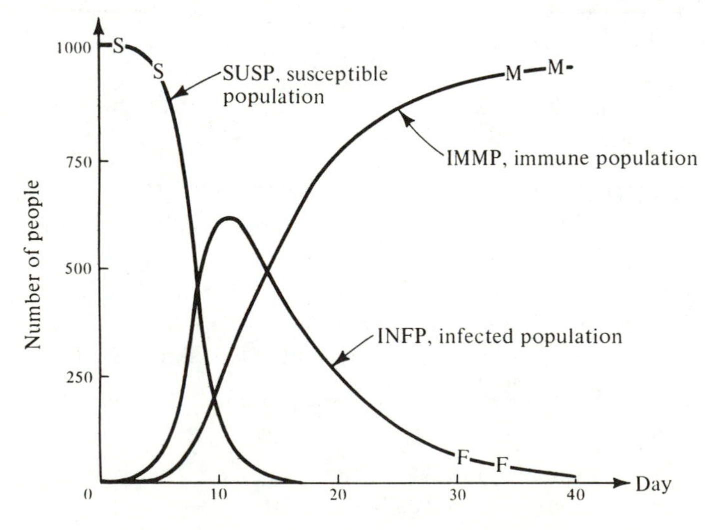

Figure 9-41 Results of Epidemic Simulation

specifies a "delta *t*" of 0.1 day over a total time interval of 40 days for the simulation; PLTPER gives the frequency with which the information is to be plotted (namely, points on the curves), once every day in the present example. PRTPER gives the frequency with which the information is to be printed out. No print out is requested in the present example, therefore zero (for zero frequency) follows PRTPER.

Results of the Simulation. The results of the simulation described by the program in Table 9-1 and reported by Henize [88] are shown in Fig. 9-41. These results show a rapid rise in the epidemic momentum from the fifth to the tenth day, then a decline as more and more people become immune and the epidemic has few susceptible people to feed on.

Critical Number SUSP. From Eq. 9-25(b) it is seen that in order for an epidemic to develop, the infected population, INFP, must increase, namely, the infection rate IR must be larger than the recovery rate RR, or

$$\text{IR} > \text{RR} \tag{9-29}$$

Using Eqs. 9-26 and 9-27, we have

$$\text{F*SUSP*INFP} > \text{INFP/PD}$$

or, dividing both sides by the product F*INFP, we have

$$\text{SUSP} > 1/(\text{F*PD}) \tag{9-30}$$

This implies that no epidemic will occur for

$$\text{SUSP} < 1/(\text{F*PD}) \tag{9-31}$$

The critical number of susceptible people, $\text{SUSP}_{\text{critical}}$, to start an epidemic is, therefore,

$$\text{SUSP}_{\text{critical}} = 1/(\text{F*PD}) \tag{9-32}$$

In the example considered here,

$$\text{SUSP}_{\text{critical}} = 1/(0.001 \times 8) = 125 \text{ people}$$

Figure 9-42, reported in [88], verifies this conclusion by showing the results of a simulation using the model of Table 9-1 with an initial susceptible population SUSP = 100 persons, namely, SUSP smaller than $\text{SUSP}_{\text{critical}}$.

> NOTES. The concept of a critical number of susceptible people to "feed" an epidemic suggests preventive measures to reduce SUSP by immunization procedures or to increase the level of $\text{SUSP}_{\text{critical}}$. A decrease in IR can be accomplished by reducing the number of possible paired contacts SUSP*INFP through quarantine isolation of the infected population. It is interesting to note that the concept of quarantine originated during

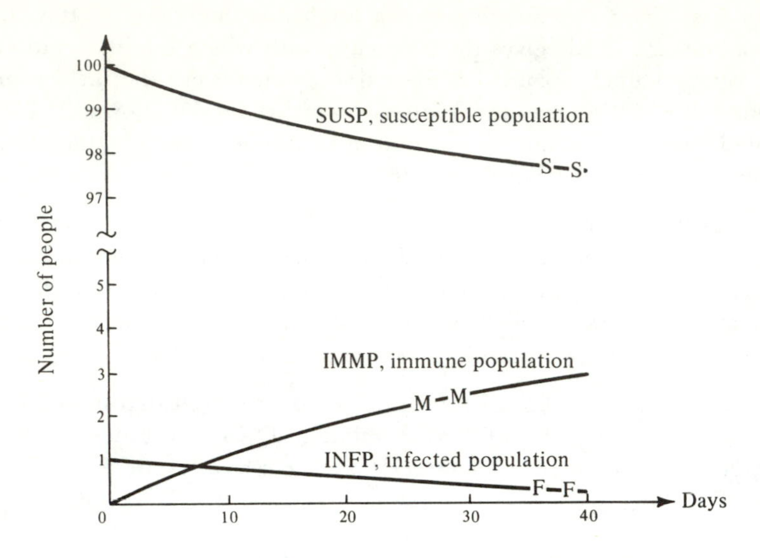

Figure 9-42 Result of Simulation for SUSP < $SUSP_{critical}$

the black death epidemic which swept Europe between 1348 and 1350 and left one-third of Europe's population dead [89]. In some crowded cities the devastation was even worse; Florence was reduced in population from 90,000 to 45,000 and Siena from 42,000 to 15,000. The word quarantine comes from the Latin word for forty (*quadraginta*), which was a period of detention or isolation imposed on ships or people arriving at a port when suspected of bringing an infectious disease. This practice goes back to biblical times, but was revived and rediscovered during the black death. The black death brought in its wake a fundamental change in man's model of the world by creating evidence that all are equally vulnerable to disaster, because the epidemic made no distinction between rich and poor, educated and ignorant, pious and heretic. In fact, the death rate among the clergy was even higher than the average (50% as compared to 30%) because of the crowded quarters in monasteries and the general tendency of the clergy to offer help and, thus, become more vulnerable to infection.

We now know that the epidemic, also known as plague (or bubonic plague because of swelling of the lymph glands) is spread by fleas that live mainly on black rats. The name "black death" comes from the dark blotches produced by hemorrhages in the skin.

9-9 A NOVEL APPLICATION IN THE MAKING

Preview. The disease known as *diabetes mellitus* is quite common. In some cases it occurs as a result of a deficiency in the amount of the hormone insulin secreted by the pancreas to regulate blood sugar metabolism.* The disease was known in ancient times and there is reference to it in the ancient literature of Egypt, China, and India. Insulin was first successfully extracted from the pancreas in 1921, and first administered to human patients in 1922. Before insulin became available, less than 20% of the patients with a severe diabetic condition survived more than ten years. Children, who usually have a severe form of the disease, seldom survived for more than one year. At present, a person suffering from diabetes can maintain a normal life style provided the insulin, which the body fails to produce internally, is supplied from an external source. Progress in understanding the nature of dynamic system behavior and man's ingenuity in contriving artificial sensing and feedback devices may ease the process of caring for this common disease. Some elements of the dynamic system, its behavior, control, breakdown, and potential repair are discussed briefly in this section.

Blood Sugar Concentration. The human body operates as a dynamic system with energy input and output. The weight of the body remains the same when input is equal to output. The output varies greatly between individuals. At a state of complete rest the average person's energy output is about 1700 Calories† per 24-hour period. Office work dissipates between 2000 and 2500 Calories a day, and vigorous efforts may dissipate 4000–5000 Calories per day. Signals to and from the hypothalamus, the brain switchboard, control the food intake for energy input. There are three general categories of food intake: carbohydrates, proteins, and fats. The food intake becomes a source of energy. The most readily available energy source to the brain and muscles is glucose in the blood.

The digestive system extracts sugar from certain foodstuffs, particularly starch, and converts it to glucose which enters the blood stream [90]. Various tissues remove glucose from the blood stream, oxidize it by using the oxygen delivered by the blood, and use the energy derived from the oxidation to perform their tasks. The central nervous system also needs glucose readily available so that it can respond rapidly to energy demands. Each milliliter of blood contains, under normal conditions, about 0.8 mg of glucose. This concentration amounts to about 1 teaspoon of sugar in the total blood volume.

---

*Metabolism is the process for generating energy for the vital activities of the body, and building material for the body structure.

†One Calorie is the quantity of heat required to raise the temperature of 1000 grams of water 1 degree centigrade. The quantity is usually determined at about 16°C which is 60.8°F.

*Control.** Normally, we eat large meals two or three times a day with stops in between, so there are two or three intervals of large sugar inputs into the blood. Since the blood must not contain excessive sugar, a control mechanism acts to store or remove the excess. The primary control elements are the kidneys, pancreas, liver and the central nervous system. The sensors are cells in the pancreas and the brain which are sensitive to glucose concentration in the blood. There are other elements in the control process not mentioned here [77, 91].

The input of glucose into the blood is accomplished as follows. The human body can draw glucose from the ingested food or from storage rooms or depots in the liver and muscles. These contain glycogen which can be converted to glucose. Also, protein in the body can be converted directly to glucose and put into the blood.

The removal of blood glucose may take the form of energy output, it may be excreted in the urine, it may be converted to glycogen and stored, or converted to fat. Normally, no glucose is converted to fat. When the concentration rises above 1.60 mg per ml, glucose is excreted in the urine. Fat is deposited only when the glucose exceeds that which is required for energy and for ample storage in the form of glycogen.

The *kidneys* as a separate unit can be considered an open loop controller that removes, through the urine, a certain amount of sugar from the blood, depending on the input concentration. Since no sugar is removed through urine for concentrations of less than about 1.6 mg/ml, and the maximum possible amount removed (i.e., the kidney's ability is saturated) is for concentrations of about 2.7 mg/ml, the input-output relation of the kidney system is approximately as shown in Fig. 9-43.

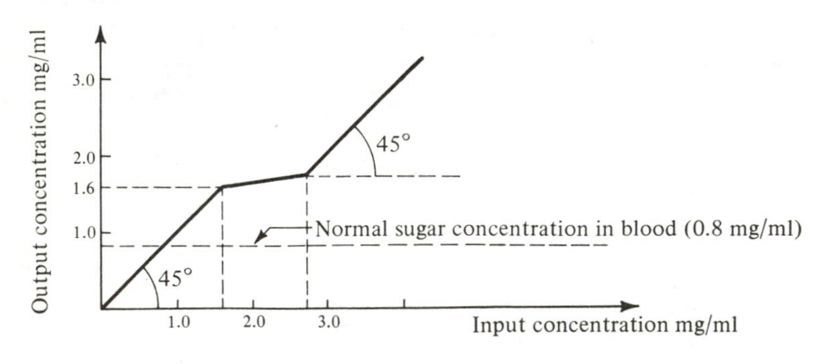

Figure 9-43 Blood Sugar Concentration through Kidney

*It is interesting to note that the investigations in the area of blood glucose metabolism were partially responsible for the discovery of homeostasis.

A second control mechanism involves the *pancreas* and *liver*. The pancreas secretes insulin, causing the liver to remove sugar from the blood by converting it into glycogen. Insulin from the pancreas also enters other tissues increasing the consumption of blood sugar. Figure 9-44 is a schematic simplified block diagram of the controlled removal of excess blood sugar. The diagram can be extended to include the central nervous system as part of the control mechanism. Most of the day the sugar concentration is at about the normal level of 0.8 mg/ml. When food with a large amount of sugar in it is consumed, the digestive system quickly transfers this sugar to the bloodstream, and the control mechanism acts to remove the excess.

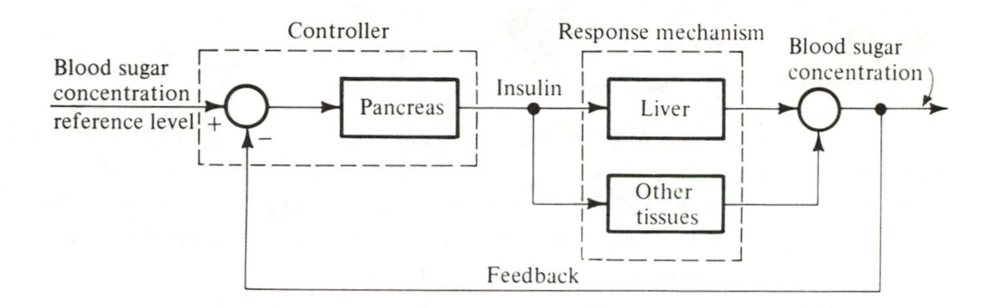

Figure 9-44 Block Diagram to Control Blood Sugar Concentration (Reduction of High Levels to Normal, 0.8 mg/ml)

Breakdown in Control. In some humans, the pancreas (a main element in the control system), fails to produce enough insulin to remove the excess sugar and lower the sugar concentration in the blood. This condition is called *diabetes mellitus*. The high sugar concentration leads to many problems. The presence of a high level of sugar in the urine is a rough indicator of the presence of the disease since, as indicated earlier, no sugar appears in the urine unless sugar concentration is about two times the normal level, i.e., 1.6 mg/ml (normal = 0.8 mg/ml).

A common treatment for diabetes is the injection of insulin. The person afflicted with the disease often acts as his own open loop controller. He measures his food intake and distributes the sugar input to his system in a way compatible with need. He then injects insulin in an amount, based on his experience, sufficient to control his problem. Without a laboratory, however, he cannot measure the actual blood sugar concentration and, hence, has only indirect feedback on whether his injection is of the proper amount. Too much insulin will lower the sugar level below normal, and too little will leave the level above normal.

A Novel Application of a Control System [92, 93]. A novel man-made procedure is now being developed for dealing with the diabetic problem. In this procedure a mechanism for sensing the sugar-level concentration will be implanted in the body of a diabetic patient and connected to an insulin supply unit, also implanted in the body. Through a regulation mechanism, the concentration of blood sugar will be automatically controlled for long periods of time. The higher level of control will again be man who will use a measurement to establish when the insulin unit must be replenished. This novel application is a number of years from the implementation stage. However, it serves to illustrate again the ubiquity and usefulness of dynamic system models.

9-10 SUMMARY

A *system* is a collection of elements aggregated by virtue of the links of form, process, or function which tie them together and cause them to interact. A system model is constructed to help *understand*, *predict*, and *control* its behavior. A system is described by *state variables* or *levels*. In a *dynamic system*, the state variables depend on the input history.

There are two configurations of control: *open* and *closed loop*. In open loop control, the controller selects an input, but there is no flow of information from output to new input. In closed loop or *feedback* control, information flows from output affecting the new input into the system.

Control can be exercised at various levels, and it can be voluntary or involuntary. The ubiquity of control in the natural and man-made world is manifested in its presence in the human body, in man's problem-solving activities, in government, in automation, in the ecological system, etc.

Cybernetics is the name for the field of control theory. It is derived from the Greek word for helmsman, *Kybernet*.

Block diagrams describe dynamic systems by using four elements: arrows, branch points, blocks, and comparators. A block represents a *transfer function* which prescribes the functional relationship between input and output. A comparator can be used to measure the *error e* which is the difference between a desired and actual output.

Models of dynamic systems date back to Ctesibius's waterclock in the 3rd century B.C. in Alexandria. The present-day toilet flush tank had its origin in Heron's "Inexhaustible Goblet."

A general block diagram for a dynamic system with feedback is shown in Fig. 9-24.

Homeostasis is biological control in living organisms; it is regulation of the internal environment in the face of disturbances from the external environment and changes in the system itself. Control of blood pressure and body temperature are examples of homeostasis.

For each dynamic system, a region of *controllability* must be established before control is attempted. Within the controllable region, feedback control has advantages over open loop control in that it is *goal seeking* and can correct for disturbances from the environment and for changes in the controlled system itself. A feedback configuration can achieve the same results as an open loop configuration with a more vulnerable (subject to change) system, namely, less costly than in open loop. Or better control is achieved for the same system by using feedback. The disadvantages of feedback are the added costs of the sensing and feedback, and the possible introduction of instability.

A *transfer function* of a dynamic system is characterized by *magnification* and *phase*. The magnification represents the ratio of peak magnitudes of output to input. The phase is related to the *time delay* in system response. The output normally lags behind the input, reaching its peak values after the input reaches its peaks. The higher the frequency of the input, the more delayed the response. *Transient response* is the initial erratic response as the system adjusts to an input. The *steady state* response is the well-behaved response which persists after the transient response decays. An *overdamped* system does not oscillate in its response, while an *underdamped* one does. *Critical damping* is the amount of energy dissipation at the transition between overdamped and underdamped response.

Phase or time delay between input and output can lead to *negative* or *positive feedback*. *Negative feedback* occurs when the input and output are 180° out of phase or half a period delayed. Positive feedback occurs when input and output are in phase or one period delayed. *Positive feedback* reinforces deviations of output from a reference level. *Negative feedback* diminishes output deviation from a reference level, provided there is no overcorrection. Overcorrection may lead to progressively increasing output deviations in successive cycles.

Input R and output O in a feedback control system with controlled system S, controller C, and feedback H are related by

$$O = \frac{CS}{1 + CSH} R \tag{9-10}$$

When disturbance D is included, the relation becomes

$$O = \frac{CS}{1 + CSH} R + \frac{1}{1 + CSH} D \tag{9-19}$$

For open loop control, the relationships are

$$O = SCR \tag{9-13}$$

and with disturbance D,

$$O = SCR + D \tag{9-18}$$

The above four equations apply to a *linear control system* in which effect is proportional to cause. Nonlinear systems can be modeled as linear systems over a small range in variation.

A system is *stable* when a disturbance from an equilibrium state creates in its wake a tendency to return to equilibrium after the disturbance stops. When equilibrium is not restored after the disturbance stops, the state of equilibrium is *unstable*. A *neutrally stable* state of equilibrium is a sustained oscillation of constant amplitude with respect to a reference state.

It can be shown quantitatively that feedback control systems are less sensitive to controlled system variations and to environmental disturbances than open loop control. However, the introduction of feedback may cause instability in a system which otherwise is stable.

Optimal control is an optimization of control with respect to an index of performance (IP) such as minimum *Mean Square Error* of deviation of actual output x_a from desired output x_d, over the time T in which the system is in use:

$$\text{Mean Square Error} = \frac{1}{T} \sum_i e^2(t_i)\Delta t_i \tag{9-23}$$

or

$$\text{Mean Square Error} = \frac{1}{T} \int e^2(t)\, dt \tag{9-24}$$

in which

$$e = x_a - x_d$$

There are other indices of performance, such as minimum time for setting error e to zero.

Adaptive control is a second feedback loop nested in the first one to introduce changes in the controller so it can adapt to changes in the controlled system and to disturbances from the environment. A block diagram for adaptive control is shown in Fig. 9-39. Adaptive control is finding many areas of applications in pattern recognition, artificial intelligence, weather prediction, and learning machines in general.

An important form of adaptation is a change in the reference goal as the need arises. This is a higher level of adaptive control than changing the controller to achieve a prescribed goal.

Simulation of a dynamic system can provide knowledge for understanding a system's characteristics in terms of amplitudes, time delays, and possible conditions of instability. The system behavior can also be studied over a wide range of environmental disturbances and changes in the system itself. The results of a simulation may also lead to questions regarding the reference goals, and cause them to be changed. DYNAMO is a computer simulation language. Using this language, we can get answers to the *what if . . .* type of questions of concern, and "act out" many years of a dynamic system. As an example, the simulation of an influenza epidemic is shown.

A novel application of control in dynamic systems is discussed in connection with blood sugar control for the disease diabetes.

EXERCISES

NOTE. The first four exercises refer to your project of Exercise 1-1.

9-1 Use a block diagram to describe your project as a dynamic system. Clearly identify the components such as input, desired output, etc.

9-2 Describe the specific role of feedback in your model.

9-3 Expand your block diagram to include adaptive control. How might this control help or hinder you in the solution of your project?

9-4 Identify disturbances that could affect your project. How sensitive is your model to these disturbances?

9-5 Put yourself in the position of a teacher who is offering a course in creative problem solving. State what your objective would be. Proceed to answer Exercises 9-1 through 9-4 with regard to this problem.

9-6 Discuss problems that might be associated with the measurement of the output in a dynamic system. Specifically state instances in which differences between measured and actual output can be attributed to errors in measurement. How might this relate to the story in Chapter 1 regarding measurement (tests) of the reading competence of students?

9-7 Draw a block diagram to show a solution process for the tile puzzle of Exercise 1-15.
(*Hint:* Consider the following index of performance. For an intermediate stage, the "distance" from the goal state is the *sum* of the individual distances of the tiles from their respective goal positions. These individual distances would equal the number of moves (following the shortest path) which will put a tile into its goal position. For example, in the initial state (of Exercise 1-15), tile 10 is a distance of 3 from its goal position.)

9-8 Construct a block diagram of the modeling process of Chapter 5 (include validation of the model).

9-9 Water flows into an empty tank at a constant rate until the water level in the tank reaches the level of a float at some initial height, h_0. From then on, the water enters the tank at an ever-decreasing rate, as the float rises. Draw a block diagram to model the system.

9-10 One-pound packages of detergent are filled automatically by machine. One package of every ten is removed from the filling line and when twelve such boxes are taken, they are all weighed together. If the boxes together weigh 12 pounds, the filling process continues as before. However, if the boxes weigh over 12 pounds, a valve is closed slightly, decreasing the amount of detergent going to each box. Similarly, if the boxes weigh less than 12 pounds, the valve is opened slightly. Consider the fact that the weight of detergent in any one box is distributed normally with mean of 1 pound and variance of 2 oz. Discuss problems associated with running the machine in this manner.

NOTE. As we saw in Sec. 9-7, we may transform complicated block diagrams into simpler ones by combining some blocks. For example, if we have inputs X and D, and block transfer functions C, S, and H, we can construct equivalent systems, as in Fig. 9-45, to find output Y. (Note that C, S, and H can represent *any* type of transfer function.)

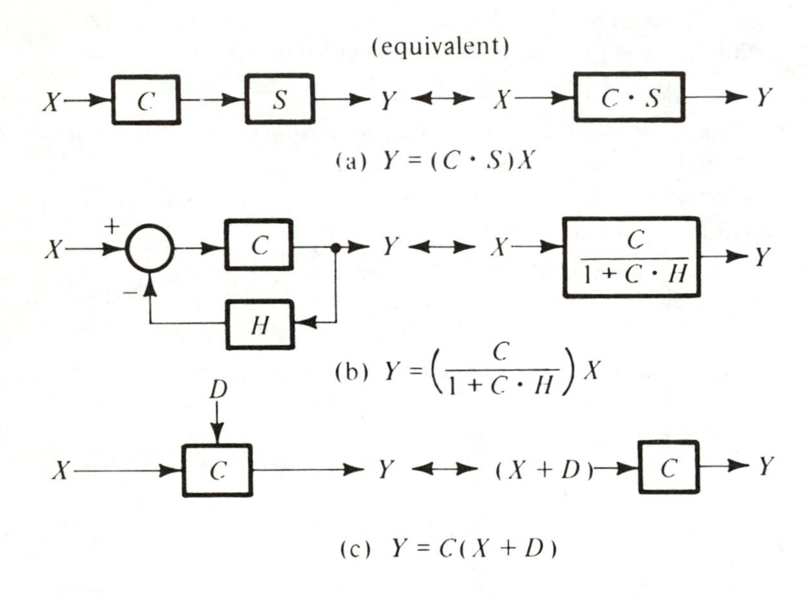

Figure 9-45

9-11 Simplify the systems in Fig. 9-46 by combining blocks, and find the transfer functions.

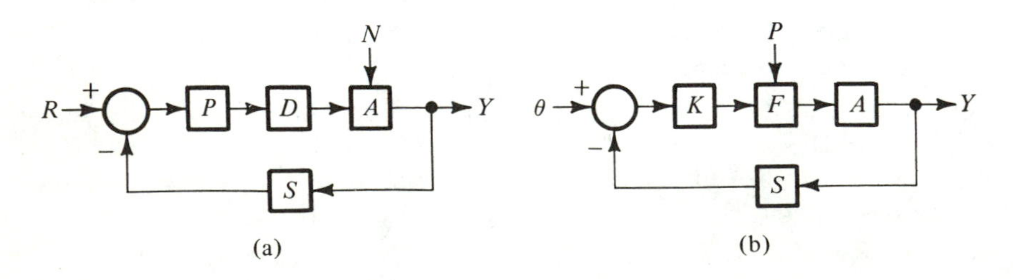

Figure 9-46

9-12 Let $P = 10$, $D = 2$, $A = 1$, $N = 0.5$, and $S = 0.01$ in Exercise 9-11.

(a) If $R = 50$, what will be the output Y?

(b) What must the value of R be to yield an output of 100?

(c) If N rises to 2 with $R = 50$, what will be the output Y?

9-13 Expand each block in the block diagram for the epidemic (Fig. 9-40) to represent Eqs. 9-25, 9-26, 9-27, and 9-28. In each block, show the input-output relationship and label the input and output variables.

9-14 Referring to the epidemic model of Sec. 9-8, discuss problems that might arise as DT is made increasingly larger. Discuss also how changes in other variables would affect the output.

9-15 In Exercise 9-9, suppose the tank is 20 inches long and 5 inches wide (thus, the level of water in the tank increases 1 inch for every 100 cubic inches of water entering the tank). Suppose that at time $t = 0$, the tank is empty and the float sits 5 inches above the bottom of the tank. Water enters the tank at a rate of 100 cubic inches per second until it reaches the level of the float. From then on, water enters at a rate of $5(25 - h)$ cubic inches per second (where h is the height of the float above the bottom of the tank). Assume this rate is constant for every 1-second interval and plot a graph of the amount of water in the tank versus time.

9-16 Discuss the arms race between world powers in terms of a dynamic system with feedback. How does such a system react to disturbances? How can the system be controlled to prevent instability? How does this relate to nonzero-sum games, as discussed in Chapter 7?

9-17 In a developing nation there is presently a need for 10,000 engineers, and this need will grow by 3000 engineers per year. Currently there are 5000 engineers and 4000 students who will become engineers in 2 years, but no students who will become engineers in 1 year. Further assume that the number of students with 2 years to go before becoming engineers is equal to the *preceding year's* shortage of engineers. (For example, next year there will be 5000 students in this category since this year's shortage is 10,000 − 5000 = 5000 engineers.) If in the preceding year there was a surplus of engineers, then assume no students enter this category.

(a) Draw a block diagram to describe this system.

(b) Calculate and plot on a graph the surplus (or shortage) of engineers, year by year, for the next 30 years. (You may assume all students with 2 years of study remaining, graduate and become engineers, and no people drop out of engineering in the next 30 years.)

9-18 Figure 9-44 shows a block diagram of the control system for blood sugar concentration in the human body. Modify this diagram to show the proposed novel control system described on page 466.

9-19 A group of foxes and rabbits live on an island capable of supporting a large number of rabbits, but providing the foxes with only the rabbits as food. Initially, there are 1000 foxes and 1000 rabbits.

Let R = number of rabbits $\qquad A$ = birth rate of foxes

F = number of foxes $\qquad B$ = birth rate of rabbits

The number of rabbits eaten (RE) by foxes is proportional to the product of the number of foxes and the number of rabbits. That is, $RE = K \cdot R \cdot F$. (K can be thought of as the probability of a rabbit being eaten when he comes in contact with a fox.)

The number (FS) of foxes that die of starvation is proportional to the number of foxes times the ratio of foxes to rabbits. That is, $FS = C \cdot F^2/R$. ($C =$ constant).

The number of foxes born each season is $A \cdot F$ and the number of rabbits born each season is $B \cdot R$.

(a) Draw a block diagram of the system.
(b) Let $K = 10^{-3}$, $C = 0.4$, $A = 0.4$, and $B = 1.0$. Show that the system is in equilibrium (that is, show that after each season, there are still 1000 rabbits and 1000 foxes.)
(c) Assume a natural disaster kills 200 rabbits and 200 foxes. Find the number of rabbits and foxes for each of the next 20 seasons to see whether or not the system is stable. (Is the system returning to a state of 1000 rabbits and 1000 foxes?)
(d) In the equation for "rabbits eaten," RE is proportional to the product of the number of foxes and the number of rabbits. Why is this product a reasonable measure for RE?

9-20 Referring to Exercise 9-19, assume that initially there are 1000 rabbits and 1000 foxes Let $K = 1.5 \times 10^{-3}$, $C = 0.4$, $B = 1.5$, and $A = 0.4$.
(a) As in 9-19(b), show that the system is in equilibrium.
(b) Assume the death of 200 rabbits and 200 foxes and, as in 9-19(c), show whether or not the system is stable.

9-21 The feedback system shown in Fig. 9-47 represents the payment scheme for a mortage loan. R is the fixed monthly payment, r is the monthly interest rate (equal to yearly rate divided by 12), and I_n is the interest payment to be made on the nth payment date. P_n is the principal remaining after n payments.

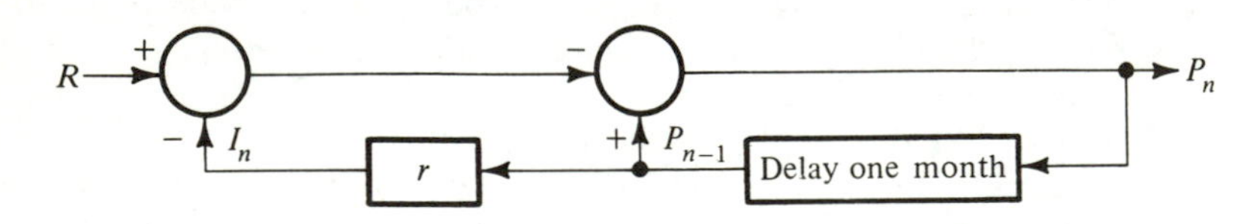

Figure 9-47

Given a $50,000 loan ($P_0 = 50{,}000$) at 6% per year, with a $300 monthly payment, compute the interest paid in the first year.

9-22 As outlined by Goldberg [94], Samuelson's economic model for the national income X_n in any given year n can be expressed as the sum of government spending in that year, G_n, consumer expenditures C_n during that period, plus private investments for capital equipment, I_n.

The consumer expenditure C_n is assumed to be proportional to the national income of the preceding year, thus $C_n = \alpha X_{n-1}$.

The private investment in turn is proportional to the *increase* in consumption from one year to the next. Thus, $I_n = \beta(C_n - C_{n-1})$.

Government expenditure is an arbitrary number and may be considered as input to a system where the output is fed back with appropriate delays.

(a) Draw a block diagram for the economic model.

(b) Assuming $\alpha = 0.5$, $\beta = 1$, $G_n = 1$, $X_0 = 1$, and $X_1 = 2$, make calculations to determine the general nature of the fluctuation of national income.

(c) Do the fluctuations seem to be approaching any limit? If so, what is the limit?

(d) Repeat part (b) for $G_n = 0.5X_{n-1}$.

9-23 This problem again refers to Samuelson's economic model (see Exercise 9-22), but the parameters have been changed so as to result in an unstable economy. Again, assume $X_0 = 1$ and $X_1 = 2$, and let $\alpha = 0.8$ and $\beta = 2$, with $G_n = 1$. Make enough calculations to establish a pattern from which the instability will be evident.

10

VALUES AND MODELS OF BEHAVIOR

10-1 INTRODUCTION

At the very beginning of the first chapter we introduced the concept of values as the central element in problem solving. A person's *value system* constitutes a framework which influences his models of reality, his appreciation of what is worth striving for, and his choice of actions to bridge the gap between what is his perceived present state and the desired or preferred goal state. Two people using the same rational tools of problem solving may arrive at different solutions because they operate within different frames of values, and therefore their behavior is different. When a number of people face a problem which they must solve as a group, consensus must be formed to establish the group values.

The tools of problem solving and decision making can guide us to achieve desired goals, or lead us to the realization of our values. But what might the values be in the first place? How do values get established? Is a rigid or flexible approach to values desirable?

In the past, it was common for successive generations to subscribe to the same values. At present, it is common for values to change more than once in a single generation. The young no longer subscribe to the values of their elders, and as time marches on they will continue to change their values. In an era of rapid social and technological change, an open-ended, flexible, and unsettled approach to life is emerging as a major characteristic of a healthy life style compatible with changing values. Education should strive to develop such an attitude. Instead of attempting to develop a false sense of security anchored in the realization of fixed values, we should strive to cultivate a mechanism for effectively coping with insecurities which may result from change in values.

What are values? The following definitions may be helpful:

> A value is anything which is the object of interest.
> Values are end states to guide human endeavor.
> Values are normative standards which influence human behavior.
> A value is that which permits us to judge what matters from what does not matter.
> A value is that which furthers the cause of human survival.

Each of these definitions contains some of the attributes of values. Therefore, rather than subscribe to any one definition as all encompassing, let us list some attributes of values and discuss them.

> Values evolve with culture and are mostly man-created.
> Values change; the rate of change differs.
> Some values are end values.
> Some values are means-to-an-end values.
> Subscription to value by people can take the form of professing or action, i.e., behavior.
> Values for an individual may conflict with values of a group.
> Values of one group may differ or conflict with those of another group.
> Values and knowledge are interrelated.
> Values can be classified in a number of ways.
> A hierarchy of values can be established.
> Values are benefit oriented and can be assessed by a cost-benefit function in which cost represents effort and energy expended for the realization of a value.

In this chapter we discuss some of the above attributes. Our discussion will serve only as a foundation for further thought and study by the reader. In the spirit of this book you are encouraged to read with a will to doubt, challenge what is not in agreement with your own experiences, and construct rational arguments for your point of view.

I believe that values, namely, the questions and the attending search for what ought to be the values of human beings as individuals and the values of society, constitute the most appropriate subject for true dialogue in education. Since no longer do parent and child, or student and teacher necessarily subscribe to the same values, communication through dialogues, in which for example student is not distinguished from teacher, is essential in order to reach understanding and form consensus.

Value Theory. In daily language we use the word "value" in a sense which is benefit oriented. We speak of economic value, aesthetic value, moral value, ethical value. The benefit associated with these values may be individual or societal in character, or both. Thus, efficiency and thrift may be of economic value to both individuals and society, while certain aesthetic values in art form may be oriented only to benefit an individual.

The general theory of value, known as *axiology*, encompasses all aspects of values. Rudolf Lotze (1817-1881) is considered the father of the general theory of value, in which value is defined as:

> A thing—anything—has value or is valuable, in the original and generic sense, when it is the object of an interest—any interest [95].

Axiology has sparked an interest and activity in values in economics, psychology, sociology, social psychology, and philosophy. Thus, it embraces the study of economic, aesthetic, ethical, and moral values.

10-2 ROLE OF VALUES IN PROBLEM SOLVING

The systems approach to problem solving can be described in terms of three fundamental stages:

Stage 1. A description of the present state of *what is*
Stage 2. A description of the desired goal state of what ought to be or *what is desired*
Stage 3. A description of a process to bridge the gap between what is and what we want; namely, a prescription of *what to do and how*.

The greatest amount of effort in systems analysis has been devoted to the third stage of *what to do*, and the least effort on the description of the goal state, namely, what we want. It is commonly considered that values enter the goal state. This is true. However, values enter stages one and three as well and, therefore, all three stages cannot be dealt with independently and must be considered as an integrated whole. To support this contention let us consider the three stages.

Stage 1—Description of the Present State

When we try to describe what is a present state, we resort to a language of representation. The language may take on different forms: natural language, art, mathematics, etc. This is a modeling process in which we attempt to abstract reality with some detail sacrificed for the sake of describing a working model that is useful. What is included in the model depends on what we consider important and relevant, and this, in turn, depends on our values.

For example, a person who values education highly may describe the billions of dollars associated with all forms of education in the United States as an *investment*. The same sum is likely to be described as a *cost* or *expenditure* by a person who puts a low value on education.

Vickers [96] points out that a crime, for instance, may be described as a disturbance of the social order by some, and as a protest by a distressed individual by others. The choice of description depends on the values of the person and his world view. A policeman, a social worker, an engineer may differ greatly in their description of an event or an observed state if their values are different. However, differences in description are not necessarily based on different world views represented by different professions. Two people of the same profession may also differ in their model of *what is* when their values are different. For example, a city planner who values meaningful social interactions in a community more highly than economic well-being, will describe the same community differently than a city planner who ranks these values in reverse order [97].

The aspects of reality which we grasp and describe in our models are strongly influenced by our biological tools of perception and by factors which relate to both language and culture in the environment in which we reside and strive for survival. Thus, for example, the Eskimos distinguish 30 different kinds of snow [98] while we distinguish between scores of automobiles. These distinctions are likely to enter the model which describes reality. To an Eskimo, a different kind of snow may be an important factor contributing to the value of survival. To an American, the particular kind of automobile may be an important factor contributing to the value of prestige.

The value and importance of the individual, the significance attached to the life here and now as contrasted with the concept of an eternity as manifested in different cultures, also enters the description of *what is*. For example, Bertalanffy [98] observed that Japanese artists never painted shadows because to them a shadow was not an important aspect of reality, but only a temporary short-lived change in appearance. Along the same lines, when the Dutch treatises on perspectives were introduced in Japan in the late 18th century, Japanese artists adopted a different form of it which was representative of what they perceived as an important aspect of reality. While European paintings considered reality as viewed by an observer and, therefore, had a central perspective with parallel lines, progressively converging away from the observer, the Japanese used parallel lines in the paintings with convergence in infinity. Again, the position of the individual observer was not considered important, it was only a detail and therefore not to be included in the model of reality.

Models in science are primarily concerned with furthering our understanding of *what is*, which in turn may enhance our ability to predict future outcomes. Honesty, comprehensiveness, and order are some values of importance in models of science. Order is either imposed, or sought, in quest for harmony, elegance, and simplification, and it attempts to explain more and more phenomena in terms of fewer key relevant parameters. This is true of Kepler's model, Newton's model, Maxwell's model, etc. This quest for order and unification, aggregates more and more phenomena in a single model. The model of relativity encompasses more than Newton's model, and much effort has been devoted in search for further unification through a unified field theory which will place electromagnetic and gravitational fields within a single model of reality.

Stage 2—Goal State

The political process is identified as the arena for generating societal goal states and for implementing what is possible. This stage of establishing *what we want* or what we consider *ought to be* is most difficult. It requires a learning process to gain insight into our own value system. An architect who works closely with a person in the design of his house must help the person identify his own values if the finished product is not to be a disappointment. Most of our education is concentrated on know-how of *what to do* and *how to do* when a goal is specified, but too little is done to cultivate a learning process to help us identify our value system and modify it as the need arises.

Stage 3—The Process, What to Do and How

Technology is primarily concerned with models of the process for bridging the gap between *what is* and *what is desired.* Building on the knowledge gained from science, technology attempts to construct models for achieving desired goals. Models of the process are also influenced by values. Efficiency and economy are two important values in models of processes to achieve goals. However, while traditionally technology has been primarily concerned with the process of *what to do* and *how to do it,* guided by its own values such as efficiency and economy, recent history has made it clear that there are competing values in the goal state which must be seriously considered, and a compromise of efficiency and economy may be required. For example, the efficiency and effectiveness in the technology of data storage and retrieval, which is important in business, industry, and government, is in competition with the social value of privacy. At some point efficiency will have to yield to privacy [99].

10-3 VALUE CLASSIFICATION

One of the most fundamental activities in problem solving is classification. The problem of values is no exception, so let us consider some possible classifications.

End Values and Means Values

Aristotle argued that there must be a limit in a series of values which form a chain in which each value is a means to the realization of a following value. The end value is the limit of the chain and is not for the sake of something else, but rather "for whose sake something else is." The end values are intrinsic values, values for their own sake. Intermediate values which are not ends in themselves are means values.

I propose that we consider this classification in a hierarchy of substructures at successive levels, in which each substructure consists of means values which terminate in an end value. The end values of the substructures are then linked and assume the role of means values to achieve the realization of an end value of a higher level in the hierarchy of levels, until we reach a truly intrinsic limiting end value.

For instance, we may consider economic welfare of a community (or a society) as an end value within a substructure which contains thrift, efficiency, progress, and growth as means values. Economic welfare may then be considered as one of a number of means values to the realization of survival as an intrinsic limiting end value.

This classification is useful as a possible explanation for some of the difficulties with our present value system. The subscription to means values, such as progress and growth, can become so intense that they take on the role of end values, namely, they become ends in themselves. Such intensity of subscription is at the expense of the realization of other values. A periodic assessment of values and their relative importance in the hierarchy of values is essential in order to put them in the proper perspective, and distinguish means from end values.

Classification According to Grouping

Values can be classified according to the social grouping to which they apply. For example, values such as individual success, family honor, unity of purpose (in the community), professional reputation, patriotism, and social justice belong, respectively, to these six groupings:

Individual
Family
Community
Professional Group
Nation
Society

Competitive and Cooperative Values

Values can be classified as competitive or cooperative. Power, status, and economic success are examples of competitive values, while aesthetics, safety, and health are cooperative values.

The above three classifications are not the only classifications or the best ones. For example, a fourth mode of classification may list values according to levels of importance, and a fifth classification may list related values such as comfort and health or courtesy and justice. A sixth may consider a grouping of values as related to things, to the environment, to the individual, the community, and society. A seventh classification may consider individual rights values, personal character values, and life-setting values in terms of the political arena and the physical environment. The purpose of this section is to show that values can be classified. The mode of classification chosen will depend on the problem considered.

10-4 VALUE JUDGMENT

Acts of judgment are inseparable from the fundamental act of matching. The doctor matches a medicine to a disease, the judge matches a punishment to an offense, and in common speech we match words to objects, experiences, thoughts, etc. Indeed, judgment as a matching activity is closely related to the concept of classification, because the act of judging what is like and what is unlike is at the very foundation of human thought process in classification.

Value judgment is the act of matching behavior, or a course of actions, to values. To be more specific, when we claim that activity or policy a is good for the purpose of realization of value v, we have performed an act of judgment in the sense of matching an activity or a policy to a value. It is conceivable, however, that an act may contribute to the realization of a number of values v_j ($j = 1, 2, \ldots, m$). In that case, we must use judgment to match a number with the relative intensity of the contribution of the act to each value.

Since values can be viewed as means values which contribute to the realization of an end value, we may use judgment to assign a rating r_j in the form of a number which matches the importance of the means value v_j, or how much v_j matters, to the realization of the end value.

Consider, for example, the design of an urban transportation system. Suppose the model consists of n alternatives a_i, and that the hierarchy of values is as shown in Fig. 5-4. Namely, the end value is the "quality of life" and the means values v_j ($j = 1, 2, 3$) are convenience, safety, and aesthetics. We note that the list of values could be expanded and that each may be considered as an end value within a substructure of lower-level values. To assess the utility of each alternative, we write:

$$u(a_i) = \textstyle\sum_{j=1}^{m} c_{ij} r_j \qquad i = 1, 2, \ldots, n \tag{10-1}$$

in which r_j is the importance rating of value v_j, and c_{ij} is the relative contribution of alternative a_i to the realization of value v_j.

In matrix form Eq. 10-1 becomes

$$\{u(a_i)\} = [c]\{r\} \tag{10-2}$$

The criterion for selecting an alternative can be max $\{u(a_i)\}$, i.e., the alternative which yields the maximum utility.

This model is, of course, very simple. It is not stipulated here that the linear model of Eq. 10-1 is the most descriptive or appropriate. In addition, the assignment of values r_j may be quite a complicated task if we consider interactions of the following nature. Consider four values v_j. It is conceivable that the assignment of a particular importance rating r_k will depend on the probabilities and extents of realization of the other three values. The number of combinations is infinite because both the probabilities and the extents of realization are on a continuous scale. In Chapter 7 we considered problems in which a utility was generated for a particular outcome B by the mechanism of a conceptual lottery. The most desired outcome A was assigned a utility of one, the least desired outcome C was assigned a utility of zero, and the utility $u(B)$ assigned to B was the probability p in the following lottery:

$$\{p, A; (1 - p), C\}$$

Efforts have been made to extend this fundamental approach to situations in which multiple attributes are involved [100].

Example—A Model for Value Judgment in Resource Allocations

Public and private agencies are often confronted with the problem of allocating resources to alternative programs of research and development. When values v_j can be stipulated and their importance rating r_j assessed, Eq. 10-1 can be used in a model of value judgment among alternative programs. Since each program may require a different level of resource allocations, a ratio of benefit, or accrued expected utility, to cost can be generated and used as an index of desirability for each program.

Cetron [101] employed such an approach in comparing the desirability of the following two research and development programs for the U.S. Navy.

Program a_1: Enhance *Tensile Strength* of Metals

Program a_2: Improve *Metal Corrosion* Resistance (in sea water)

The programs were compared in terms of their value to the country, their cost, and their probability of success, using these steps:

1. The national values v_j or goals and their ratings r_j were established.
2. The contributions c_{ij} of programs a_1 and a_2 to the realization of values v_j were assessed.
3. A probability was assigned to the likelihood of success for each program in meeting its objective (say, a certain level of tensile strength for program a_1, and a certain level of corrosion resistance for program a_2).
4. The costs for each program were estimated at three levels of funding:
   A threshold level below which the program should not be undertaken.
   A maximum level which represents the amount of funding for the program under conditions of unlimited resources, i.e., the program can progress at a rapid pace to meet the stipulated objective.
   An optimum funding level to proceed at an optimum rate which is compatible with normal but productive procedures.

Table 10-1 Contributions c_{ij} of Two Programs to the Realization of National Goals (Values) and their Ratings r_j [101, 102]*

			Contribution c_{ij} to Realization of Goal	
j	National Goals (values, v_j)	Rating r_j	c_{1j}, Program a_1 Tensile Strength	c_{2j}, Program a_2 Corrosion Resistance
1	National Defense	14	0.8	0.9
2	Social Welfare	12	0.3	0.0
3	Education	8	0.1	0.0
4	Urban Development	8	0.9	0.0
5	Health	7	0.1	0.0
6	Area Redevelopment	7	0.7	0.1
7	Housing	6	1.0	0.3
8	Agriculture	6	0.2	0.4
9	Manpower Retraining	6	0.0	0.0
10	Transportation	5	0.9	0.6
11	Consumer Expenditure	5	0.2	0.0
12	Private Plant and Equipment	4	0.8	0.0
13	Space	4	0.6	0.1
14	Research and Development	3	1.0	0.9
15	International Aid	3	0.0	0.0
16	Natural Resources	2	0.9	0.8

*Adapted by permission of John Wiley and Sons, Inc. from Marvin J. Cetron and Christine A. Ralph *Industrial Applications of Technological Forecasting*, Wiley, N.Y., 1971.

Table 10-1 lists the values,* their ratings r_j, and the contribution coefficients c_{ij} of the two programs. The ratings r_j were assigned on the basis of

$$\sum_j r_j = 100$$

and $$0 \leq r_j \leq 100$$

From the information in Table 10-1, the national utilities $u(a_i)$ of the programs were calculated by applying Eq. 10-1:

$$\begin{aligned} u(a_1) &= \textstyle\sum_{j=1}^{16} c_{1j} r_j \\ &= 0.8 \times 14 + 0.3 \times 12 + \ldots + 0.9 \times 2 \\ &= 51.5 \text{ utiles for program } a_1 \\ u(a_2) &= \textstyle\sum_{j=1}^{16} c_{2j} r_j \\ &= 0.9 \times 14 + \ldots + 0.8 \times 2 \\ &= 25.2 \text{ utiles for program } a_2 \end{aligned}$$

The probabilities of success were 0.9375 and 0.75 for programs a_1 and a_2, respectively. These values were calculated on the following basis. Four *independent* efforts† were considered in progress in program a_1, each with

*The values are from Lecht [102]. The ratings r_j and coefficients c_{ij} are those used by Cetron [101].

†Possibly different approaches to achieve the same objectives of the program.

probability 0.5 of success. Since the success of one effort or more is equivalent to the success of the program, then $p_{\text{success}}(a_1) = 1 - p_{\text{failure}}(a_1) = 1 - (0.5)^4 = 0.9375$. Two independent efforts were considered for program a_2, each with a probability 0.5 of success; therefore, $p_{\text{success}}(a_2) = 1 - p_{\text{failure}}(a_2) = 1 - (0.5)^2 = 0.75$.

Thus, the expected utilities of the programs were calculated as

$$\begin{aligned} E[u(a_1)] &= p_{\text{success}}(a_1) \times u(a_1) \\ &= 0.9375 \times 51.5 \\ E[u(a_2)] &= p_{\text{success}}(a_2) \times u(a_2) \\ &= 0.75 \times 25.2 \end{aligned}$$

The optimum fundings for the programs were 1 billion dollars for program a_1 and 0.516 billion dollars for program a_2. Therefore, the expected utility per billion dollars, or the index of desirability D_i, was calculated for each program as follows:

$$\text{Program } a_1\text{: } D_1 = \frac{0.9375 \times 51.5}{1} = 48.3 \text{ utiles/billion dollars}$$

$$\text{Program } a_2\text{: } D_2 = \frac{0.75 \times 25.2}{0.516} = 36.6 \text{ utiles/billion dollars}$$

10-5 KNOWLEDGE AND VALUES

It has been asserted that our knowledge concerned with natural law deals with matters of fact and, therefore, has no place for values. Models of natural law can predict outcomes on the basis of inputs and knowledge of the current state of the system under consideration. These models are instrumental in problem solving in the sense that they guide us to select a behavior which will lead to a desired outcome or the realization of a value. However, knowledge of natural law does not tell us what we ought to desire or what ought to be our values.

The separation of knowledge of natural law from human values is not fruitful. For one thing knowledge of natural law may be essential for the purpose of realizing values. For example, we may have the best intentions for curing an epidemic in a community on a remote island because we subscribe strongly to the value of physical well-being of people and value suffering negatively. However, our good intentions will do little to further our cause if, because of ignorance, we administer medication which not only does not stop the epidemic but worsens conditions and intensifies suffering. Here we may take comfort in our good intentions; however, knowledge of the cure would certainly be more effective in leading us to the realization of our values.

There is, then, a strong connection between knowledge of natural law and values. Certainly, such knowledge can be used in malice, but I believe we should learn to live with such possibilities and make every effort to cultivate a life style in which respect for knowledge and values is positively valued.

It has also been suggested that science may have done more than merely provide the knowledge which can help us realize our values. Lindsay [103] suggests that the second law of thermodynamics gives us a strong clue as to what human beings ought to do by identifying a value which is central to reverence for life and its sanctity. He suggests that man ought to behave so as to decrease entropy or increase the consumption of entropy. Before we present this point of view in more detail, let us digress for a brief discussion of the first and second laws of thermodynamics and review the concept of entropy.

Digression: First and Second Laws of Thermodynamics

The first law of thermodynamics is the conservation of energy. Energy can be transformed from one form to another. While energy is conserved in all transformation processes, it is accompanied by a loss of availability of energy to perform work in the future. Thus, every transformation reduces the possibility of future transformations, or the process is irreversible, i.e., directed one way. This is the second law of thermodynamics. Entropy is used to describe the loss in available energy in all naturally occurring transformations.

Digression Continued: Entropy, A Revisit (remember Chapter 4?)

In Chapter 4 we considered entropy as a lower bound on the average number of binary questions that must be asked to identify an event from E_i ($i = 1, 2, \ldots, n$) possible events with probabilities $p(E_i)$ of occurrence. *The entropy is maximum for events E_i when the probabilities of occurrence $p(E_i)$ are identical.* Let us refer to a set of events E_i with associated probabilities $p(E_i)$ as a *state*. When the probabilities change, the state changes. A state in which the probabilities $p(E_i)$ are equal for all events, i.e., $p(E_i) = 1/n$ for $i = 1, 2, \ldots, n$, is a state of maximum entropy. This state represents the state of highest disorder because to identify events E_i requires on the average a larger number of binary questions than for any other assignment of probabilities to events E_i. Thus, the larger the entropy, the higher the degree of disorder of the state, i.e., the more binary questions are required to identify the events E_i. Conversely, the smaller the entropy, the higher the degree of order of the state, i.e., the fewer binary questions are required to identify events E_i. "Order" as used here is, therefore, inversely related to randomness; the less randomness, the more order.

The second law of thermodynamics asserts that the entropy in the universe is increasing because highly ordered states of low entropy tend toward states of lower order and higher entropy, in which the events are more equally probable. For example, consider a room of dimensions 10 ft × 10 ft × 10 ft, i.e., 1000 cubic ft. Suppose we identify *one molecule of oxygen* in the room. If we divide the volume of the room into 1000 separate cubic feet, then we can speak of an event E_i ($i = 1, 2, \ldots, 1000$) as the event of finding our identified oxygen molecule in the ith cubic foot of volume. The most natural state appears intuitively to be the one of highest entropy, namely, $p(E_i) = 1/1000$ for all i. However, we can decrease the entropy locally by pumping all the air in the room into a container of one cubic foot in volume and placing it in a corner of the room where $E_i = E_1$. Now $p(E_1) = 1$, and all $p(E_i) = 0$ for $i \neq 1$. The entropy is zero for the new state in terms of the one thousand events E_i of interest to us. This state is a much less probable state than the original state in which $p(E_i) = 1/1000$ for all i. Indeed, if we open the container, the molecules of air will tend to return to the original state which is the most probable state in the absence of an external influence. Thus, the most probable states in nature are the states of highest entropy or highest disorder. When fast-moving particles come in contact with slow-moving particles, a new average velocity is reached, and this new state is more probable and thus of higher entropy.

But note that energy can be extracted in a transition from a lower to a higher entropy state. Therefore, once energy is extracted in the transition process, or the transformation to a higher entropy state, fewer transformations of such type will be available in the future. This is so because the high entropy state has a very, very small probability (if it is at all different from zero) of going in the reverse direction on its own, i.e., going to a lower entropy state. Can you ever imagine all the molecules in the 10 ft × 10 ft × 10 ft room "getting together" and "agreeing" to occupy 1 cubic foot at E_1?

This is, in rather simple terms, the nature of the irreversibility of thermodynamic processes.*

---

*To point out the connection between the concept of information and the second law of thermodynamics, physicist Leo Szilard used the "Maxwell Demon" thought experiment. In this experiment a chamber containing gas in equilibrium (a state of highest entropy) is divided into two compartments, E_1 and E_2, with a swinging door between them. The door is held by a demon. When a molecule from E_1 approaches the door, he lets it through to E_2, but no molecules are permitted to go from E_2 to E_1. Eventually, all the molecules in the chamber are in E_2. The entropy of the new state is lower than that of the original state. The new state is, of course, less probable and, indeed, if the demon disappears from the chamber, the system will return on its own to the original more probable state, beginning with a "rush" of molecules through the free swinging door.

The transition to the new state required a demon who "consumed" information concerning the motion of the molecules. This information is equivalent to the difference in entropy between the original and new state of lower entropy. Since such a demon cannot be contrived, the process of transition from high to low entropy states without external interference is considered very unlikely, and thus follows the conclusion regarding the irreversibility of thermodynamic processes in the universe, and the trend toward increase in entropy.

Entropy and Life

A local decrease in entropy takes place when atoms of oxygen, hydrogen, nitrogen, carbon, phosphate, and other elements become arranged in a highly complex but orderly fashion to produce a living organism. Life represents a transition from disorder to order, or entropy consumption. While the creation of life may be considered as an unconscious consumption of entropy, the development of civilization and evolution of culture accompanied by a man-made world represent a conscious consumption of entropy. All man-made creations from the building of a primitive hut to the construction of a sophisticated space vehicle consume entropy; they represent the creation of more orderly states from less orderly states.

There are of course deviations from a total commitment to entropy consumption by living things. Both man and animal kill, and in so doing may reduce order to disorder. We can point to many examples of increase in local entropy in society as they are manifested in wars, crime, and destruction of elements of the natural and man-made world.

Back to Values: A Model of Ethical Behavior

Lindsay [103] suggests that, although eventually every living organism dies, and thus a highly developed order of organism is reduced to a random disorderly collection of atoms, during our lifetime *we ought to behave in a way which will produce as much order in our environment as possible*, or *maximize the consumption of entropy*.

This he suggests as a normative principle, a basis for ethical behavior in society. Adherence to this principle as a value presupposes a supreme value in "reverence for life" inasmuch as all life is entropy consuming.

The suggestion has some interesting features, but (as Lindsay himself recognizes) it is also possible to derive a diametrically opposed value from the second law of thermodynamics. Namely, if nature tends to go from order to disorder, so should man, with no effort to fight the second law. Or the lesson of the second law of thermodynamics is that anarchy is a principal value and a basis for ethical human behavior. However, it may be more appealing, on the basis of our observations and the study of man, to consider life and man's behavior as a continual challenge to the second law and make the striving to consume entropy a principal value.

I do not propose that the above rationalization to regard entropy consumption as a principal value served as the basis for a strong commitment to order in industry, in government, in education, in business, and in other institutions of our society. But commitment to order did indeed become most pronounced with the advent of the industrial revolution which also carried in its wake great emphasis on the values of efficiency, discipline, progress, and growth. Huxley [104] uses the phrase *Will to Order* to describe over-organization and points out its dangers when applied to society in the realms of politics and economics.

"A Little Entropy is Good for You"

I believe there is danger in over-organization and order. Over-planned order tends to transform man from the "World of Thou" to the "World of It," in the language of Buber [105].

According to Buber's model, in the World of Thou, there are two centers of consciousness that meet. Neither one can absorb or appropriate the other. Neither can become an object to the other. The meeting of the two is the core of the relation between them.

In the World of It there is only one center of consciousness, the one in which a person experiences an object and appropriates the object to his own use. Objects are tools that can be used to further our interests, or they may be obstacles to be removed from the path. When we know an object, it means that we know how to use it for our own benefit. Man loses his self in the process of over-order and becomes the object in the World of It.

Buber makes a distinction between *Collectivity* and *Community*, which is enlightening and relevant to our discussion of a society totally committed to planned order as contrasted with a society which is flexible, and permits meaningful relations between human beings to develop on their own by merely creating the environment and climate conducive to the development of such relations. In Buber's language, collectivity is grouping in the World of It, while community is grouping in the World of Thou. In his words:

"Collectivity is not a binding but a bundling together: Individuals packed together, armed and equipped in common, with only as much life from man to man as will inflame the marching step. But community, growing community, is the being no longer side by side but with one another of a multitude of persons. Community requires communication, dialogue, a flowing from I to Thou. *Community cannot occur by deliberate planning. Community is where community happens.*"

Perhaps the last sentence is most descriptive of the danger in over-organization and a complete commitment to a *Will to Order*. Such a commitment smacks of the ultimate transformation of man to an object, a tool to be used, or an obstacle to be removed from the path by those who are the masters of order. Who are to be the masters of order or the benevolent dictators in a society committed to the Will to Order?

Plato had his model of social order in *The Republic*. Skinner [106] argues in *Beyond Freedom and Dignity* that human beings are hardly free in the sense of the songs of poets. They are manipulated by the inputs from the environment over which they have little if any control. With this hypothesis that we are the objects of external manipulation as a basis, he proposes that we should delegate the manipulation to the behavioral scientists who are best equipped to manipulate human behavior so as to produce an orderly society which is best or at least good for all. But who is to decide what is good or best for all? Is a cult of behaviorist "priests" to decide? Can we ever regain control of our destiny, once we give it up? Who will decide who is "priest" and who is subject?

Perhaps a more central question than the one focusing on "What kind of environment do we want?" is: "What kind of man do we want?" Are freedom and dignity, indeed, no longer relevant values in our society?

Figure 10-1 shows schematically a dynamic system with feedback in the spirit of Chapter 9. The central question before us is: "Where are we in the loop?" Are we the super-controllers who establish the Reference of what ought to be? Are we the policing authority of the State which acts as the obedient Controller? Are we the subjects of control, i.e., the Controlled System? Are we the obedient Feedback agents of the State measuring behavior, or response of subjects? Where are we?

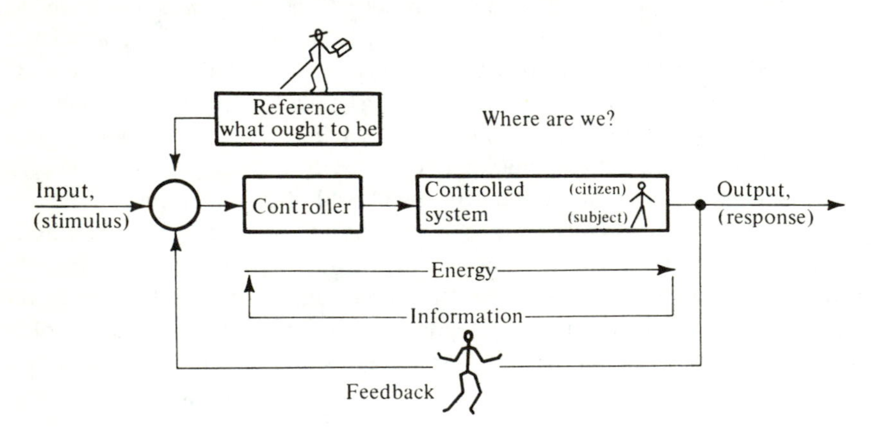

Figure 10-1 Schematic Diagram of a Dynamic Control System of Human Behavior in Society

I wish to believe that it is possible for us to be anywhere in the loop, and that no elite group can effectively get hold of the super-reference control for very long. We may not desire to assume each position in the loop, but it should be possible. I firmly believe it is worth the price of a less orderly society, because it marks the difference between being subjects of a state, or citizens with some measure of freedom and dignity.

10-6 A MODEL OF ETHICAL BEHAVIOR

In Chapter 4 we discussed rational behavior as the habit of considering all relevant desires, and not only the one which is strongest at a given moment. The probabilities of satisfying the various desires must be weighed. To achieve such rational behavior, it is necessary to have some ability to predict consequences of behavior. Prediction is facilitated by the construction of a model. The model may be deterministic or probabilistic in character.

In the spirit of problem solving discussed to this point, we can consider a model in which a desired goal is prescribed and we wish to predict the best path of a system to achieve the goal. Or we may wish to predict the outcome of a system on the basis of an input and an initial state. For example, in physical science we can construct two such models for the motions of rigid bodies in a way which will reflect prediction for a future state from an initial state, or the prediction of the path from an initial state to a final state. Newton's laws of motion, for example, lead to a model in the form of differential equations in which forces acting on a rigid particle are expressed in terms of position, velocity, and time. Given the state of the rigid particle at any time, the solution of the equations will predict the state of the particle at any other time. On the other hand, a formulation of the same problem in terms of a model of an integral equation, representing energy, stipulates that, of all possible paths between an initial and final state, the predicted path is that which renders the energy stationary (minimum for stable systems). This is a teleological or purposive model; it predicts the behavior of the system on the basis of a goal *that must be satisfied.*

In physical science our models are based on matters of fact and observations; *we do not select, but rather discover goals* or objectives such as minimum energy, increase in entropy, and invariants or constraints such as conservation of energy and conservation of momentum. But what about models of human behavior? Can we identify a value, which is not a matter of human choice, to serve as the prime indicator or index of performance to predict human behavior? That is, can we construct a model which will predict the pattern of behavior so as to maximize the probability of realizing a supreme human value which is not, most generally, a matter of human choice? Can such a model be considered a *model of ethical behavior* or *a model of moral law?*

It has been suggested that survival constitutes such a supreme end value [107, 108, 109]. It is indeed a matter of observation that, in general, human beings behave in a way which tends to maximize the probability of survival. It is a good working hypothesis to treat survival as a supreme value of human beings in the sense that it serves as the best criterion to predict behavior in a gross general way. For example, it is quite reasonable to state that as an audience leaves a room after a lecture, their behavior will, in general, be characterized by an index of performance which is the maximization of probability of survival. Thus, they will cross the streets, drive their cars, in general, in a way which renders validity to this index of performance. Many religious doctrines have subscribed to survival and the sanctity of life as a supreme end value, and have defined ethical and moral behavior as that behavior which furthers the cause of survival. This is also reflected in the Ten Commandments. It is possible to interpret the values of justice, equality, freedom, and others as means values which further the cause of survival. In many respects, in the history of civilization man's values evolved on the basis of their contribution to survival. Whatever served this supreme goal became a value (a means value to this end). Values became integrated into man's tools for adaptation, but developed a momentum of their own which made their prime purpose obscure and, therefore, became at times a detriment that impaired the ability to adapt to new conditions which man has created in his man-made world.

A model of ethical behavior can be constructed according to the hierarchy of Fig. 10-2. The quality-of-life values would be all those values which make life worthwhile, so as to enhance the value of survival. These values vary from culture to culture and change in time within a given culture. In a society dominated by scarcity, the basic needs* of food, shelter, and sex overshadow other values which an affluent society considers of great importance. In an affluent society, the search for relevant values to make life worth sampling becomes more difficult, and consensus on values is more difficult to establish. *Survival appears as one possible end value that is innate, one which is not entirely subject to man's choice.*

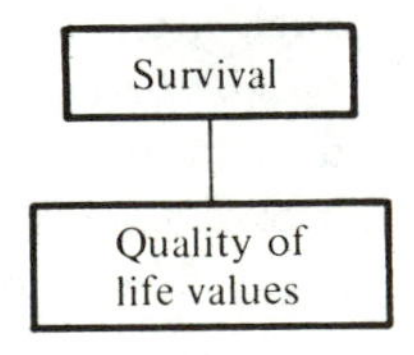

Figure 10-2

Survival has two aspects: survival of the individual and survival of the group. The group may be the family, the community, the country, or the species on a grand scale. Man is a gregarious but not completely social creature, more like a wolf than a bee or an ant. Man creates organization, but not a social organism in the sense of the bee hive or the ant heap. A community, in the sense that Buber describes it, is the closest man can come to a social organism. It is, therefore, again a matter of choice or value judgment when man considers the consequences of his behavior in terms of its contribution

*Note that these needs are associated with pleasure as an intermediate means value to further the cause of survival.

to his survival and the survival of the group. We can describe this in terms of a linear function for an index of survival, S, in which a factor α, $0 \leq \alpha \leq 1$, must be selected

$$S = \alpha \text{ Prob(individual survival)} + (1 - \alpha) \text{ Prob(group survival)} \quad (10\text{-}3)$$

α and $(1 - \alpha)$ are weighting factors of the relative importance attached to individual and group survival, respectively. α may vary considerably, depending on circumstances, culture, total experience, world views, and value commitment. For bees and ants, α is small. Man is burdened with the permanent task of making a choice for the value of α.

The commitment of people to the value of survival and the sanctity of life is reflected in the political arena. The United States has rejected repeatedly the policy of a first strike or preventive war since World War II. Had this not been the case, then delay could be interpreted only as an unwise strategy because the military superiority of the U.S. could only decline with the prospects that other nations would discover first the A-bomb, and next the H-bomb.

Considering survival as the supreme end value, it can be rationalized that societal values, such as love, human relatedness, freedom, social justice, economic justice, and mutual respect, will best serve the cause of survival for all. Education could then concentrate on such essential kernels of societal values by recognizing that reverence for one's own life will lead to reverence for the life of others, just as respect for oneself will lead to respect for others.* It is, therefore, important that we create in our educational systems an environment which is conducive to the development of self-respect by striving to let man develop to the fullest of his abilities.

Other Supreme End Values as Indicators of Human Behavior

Albert Schweitzer believed that the supreme human value was "Reverence for Life" or the "Will to Live" similar to the value of survival discussed above. Freud introduced into psychology the pleasure principle or the "Will to Pleasure." Adler considered the "Will to Power" as central in human behavior. Viktor Frankl [110] a one-time Freudian psychiatrist, survived the Nazi death camps of World War II to claim, on the basis of his experiences, that man is dominated neither by the will to pleasure nor by the will to power, but rather by a deeply rooted striving for meaning, namely, a "Will to Meaning." Erich Fromm and others would feel equally strong about the

*The time-honored saying, "Love Thy Neighbor as Thy Self," is written in Hebrew in a form which permits the translation "Love Thy Neighbor, He is Like Yourself." Having respect for yourself, having reverence and love for life as it is manifested in yourself, you will do the same to your fellow man because he is like you.

"Will to Love" and the "Will to Relatedness." Thus the search for a descriptive value or values which are most compatible with human nature goes on. Descriptive models of human behavior can easily become normative models which attempt to prescribe what man's values ought to be. We must exercise caution not to be caught in a blind subscription to a supreme value. While we can have faith before all doubt has been dispelled, we should retain a will to doubt so we can be open to new experiences and continue to sample life in a search for meaning and values which will lead to fulfillment.

10-7 VALUES OF THE PRESENT

In a study with a group of students at UCLA, Dalkey [111] has considered the means values which contribute to the quality of life. Dalkey used the Delphi method* for forming consensus to generate a list of values and assess their relative importance in terms of their contribution to the quality of life. The values and their relative ratings in order of importance are shown in Table 10-2. Conspicuous by their absence are the "old fashioned" values of honesty, thrift, respect for others, dignity, friendship, charity, patriotism, truthfulness, loyalty, social order, discipline, responsibility, and accountability.

Table 10-2 Quality of Life*

	Relative Importance
1. Love	15.0
2. Self-Respect	11.5
3. Peace of Mind	10.0
4. Sex	9.5
5. Challenge	8.0
6. Social Acceptance	8.0
7. Accomplishment	7.0
8. Individuality	6.0
9. Involvement	6.0
10. Well Being (economic, health)	6.0
11. Change	5.0
12. Power (control)	3.5
13. Privacy	2.0

*Reproduced by permission of the RAND Corporation and the publisher, from *Studies in the Quality of Life: Delphi and Decision-Making* by Norman C. Dalkey, *et al.* (Lexington, Mass.: Lexington Books, D. C. Heath and Co., 1972) Copyright 1972 by the RAND Corporation.

Table 10-3 shows a second study by Dalkey with the same group on the relative importance of education goals. Note the low rating of the "old fashioned" value of career skill.

---

*The Delphi method is discussed in Sec. 10-14.

Table 10-3 Education Goals*

	Relative Importance
1. Learning to Learn	12.0
2. Broad Outlook	10.0
3. Greater Creativity	8.0
4. Social Awareness	8.0
5. Communication Skills	7.0
6. Understanding of Others	6.0
7. Self-Awareness	6.0
8. Self-Confidence	6.0
9. Concern for Society	5.0
10. Impractical Education	5.0
11. Career Skills	5.0
12. Motivation (develop goals)	5.0
13. Involvement	1.0
14. Dependency (prolonged youth)	0.0

*Reproduced by permission of the RAND Corporation and the publisher, from *Studies in the Quality of Life: Delphi and Decision-Making* by Norman C. Dalkey, *et al.* (Lexington, Mass.: Lexington Books, D. C. Heath and Co., 1972) Copyright 1972 by the RAND Corporation.

10-8 VALUES OF THE FUTURE

The following values hold a prominent place in the first set of working papers of a committee on "Values and Rights" of the Commission on the Year 2000, American Academy of Arts and Sciences [112]:

Privacy
Equality
Personal integrity
Freedom
Law and order
Pleasantness of environment
Social adjustment
Efficiency and effectiveness of organizations
Rationality
Education
Ability and talent

These values can be classified as values of individual rights, values of personal character, and values pertaining to outer environment (law and order, environment pleasantness). Social values such as equality, privacy, freedom, and order represent the social setting which contributes to a greater utility for personal values of ability, talent, and adjustment. Again, conspicuious by their absence are the "old fashioned" values. What happened to personal responsibility and accountability? It appears that personal responsibility and accountability have been eroded progressively in the past decades with the advent of, and ever-increasing, social accountability. Accountability of the individual is not enhanced when such concepts as no-fault insurance are promoted. The new procedures may constitute an improvement of current practices of insurance settlements. However, the mere choice of the words, "no-fault insurance," carries a hidden message leading to reduced personal accountability. This reminds me of the contractor who sought my advice when he was about to construct a tall building on a slope adjacent to an existing structure. When I pointed out the danger to the workers in the course of the construction as planned, and suggested a more costly but safer procedure, he refused to revise the plans on the grounds that he was insured for such accidents. When I also pointed out that the sidewalk adjoining his property might collapse during the excavations with possible loss of life to innocent people walking the sidewalk, he assured me that he was insured for that too. . . .

Yet, perhaps as Reich [113] suggests in *The Greening of America*, the new consciousness is searching for ways to find the self. In the process of self-discovery, self-respect will promote respect for others, and an appreciation and reverence for one's own life will promote a reverence for the life of others.

Some values of the past appear definitely out. For example, it is difficult to rationalize great importance for charity in a society which is on the brink of a guaranteed annual income or a welfare state.

10-9 DYNAMIC CHANGE IN VALUES AND VALUE SUBSCRIPTION

Values change due to various causes. A society dominated by scarcity has a different scale for the value of economic well being than an affluent society. At a certain stage in life a person may begin to subscribe to a new value or he may abandon a value to which he had subscribed. Values change with age. As a person matures psychologically and becomes more conscious of himself and his behavior, he may begin to recognize the distinction between values which originate within himself in his inner environment and those which are promoted outside him by the outer environment. Such maturity leads to change in values. Hence, the list of relevant values changes. Values may become more important or less important in terms of the degree to which they are subscribed by an individual or society.

Here is a list of some other ways in which change in values can occur [112]:

New knowledge. When we learn that a means value will not lead to the realization of a desired end value we may abandon it. For example, if we were to learn that frugality will not enhance the prospects of economic prosperity, then we would abandon it as a means value to prosperity.

Erosion of relative importance. The intensity of subscription to a value may change because of a shift in its relative importance. For example, intensity of subscription to the value of economic prosperity will be eroded during times of great anxiety when the survival of man is threatened (say, the threat of thermonuclear war). The relative importance of economic prosperity is greatly diminished under such circumstances.

A shift in weighting factor between individual and group values. If we consider, for example, economic prosperity, we may use an equation similar to Eq. 10-3 as a simple model for a person's behavior when he recognizes a relationship between his personal prosperity and the national or group prosperity. He could then behave so as to maximize the following objective function in which E is an index of economic prosperity and $0 \leq \alpha \leq 1$:

$$E = \alpha \text{ Prob(personal prosperity)} + (1 - \alpha) \text{ Prob(group prosperity)} \qquad (10\text{-}4)$$

As in the case of survival, here, too, the choice of α will depend on the circumstances and the relative prospects for the realization of prosperity by the individual and the group. Consider a worker with a strong subscription to the values of productivity and efficiency because he is convinced that they are instrumental as means values to the realization of prosperity for him and society. However, should his job become threatened by automation, which will enhance productivity and efficiency, he is most likely to become concerned with his own fate and make α close to 1 because the probability of his personal prosperity has been reduced greatly. His subscription to efficiency and productivity as values is likely to be eroded greatly in the light of such a development of new circumstances.

Shift in responsibility. For a society in which responsibility shifts from the individual to the group, old virtues and values change. Charity has been eroded in importance as we progress toward a welfare state.

New politics. A strong emergence of new political movements with the attendant indoctrination through demagogy and propaganda may cause a change in values. Effectiveness of such a political movement may, for example, elevate patriotism to new heights as a value, or may do the reverse. It may promote equality or the reverse. It may promote law and order or increase subscription to anarchy.

Total influence of environment. Technological changes, cultural and social trends, economic cycles, population mix (demography)—all contribute to our view of the world and to our values. Some features of the contribution of these factors to value change are discussed in the next section.

10-10 COST-BENEFIT ASSESSMENT OF VALUES

Values are benefit oriented in the sense that we believe that their realization will be good for a purpose. The purpose or end value (or intermediate end value) may be pleasure, meaning, power, "quality of life," "the good life," or ultimate survival. Values may be benefit oriented in the sense that there is benefit in the act of doing, namely, in the pattern of behavior which leads to the value, rather then the benefit associated with the realization of the value. Thus, pleasure may perhaps be induced one day by stimulating the appropriate pleasure nerves, but the benefit may be viewed as much smaller than that derived from striving and achieving the same pleasure as that induced instantaneously by artificial means. Benefit may be difficult to assess, in particular when we consider intrinsic end values, such as learning for its own sake.*

The cost associated with striving for the realization of a value, or with maintaining a value once realized, can take a number of forms. Cost in terms of money is only one direct form of such cost. In our discussion here, we consider as costs the expenditure of energy, the devotion of time, as well as the taking of risks which may attend various efforts. The risks may contribute to a lower probability of realizing other values of interest to the subscriber, and may even contribute to a reduction in probability of survival in some cases [42].

Both costs and benefits associated with values may change. Thus, the cost of cleanliness in modern cities is lower now than in medieval times, while the cost of privacy and serenity in an uncongested environment is higher now than in the past. There is less benefit associated with the value of wealth in a welfare society in which a guaranteed annual income is in effect. There is more benefit associated with knowledge and learning in a just and free society which rewards a man for competence and abilities, the truly relevant factors, rather than on social status, political affiliation, and other extraneous factors.

The concept of benefit may be viewed in the context of the concept of utility discussed in Chapter 7. Utility is a subjective quantity and we may change our minds on the measure of utility or benefit as our values change. We may oversubscribe to a value, or undersubscribe to it, in the sense that our judgment of benefit and cost may be wrong, or the judgment required to match a cost to a benefit may be wrong.

Values, like friends, can be perceived as true or false. Values which deceive us are false values. Means values, which we hold in esteem because we have faith in their contributions to the "good life," may turn out to be deceptive. Such values are false.

---

*But even for such seemingly benefit-free values, we may consider potential benefits although they may not be realized.

*10-11 SOCIAL PREFERENCES—AN AXIOMATIC APPROACH

One of the main tasks of social order is to establish a criterion for fusing individual preferences for values into consensus for social or group values. A number of models have been considered for this purpose. In this section we discuss an axiomatic approach to preference ranking of values by introducing a ranking matrix and the concept of distance between rankings. The axiomatic approach is discussed by Kemeny and Snell [114] and only a brief summary is included here. An alternate similar approach, which employs a ranking vector and a measure of distance between rankings, is introduced in Sec. 10-12. In Sec. 10-13 we discuss a metric approach for social fusion of individual preferences, and Sec. 10-14 describes the Delphi method for forming consensus.

Question. Suppose that a group is asked to form consensus on ranking n values. How do we use the individual rankings to arrive at a consensus ranking? The following discussion attempts to answer this question.

Digression

In statistics, when we make a claim that the mean height $\bar{h}$ of m people in a room is 5′6″, it is quite possible that not a single person in the room is exactly this tall. However, we generally accept the idea that the average is most representative of group heights in the following sense. Of all possible heights which we could have selected, the mean height is the least distant from the individual heights. As a criterion for distance D of the group mean height $\bar{h}$ from the individual heights h_i, we use the sum of the squares of the deviations:

$$D = \textstyle\sum_{i=1}^{m} (h_i - \bar{h})^2 \tag{10-5}$$

The value of $\bar{h}$ which minimizes D can be computed by setting to zero the derivative of D with respect to $\bar{h}$:

$$\frac{dD}{d\bar{h}} = \textstyle\sum_{i=1}^{m} 2(-1)(h_i - \bar{h}) = 0$$

$$\textstyle\sum_{i=1}^{m} (h_i - \bar{h}) = 0$$

$$\textstyle\sum_{i=1}^{m} h_i = m\bar{h}$$

$$\bar{h} = \frac{1}{m} \textstyle\sum_{i=1}^{m} h_i \tag{10-6}$$

Back to Ranking

The concept of distance in the above example suggests a similar concept as a criterion for determination of consensus in ranking. A distance between rankings will be defined, and the consensus ranking will be least distant from the individual rankings in a sense analogous to that of the distance of the mean height $\bar{h}$ from individual heights h_i.

Kemeny and Snell [114] suggest the following axiomatic approach for the determination of such consensus of individual preferences. A number of conditions are postulated in the form of four axioms which a distance function $D(A, B)$ between rankings A and B must satisfy.

Consider rankings A and B of three objects, a_1, a_2, a_3, for example. Then, if A is $a_1 > a_2 > a_3$ and B is $a_1 \sim a_2, a_3 > a_1$, we can write these rankings in ranking columns or ordering matrices as follows:

Ranking columns.

$$A = \begin{Bmatrix} a_1 \\ a_2 \\ a_3 \end{Bmatrix} \qquad B = \begin{Bmatrix} a_3 \\ a_1, a_2 \end{Bmatrix}$$

Ordering matrices. The above column rankings can also be represented by *ordering matrices.* Consider the ranking A of n objects $a_1, a_2, \ldots, a_n$. We construct a square $n \times n$ matrix as follows:

$$\begin{array}{c} \\ a_1 \\ a_2 \\ \cdot \\ \cdot \\ \cdot \\ a_k \\ \cdot \\ \cdot \\ \cdot \\ a_n \end{array} \begin{array}{c} \begin{array}{ccccccc} a_1 & a_2 & \cdots & a_l & \cdots & a_n \end{array} \\ \begin{bmatrix} a_{11} & a_{12} & \cdots & a_{1l} & \cdots & a_{1n} \\ a_{21} & a_{22} & \cdots & a_{2l} & \cdots & a_{2n} \\ \cdot & \cdot & & \cdot & & \cdot \\ \cdot & \cdot & & \cdot & & \cdot \\ \cdot & \cdot & & \cdot & & \cdot \\ a_{k1} & a_{k2} & \cdots & a_{kl} & \cdots & a_{kn} \\ \cdot & \cdot & & \cdot & & \cdot \\ \cdot & \cdot & & \cdot & & \cdot \\ \cdot & \cdot & & \cdot & & \cdot \\ a_{n1} & a_{n2} & \cdots & a_{nl} & \cdots & a_{nn} \end{bmatrix} \end{array}$$

Each element a_{kl} in row k and column l of the matrix can have a value of 0, 1 or -1 in accordance with these rules:

$$\begin{aligned} a_{kl} &= 1 && \text{when} \quad a_k > a_l \\ a_{kl} &= -1 && \text{when} \quad a_k < a_l \\ a_{kl} &= 0 && \text{when} \quad a_k \sim a_l \end{aligned}$$

The ordering matrix then has the following properties:

1. $a_{kl} = 1, 0, -1$; $a_{kk} = 0$.
2. $a_{kl} = -a_{lk}$.
3. If $a_{kl} > 0$ and $a_{lm} > 0$, then $a_{km} > 0$.
4. If $a_{kl} = 0$ and $a_{lm} = 0$, then $a_{km} = 0$.

For example, the rankings $A = \begin{Bmatrix} a_1 \\ a_2 \\ a_3 \end{Bmatrix}$ and $B = \begin{Bmatrix} a_3 \\ a_1, a_2 \end{Bmatrix}$ take the form of the following ordering matrices:

$$[a_{kl}] = \begin{bmatrix} 0 & 1 & 1 \\ -1 & 0 & 1 \\ -1 & -1 & 0 \end{bmatrix}; \quad [b_{kl}] = \begin{bmatrix} 0 & 0 & -1 \\ 0 & 0 & -1 \\ 1 & 1 & 0 \end{bmatrix}$$

Distance Function $D(A, B)$

A distance function $D(A, B)$, which provides a measure of distance between rankings A and B, is generated in Kemeny and Snell [114] in the form of

$$D(A, B) = \frac{1}{2} \sum_{k,l} |a_{kl} - b_{kl}| \tag{10-7}$$

The vertical lines indicate absolute value. For example, the distance $D(A, B)$ between the rankings

$$[a_{kl}] = \begin{bmatrix} 0 & 1 & 1 \\ -1 & 0 & 1 \\ -1 & -1 & 0 \end{bmatrix} \quad \text{and} \quad [b_{kl}] = \begin{bmatrix} 0 & 0 & -1 \\ 0 & 0 & -1 \\ 1 & 1 & 0 \end{bmatrix}$$

is computed as

$$\begin{aligned} D(A, B) &= \frac{1}{2}(\overset{\text{row 1}}{0 + 1 + 2}, \quad \overset{\text{row 2}}{1 + 0 + 2}, \quad \overset{\text{row 3}}{2 + 2 + 0}) \\ &= 5 \end{aligned}$$

The distance function is unique and satisfies the following four postulated axioms.

AXIOM 1.

$$\begin{aligned} &D(A, B) \geq 0 \\ &D(A, B) = D(B, A) \\ &D(A, B) + D(B, C) > D(A, C) \end{aligned}$$

except

$$D(A, B) + D(B, C) = D(A, C)$$

when ranking B is between A and C.

Here *between* means that for each pair of objects, a_k and a_l, the judgment of B either agrees with A or agrees with C; or A prefers a_k, C prefers a_l, and B claims a tie. Namely,

$$a_{ij} \leq b_{ij} \leq c_{ij} \qquad \text{or} \qquad a_{ij} \geq b_{ij} \geq c_{ij} \quad \text{for all } i \text{ and } j$$

AXIOM 2. If the same permutation to preference ordering is applied to rankings A and B, yielding new rankings A' and B', then

$$D(A, B) = D(A', B')$$

For example, the above equality of distances holds for

$$A = \begin{Bmatrix} a_1 \\ a_2 \\ a_3 \end{Bmatrix} \qquad B = \begin{Bmatrix} a_3 \\ a_2 \\ a_1 \end{Bmatrix}$$

and

$$A' = \begin{Bmatrix} a_2 \\ a_3 \\ a_1 \end{Bmatrix} \qquad B' = \begin{Bmatrix} a_1 \\ a_3 \\ a_2 \end{Bmatrix}$$

in which the transformation from A and B to A' and B', respectively, is achieved according to the scheme of Fig. 10-3. An object a_k at the head of an arrow replaces the object at the origin of the arrow, namely a_1 is changed to a_2, a_2 to a_3, and a_3 to a_1. This operation can be viewed as merely an assignment of new names to the objects being ranked; hence distances should be preserved.

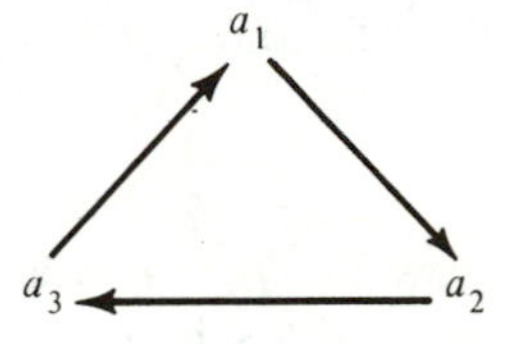

Figure 10-3

AXIOM 3. If rankings A, B agree completely at the beginning or/and at the end of a list of ranked objects, but differ only in the ranking of k objects in the middle, then $D(A, B)$ may be computed as if these were the only objects being ranked. For example, for the rankings

$$A = \begin{Bmatrix} a_1 \\ a_2 \\ a_3 \\ \hdashline a_4 \\ a_5 \\ a_6 \\ \hdashline a_7 \\ a_8 \end{Bmatrix} \qquad B = \begin{Bmatrix} a_1 \\ a_2 \\ a_3 \\ \hdashline a_6 \\ a_4 \\ a_5 \\ \hdashline a_7 \\ a_8 \end{Bmatrix}$$

A and B differ in

$$A^* = \begin{Bmatrix} a_4 \\ a_5 \\ a_6 \end{Bmatrix} \qquad B^* = \begin{Bmatrix} a_6 \\ a_4 \\ a_5 \end{Bmatrix}$$

Hence, $D(A, B) = D(A^*, B^*)$.

AXIOM 4. The minimum positive distance is 1.

Kemeny and Snell [114] show that:

(a) These four axioms are consistent (no contradictions).
(b) Equation 10-7 is the unique distance function satisfying the four axioms.
(c) No additional assumptions are required.

Consensus

We now need a criterion for generating a consensus ranking $\bar{R}$ for the individual rankings R_i of m people. The criterion of the mean provides a consensus ranking $\bar{R}$ for which $\sum_{i=1}^{m} D(R_i, \bar{R})^2$ is minimum. For example, for

$$R_1 = R_2 = \begin{Bmatrix} a_1 \\ a_2 \\ a_3 \end{Bmatrix}, \text{ and } R_3 = \begin{Bmatrix} a_2 \\ a_1 \\ a_3 \end{Bmatrix}; \bar{R} = \begin{Bmatrix} a_1, a_2 \\ a_3 \end{Bmatrix},$$

and $\sum_{i=1}^{3} D(R_1, \bar{R})^2 = 3$.

Verify.

For complete disagreement in ranking by three people:

$$R_1 = \begin{Bmatrix} a_1 \\ a_2 \\ a_3 \end{Bmatrix}, \quad R_2 = \begin{Bmatrix} a_2 \\ a_3 \\ a_1 \end{Bmatrix}, \quad R_3 = \begin{Bmatrix} a_3 \\ a_1 \\ a_2 \end{Bmatrix}$$

$$\bar{R} = \{a_1, a_2, a_3\}, \text{ i.e., } a_1 \sim a_2 \sim a_3$$

The justification for the mean (the median may also be used) as a criterion for determining consensus is derived from the methods of statistical analysis. If we wish to establish the consensus ranking of a large population, we may use a random sample and generate from it a ranking with a guarantee that for large samples the probability is nearly one that the sample consensus will approach the population consensus.

10-12 ALTERNATE PROCEDURE FOR CONSENSUS DETERMINATION

In Sec. 7-9 we discussed the Arrow paradox. Arrow [67] has shown that if a group consensus ranking is to satisfy certain plausible requirements, then we cannot count on the firmly rooted majority rule of our culture to produce a social preference function. Arrow demonstrated that his six plausible and reasonable requirements are mutually incompatible, and no group consensus can be generated to satisfy them all. This paradox or blocking for a method to fuse individual preferences into a social consensus is known as the *Arrow barrier*.

Rescher [112] suggests two possible ways to generate consensus which circumvent the Arrow barrier. One is a procedure, similar to that discussed in Sec. 10-11, which is discussed in this section, and the second (treated in Sec. 10-13) is an averaging-out process of individual preferences which have been quantized by assigning a figure of merit or a utility to each object.

Consensus of Ranking

Consider, for example, these rankings:

$$R_1 = R_2 = \begin{Bmatrix} 1 \\ 2 \\ 3 \end{Bmatrix}, \quad R_3 = \begin{Bmatrix} 2 \\ 1 \\ 3 \end{Bmatrix}$$

Here each position in the column refers to a different object. The number indicates the rank of the object identified by its position (row) in the column. Three objects, x_1, x_2, and x_3, are ranked above as

$$x_1 > x_2 > x_3 \quad \text{for} \quad R_1 \text{ and } R_2$$

and

$$x_2 > x_1 > x_3 \quad \text{for} \quad R_3$$

For a ranking $\begin{Bmatrix} 1 \\ 1 \\ 2 \end{Bmatrix}$, $x_1 \sim x_2$, $x_1 > x_3$.

Rescher [112] suggests a consensus ranking $\bar{R}$ such that for m individual rankings R_i,

$$\textstyle\sum_i (R_i - \bar{R})^2$$

is a minimum, namely, a criterion similar to the mean criterion of Sec. 10-11. In the above expression, each term $(R_i - \bar{R})^2$ stands for the sum of the squares of the differences between corresponding numbers in the columns for R_i and $\bar{R}$. For example, for

$$R_i = \begin{Bmatrix} 2 \\ 1 \\ 3 \end{Bmatrix} \quad \text{and} \quad \bar{R} = \begin{Bmatrix} 1 \\ 2 \\ 3 \end{Bmatrix}$$

$$(R_i - \bar{R})^2 = (2-1)^2 + (1-2)^2 + (3-3)^2$$
$$= 2$$

A difficulty with the above procedure is that it does not always lead to a Pareto optimal consensus. A consensus is *Pareto optimal* if it is not possible to change it and make one individual better off without, at the same time, making another individual worse off. For example, consider the rankings

$$R_1 = R_2 = R_3 = \begin{Bmatrix} 1 \\ 1 \\ 1 \end{Bmatrix}, \quad \text{i.e., } x_1 \sim x_2 \sim x_3$$

$$R_4 = \begin{Bmatrix} 1 \\ 2 \\ 3 \end{Bmatrix}$$

The consensus ranking $\bar{R}$ which minimizes $\sum_{i=1}^{4} (R_i - \bar{R})^2$ is

$$\bar{R} = \begin{Bmatrix} 1 \\ 1 \\ 1 \end{Bmatrix}, \quad \min \textstyle\sum_i (R_i - \bar{R})^2 = 5$$

However, since R_1, R_2, R_3 are indifferent to the ranking of x_1, x_2, and x_3, they might as well accommodate R_4 and accept R_4 as the consensus. $\bar{R}$ above is, therefore, not Pareto optimal because it can be changed with the result of making one individual better off without making anyone else worse off. In a similar manner, the rankings

$$R_1 = \begin{Bmatrix} 1 \\ 2 \\ 3 \end{Bmatrix}, \quad R_2 = \begin{Bmatrix} 1 \\ 2 \\ 2 \end{Bmatrix}, \quad R_3 = \begin{Bmatrix} 1 \\ 1 \\ 1 \end{Bmatrix}$$

yield

$$\bar{R} = \begin{Bmatrix} 1 \\ 2 \\ 2 \end{Bmatrix}$$

which makes $\sum_i (R_i - \bar{R})^2$ a minimum. However, replacing $\bar{R}$ by R_1, as the consensus, will make R_1 better off without making R_2 and R_3 worse off.

*10-13 METRIZATION OF PREFERENCES

Consider m people $P_1, P_2, \ldots, P_m$ who assign an importance rating or a utility to each of n objects $x_1, x_2, \ldots, x_n$. Let the utility assignment of person P_i to object x_k be designated by M_{ik}. How are we to combine the individual assignments and form a group assignment of importance ratings to the n objects? In mathematical notation, we wish to find a function $F(M_{ik})$ which will transform the importance ratings M_{ik} of any object x_k into a group importance rating M_k for the object:

$$M_k = F(M_{1k}, M_{2k}, \ldots, M_{mk}) \tag{10-8}$$

It can be shown that, by requiring the function F to satisfy some plausible requirements, the result is the arithmetic mean. Namely,

$$M_k = \frac{1}{m} \sum_{i=1}^{m} M_{ik} \quad \text{for all } k \tag{10-9}$$

The requirements which function F must satisfy [112] can be summarized as these four:

1. *Symmetry.* The same group rating is obtained if any two individuals r and t exchange their ratings M_{rk} and M_{tk} for an object x_k. That is,

$$F(M_{1k}, M_{2k}, \ldots, M_{rk}, \ldots, M_{tk}, \ldots, M_{mk}) \\ = F(M_{1k}, M_{2k}, \ldots, M_{tk}, \ldots, M_{rk}, \ldots, M_{mk}) \tag{10-10}$$

Namely, F takes account of each individual the same way with no preferential treatment of anyone.

2. The same scale is used for all importance ratings by all people. Or for any incremental change in rating by an amount δ, for object x_k, the same group rating is obtained regardless of whether the change is introduced by person P_r or P_t.

$$F(M_{1k}, M_{2k}, \ldots, M_{rk} + \delta, \ldots, M_{tk}, \ldots, M_{mk}) \\ = F(M_{1k}, M_{2k}, \ldots, M_{rk}, \ldots, M_{tk} + \delta, \ldots, M_{mk}) \quad (10\text{-}11)$$

3. Requirement (2) leads to the result that for any object x_k

$$F(M_{1k}, M_{2k}, \ldots, M_{mk}) = F(\textstyle\sum_{i=1}^{m} M_{ik}, 0, 0, \ldots, 0) \qquad (10\text{-}12)$$

This is obtained by first recording M_{1k} on each side of Eq. 10-11 with all other entries zero; next, setting δ equal to M_{2k} on the left, but adding it to M_{1k} on the right; then, setting δ equal to M_{3k}, etc., as shown here:

$$\begin{aligned} F(M_{1k}, 0, \ldots, 0) &= F(M_{1k}, 0, \ldots, 0) \\ F(M_{1k}, 0 + M_{2k}, 0, \ldots, 0) &= F(M_{1k} + M_{2k}, 0, \ldots, 0) \\ F(M_{1k}, M_{2k}, 0 + M_{3k}, 0, \ldots, 0) &= F(M_{1k} + M_{2k} + M_{3k}, \ldots, 0) \\ \vdots \qquad & \qquad \vdots \\ F(M_{1k}, M_{2k}, \ldots, M_{mk}) &= F(\textstyle\sum_{i=1}^{m} M_{ik}, 0, \ldots, 0) \end{aligned}$$

Hence, for any object x_k function F is a sum of the parameters M_{ik}. The simplest such function is a linear function of the form

$$F(M_{1k}, M_{2k}, \ldots, M_{mk}) = \alpha \textstyle\sum_i M_{ik} + \beta \qquad (10\text{-}13)$$

in which α and β are constants.

4. For identical rating M_{ik} for all i of a given object x_k, the group rating of x_k should also be M_{ik}:

$$F(M_{ik}, M_{ik}, \ldots, M_{ik}) = M_{ik} \qquad (10\text{-}14)$$

Generating Equation 10-9. Constrained by these four requirements, Eq. 10-9 is generated as follows. Setting in Eq. 10-14 $M_{ik} = 0$ for all i, then $F = 0$. Substituting this into the linearity stipulation of Eq. 10-13 yields $\beta = 0$. For $M_{ik} \neq 0$, say, $M_{ik} = x$ for all i, then from requirement (3) (Eq. 10-13),

$$F(x, x, \ldots, x) = \alpha \textstyle\sum_{i=1}^{m} x$$

But from requirement (4) (Eq. 10-14),

$$F(x, x, \ldots, x) = x$$

hence, $x = \alpha(mx)$

and $\alpha = \dfrac{1}{m}$

Thus, we are led to Eq. 10-9, rewritten here for completeness:

$$F(M_{ik}) = M_k = \frac{1}{m} \sum_{i=1}^{m} M_{ik} \quad \text{for all } k \qquad (10\text{-}15)$$

Some objections can be raised against the procedure suggested by Eq. 10-15. In addition, even the assignment of importance rating or utilities by an individual to an outcome with multiple attributes which are not mutually exclusive is no simple task, and is the subject of research[100].

Deficiencies in the Procedures for Consensus

The procedures for consensus in ranking discussed in the last three sections have a number of deficiencies. In each case, a criterion for consensus is stipulated such as minimum sum of squares of distances from the consensus. This in itself requires general acceptance. However, of even greater importance is the inherent assumption that all participants of the group share the same list of values. To arrive at such a state normally requires a great deal of dialogue in our democratic society. For small groups, we can visualize how such a dialogue is likely to take place, but for a very large community a dialogue is much more difficult to achieve. The Delphi method discussed in the next section offers an opportunity for progress toward community dialogue.

10-14 THE DELPHI METHOD*

Historical Preview to the Subject of Consensus

In the years between 285 B.C. and 246 B.C., King Ptolemy ruled in Egypt. Ptolemy was a strong supporter of Greek culture. Story has it that when he decided to undertake translation of biblical writings into Greek, he approached the high priest in Judea and requested the assistance of scholars who were well versed in both Hebrew and Greek. The high priest sent 70 scholars,† and the translation, therefore, later became known as the *septuagint* after the word *septuaginta*, seventy in Latin.

The story continues, then, to tell us that Ptolemy brought the scholars to Alexandria, placed each one in a separate room in isolation so that they could not communicate with each other or anyone in the outside world. The scholars set about their translation job without interruption, completed their work on schedule, and presented their translations to Ptolemy. An independent review committee inspected the 70 translations, and legend has it that

---

*The name Delphi comes from the site of an ancient Greek temple where the gods of Greek mythology gathered to profess their prognosis of the future.

†Another version claims that he actually sent 72 scholars, 6 from each of the 12 tribes.

they were found to be identical to the iota. This was such an unbelievable coincidence that the whole world known at that time was astonished to learn the story. The story traveled from city to city, from market place to market place, from gymnasium to gymnasium, until it finally reached Judea and got to a little town that was the home of one of the scholars who participated in the translation.

A young resident of the town, having heard the story from a traveler, rushed into the Rabbi's home and, full of excitement, related to him the miracle of the identical translations. The Rabbi, an old experienced man, listened with great patience and then turned to the young man: "Seventy scholars in separate rooms, and this you call a miracle? Put them in one room and get the same translation—this is a miracle."

In Sec. 7-9 we considered group decision making and the resulting difficulties as compared to individual decision making. In this section, we discuss a "modern version of Ptolemy's procedure" for group decisions known as the Delphi method developed by Helmer [115].

The Delphi Method—A Forum for Group Communication

The emergence and development of the Delphi method in recent years offers an opportunity for achieving progress toward dialogue and effective communications to gain a clearer understanding of the process of generating consensus. The method is a dynamic procedure for forming consensus through the use of questionnaires (or direct on-line computer consoles) with no face-to-face discussion (or even identification) by the members of the group.

The Delphi method has been used effectively in a number of areas [116–123]. For example, Dalkey [124] has applied the Delphi procedure to rate quality-of-life factors (see Table 10-2) in which "quality of life" was defined as the sense of well-being, satisfaction or dissatisfaction with life, happiness, or unhappiness. The study consisted of three stages. The first two were devoted to generating a group-aggregated list of quality-of-life factors, and the third stage was used to place the items of the list on a scale of importance. In the first stage, the participants (a group of 90 UCLA students) were each asked to make up a list of 5 to 10 factors relevant to the quality of life and to rank them in importance. About 250 different items were submitted. These were sorted into 48 categories by virtue of their similarity. The individual rankings of items by each participant as well as the following two criteria were used as guides in the aggregation. (1) No more than 50 categories were to be formed. (2) The difference between any pair of items within a category was to be smaller than the difference between any pair of items coming from two different categories. The 48 categories formed this way by the Delphi study team were given new labels which best represented the aggregates of items. In the second stage, each participant was informed of the new labels and was given a list of 1128 pairs of labels that could be formed with the 48 aggregates. For example, money-freedom, aggression-freedom, status-money, good health-money, ambition-honesty, fear-money, ambition-self respect, etc. (The order of the pairs was random.) The participants were asked to rate the similarity of the factors in each pair on a scale 0–4 signifying the strength of similarity:

0 unrelated
1 slightly related
2 moderately related
3 closely related
4 practically the same

Dissimilarities were given negative ratings, with —4 signifying extreme opposites.

The Delphi study team computed the means of the absolute values of the similarity ratings for each pair, and the 48 labels were then clustered into 13 according to the similarities between them, using Johnson's hierarchical clustering scheme [125].

In the third stage, the participants rated the 13 labels in importance by assigning a numerical value to each. These ratings were used to compute the median* and quartiles† of the ranking for each label. These results were reported to the individual group members who were asked to revise their ratings. This procedure can be repeated with revisions in ratings accompanied by the impersonal debate. Eventually, convergence may be reached and consensus formed.

Thus the Delphi method establishes a forum for group communication, which is most useful when the group is too large for meaningful face-to-face interaction. But perhaps most importantly, the group can start from strongly different views held by individuals and attempt to focus on areas of agreement rather than disagreement. In this sense, it brings us closer to the key feature in modeling which has been marked by so much success in other areas.

The method need not be limited to consensus on values. Consensus may be sought regarding relevant parameters for models of reality (*what is*) as well as models of *what to do* and *what is possible.* Since all these stages are not value free, the communication framework can serve as a learning experience in which insight is gained into issues, and values are identified in the process, and what is desirable is considered in the light of what is possible. The whole

*The median is the measurement chosen so that half the measurements are above it and half below it.

†A quartile is a measurement chosen so that one-quarter of the measurements are on one side of it and three-quarters are on the other.

process is much like that of a man who discovers what kind of house he wants in the light of what he can afford as he continues a meaningful communication with the architect.

But who is the "architect" in a Delphi communication scheme? The Delphi method has its roots in the jury system. The jury explores the issue, individuals convey their thoughts and feelings and reach an understanding of what is important. Finally, a group decision is sought in the form of a verdict. The jury is guided by instructions of a judge. The equivalent of the judge in the Delphi exercise is the agency that asks the questions, proceeds with the analysis of the inputs, provides feedback, and in general structures the framework for the communication. This may strongly influence the ultimate results and it is, therefore, an important element which must be explored further. The method shows great promise as an effective way to introduce dialogue from afar leading to better insight, continued learning, and ultimate formation of consensus in problems of a socio-technical nature, in which *what we want* and *what we do* can be treated as an integrated whole.

10-15 USE OF DELPHI METHOD TO DEVELOP AN INTERDISCIPLINARY COURSE

This section summarizes a study [126] in which the Delphi method was used to develop the content of an interdisciplinary course in problem solving which discusses tools, concepts, and philosophies underlying the formulation and solution of problems relevant to society and technology. The study formed the basis for the course, Patterns of Problem Solving, mentioned in the preface of this book.

Objective

The objective of the study was to maximize the benefit derived by a class of students with diverse interests. The Delphi technique of eliciting and refining expert opinion was used to model the benefit functions, and dynamic programming was employed to optimally allocate class time to the various topics of the course.

Considering n main topics, we wish to allocate time to these topics so as to maximize the benefit derived from the course. The model of the problem is stated in the form:

$$\text{maximize } F = \textstyle\sum_{i=1}^{n} w_i f_i(t_i) \tag{10-16}$$

subject to these constraints:

$$\textstyle\sum_{i=1}^{n} t_i \leq T \text{ and } t_i^L \leq t_i \leq t_i^H, \quad i = 1, 2, \ldots, n \tag{10-17}$$

in which $f_i(t_i)$ = benefit function of the ith topic (discussed below).
w_i = weighting factor for the ith topic.
F = sum of the benefits derived from the n topics.

t_i^L and t_i^H represent, respectively, the lower and upper limits on the time that should be devoted to the ith topic, and T is the total time, in hours, available for the course.

The general recursive relationship for the solution has the form

$$F_n(T) = \max_{t_n^L \leq t_n \leq t_n^H} f_n(t_n) + F_{n-1}(T - t_n) \tag{10-18}$$

This is a dynamic programming model (see Chapter 8). Namely, the benefit from n topics is the sum of the returns from the nth topic by optimal allocation of t_n hours to it, and the optimal allocation of $(T - t_n)$ hours to the remaining $(n - 1)$ topics.

Benefit Function

To obtain the benefit functions $f_i(t_i)$, a panel of experts was asked to estimate t_i^L and t_i^H for each topic. t_i^L represents the number of hours below which the adoption of topic i would be pointless, and t_i^H the number of hours above which the marginal benefits would be so small as to make allocation of additional time wasteful. A cumulative beta distribution function [127] in the shape of a sigmoid was then fitted through the points (0, 0), $(t_i^L, 0.1)$ and $(t_i^H, 0.9)$, as shown in Fig. 10-4.

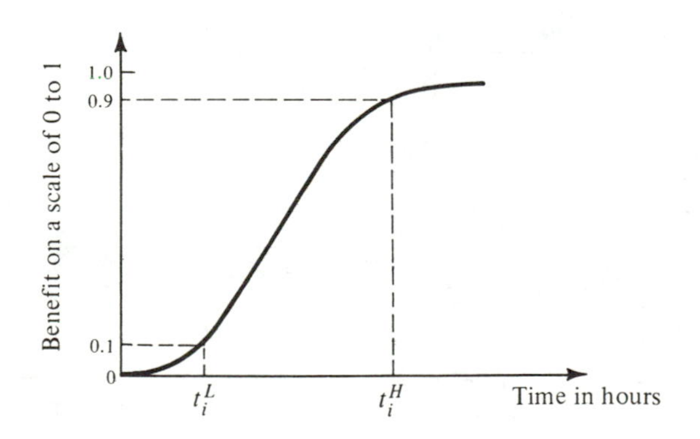

Figure 10-4 Benefit Function

Delphi Method to Generate Consensus*

The questionnaire for the study was designed to obtain the best topics and to decide how much time should be devoted to each. The 23 respondents

---

*The application of the method was done with the help of Dr. N. Dalkey of the RAND Corporation, Santa Monica, CA, Adjunct Professor, Engineering Systems Department, UCLA.

Table 10-4 Summary of Total Group Response

Topic	First Questionnaire t^L(hrs) Q_1	Q_3	M	t^{II}(hrs) Q_1	Q_3	M	Average Self-Rating (based on a maximum of 5)	Average Ranking of Topics (8 = most important 1 = least important)
(i) Modeling	1.5	4	2.4	4	10	7.6	3.8	6.54
(ii) Decision Making Elements and Utility Theory; Game Theory	3	6	4.8	10	20	13.6	4.0	6.3
(iii) Important Tools and Concepts in Probability	2	4	3.2	6	12	9.3	3.8	5.4
(iv) Examples of the Use of Certain Well Known Functions in Modeling in Physics, Economics and Sociology	2	4.5	3.3	5	15	11	3.8	5
(v) Dynamic Behavior of Systems	2	4.5	3.3	4	16	10	3.1	3.8
(vi) Optimization: Linear and Dynamic	2	4	2.8	4	14	9.8	3.3	3
(vii) Feedback and Stability—Examples	1.5	4	2.3	4	13	7.7	2.7	3.1
(viii) Systems Theory*								

*This topic was added after the first questionnaire.

Topic	Second Questionnaire t^L(hrs) Q_1	Q_3	M	t^{II}(hrs) Q_1	Q_3	M	Average Weighting of Each Topic (scale of 0 to 1 1 = most important)	Third Questionnaire t^L(hrs) Q_1	Q_3	M	t^{II}(hrs) Q_1	Q_3	M	Average Weighting of Each Topic (scale of 0 to 1 1 = most important)
(i)	2	4	3.4	6	10	7.6	0.89	3	4	3.33	6	10	7.83	0.85
(ii)	3	5	3.7	6	12	9.5	0.82	3	5	4.5	7	11	9.4	0.82
(iii)	4	5	4.3	9	12	10	0.85	4	5	4.3	8	11	9.4	0.78
(iv)	2	3.5	2.6	6	11	10	0.62	2	3	2.6	4	9	6.6	0.58
(v)	3	4	3.6	8	14	9.8	0.61	2.5	3.5	3.1	5	9	7.4	0.60
(vi)	3	5	3.8	6	12	9.3	0.58	3.5	4.5	4.1	6	11	8.67	0.67
(vii)	1.5	4	2.5	4	10	7.6	0.56	1.5	3	2.92	4	8	6.0	0.45
(viii)	1	6	3	5	10	8.4	0.47	1.5	3.5	3.2	4	11	7.0	0.52

knowledgeable in the topics of the proposed course were asked to:

(a) Rank the importance of the topics.
(b) Fill in the t_i^L and t_i^H values for each topic.
(c) Rate their own knowledge of each topic.

The response to (c) provided a criterion for selecting more accurate subgroups of respondents. Only the upper subgroup with ratings of 3 and higher, on a scale of 1 to 5, was used for analyses of the data obtained from the final questionnaire. The feedback after each round was given as the average of the total group.

Tables 10-4 and 10-5 summarize the computed consensus. Table 10-4 contains the estimates of the total group while Table 10-5 lists the estimates

Table 10-5 Summary of Upper Subgroup Response Only (Third Questionnaire)

Topics	t^L(hrs) Q_1	Q_3	M	t^H(hrs) Q_1	Q_3	M	Average Self-Rating (based on a maximum of 5)	Average Weighting of Each Topic (scale of 0 to 1 1 = most important)	Number of Members in Upper Subgroup (self-rating of 3 and above)
Modeling	3	4	3.4	6	11	8	4.4	0.86	22
Decision	3.5	5.5	4.6	7	11	9.5	4.5	0.92	18
Probability	4	5	4.8	10	12	10.5	4.7	0.88	16
Functions, Use of	2	4	2.9	8	10	9	4.4	0.61	12
Dynamic Behavior	3	5	3.6	7	11	9.4	4.4	0.71	10
Optimization	3	5	4.1	7	11	8.6	4.3	0.68	16
Feedback, Stability	2.8	5	3.7	5	9	7	4.3	0.58	8
Systems Theory	1	5.5	3.1	4	10	7.1	4.6	0.48	20

for the upper subgroup only. Q_1 and Q_3 in these tables stand, respectively, for the first and third quartiles and M is the median. Most of the questions put to the panel members were asked more than once to determine the degree of convergence in the process of interrogation. The quartile range (QR) of the responses was used to measure the spread of opinion. The quartiles Q_1, Q_3 and median M were determined for t_i^L and t_i^H of each topic so that each of the four intervals formed by these three points contained one-quarter of the responses. Figure 10-5 shows the process of convergence for the t^L value in the topic "modeling".

Optimal Allocation of Class Time

The optimal allocation of class time for a 40-hour course is shown in Table 10-6. The upper subgroup had less dispersion in their responses than the total group.

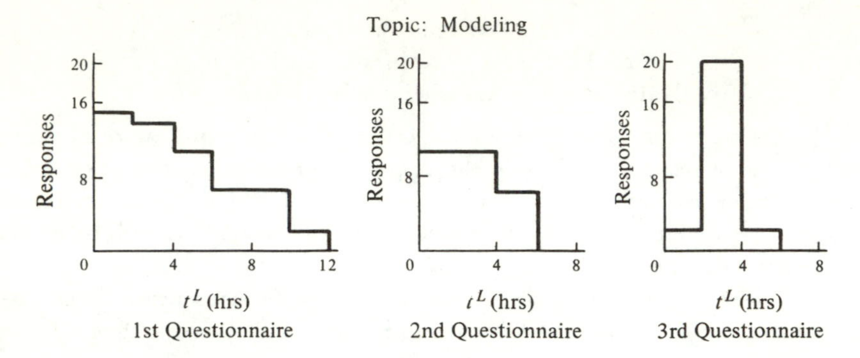

Figure 10-5 Progressive Convergence of Opinion (t^L Estimate)

Table 10-6 Optimal Allocation of Time (40 hours total)

Topic	Total Group	Upper Subgroup
Modeling	8	8
Decision	6	9
Probability	4	5
Functions, Use of	6	3
Dynamic Behavior	3	3
Optimization	7	4
Feedback and Stability	3	5
Systems Theory	3	3

NOTES. This study is an example of how a systems approach which incorporates the Delphi method can be used to design an interdisciplinary course. However, the fundamental approach developed here can be extended to the design of an entire curriculum for a school.

The linear form of the total benefit function (Eq. 10-16) is a simple model. It is important to establish the significance and strength of interaction between the various topics in the course. To include any interaction in the model may be comparable to generating a multidimensional nonlinear utility function. Such a function could be generated by using the Delphi method, and could be the subject of further study.

It is important to devote part of the analysis to problems of acceptability and implementation. The participants in the study should be made aware of the final results and their cooperation sought in implementation. This phase is important if the analysis is to realize its full potential.

10-16 CONSENSUS ON VALUES AND CONSENSUS ON WHAT TO DO

Manheim [128] discusses a plan for community involvement in reaching decisions about projects of a socio-technical nature. The strategy he proposes consists of four phases:

1. Initial survey
2. Issue analysis
3. Design and negotiation
4. Ratification

The first phase is equivalent to a study of *what is*. Phase two is concerned with gaining insight into the goal state as part of a learning stage while alternatives are explored. Phases 3 and 4 represent the procedure for forming consensus on the process to achieve a goal which is most likely to represent what the community wants. Manheim's proposal is an important contribution to the new age of technological projects which assess the social consequences beforehand. The community involvement in all phases is stipulated from the outset.

However, each phase above is laden with values; the description of *what is*, what the needs are, the objectives, and the alternative solutions to achieve objectives. Manheim states the following basic premise in his approach: "The community consists of all those individuals and groups who will potentially be affected, positively or negatively, by any of the courses of action being considered. A basic premise of our approach is that the community so defined is composed of diverse groups with very different values."

This basic premise is plausible; however, the conclusion which follows is questionable: "Therefore it is infeasible to get agreement on a statement on values; it is more feasible to get agreement on a course of action."

In principle, the statement may be true. However, too often we agree on *what to do* without questioning *what we want* in the first place. The general spirit of such an approach is the subordination of end values to means values. In the process, the instruments for achieving goals become end goals. *What to do* should reflect what is possible, and although values cannot always be achieved, they should definitely be considered in the light of what is possible.

It is quite possible that the "freeway revolt" of the San Francisco citizens who objected on aesthetic grounds to the completion of the Embarcadero Freeway (because it would spoil the view of the San Francisco Bay) could have been avoided, had values been considered more thoroughly and seriously in the light of what was possible to do. Construction of the partially completed freeway was abruptly halted in 1958, and the "monument" still stands as a reminder of the failure to consider an important community value: aesthetics.

It is important to gain insight into values of a community as defined by Manheim because such insight will lead to improved procedures for consensus on values, and as a result will facilitate a better understanding of *what to do*. A basic notion which may then emerge, a kind of meta-value, may convince a community that to arrive at consensus, which nearly always involves compromise, is of a higher value than arriving at no consensus.

10-17 SUMMARY

Values are a central component of problem solving. Values guide human behavior. They are the tools of man's adaptability and contribute to his survival. A person's value system influences his models of reality, his goals, and his actions to achieve them.

The general theory of value is known as *axiology*, and it encompasses all aspects of values: economic, aesthetic, moral, and ethical.

Values enter all stages of the systems approach to problem solving:

Stage 1. Description of "what is."
Stage 2. Description of "what is desired."
Stage 3. Description of "what to do and how" in going from "what is" to "what is desired."

Values can be classified in many ways; for example, as end values or means values, as competitive or cooperative values, as personal character values, life-setting values, etc.

Value judgment is an act of matching: matching importance ratings r_j to values v_j, and matching relative contributions c_{ij} of an action a_i ($i = 1, 2, \ldots, n$) to a value v_j. The utilities $u(a_i)$ of alternative actions a_i can be assessed from

$$u(a_i) = \textstyle\sum_j c_{ij} r_j; \quad i = 1, 2, \ldots, n$$

or

$$\{u(a_i)\} = [c]\{r\}$$

The criterion for selecting a course of action can be $\max\{u(a_i)\}$, i.e., the action which yields the maximum utility.

Knowledge and values are related. Knowledge helps us realize our values. Knowledge may give us clues to possible intrinsic values which are most compatible with human nature. Entropy consumption? . . .

Entropy is a measure of disorder. For a state with E_i ($i = 1, 2, \ldots, n$) possible events, the larger the entropy, the higher the degree of disorder of the state, or the larger the number of binary questions which must be asked, on the average, to identify the events in the state. The state of highest entropy occurs when $P(E_i) = 1/n$ for all $i = 1, 2, \ldots, n$. Highly ordered states have a small probability of occurrence, and they tend to change to more probable states of less order or higher entropy. Therefore, the entropy in the universe is increasing. Entropy can be decreased locally by exerting an external influence.

While life is an entropy-consuming process, too much emphasis on order or a total commitment to a *Will to Order* goes against human nature.

Models of ethical behavior or moral law have been suggested. Such models attempt to predict human behavior. Teleological or purposive models are considered based on an end value of *Survival*, *Reverence for Life* (Schweitzer), *Will to Pleasure* (Freud), *Will to Power* (Adler), *Will to Meaning* (Viktor Frankl).

It is a good working hypothesis to consider survival as a supreme value in the sense that it serves to predict human behavior in a gross general way. With this hypothesis, a model of behavior can be stated in this form: human beings generally behave so as to maximize the following objective function of survival S, in which $0 \leq \alpha \leq 1$,

$$S = \alpha \text{ Prob(individual survival)} + (1 - \alpha) \text{ Prob(group survival)}$$

Behavior consistent with this objective may be considered as ethical behavior.

Values of the present and predicted values of the future are different from the traditional values of the recent past. Conspicuous by their absence are honesty, dignity, charity, loyalty, discipline, responsibility, and accountability.

Values change dynamically with the age and psychological maturity of an individual. Values change when affluence replaces economic scarcity, when new knowledge becomes available, and when new developments in technology or in the political arena cause shifts in emphasis and importance of values.

Values are benefit oriented. The cost-benefit relationship as applied to values attempts to assess the extent of cost in terms of energy, time, money, and the undertaking of risks for the purpose of realizing a value. Values can be true or deceptive, i.e., false.

Social order strives to fuse individual values into group values. Mathematical procedures can be developed for the formation of group consensus. Group consensus for ranking of preferences and metric rating of preferences are considered. The Delphi method provides a means for forming group consensus by an impersonal debate. A number of iterations are used through a feedback scheme which leads to convergence of opinion and the formation of consensus.

Often we agree on "what to do" without questioning "what we want." In the process, instruments for achieving goals become goals. What we want as a group must be based on a consensus of group or community values.

TOPICS FOR DISCUSSION WITH THE FOCUS ON VALUES

1. The book of Genesis, Chapter 23, in the Scriptures describes the purchase of the cave of Machpelah by Abraham from Ephron the Hittite, as a burial ground to bury his wife Sarah. The meeting between the two "tycoons" of the day takes place outside the city gates in the presence of the city dwellers as well as merchants and visitors from out of town. Abraham offers to pay the full price of the cave, but Ephron expresses his generosity and says:

> "Nay, my lord, hear me: the field give I thee, and the cave that is therein, I give it thee; in the presence of the sons of my people give I it thee; bury thy dead."

and the story continues:

> "And Abraham bowed down before the people of the land. And he spoke unto Ephron in the hearing of the people of the land, saying: 'But if thou wilt, I pray thee, hear me: I will give the price of the field; take it of me, and I will bury my dead there.' And Ephron answered Abraham, saying unto him: 'My lord, hearken unto me: a piece of land worth four hundred shekels of silver, what is that betwixt me and thee? bury therefore thy dead.' And Abraham hearkened unto Ephron; and Abraham weighed to Ephron the silver, which he had named in the hearing of the children of Heth, four hundred shekels of silver, current money with the merchant."

The sum of four hundred shekels was a very substantial and a highly inflated price. Six shekels were the average yearly wages of a worker in those days.

Identify and discuss the role played by values which guided Ephron to achieve his goal of receiving a high price for his property and led Abraham to overpay.

2. The following epilogue appears in *Teacher and Child* by Haim G. Ginott, published by The Macmillan Co., 1971, page 317:*

> On the first day of the new school year, all the teachers in one private school received the following note from their principal:
>
> Dear Teacher:
>
> I am a survivor of a concentration camp. My eyes saw what no man should witness:
>
> Gas chambers built by *learned* engineers.
>
> Children poisoned by *educated* physicians.
>
> Infants killed by *trained* nurses.
>
> Women and babies shot and burned by *high school* and *college* graduates.

*Reprinted by permission of the publisher and author.

> So, I am suspicious of education.
>
> My request is: Help your students become human. Your efforts must never produce learned monsters, skilled psychopaths, educated Eichmanns.
>
> Reading, writing, arithmetic are important only if they serve to make our children more humane.

Discuss the implication of this message in relation to the role of values in education.

3. The table on page 518 is the result of a survey on educational goal-setting activities conducted by the junior high school attended by my daughters. Each of the eighteen activities was assigned a value of importance from the following scale:

	1	2	3	4	5	
least important						most important

The teachers, students, and parents are not always in agreement. Identify activities of largest disagreement and discuss the role of values which may account for the different importance ratings.

4. When it became apparent that ordinary tomatoes could not withstand the rough treatment of mechanical harvesting, a new breed of tomato was developed. It is elongated in shape, has tough skin, and all tomatoes ripen at the same time so that mechanical harvesting can be done in a single run. Sommer [129] uses this development of a new tomato as an analogy to the "new man" of the future who may adapt to crowding, pollution, and crime. He then claims that good designs of space become meaningless if we consider that man can be reshaped to fit whatever environment he creates, because the content is made to fit the form or the object is shaped to fit the instrument. To avoid this, we must establish criteria for two fundamental questions:

1. What kind of man do we want?
2. What kind of environment do we want?

What implications do these questions have regarding values of a society?

5. Discuss the relationship between culture and values that can be identified in the Ahmed story of Sec. 1-1.

6. Discuss the change in values which can be attributed to automation and the advent of computers.

7. What changes in values are likely to take place if computer-aided instruction becomes widely used at all levels of learning? Discuss both positive and negative attributes of computer-aided instruction and learning.

8. Discuss the role of humor in the realization of human values.

9. Discuss the relationship between individual and group values in Model 2 of Sec. 7-2.

OLIVER WENDELL HOLMES JUNIOR HIGH SCHOOL
NORTHRIDGE, CALIFORNIA

*SUMMARY OF EDUCATIONAL GOAL-SETTING ACTIVITIES**

Consensus of:	*Teachers*		*Students*		*Parents*	
Goals	Pt. Av.	Rank	Pt. Av.	Rank	Pt. Av.	Rank
1. To gain a general education	4.21	2	4.5	1	4.65	1
2. Develop skills in reading, writing, speaking, and listening	4.36	1	4.0	2	4.60	2
3. Learn how to examine and use information	3.36	5	2.7	9	4.17	4
4. Develop a desire for learning now and in the future	2.83	8	3.3	3	4.21	3
5. To learn about and try to understand the changes that take place in the world	2.14	11	3.08	4	3.85	7
6. Help students develop pride in their work and a feeling of self-worth	4.0	3	2.83	8	4.08	5
7. To develop good character and self-respect	3.07	7	2.3	11	3.89	6
8. To help students appreciate culture and beauty in their world	2.07	12	1.33	15	3.42	11
9. To learn to use leisure time	2.0	13	1.6	13	2.93	15
10. Develop the ability to make job selections	1.07	17	2.9	6	3.09	14
11. To learn to respect and get along with people with whom we work and live	3.54	4	2.8	7	3.85	7
12. To learn how to be a good citizen	3.29	6	2.0	12	3.76	8
13. Understand and practice democratic ideas and ideals	2.36	10	1.3	15	3.61	9
14. To learn how to respect and get along with people who think, dress, and act differently	2.79	9	3.0	5	3.41	12
15. Practice and understand the ideas of health and safety	2.0	13	1.5	14	3.50	10
16. To prepare students to enter the world of work	1.54	15	2.9	6	3.21	12
17. To understand and practice the skills of family living	1.58	14	1.3	15	2.92	16
18. To learn how to be a good manager of time, money, and property	1.45	16	2.4	10	3.21	12

*Reproduced by permission of Oliver Wendell Holmes Junior High School, Northridge, California.

10. Discuss the relationship between individual and group values in Model 3 of Sec. 7-2.

11. Describe a situation in which the farmer of Model 1, Sec. 7-2 may have to consider personal versus societal values in his decision.

12. Discuss the differences in both individual and societal values in a society in which schooling is compulsory as compared with one in which it is voluntary.

13. Consider the implication on values in our society if we were to reach a state in which virtually all adults had a university education.

14. Some claim that the word "civilization" introduced in the 18th century, before the industrial revolution, referred to ethical, moral, and social progress. How did the change in values which attended the industrial revolution change the meaning of this word?

15. It is said that man, not knowledge, is the cause of violence. Yet after Hiroshima, the famous physicist Robert Oppenheimer said: "For the first time, the scientist has known sin." Oppenheimer made reference to the scientist who contributed to the knowledge which made Hiroshima possible. What is the role of knowledge in human values?

16. What values did Lawrence of Arabia exploit when he used existing Arab tribes as units in forming an army to revolt against the Turks?

17. Microminiaturization of electronics and developments in computer technology have been blamed for contributing to the emergence of a new era of a depersonalized machine culture. What values have been compromised or eroded as a result of these developments?

18. What is the role of values in the nonzero-sum game models summarized in Figs. 7-25 and 7-26?

19. What do you think would have been the relative importance ratings of Tables 10-2 and 10-3 had the consensus been taken in the year 1900?

20. The increase in literacy and the availability of books on every subject has transformed our society to what Peter Drucker [130] calls the knowledge society. Gutenberg's printing press, invented in the 1450s, marked the beginning of a new era in dissemination of information through the written word. Instead of a slow, painstaking process of copying by hand, three hundred pages a day (with impressions on each side) could be produced [131]. This rate was virtually unchanged until 1813 when Fredric Koenig, a Saxon printer, invented a metal rotary press which he ran with steam one year later in 1814 and thus increased the rate to 550 pages an hour. By the 1850s, it increased to 10,000 pages per hour.

How did this exponential growth in the availability of printed material affect social values?

21. A local newspaper article related the following story:

> A 12-year-old girl lost a $900,000 lawsuit when a jury decided a hospital gave her oxygen to save her life.
>
> The oxygen damaged her sight at the time of her premature birth, the jury was told.
>
> A physician testified during the trial that a premature baby who is not given oxygen may die, but the grim alternative is that an eye disease may develop from the extra oxygen.
>
> A jury reached a unanimous verdict in favor of the hospital.
>
> "It was a difficult decision," said the jury foreman. "We took seven polls before deciding that the hospital was not negligent. They had to give her oxygen so she would survive."
>
> She weighed 2 pounds 10 ounces at birth and received oxygen in an incubator for seven days.

What values do you think were involved in the deliberations of the jury?

22. Discuss how the values of work, frugality, respect for elders, charity, honesty, brotherhood, and religious faith contributed to survival in a society dominated by scarcity.

23. What changes in values are likely to accompany a transition to a society in which a guaranteed annual income is sufficient to provide for a person's basic needs of food and shelter?

24. What changes in values can be attributed to the availability of various forms of insurance policies on almost anything?

25. The institution of an ombudsman began in Sweden in the early 1800s (c. 1810) and has been adopted in other countries. The ombudsman* acts as an agent of Parliament to investigate citizens' grievances against bureaucracy. What contribution can this institution make to human values? To what extent can the ombudsman become the Tribune of the People?

26. Propose a list of seven most important values which are likely to prevail in a society dominated by material abundance for all, with adequate health-care, open space, etc. Compare your list with what you consider as the seven most important values in a society dominated by scarcity.

27. Discuss the importance of certain ethical obligations expected of those selected for judicial office.

28. Discuss the values which are likely to be considered in the deliberations of a medical society whose membership is about to go on a nationwide strike.

---

*The word means, in Swedish, agent or representative.

29. Discuss the influence on human values which may be attributed to the entertainment provided by mass media.

30. Discuss the value of work in our society. To what extent is the following biblical injunction still relevant, "If any would not work, neither should he eat"?

31. Discuss the values implicated when a reporter is faced with the option to reveal information which supports his personal commitment to search for and tell the truth, but which could endanger national security.

32. Suppose you are offered an important job within the ranks of a political campaign, with the opportunity for experience and high pay. However, you do not agree with the basic philosophy of the candidate and would not vote for him. What values are implicated in the decision to accept or reject the offer?

33. What are the values behind different admission requirements to the same institutions for different individuals?

34. Discuss conditions under which personal and societal values have changed in your experiences.

References

1. SKINNER, B. F., "An Operant Analysis of Problem Solving," in *Problem Solving: Research, Method, and Theory*, ed. Benjamin Kleinmuntz. New York: John Wiley and Sons, 1966.
2. SIMON, HERBERT, *The Sciences of the Artificial*. Cambridge, MA: M. I. T. Press, 1969.
3. GERHOLM, TOR RAGNAR, *Physics and Man*. Totowa, NJ: The Bedminster Press, 1967.
4. ERNST, GEORGE W., and ALLEN NEWELL, *GPS: A Case Study in Generality and Problem Solving*. New York: Academic Press, 1969.
5. MILLER, GEORGE, "The Magical Number Seven, Plus or Minus Two," *Psychological Review*, Vol. 2, 63 (1956), 81–97.
6. DUNCKER, KARL, "On Problem Solving," *Psychological Monographs*, Vol. 5, No. 5 (1945), 270.
7. CHURCHMAN, C. WEST, "The Role of Weltanshauung in Problem Solving and Inquiry," in *Theoretical Approaches to Non-Numerical Problem Solving*, eds. R. B. Banerji and M. D. Mesarovic. Berlin and New York: Springer-Verlag, 1970.
8. HYMAN, RAY, and BARRY ANDERSON, "Solving Problems," in *The R & D Game*, ed. David Allison. Cambridge, MA: M.I.T. Press, 1969.
9. LEVENSTEIN, AARON, *Use Your Head*. New York: Macmillan, 1965.
10. ELLIOTT, ROBERT S., *Electromagnetics*. New York: McGraw-Hill, 1966.
11. MORRIS, CHARLES, "Foundations of the Theory of Signs," in *International Encyclopedia of Unified Science*, Vol. 1, No. 2. Chicago: University of Chicago Press, 1938.
12. ———, *Signs, Language and Behavior*. New York: George Braziller, 1955.
13. KRAUSS, ROBERT M., "Language as a Symbolic Process in Communication", *American Scientist*, Vol. 56, No. 3 (1968), 265–278.
14. WALKER, MARSHALL, *The Nature of Scientific Thought*. Englewood Cliffs, NJ: Prentice-Hall, Inc., 1963.
15. GLEASON, HENRY A., JR., *An Introduction to Descriptive Linguistics*. New York: Holt, Rinehart, and Winston, 1961.

16. DIRINGER, DAVID, *The Alphabet* (3rd edition). London: Hutchinson and Co., 1968.
17. SCHOLEM, GERSHOM G., *Jewish Mysticism* (3rd ed.). New York: Schoken, 1960.
18. DANTZIG, TOBIAS, *Number: the Language of Science*. New York: Macmillan, 1954.
19. KARPLUS, W. J., *Analog Simulation*. New York: McGraw-Hill, 1958.
20. ———, and W. W. SOROKA, *Analog Methods* (2nd ed.). New York: McGraw-Hill, 1959.
21. ARDEN, BRUCE W., *An Introduction to Digital Computing*. Reading, MA: Addison-Wesley, 1962.
22. DESMONDE, W. H., *Computers and Their Uses*. Englewood Cliffs, NJ; Prentice-Hall, Inc., 1964.
23. HULL, T. E. *Introduction to Computing*. Englewood Cliffs, NJ: Prentice-Hall, Inc., 1966.
24. WRUBEL, M. H., *A Primer of Programming for Digital Computers*. New York: McGraw-Hill, 1959.
25. FISHER, F. P., and G. F. SWINDLE, *Computer Programming Systems*. New York: Holt, Rinehart, and Winston, 1964.
26. MCCRACKEN, D. D., *Digital Computer Programming*. New York: John Wiley and Sons, 1957.
27. ———, *A Guide to FORTRAN Programming*. New York: John Wiley and Sons, 1961.
28. GERMAIN, C. B., *Programming the IBM 1620*. Englewood Cliffs, NJ: Prentice-Hall, Inc., 1962.
29. CHAPIN, NED, *Computers, a Systems Approach*. New York: Van Nostrand Reinhold, 1971.
30. ORTEGA Y GASSET, JOSÉ, *What Is Philosophy?* New York: W. W. Norton, 1967.
31. TRIBUS, MYRON, *Rational Description, Decision and Design*. New York: Pergamon Press, 1969.
32. RUSSELL, BERTRAND, *The Will to Doubt*. New York: Philosophical Library, 1958.
33. POLYA, G., *Patterns of Plausible Inference*. Princeton, NJ: Princeton University Press, 1954.
34. ABRAMSON, NORMAN, *Information Theory and Coding*. New York: McGraw-Hill, 1963.
35. PIERCE, J. R., *Symbols, Signals and Noise*. New York: Harper & Row, 1965.
36. TOULMIN, STEPHEN, *Foresight and Understanding*. Bloomington, IN: Indiana University Press, 1961.
37. MEADOWS, D. H., D. L. MEADOWS, J. RANDERS, and W. W. BEHRENS III, *The Limits to Growth*. New York: Universe Books, 1972.
38. ALEXANDER, CHRISTOPHER, *Notes on the Synthesis of Form*. Cambridge, MA: Harvard University Press, 1970.
39. MANHEIM, M. L., and F. L. HALL, "Abstract Representation of Goals," in *Transportation: A Service*. New York: New York Academy of Sciences and American Society of Mechanical Engineers, 1967.
40. SIMON, HERBERT A., and ALLEN NEWELL, "Models: Their Uses and Limitations," in *The State of the Social Sciences*, ed. Leonard D. White. Chicago: University of Chicago Press, 1955.
41. WHITEHEAD, ALFRED NORTH, *Science and the Modern World*. New York: New American Library, 1953.

42. STARR, CHAUNCEY, *Social Benefits vs. Technological Risk*, UCLA Report No. 62, 1969.
43. HESSE, MARY B., *Models and Analogies in Science*. Notre Dame, IN: University of Notre Dame Press, 1966. Also see [40].
44. HUNT, EARL, "What Kind of Computer Is Man?" *Cognitive Psychology*, Vol. 2, (1971), 57–98.
45. FORRESTER, J. W., *Industrial Dynamics*. Cambridge, MA: M.I.T. Press, 1961.
46. DE NEUFVILLE, RICHARD, and JOSEPH H. STAFFORD, *Systems Analysis for Engineers and Managers*. New York: McGraw-Hill, 1971.
47. RABINOWITZ, HOSEA, "Analysis for Automatic Extraction of Causal Relations in Empirical Data." A thesis submitted in partial satisfaction of the requirement for the degree of Master of Science, UCLA, 1972.
48. SPENGLER, OSWALD, *The Decline of the West*. New York: Knopf, 1945.
49. TOYNBEE, ARNOLD, *A Study of History*. New York: Oxford University Press, 1954.
50. GAMOW, GEORGE, *The Creation of the Universe*. New York: Viking, 1952.
51. HOYLE, FRED, *Frontiers of Astronomy*. New York: Harper & Row, 1955.
52. SIMON, HERBERT A., and ALLEN NEWELL, "Information Processing in Computer and Man," *American Scientist*, Vol. 52 (Sept. 1964), 281–299.
53. MILLER, GEORGE, EUGENE GLANTER, and KARL PRIBRAM, *Plans and the Structure of Behavior*. New York: Holt, Rinehart and Winston, 1960.
54. MOOD, ALEXANDER MCFARLANE, *Introduction to the Theory of Statistics*. New York: McGraw-Hill, 1950.
55. HOEL, PAUL G., SIDNEY C. PORT, and CHARLES J. STONE, *Introduction to Statistical Theory*. Boston: Houghton Mifflin, 1971.
56. MILLER, DAVID W., and MARTIN K. STARR, *The Structure of Human Decisions*. Englewood Cliffs, NJ: Prentice-Hall, Inc., 1967.
57. HURWICZ, LEONID, "Optimality Criteria for Decision Making under Ignorance." Cowles Commission discussion paper, *Statistics*, No. 370, 1951, mimeographed; cited in Luce and Raiffa [59].
58. VON NEUMANN, JOHN, and OSKAR MORGENSTERN, *Theory of Games and Economic Behavior*. Princeton, NJ: Princeton University Press, 1947.
59. LUCE, R. DUNCAN, and HOWARD RAIFFA, *Games and Decisions: Introduction and Critical Survey*. New York: John Wiley and Sons, 1957.
60. GINSBERG, A. S., and F. L. OFFENSEND, "An Application of Decision Theory to a Medical-Diagnosis Treatment Problem," *IEEE Transactions*, Vol. SSC-4, No. 3 (Sept. 1968), 355–362.
61. SPRINGER, C. H., R. E. HERLIHY, R. T. MALL, and R. I. BEGGS, "Statistical Inference," in *Mathematics for Management*, Vol. III. Homewood, IL: Richard D. Irwin, Inc., 1966.
62. NORTH, D. WARNER, "A Tutorial Introduction to Decision Theory, System Science and Cybernetics," *IEEE Transactions*, Vol. SSC-4, No. 3 (Sept. 1968), 203.
63. SWALM, RALPH O., "Utility Theory—Insight into Risk Taking," *Harvard Business Review* (Dec. 1966).
64. WILLIAMS, I. D., *The Compleat Strategyst*. New York: McGraw-Hill, 1966.

65. KASSOUF, SHEER, *Normative Decision Making*. Englewood Cliffs, NJ: Prentice-Hall, Inc., 1970.
66. BOULDING, KENNETH E., "The Learning and Reality-Testing Process in the International System," *J. of International Affairs*, Vol. XXI (1967), 1-15.
67. ARROW, KENNETH, *Social Choice and Individual Values*. New York: John Wiley and Sons, 1951.
68. ALBERTS, DAVID S., and RICHARD C. CLELLAND, "Individual and Societal Utility," *Environmental Systems*, Vol. 1, No. 1 (March 1971), 19–36.
69. DOHERTY, NEVILLE, "Increasing Practice Efficiency by Linear Programming," *J. of American Dental Association*, Vol. 85 (Nov. 1972), 1099–1104.
70. CHURCHMAN, C. West, *The Systems Approach*. New York: Dell Publishing Co., Inc., 1968.
71. FELTON, LEWIS P., and MOSHE F. RUBINSTEIN, "Optimum Structural Design," *Society of Automotive Engineers*, #680752 (Oct. 1968).
72. BELLMAN, RICHARD, *Dynamic Programming*. Princeton, NJ: Princeton University Press, 1957.
73. SASIENI, MAURICE, ARTHUR YASPAN, and LAWRENCE FRIEDMAN, *Operations Research—Methods and Problems*. New York: John Wiley and Sons, 1959, p. 281.
74. WIENER, NORBERT, *Cybernetics or Control and Communication in the Animal and the Machine*. New York: John Wiley and Sons, 1948.
75. MAYR, OTTO, *The Origins of Feedback Control*. Cambridge, MA: M.I.T. Press, 1970.
76. CANNON, W. B., *The Wisdom of the Body*. New York: W. W. Norton, 1936.
77. LANGLEY, L. L., *Homeostasis*. New York: Van Nostrand, 1965.
78. BURTON, A. C., *Physiology and Biophysics of the Circulation*. New York: Yearbook Medical Publishers, 1965.
79. BEKEY, GEORGE, and M. B. WOLF, "Control Theory in Biological Systems," in *Biomedical Engineering*, eds.: J. H. Brown, J. E. Jacobs, and L. Stark. Philadelphia: F. A. Davis, 1971.
80. HURTY, WALTER C., and MOSHE F. RUBINSTEIN, *Dynamics of Structures*. Englewood Cliffs, NJ: Prentice-Hall, Inc., 1964, p. 352.
81. BODE, H. W., *Network Analysis and Feedback Amplifier Design*. New York: Van Nostrand, 1945.
82. HOROWITZ, I., *Synthesis of Feedback Systems*. New York: Academic Press, 1963.
83. Eveleigh, Virgil W., *Adaptive Control and Optimization Techniques*. New York: McGraw-Hill, 1967.
84. PUGH, ALEXANDER L., III, *DYNAMO II User's Manual*. Cambridge, MA: M.I.T. Press, 1971.
85. FORRESTER, J. W., *Urban Dynamics*. Cambridge, MA: M.I.T. Press, 1969.
86. ———, *World Dynamics*. Cambridge, MA: Wright-Allen Press, Inc., 1971.
87. KADANOFF, LEO P., "From Simulation Model to Public Policy," *American Scientist*, Vol. 60 (Jan.–Feb. 1972), 74–79.
88. HENIZE, JOHN, "System Dynamics Memorandum," D-1598, course material for "Principles of Dynamic Systems I," M.I.T., 1971.
89. LANGER, WILLIAM, "The Black Death," *Scientific American*, Vol. 210, No. 2 (Feb. 1964), 114–121.

90. Tuttle, W. W., and Byron A. Schottelius, *Textbook of Physiology* (15th ed.). St. Louis, MO: C. V. Mosby Co., 1965.
91. Milhorn, H. T., *Application of Control Theory to Physiological Systems*. Philadelphia: W. B. Saunders Co., 1966.
92. Bessman, S. P., and R. D. Schultz, "Sugar Electrode Sensor for the 'Artificial Pancreas'," *Horm. Metab. Res.*, 4 (1972), 413–417.
93. ———, "Prototype Glucose-Oxygen Sensor for the Artificial Pancreas," *Trans. Am. Soc. Artificial Internal Organs*, Vol. XIX (1973), 361.
94. Goldberg, S., *Introduction to Difference Equations*. New York: John Wiley and Sons, 1958.
95. Perry, Ralph B., "The Definition of Value," *J. of Philosophy*, Vol. 11 (1914), 141–162.
96. Vickers, Sir Geoffrey, *Values Systems and Social Progress*. New York: Basic Books, 1968.
97. Berson, James H., "The Human Values of City Planning," *J. Environ. Sys.*, Vol. 1, No. 3 (Sept. 1971), 283–287.
98. Von Bertalanffy, Ludwig, *General System Theory*. New York: George Braziller, 1968.
99. *Records, Computers and the Rights of Citizens*. Report of the Secretary's Advisory Committee on Automated Personal Data Systems, U. S. Department of Health, Education, and Welfare, July 1973. DHEW Publication No. (OS) 73–94.
100. Raiffa, Howard, "Preferences for Multi-Attributed Alternatives." Memorandum RM 5868-DOT/RC. Santa Monica, CA: The RAND Corporation, April 1969.
101. Cetron, Marvin J., "A Method for Integrating Goals and Technological Forecasting into Planning," *Technological Forecasting*, Vol. 2, No. 1 (1970).
102. Lecht, Leonard A., "Manpower Requirements for National Objectives in the 1970s," Center for Priority Analysis, National Planning Association (1968).
103. Lindsay, R. B., "Entropy Consumption and Values in Physical Science," *American Scientist*, Vol. 47, No. 3 (Sept. 1959), 376–385.
104. Huxley, Aldous, *Brave New World Revisited*. New York: Harper & Row, 1959.
105. Buber, Martin, *I and Thou* (1st ed.), translated from the German by Ronald Gregor Smith. Edinburgh: T. & T. Clark, 1950; (2nd ed.; with P. S. by Buber added), also translated by Ronald Gregor Smith. New York: Scribner, 1958; Edinburgh: T. & T. Clark, 1959.
106. Skinner, B. F., *Beyond Freedom and Dignity*. New York: Bantam Books, 1971.
107. Ice, Jackson Lee, *Schweitzer*. Philadelphia: Westminster Press, 1971.
108. Rogers, Carl R., *Freedom to Learn*. Columbus, OH: Charles E. Merrill Publishing Co., 1969, p. 254.
109. Walker, Marshall, *The Nature of Scientific Thought*. Englewood Cliffs, NJ: Prentice-Hall, Inc., 1963.
110. Frankl, Viktor E., *The Will to Meaning*. New York: World Publishing Co., 1969.
111. Dalkey, N. C., D. L. Rourke, R. Lewis, and D. Snyder, *Studies in the Quality of Life*. Boston: D. C. Heath, 1972.
112. Rescher, Nicholas, *Introduction to Value Theory*. Englewood Cliffs, NJ: Prentice-Hall, Inc., 1969.

113. Reich, Charles A., *The Greening of America*. New York: Random House, 1970.
114. Kemeny, John G., and J. Laurie Snell, *Mathematical Models in the Social Sciences*. Waltham, MA: Blaisdell Publishing Co., 1962.
115. Helmer, Olaf, *Social Technology*. New York: Basic Books, 1966.
116. Dalkey, Norman C., and Olaf Helmer, "An Experimental Application of the Delphi Method to the Use of Experts," *Management Science*, Vol. 9 (1963), 458.
117. Dalkey, Norman C., "An Experimental Study of Group Opinion," *Futures* (Sept. 1969), 408–426.
118. Gordon, T. J., and Olaf Helmer, "Report on a Long-Range Forecasting Study," The RAND Corporation, P-2982 (Sept. 1964).
119. Campbell, R., "A Methodological Study of the Utilization of Experts in Business Forecasting." Ph.D. thesis, University of California, Los Angeles, 1966.
120. Martino, J., "An Experiment with the Delphi Procedures for Long-Range Forecasting," *IEEE Transactions on Engineering Management*, Vol. EM-15 (Sept. 1968), 138–144.
121. North, Harper Q., "Delphi Forecasting in Electronics." Presented in a panel on *Input of Future Technology*, Wincon (Feb. 15, 1968).
122. Bender, A. Douglas, Alvin E. Strack, George W. Ebright, and George von Haunalter, "Delphic Study Examines Developments in Medicine," *Futures* (June 1969).
123. Dalkey, Norman C., B. Brown, and S. Cochran, "The Delphi Method, III: Use of Self Ratings to Improve Group Estimates," The RAND Corporation, RM-5883 (Aug. 1969).
124. Dalkey, Norman C., and Daniel L. Rourke, "Assessment of Delphi: Procedures with Group Value Judgments." Santa Monica, CA: The RAND Corporation, 1971, pp. 8–24.
125. Johnson, Stephen C., "Hierarchical Clustering Schemes," *Psychometrika*, Vol. 32, No. 3 (Sept. 1967), 241–254.
126. Batra, Pradeep, and Moshe F. Rubinstein, "Systems Approach to the Design of an Interdisciplinary Course," *J. of Engr. Education*, Vol. 63, No. 4 (Jan. 1973), 283–285.
127. Pearson, Karl, *Tables of the Incomplete Beta Functions*. London: Biometrika Office, University Press, 1948.
128. Manheim, Marvin L., "Reaching Decisions about Technological Projects with Social Consequences: A Normative Model," in *Analytic Formulation of Engineering Problems: Case Studies in Systems Analysis*, Vol. III: Evaluation, eds. Richard de Neufville and David H. Marks. Cambridge, MA: M.I.T. Dept. of Civil Engineering (Sept. 1972).
129. Sommer, Robert, *Personal Space*. Englewood Cliffs, NJ: Prentice-Hall, Inc., 1969, pp. 171–172.
130. Drucker, Peter F., *The Age of Discontinuity*. New York: Harper & Row, 1969.
131. Bagdikian, Ben H., *The Information Machines*. New York: Harper & Row, 1971, p. 9.
132. Hamming, R. W., "Error Detecting and Error Correcting Codes," *Bell System Tech. J.*, Vol. 29 (1950).

133. Dalkey, Norman C., "Delphi Research: Experiments and Prospects," *Proceedings International Seminar on Trends in Mathematical Modeling,* Venice (Dec. 13–18, 1971).

134. Pratt, John W., Howard Raiffa, and Robert Schlaifer, *Introduction to Statistical Decision Theory.* New York: McGraw-Hill, 1965.

135. Raiffa, Howard, *Decision Analysis.* Reading, MA: Addison-Wesley, 1968.

136. Dantzig, G. B., "The Simplex Method," RAND Corporation Report P-891 (1956).

137. English, J. Morley, "Some Investment Concepts in Engineering Systems with Particular Emphasis on Long-Range Investment," in *Economics of Engineering and Social Systems,* ed. J. Morley English. New York: John Wiley and Sons, 1972, pp. 74–77.

138. Duda, R. O., "Elements of Pattern Recognition," in *Adapters Learning and Pattern Recognition Systems,* eds. J. M. Mendel and K. S. Fu. New York: Academic Press, 1970, pp. 3–33.

139. Mosteller, Frederick, Robert E. K. Rourke, and George B. Thomas, Jr., *Probability and Statistics.* Reading, MA: Addison-Wesley, 1961.

Answers to Selected Exercises

Chapter 1

1-19. 15 days
1-21. 255 steps
1-22. 2, 2, 9

Chapter 2

2-9. Black
2-10. 5
2-11. Not valid
2-12. (a) Erroneous bit is bit #5
2-17. Valid
2-19. 22
2-22. 20 paths

Chapter 3

3-13. 6 bits
3-16. $111110_{\boxed{2}} = 62_{\boxed{10}}$; $10101_{\boxed{2}} = 21_{\boxed{10}}$; $100_{\boxed{2}} = 4_{\boxed{10}}$; $11_{\boxed{2}} = 3_{\boxed{10}}$
3-17. (a) 85.625; (b) 88; (c) 4.625; (d) 11; (e) 64.125
3-18. (a) 101111; (b) 1000000; (c) 1110; (d) 1111101; (e) 111001101010
3-19. (a) 57; (b) 100; (c) 16; (d) 175; (e) 7152
3-20. $7325_{\boxed{8}} = 3797_{\boxed{10}}$; $11060_{\boxed{8}} = 4656_{\boxed{10}}$; $521_{\boxed{8}} = 337_{\boxed{10}}$; $1166_{\boxed{8}} = 630_{\boxed{10}}$
3-21. $100011_{\boxed{2}} = 35_{\boxed{10}}$; $11001_{\boxed{2}} = 25_{\boxed{10}}$
3-22. $156_{\boxed{8}} = 110_{\boxed{10}}$; $372_{\boxed{8}} = 250_{\boxed{10}}$
3-23. $11101_{\boxed{2}} = 29_{\boxed{10}}$

Chapter 4

4-1. 0.4116
4-2. (a) \$0.45; (b) −\$0.15
4-3. \$0.78
4-5. (i) 0.86; (ii) 0.047; (iii) 0.0025
4-6. (f) 0.0456

4-7. (a) 0.4; (b) 0.9
4-8. 0.6226
4-9. (a) 0.25; (b) Democrat
4-10. 0.429
4-13. (a) 2; (b) $\text{Log}_2 10$
4-14. (a) 14/8
4-15. 8/3
4-18. (a) 0.1, 0.2; (b) 0.02; (c) 0.98
4-20. (a)0.63; (b) 0.37
4-26. 4/9
4-30. (a) 1/4; (b) 1/13; (c) 1/52; (e) 4/13; (f) Both 1/13; (g) 0; (h) 2/13; (i) 8/52
4-31. $0.33
4-32. 1/10
4-33. (a) 8/90; (b) 1/9
4-34. $0.17
4-35. 2/3
4-36. 19/40
4-37. (a) 1/54, 145; (b) 13/54, 145
4-38. (a) 16/75; (b) 24/91
4-39. (a) 13.5¢; (b) $f = 0.8$

Chapter 5
5-13. 500 units
5-14. C is most profitable
5-15. 15 hours and 20 minutes
5-17. (a) 15,000 years; (b) 0.014% per year
5-18. 13.86 years
5-19. Annex now
5-23. 2 cities, A and B

Chapter 6
6-1. (a) 1,150,000; (b) 1,150
6-2. (b) 9.0
6-4. (a) 4,600,000; (b) 1,900,000
6-8. 0.4116
6-9. (a) $n = 7$, $x = 4$ or 3; (b) 35
6-10. 0.868
6-11. 0.704
6-12. 0.0179
6-13. 0.2
6-14. (a) 0.188; (b) 73/80
6-15. (a) 0.0228; (b) ≈ 1; (c) 0.9772; (d) 0.9772
6-16. (a) Both are 2 standard deviations away; (b) 0.954; (c) (i) 0.682;
(c) (ii) Between 5 and 15
6-17. (a) (i) 0.3821, (ii) 0.8185; (b) (i) 0.3821, (ii) 0.8185
6-18. 0.997, 0.954, 0.682
6-19. 0.9544
6-20. (a) (i) 0.023, (ii) 0.7881; (b) $(0.023)^3 = 0.0000122$; (c) 0.9983

6-21. 9.85%
6-22. $X_{upper} = 119.6$, $X_{lower} = 80.4$
6-23. (b) 0.0635
6-24. (a) 7.512 ounces; (b) Aleph
6-25. $-2/38$
6-26. 8
6-27. 255.5
6-28. 0.6915
6-29. 0.75

Chapter 7
7-1. (a) S_4; (b) S_1; (c) S_2
7-2. (a) Watch TV; (b) Skim notes; (c) Study hard
7-4. Company should insure
7-7. Both games have $EV = \$42$. No preference for one game over another.
7-8. (a) Doctor #1 will treat; (b) Doctor #2 will wait
7-11. (a) Strategy for A is $[3/11, A_1; 8/11, A_3]$; (b) 57/11
7-12. Always use S_2
7-13. (a) Doctor will not give the pill
7-17. (b) Strategy for A is $[5/8, H; 3/8, T]$; (c) A will lose 12.5 ¢/game

Chapter 8
8-1. (a, b) Nonlinear; (c, d, e, f) Linear
8-2. (a) Convex; (b) Convex; (c) Nonconvex; (d) Nonfeasible; (e) Convex; (f) Nonconvex; (g) Convex; (h) Nonfeasible
8-4. (a) 4; (b) 24; (c) 12
8-5. (a) 4; (b, c) Unbounded
8-6. (d) Maximum = 340
8-8. (c) Maximum = 260
8-10. Minimum cost = \$10,100
8-13. (a) $F_{max} = 2$; (b) $F_{max} = 8$; (c) $F_{max} = 80$; (d) $F_{max} = 98$; (e) $F_{max} = 100$
8-16. $L = W = 15$
8-18. 25
8-19. (a) 17 units; (b) 70
8-20. (a) 14

Chapter 9
9-12. (a) $Y = 833.75$; (b) $R = 5.96$; (c) $Y = 835$

9-15.

Time (seconds)	*Height of Water (inches)*
1	1.00
2	2.00
3	3.00
.	
.	
.	
29	19.16
30	19.45

9-17. (b)

Year	*New Students*	*No. of Engrs.*	*No. of Engrs. Needed*	*Surplus*
0	4	5	10	−5
1	5	5	13	−8
2	8	9	16	−7
3	7	14	19	−5
.				
.				
.				
29	0	100	97	+3
30	0	100	100	0

9-19. (c)

Season, n	R_{n-1}	F_{n-1}	*RE*	*FS*	$R_{n-1} + RB$	$F_{n-1} + FA$	R_n	F_n
1	800	800	640	320	1600	1120	960	800
2	960	800	768	267	1920	1120	1152	853
3	1152	853	985	252	2304	1192	1319	940
.								
.								
.								
19	754	869	654	400	1508	1218	854	818
20	854	818	697	314	1708	1148	1011	834

9-21. $2983.21

INDEX

A

Absolute space, 26-27
attributes, existence, and mechanics of, 27
Accountability, personal and social, 494
Adaptive control, 453, 468
decision in, 453
goal change and, 454
identification in, 453
learning machines and, 454
modification in, 453
Adaptive learning, 454
Address, in computer, 95
Aggregates, of random variables, 264
Aggregation, criteria for, 506
Delphi method procedure for, 506-7
of measurements, 251
Aleph, origin of, 47
Allocation, optimal, by Delphi method, 509
Alpha, origin of, 47
Alphabet, Hebrew, origin of, 49
Ambiguity, in language, 44
Amplitude, 232, 241
dynamic system response, 437
Analogies, in problem solving, 20
Analysis, of system, 224
AND circuit, 65
Approaches to problem solving, 3
behaviorist, 2, 30
Arabic numerals, 50
Arguments, valid, 61, 83
check of, 61
Arithmetic unit, in computer, 99, 129
Arrow barrier, 501 (*See also* Arrow paradox)
Arrow paradox, 350
Artificial intelligence, 5
use of computer in, 128
Assembly language, 82
Associative laws:
Boolean algebra, 66, 85
sets, 73, 85
symbolic logic, 64, 85
Atom, models of, 217
Atomic number, 202
Axiology, 476, 514
value defined, 476

B

Babel, tower of, 37
Baroreceptors, 432
Base, in number system, 109
Basic solution, 365
Basic variable, 364, 366
Bassa language, 43
Bayes' equation (*See* Bayes' theorem)
Bayes' theorem, 149
applications, 151, 156
general form, 150, 181
relevance of information and, 166
tree structure and, 158, 161
validation of Newton's model and, 215
Behavior, ethical, model of, 486, 489
Benefit function, 509
Bernoulli trials, 254
Biconditional, in logic, 60
Binary decision, 389
Binary number system, 110, 112
division in, 134
multiplication in, 133

Binomial distribution, 254, 300
  normal approximation of, 272
  normal distribution model for, 271
  number of routes computed by, 306
Binomial model, 229, 240
Bit:
  (binary digit), 75, 106
  of information, 171, 182
  time-delay box for, 118
Black death, 462
Block diagram(s):
  for drinking straw, 424
  of dynamic system, 415
  for epidemic, 459
  equivalent, 470
  of Heron's floating syphon, 422
  for feedback, 445
  general, 429
  for glucose control, 465
  of governor, 426
  of half-adder, 415
  of house heating, 425
  of Heron's inexhaustible goblet, 418
  of pricing model, 428
  in problem solving, 429
  symbols in 415, 416
  unit delay in, 417
  of Ctesibius water-clock, 423
Blood, 430, 431
Blood glucose, 463
Blood pressure, 430-433
Blood sugar, 463-64 (*See also* Glucose)
Bohr model, 217
Boolean algebra, 66, 68, 85
Brain, model of, 8, 220
  and problem solving, 221
Break-even point, 242

C

Calorie, defined, 463
Carbon dating, 237
Cartesian coordinates, 23
Central limit theorem, 248, 269, 302
  applications, 269
Channel, communication, 74
χ^2 distribution, 285
Chunking, 8, 9
Circle, equation of, 23
Circuit,
  AND, 65
  NOT, 65
  OR, 65
Classification, sets and, 68
Classification model, 228, 240
Code:
  for aggregated events, 178
  extension, 178
  instantaneous, 175
  length of and entropy, 175
  Morse, 177
  prefix, 175
Collectivity, Buber's definition of, 487
Communication:
  channel, 74
  community and, 36
  language and, 36, 40
  message, 75
  noise in, 74
  redundancy in, 75
Communication systems, 74
Community, Buber's definition, 487
  communication and, 36
Commutative laws:
  Boolean algebra, 66, 85
  sets, 73, 85
  symbolic logic, 64, 85
Comparator, in dynamic system, 417
Compiler, 121, 130
Compiler language, 82
Complement, of set, 69
Complement laws:
  Boolean algebra, 67, 85
  sets, 73, 85
  symbolic logic, 64, 85
Complex number, 438
Computer:
  address in (location in), 95
  analog, 95
  arithmetic unit, 99, 129
  command counter, 99, 120, 129
  command register, 99, 120, 129
  components of, 99, 129
  computations in, 113
  digital, 95
  ENIAC, 93
  full adder (*See* Full-adder)
  half adder (*See* Half-adder)
  hardware, 91
  history, 92
  hybrid, 95
  input, 99, 108, 129
  instruction, 97-98
  language, 109
  machine word in, 98
  memory (storage), 99, 105-107, 129
  model of, 95
  output, 99, 108, 129
  parallel adder, 119, 130
  program, 94, 98, 129
  programming, 120, 121, 130
  register in, 95
  software, 91
  types, 120
  UNIVAC, 94
  uses of, 127-128, 131
Computer language, 81, 82
Computer program:
  address modification in, 103

to add N numbers, 100
counter in, 102
definition, 98, 129
loop in, 103
Concepts, and words, 41
Condition, necessary, 60
necessary and sufficient, 60
sufficient, 59
Conditional, in logic, 59
Confidence intervals, 280, 303
Confidence level, 281, 303
Confidence limits, 281, 303
Consensus:
criterion for, 501
deficiencies in procedures, 505
by Delphi method, 507
on educational goals, 518
equations of, 362
pareto optimal, 502
in problem solving, 10-13, 30
on ranking, 497, 502
of ratings, function for, 503
on values, 513
on what to do, 513
Control, 413-414, 466
adaptive (*See* Adaptive control)
blood pressure, 433
blood sugar (*See* Glucose control)
closed loop, 413, 434, 466
diabetes and, 466
feedback, 413, 434, 466
flush tank, 420
glucose (*See* Glucose control)
open loop, 413, 434, 466
optimal, 452, 468
in organism, 430, 466
system, 224
Controllability, 434, 467
example of, 435
Controlled system, sensitivity of output, 446
Controller, 430
Control variables, 372
Convex set, 373
Coordinates, cartesian, 23
inertial, 26
natural, 26
Cosine model (function), 232
Cost-benefit, values, 496
Critical damping, 441, 467
Critical region, in test of hypotheses, 289
Culture, language and, 42
role in problem solving, 1
Cybernetics, 414, 466

D

Damping, critical, 441, 467
Decision(s):
binary, 389
emotional, 311
rational, 311
sequential with random outcomes, 391
Decision maker, 326
types of, 318-320
Decision making:
under certainty, 316
under conflict, 337
group, 349
under risk, 317
under uncertainty, 317-320
utility of money and, 330
Decision model(s), 311, 350
classification, 315-316, 351
decision rule in, 315
doctor, 314
farmer, 312
jury, 313
main elements of, 315, 350
utility assignments and, 335
Decision policy, in dynamic programming, 387
Decision rule, in decision theory models, 315
in test of hypotheses, 287
Decision theory, 311
Decision variables, 372
Decomposition and complexity, 200
Definitions, circular, 41
Delphi method, 505-506
aggregation procedure in, 506-507
for allocation, optimal, 509
consensus by, 507, 512
course design by, 508
group communication in, 506
quality of life study by, 506
questionnaire, example of, 509, 510
uses of, 506
DeMorgan's laws:
Boolean algebra, 67, 85
sets, 73, 85
symbolic logic, 64, 85
Descartes, 22
coordinates and, 23
problem solving and, 22
Descriptive utility, 335, 336
Desirability, index of, 481
Diabetes mellitus, 463, 465
control system applied to, 466
Digital computer (*See* Computer, digital)
Distance function, in ranking, 499
Distribution(s), 258, 301
binomial, 254, 300
χ^2, 285
cumulative, 253
hypergeometric, 256
normal, 248, 256, 300
Poisson, 255
of proportion $\overline{p}$, 279, 302
student t, 284
Distributive laws:
Boolean algebra, 67, 85

Distributive laws, (*con't.*)
sets, 73, 85
symbolic logic, 64, 85
Dominance, 344
Doppler effect, 213
Doubling time, 236, 241
Doubt, will to, 17
Dual problem, in linear programming, 379, 380, 401
Dynamic programming, 386
decision policy in, 387
levels in, 393
random outcomes treated by, 393
recursive relationship of, 390
restrictions in, 400
route problem solution, 387-389
sequential decisions by, 393
state in, 387
working backwards and, 389
Dynamic system(s):
block diagrams of, 415, 429
comparator, 417
control (*See* Control)
energy flow in, 412
examples of, 411
glucose control and, 463
information flow in, 412
input-output relationship, 437
measurement of output in, 429
model, 409
natural frequency of, 440
neutrally stable, 450, 468
overdamped, 443, 467
response, 437, 443, 467
sensitivity of, 447, 448
simulation, 455, 457, 468
stability, 449, 450, 468
transfer function 416, 437, 467
underdamped, 443, 467
DYNAMO, 455, 458-460, 469
level equations in, 458
program for epidemic model, 460
rate equations in, 458
simulation program in, 459

E

Effector, 430
Einstein, 27-29
Element, of set, 68
Embarcadero freeway, 513
Emotional signs, in problem solving, 21
Empty set, 69, 84
Encoding, language, 40
End values, 478
supreme, 491-492
Energy, in dynamic system, 412
Engineering, models in, 222
ENIAC, 93
Entropy, 484
and code length, 175
and English, 177
and information, 171, 175, 182, 484-485
and life, 486
maximum, 484
and model of behavior, 486
and order, 484, 514
and thermodynamics, 484-485
Epidemic, black death, 462
block diagram for, 459
critical number in, 461
DYNAMO program for, 460
quarantine, 462
simulation of, 457
Equilibrium, 450, 468
Equilibrium solution, in games, 340
Equivalent forms, symbolic logic, 64
Error(s):
commission, 201, 285
mean square, 452, 468
omission, 201, 285
type I, 201, 286, 303
type II, 201, 286, 303
Estimate, of variance, 266
Estimation, population, parameters, 277, 300
Ethical behavior, model of, 486, 489
Event(s), aggregation of, 178
types of, 141-143, 180
Expected value, 171, 251, 264, 301, 309
of information, 171
operator, 264, 301
Exponential decay, carbon dating and, 237
model, 236, 241
Exponential growth, model, 234, 241

F

Fair game, 340
Feasible region, 362
Feasible solutions, 362, 356
Feedback:
advantages, 436, 451
block diagram for, 429, 445
disadvantages, 451
negative, 443, 450, 467
positive, 443, 451, 467
Feedback control, 413, 434, 466
Feedback system(s), characteristics, 444
transfer function, 444
Finite games, 338
Flexible problem solver, 19
Floating syphon, Heron's, 420-421
Flow chart, 104, 129
Flux, 224
FORTRAN, 121, 122, 126, 130, 131
compilation in, 122, 131
DO loop in, 123
object program, 122, 131
precompilation in, 122, 130
program, to add *N* numbers, 121, 125
source program, 122, 131

subprogram, 126, 131
subroutine in, 126, 131
Frequency, 232, 241
natural, 440
Full adder, 116, 130
example of use, 118
truth table for, 117

G

Game(s):
classification, 338, 351
equilibrium solution in, 340
fair, 340
finite, 338
infinite, 338
maximin criterion in, 339
model of, 339
nonzero-sum, 339, 347
n persons, 338
prisoner dilemma, 347
pure strategy in, 341
saddle point in, 340
solution, 340, 345
theory of, 337, 341
two-person, 338-339, 347
value of, 340
zero-sum, 338
Game Theory, 337, 341
General Problem Solver (GPS), theory, 4
Generalizing process, in problem solving, 19
Gimatria, 51
examples of, 53
Jehovah and, 52
theology, Christian, and, 53
Glucose, in blood:
content and function, 463
input, 464
removal, 464
Glucose control:
block diagram, 465
breakdown, 465
insulin, 465
kidneys, 464
liver, 465
man-made procedure, 466
novel application, 466
pancreas, 465
sensors in, 464
Goal-seeking system, 434
Goal state, example of, 32
Goals, study of, in education, 493
national, and value judgments, 482
Goblet, Heron's, 418
Governor, Watt's, 425-426
Grammar, definition, 38, 82
semantic theory and, 38, 82
Graphical method, linear programming by, 368
Graphical solution, of 2 × *n* games, 345
Gravitation, Newton's model, 194
Gravitons, 195
Group decision making, 349

H

Half-adder, 114, 130
AND box in, 115
block diagram of, 415
INVERTER box in, 116
OR box in, 115
truth table for, 116
Half-life, 236, 241
Hamming code, 76
Hanoi, tower problem, 34
Heart (*See* Blood, circulation)
Heron, floating syphon of, 420-421
inexhaustible goblet of, 418
History, models of, 210
analog, 211
cyclic, 211
economic, 211
evolutionary, 212
geographic, 209
psychoanalytic, 211
religious, 212
Homeostasis, 430, 466
Human affairs, models, 227
Hybrid computer, 95
Hyperbolic paraboloid, 24
Hypergeometric distribution, 256
Hypothesis:
alternate, 286
null, 286
testing, 135, 285, 287, 289, 290, 303

I

Idempotent laws:
Boolean algebra, 67, 85
sets, 73, 85
symbolic logic, 64, 85
Identification, of system, 224
Identity laws:
Boolean algebra, 67, 85
sets, 73, 85
symbolic logic, 64, 85
"Inbetweenist," in decision making, 319
Inference, 252
Infinite games, 338
Information:
average, 171
bit of, 171, 182
credibility, 136
in dynamic system, 412
and entropy, 484-485
expected value of, 171
failure to use, 8, 30
measurement of, 166, 182
mutual, 166
premise in theory of, 182
relevance, 136, 149, 166, 181
and second law of thermodynamics, 484-485
unit of, 182

Information processing, in problem solving, 3, 30
  model of the brain, 8
Initial state, example of, 32
Input, 99, 129
Input-output:
  in dynamic system, 437
  relationship, 372
Input-output model, 409
Insulin, in glucose control, 465
Insurance, 329
Intersection, of sets, 69

J

Judgment, in problem solving, 14
Judgment, of value, 480, 482, 514

K

Kepler, laws, 214
  model, 214
Kidneys, in glucose control, 464
Knowledge and values, 483
Kraft inequality, 175

L

Language, 36-41
  ambiguities and, 44
  Bassa, 43
  computer, 81, 109
  culture and, 42
  English, and entropy, 177
  metalanguage and, 39
  problem solving and, 37
  semantic theory of, 38, 82
  of sets, 68
  simulation, DYNAMO (*See* Dynamo)
  and thought, 28
  of Wintu Indians, 42
  and World, 43
  written, evolution of, 44
Large numbers, 273
Laws:
  Boolean algebra, 67, 85
  Kepler's, 214
  of large numbers, 273
  Newton's, 42, 214, 215
  of probability, 140, 148, 180
  of sets, 73, 85
  of symbolic logic, 64, 85
  thermodynamics, 484
Learning machines, 454
Letters:
  Hebrew, meaning of, 49, 52
  origin of, 47
  Phoenician, 49
  Semitic, 49
Level(s):
  of control, 413
  in dynamic programming, 393
  of dynamic system, 410
  equations of, in DYNAMO, 458
Line, equation of, 23
Linear functions, 357
Linear function(s), geometric interpretation, 359
  models of, 358
Linear operator, 264, 301
Linear programming:
  in dental practice, 366
  basic building blocks, 372
  dual problem in, 379, 401
  exposure to, 361
  general form, 400
  generalization of method, 371
  graphical solution, 368
  performance criteria, 372
  premises in, 366, 378
  primal problem in, 379, 401
  transformation to game, 368
Listening, in problem solving, 21
Liver, in glucose control, 465
Location, in computer, 95
Logic, algebra of, 64
Logic, symbolic, 53, 64, 83, 85
Lottery, in utility theory, 324

M

Machine language, 82
Machine word, 98
Magnification, response, 439, 467
Majority rule, 349-350
  group choice by, 349-350
Maximin criterion, 318, 339
Maximax criterion, 319
Maximum expected utility criterion, 318
Maxwell Demon, 485
Mean, 248, 300
  of aggregate, 266
  of proportion $\overline{p}$, 278, 302
  of population, 251, 300
  of sample, 252, 300
Mean square error, 452, 468
Means values, 478
Measurement(s), aggregation of, 251
Median, 507
Memory, 8 (*See also* Computer, memory)
Mercury, orbit of, 216
Message, in communication, 75
Metabolism, 463
Metalanguage, 39
Meta-metalanguage, 39
Metaphors, in problem solving, 20
Metaphysics, 28
Metrization of preferences, 503
Michelangelo, 51

Minimax regret, 321
Mixed strategies, in games, 340
  alternate interpretation, 343
  generating, 341
Model(s):
  and abstraction levels, 205
  analog, 207
  of atom, 217
  of behavior, 474, 486
  behaviorist, 192
  binomial, 229, 240
  Bohr's, 217
  of the brain, 8, 220, 221
  Buber's, Thou/It, 487
  causal, 208
  classification, 203, 208, 239
  for classification, 228, 240
  for competition, 334
  of computer, 95
  content, 203
  correlative, 208
  cosine function, 232
  decision (*See* Decision models)
  definition, 15, 192, 196, 238
  dynamic system, 409
  economic, 472
  in engineering, 222
  of epidemic, 460
  ethical behavior, 489
  exponential decay, 236, 241
  exponential growth, 234, 241
  form, 203, 400
  game theory, 339
  of history, 194 (*See also* History, models of)
  homomorphic, 209
  human affairs, 194, 227
  of hydrogen atom, 217
  information processing, 192
  input-output, 409
  isomorphic, 209
  Kant's, 195
  Kepler's, 214
  Limits to Growth, 197
  linear function, 358
  linear programming, 400
  of man, 196
  mathematical, 205, 228
  of matter, 219
  of national income, 472
  nature of, 196
  normal distribution, 256, 300
  optimization, 356
  parameters in, 205
  physical science, 227
  for prediction, 195
  pricing, 428
  of prisoner's dilemma, 347
  probabilistic, 246, 293
  probability distribution, 253
  of problem solving, 4
  purpose of, 193, 238, 489
  qualitative, 206
  quantitative, 206
  for remembering, 222
  resource allocation, 481 (*See also* Linear programming)
  in science, 477
  simulation, 209
  sine function, 232
  supply and demand, 425
  of tea kettle design, 198
  teleological, 489
  for understanding, 193, 194
  of the universe, 212, 213
  of urban dynamics, 455
  validation, 201, 215
  value judgment, 481
  verbal, 204, 205
  working, 223
  world population, 457
Modeling:
  and aggregation, 199, 200
  errors in, 201
  process of, 197, 239
Monte Carlo method, 293, 295
Morse code, 177

N

Natural frequency, 440
Neptune, discovery of, 216
Newton, 25
  absolute space and, 26
  inverse-square model, 194
  law of gravitation, 215
  laws of motion, 214
  model, 215
    validation, 215
  Second Law of, 42
Noise, in communication, 74
Nonbasic variable, 364, 366
Nonconvex set, 373
Nonlinear programming, 382, 386
Non-zero sum game(s), examples, 347-348
Normal distribution, 248, 256, 300
  approximation, 272
  random sample from, 298
  simulation, 297
  standard, 258, 301
Normative utility, 335
NOT circuit, 65
Null hypothesis, 286
Null set, 69, 84
Number(s):
  Arabic, 50
  atomic, 202
  complex, 438
  critical, in epidemic, 461
  Hebrew letters and, 52

Number(s), (*con't.*)
hieroglyphic, 51
Hindu, 50
origin of, 50
Sumerian, 51
Number system:
base of, 109
binary, 110
binary coded decimal, 112
decimal, 109, 111
general, 110
octal, 110
radix, 109

O

Objective, example of, 32
Objective function, 361
in linear programming, 372
restrictions in dynamic programming, 400
sensitivity of, 371, 380
Object language, 39
Octal number system, 110
multiplication in, 133
Ombudsman, 520
Operator, linear, 264, 301
Opportunity, 321
Optimal control, 452, 468
Optimality, principle of, 386
Optimal policy, 386
Optimism, index of, 319
Optimist, in decision making, 319
Optimization models, 356
Optimum solution, 356-357, 363
OR circuit, 65
Ordering matrices, 498
Organism, control in, 430, 466
Oscillation, stable, 443, 468
Output, 99, 129
Output, in closed loop control, 447
Overtones, semantic, 44

P

Pacemaker, in heart, 434
Parallel adder, 119, 130
Parameter(s):
of dispersion, 251, 300
estimates of, 252, 300
in a model, 205
of population, 252, 277, 300
of straightness ratio, 290
of system, 225
Pancreas, in glucose control, 465
Pareto optimal consensus, 502
Pascal's triangle, 231
Pattern recognition, example, 290
Payoff matrix, 313
dominant strategies in, 344
in game theory, 339
saddle point in, 340
Performance, index of, 452-453, 468
criteria in linear programming, 372
Perihelion, 216
Period, in sine and cosine model, 232, 241
Pessimist, in decision making, 318
Phase, 234, 241
dynamic system response, 437, 467
Phase-advancers, 450
Phoenician letters, 49
Photon, 218
Pivoting, in simplex algorithm, 378, 402
Planck's constant, 217
Planets, distances from sun, 216
Poisson distribution, 255
Policy, optimal, 386,
Population, defined, 250, 300
Potential, 224
Pragmatics, 38
Preferences, metrization of, 503
Premises, in linear programming, 366, 378
major, 61
minor, 61
in problem solving, 18
Prescriptive utility, 335
Present value, 243
Pricing model, 428
Primal problem, in linear programming, 379, 380-381, 401
Principle of optimality, 386, 402
Printing press, Gutenberg's, 519
Prisoner dilemma, game model of, 347
Probability:
of A AND B, 144
of A OR B, 143
a posteriori, 146, 149, 181
a priori, 146, 149, 162, 181
conditional, 145
and credibility, 164, 167, 181
definition, 137
density function, 253, 254, 301
and doubt, 135
frequency function, 253
and information measurement, 169, 182
laws of, 140, 148, 180
objective, 140, 180
personal, 140, 180
and reliability, 188
subjective, 140, 180
of symbols, in English, 178
Probability theory, 312
Problem solver(s):
Descartes as, 22
Einstein as, 27
flexible, 19
Newton as, 25
rigid, 19

Problem solving:
analogies in, 20
analysis, 7, 30, 224
attitudes in, 7
block diagram for, 429
change in representation, 15
chunking in, 8-9
constraints in, 10-13, 30
in course design, 508
role of culture in, 1
Descartes' process for, 22
difficulties in, 8, 10, 30
emotional signs in, 21
generalizing process in, 19
goal state description, 32, 478
guides to, 7, 14-21, 30
heuristic, computer and, 128
initial state, 32
and language, 37
listening in, 21
metaphors in, 20
and, model of brain, 221
models of, 4
objective, 32
premises in, 18
present state description, 476
process description, 478
schools of thought in, 2, 3, 30
"seven, plus or minus two," 8, 30
specializing process in, 19
and stable substructures, 20
stages in, 476
and stimulus-response, 2, 30
synthesis, 6, 30, 224
systems approach to, 476
talking in, 21
and values, 1, 476
working backwards, 19
Processing unit, in brain, 8-9
Program, computer (*See* Computer, program)
Programming, dynamic (*See* Dynamic programming)
Programming, linear (*See* Linear programming)
Programming, nonlinear (*See* Nonlinear programming)
Programming methods, 357
Proposition, 54
Pulmonary circulation, 431
Pure strategy, in game theory, 341

Q

Quality of Life study, 492, 506
Quanta, 218
Quantum mechanics model, of matter, 219
Quarantine, 462
Quartile, 507

R

Radix, in number system, 109
Random numbers, table of, 299
Random outcomes, sequential decisions with, 391
Random sample, 277
Random variables, 248, 264
Random walk, 273, 294
Randomness, order and, 484
Ranking(s), consensus on, 502
distance function in, 499
of values, 497
Ranking columns, 498
Ratings, consensus, 503, 507
Rational behavior, 138
Rational opinion, 138
Read-write head, in computer, 106
Reasoning, demonstrative, 164, 182
plausible, 165, 182
Red shift, 213
Redundancy, 75
Register, in computer, 95
Regret matrix, 321
Regretist, in decision making, 320
Relevance, measure of, 149, 181
Reliability, 188
Representation, in problem solving, 15
Response, of dynamic system (*See* Dynamic system, response)
magnification, 439, 467
static, 440
Reverence for Life, 491
Rigid problem solver, 19
Risk, attitudes toward, 319, 326, 330

S

Saddle point, in games, 340
Sample:
definition, 250, 300
mean of, 252, 300
random, 277, 298
variance of, 252, 300
Sample point, 141, 180
Sample space, 141, 180, 250
Sample statistics, 252
Science, models of physical, 227
Scientific method, 202
Self insurance, 329
Semantic overtones, 44
Semantic theory, grammar and, 38, 82
Semantics, 38
levels of, 40, 82
Semiotics, 38, 82
Semitic letters, 49
Sensing float, 418
Sensor, 430
pressure, 432

Sentence, structure of, grammatical, 39
Septuagint, 505
Sequential decisions, normal distribution, 397
  random outcomes, 391
  schematic description of, 393
Set(s):
  algebra of, 73
  analogy to statements, 70
  and classification, 68
  complement of, 69
  convex, 373
  definition, 68, 84
  elements of, 68
  empty, 69, 84
  intersection of, 69
  language of, 68
  nonconvex, 373
  null, 69, 84
  operations on, 68-69
  symbols, 69, 84
  union of, 69
  universal, 69, 84
"Seven plus or minus two," 8, 30
Significance level, in test of hypotheses, 289
Signs, emotional, 21
Simplex algorithm, 363, 401
  formalizing, 374
  idealized version, 369
  indicators in, 374, 401
  and linear algebra, 377
  pivoting in elimination process, 378
  pivots, 375, 402
  power of, 374
  practical version of, 369
  solution by, 369, 401
  tableaus, 374-375
Simulation:
  of dynamic system, 455, 457, 468
  of epidemic, 457
  normal distribution, 297
  of probabilistic models, 293
  of random walk, 294
Simulation language, (*See* DYNAMO)
Simulation models, 209
Simultaneity, Einstein and, 28
Sine model (function), 232
Slack variable, 362
Social preferences, 497
Solution(s), basic, 365
  feasible, 356, 362
  optimum, 363
  paths to, 19-21, 30-31
Space, absolute, 26-27
  attributes of, 27
Specializing process, in problem solving, 19
Spectrum of color, and Bassa language, 43
Sphere, equation of, 23
Stability, of dynamic system, 449, 468
Stable equilibrium, 450, 468
Stable oscillation, 443, 468
Stable system, energy in, 489
Standard deviation, 257, 300
Standard normal distribution, 258, 301
  areas under, 259
  ordinates of, 260
  use of, 258
Standard normal variable, transformation to, 262, 301
State(s):
  in dynamic programming, 387
  goal, description, 478
  of nature, in decision models, 312
  present, description, 476
  of system, 409
Statements, analogy to sets, 70
  imply, interpretation of, 59
  language of (*See* Symbolic logic)
State variables, 409
Static system, 410
Statistic, 278
Steady state response, 437, 440, 467
Stimulus-response, in problem solving, 2, 30
Stored program, in computer, 94
Straightness ratio, 290
Strategies, 312
  dominant, 344
  dominated, 344
  mixed, 340
  pure, 341
Strings, in language, 37
Student *t* distribution, 284
Subjectivist, in decision making, 318
Subroutine(s), 125, 131
  for matrix additions, 126
  for matrix multiplication, 126
Subset, definition, 69, 84
Substructures stable, 20
Summation notation, 304
Superposition, property of, 358
Supply and demand, model of, 425
Supreme value, survival as, 489
Survival, group, 490
  index of, 491
  individual, 490
Switching circuits, 64
Syllogisms, 61, 83
Symbolic logic, 53, 83
  axioms of, 54, 83
  equivalent forms in, 64
  laws of, 63
  operations in, 54, 63
  symbols, 63
  tree diagrams in, 55
Syntactics, 38
Synthesis, of system, 224
System(s):
  analysis, 224
  behavior, 225
  control, 224, 413, 466

definition, 409, 466
dynamic (*See also* Dynamic system)
level of, 410
model, 409
identification, 224
model, 409, 466
parameters, 225
state of, 409
static, 410
synthesis, 224
variables, 224
System analysis, and problem solving, 476
System approach, to problem solving, 476
Systemic circulation, 431

T

Talking, in problem solving, 21
Tautologies, 61, 83
verification, 61
Tchebysheff inequality, 274
Teleological model, 489
Thermodynamics, and entropy, 484-485
first law of, 484
irreversibility, 485
second law of, 484
Thought experiment, 485
Thought and language, 28
Time constant, 441
Time delay, 443, 467
Tower of Hanoi problem, 34
Transducer, 445
Transfer function, 437, 467
in dynamic system, 416
of feedback system, 444
Transformation, to standard normal variable, 262, 301
Transient response, 437, 440, 467
Transitivity, violation of, 350
property of, 323
Tree, and Bayes' theorem, 158, 161
for code of symbols, 172, 173
decision, 169
Tree diagrams, 55
in probability, 146
Truel, story of, 332
Truth tables, 57
for AND, OR, NOT, 67
for Boolean algebra functions, 68
for full adder, 117
for half adder, 116
interpretation of, 58

U

Uncertainty principle, 219
Union of sets, 69
UNIVAC, 94
Universal set, 69, 84
Universe, models of the, 212
Unstable equilibrium, 450, 468
Uranus, discovery of, 215
Urban dynamics, Forrester's model, 455
Utility, 286, 313
assignments, 335
cardinal, 322
descriptive, 335, 336
and doctor decision, 327
of money, 322, 330
normative, 335
ordinal, 322
prescriptive, 335
and self insurance, 329
Utility function(s), 326
conservative behavior, 330
expected value behavior, 330
gambler behavior, 330
multidimensional, 512
properties, 326-327, 351
Utility theory, 286, 312, 322, 351
axioms of, 322, 351
lotteries in, 324

V

Valid arguments, 61, 83
Value(s):
assessment of, 496
attributes of, 475
benefits, 496
change in, 494
classification, 478, 479
competitive, 479
consensus on, 497, 513
cooperative, 479
costs, 496
definitions, 475
end, 478
erosion of importance, 495
of the future, 493
group, 495, 497
individual, 495
and knowledge, 483
means, 478
of the present, 492
ranking, consensus on, 497
rating of, 480
role, in problem solving, 1, 476
subscription to, 494
survival, 489
theory of, 475
topic for discussion, 516
Value judgment, 480, 514
and national goals, 482
Value system, 474
Variable(s):
across, 224
basic, 364, 366

Variable(s), (*con't.*)
control, 372
decision, 372
dependent, 205
independent, 205
nonbasic, 364, 366
random, 248, 264
slack, 362
state, 409
transformation to standard normal, 262, 301
through, 224
Variance, 248, 300
of aggregate, 266
of population, 251, 300
of proportion $\bar{p}$, 278, 302
of sample, 252, 300
unbiased estimate of, 266
Venn diagrams, 70
Hamming code and, 76
Vulcanus, 216

W

Water-clock, Ctesibius, 422
Watt's governor, 425-426
Wave number, 234
Wavelength, 233, 241
Will to doubt, 17
basic premise of, 136
probability and, 135
Will to Live, 491
Will to Love, 492
Will to Meaning, 491
Will to Order, 487
Will to Pleasure, 491
Will to Power, 491
Will ro Relatedness, 492
Wintu Indians, language of, 42
Words, concepts of, 41
Working backwards, in dynamic programming, 389
in problem solving, 19
Written language, evolution of, 44

Z

Zero-sum games, 338